PRINCIPLES OF
PHYSICAL GEOGRAPHY

1 The Nesthorn (12,530 feet) in the Bernese Oberland, Switzerland, with part of the Aletsch Glacier in the foreground

Note the gullying on the face of the peak, the small hanging glacier, the deep gorge cut by its melt-water stream, and the several lines of moraine.

(*Eric Kay*)

PRINCIPLES OF
PHYSICAL GEOGRAPHY

F. J. Monkhouse M.A., D.Sc.

formerly Professor of Geography
in the University of Southampton

AMERICAN ELSEVIER PUBLISHING COMPANY, INC.
NEW YORK 1970

Published in the United States by
AMERICAN ELSEVIER PUBLISHING COMPANY, INC.
52 Vanderbilt Avenue, New York, New York 10017

GB 53
.M6
1970b

First published 1954
Seventh edition
Copyright © 1970 F. J. Monkhouse

Library of Congress Catalog Card Number: 79-134278
International Standard Book Number: 0-444-19622-6

Printed and bound in Great Britain

Preface

It is significant that such terms as 'basis', 'groundwork', 'elements' and 'background' are commonly applied to physical geography. It may be said that there can be no geography at all without physical geography, which concerns itself with the solid rocks, the actual shape and form of the land surface, the configuration and extent of the seas and oceans, the enveloping atmosphere without which life as we know it cannot exist, the physical processes which take place in that atmosphere, the thin vital layer of the soil and the 'green mantle' of vegetation. All these, in conjunction, comprise man's physical environment.

Each of these various aspects has an interest in its own right, and the geologist, the meteorologist, the soil scientist and the botanist are intimately concerned with the understanding of the various phenomena falling within their respective fields. From these allied natural sciences physical geography necessarily draws much of its data. But geography, physical or otherwise, is far from being a mere descriptive compilation of facts derived *en bloc* from these external sources. The geographer seeks to use this information so as to describe and attempt to explain the features of the stage on which man plays his part; description without explanation is a poor thing, unsatisfactory and unsatisfying. Wherever possible, classifications, with careful headings and sub-headings, have been used in order to bring some system and ordered relationship into the mass of material with which we are confronted.

Since this book first appeared in 1954, minor amendments and additions have been made to the text in subsequent reprintings and additions, notably to the sixth edition. University of London Press Ltd has now been so kind as to allow me to undertake another major revision, involving a re-setting of the type and consequent repagination of the book, the addition or replacement of material in the light of recent concepts in geomorphological, geophysical and meteorological research, and a number of new or modified diagrams. Opportunity has been taken to conform with the British policy of officially using the Centigrade (Celsius) scale of temperature, though in most cases the Fahrenheit equivalent is retained in brackets. Other dimensions are also given in both Imperial and metric systems, rounding off where necessary to the equivalent order of magnitude, except where precise figures need exact transnumeration.

ENNERDALE, 1969   F. J. M

Contents

CHAPTER I

The Materials of the Earth's Crust

THE solid crust of the earth consists of rocks in great variety, of different degrees of hardness, coherence and permeability. The study of the rocks is primarily the province of the geologist, but the geographer must also be interested in at least the more common types. He should be able to recognize them in the field, and he should know something of their chemical composition and of their origin, for many rocks have come into existence by means of natural processes which are still at work. The study of the rocks and the facts which they reveal are of immense help in understanding how the present landscape has come to be what it is.

Perhaps the most fundamental fact in the physical make-up of Britain is the familiar concept that a line drawn from the mouth of the Tees to that of the Exe divides the country broadly into 'Upland Britain' and 'Lowland Britain'. To the north and west are found mainly the old hard rocks—the ancient schists, gneisses and sandstones, the granite masses, the volcanic rocks and the older limestones. Most of this part of Britain, except for narrow coastal plains and valleys, is upland. These rocks stand out as hill masses— much of the Highlands of Scotland, the Southern Uplands, the Pennines, Cumbria, Wales and the South-west Peninsula. Soils are thin and poor, moorland and coarse grazing is widespread, and population over much of the area is scanty, except for widely separated lowlands of high productivity.

Lowland Britain is indeed far from flat, except in parts of East Anglia; rather it is undulating country, where lines of low hills of limestone and chalk rise above valleys worn from less resistant clays, but they rarely do so for more than 200 to 230 m (600–700 ft) above sea-level. The soils are for the most part deep and rich, and population and settlement are fairly continuous. Thus the basic geological difference between the areas of old resistant rocks and younger softer rocks is largely responsible for the different landscapes of each.

The nature of rocks Rocks consist for the most part of aggregates of mineral particles; sand, clay, slate and granite are all rocks in the correct sense of the word. About 2000 named minerals are known, though relatively few are important as rock constituents.

Some elements, like the diamond and graphite forms of carbon, and sulphur, are minerals themselves, as are metals such as gold and copper, but most rock minerals consist of compounds of two or more elements. Thus quartz is a form of the oxide of silicon (silica, or SiO_2). Estimates indicate that 59 per cent of the rocks of the earth's crust consist of silica, for in addition to its oxide form of quartz, silica combines with various other oxides to form the largest group of rock-forming minerals. These silicate minerals can be divided into two groups. The first is the ferromagnesian or *mafic* group, in which the silicates are combined with iron and magnesium, including olivine, augite, hornblende and biotite (dark mica). These are usually dark in colour, and with a high density. The second group are the non-ferromagnesian minerals: muscovite (or white mica), the felspars (orthoclase, albite and anorthite) and quartz. These are usually light in colour and have a lower density than the ferromagnesians. Other common rock-forming minerals include chlorides (such as rock-salt), sulphides (such as iron pyrites and galena), oxides of iron (magnetite and haematite) and carbonates (notably calcium carbonate, the crystalline form of which is calcite). Apart from their chemical composition, each mineral has clearly marked properties and features—the crystalline form in which it may occur, its cleavage or tendency to break along a certain plane or line of weakness, its hardness (there is a scale ranging from talc $= 1$, which can be scratched with a finger-nail, to a diamond $= 10$), its colour (which, however, is very variable), its lustre and its density. It is useful to be able to distinguish the main mineral constituents of the rocks, and a text-book of mineralogy should be available for reference.

CLASSIFICATION OF THE ROCKS

The rocks of the earth's surface may be classified according to their mode of formation. It is believed that below the earth's outer solid crust, rock material exists at immensely high temperatures, but under enormous pressure. When the pressure is relieved locally, this rock material becomes fluid, rising through lines of weakness towards the surface (Chapter 3). The *igneous rocks* have been formed by solidification in various ways of this molten rock material, known as *magma*, so comprising a group of 'heat-formed' rocks which constitute about 95 per cent of the outer 16 km (10 miles) of the earth. The *sedimentary rocks* consist for the most part of the remains of previously existing rocks, reassembled and consolidated in various forms; they are sometimes known as *derived rocks*. They also include a number of rocks formed of the organic remains of plants and animals, and others formed chemically by the precipitation of substances from solution. The *metamorphic rocks* are those in which pre-existing rocks, both igneous and sedimentary, have undergone chemical or physical changes, either by heat or by pressure, or both, to cause degrees of alteration and modification.

IGNEOUS ROCKS

The character of an igneous rock depends on two main features: the first is the chemical composition of the magma from which the rock solidified; the second is the physical circumstances under which this cooling and solidification took place.

Chemical composition Many of the igneous rocks are extremely complicated in their chemical composition; probably most, if not all, the known elements are found in rocks of this class. Actually, about nine elements make up 99 per cent of all igneous rocks. Perhaps the most useful index of classification is the amount of silica present in a rock. When there is a high proportion of silica (exceeding about 65 per cent), the rock is said to be *acid*. Where, however, the amount of silica is 55 per cent or less, and where various basic oxides (i.e. oxides of aluminium, iron, calcium, sodium, magnesium and potassium) exceed 45 per cent, it is known as a *basic rock*. There is an *intermediate* category of acidity between the two, while rocks with under 45 per cent of silica are known as *ultrabasic*. Generally speaking, the acid rocks are light in colour and in weight, the basic rocks darker and heavier. Acid rocks include granite and obsidian, intermediate rocks diorite and andesite, basic rocks gabbro and basalt, and ultrabasic rocks peridotite.

The cooling of the magma The igneous rocks solidified from the molten magma, either intruded into the crust, forming *intrusive rocks*, both in large masses and in thin sheets, or poured out on to the surface, there solidifying as *extrusive rocks*. These processes are known collectively as vulcanicity (Chapter 3). In either case there is a definite order of crystallization, or *magmatic differentiation*, with a specific temperature at which each of the nine silicate minerals is formed, beginning with olivine, which has a simple structure, and ending with quartz. A single magma may therefore produce a wide range of igneous rocks.

If the intrusive magma cools in large masses deep in the earth's crust, the process is slow, and the resulting rocks are compact, coarse in texture and large crystalled; they are known as *plutonic rocks*, and examples are granite, diorite, gabbro and peridotite. Some geologists do not now regard all granites as igneous rocks formed by direct solidification from a magma, but believe that in some cases older rocks have been altered or transformed by a process known as

3

granitization. Emanations from the magma penetrated the surrounding country rock and transformed it by chemical action into granite, without any liquid magmatic phase. At any rate, it is safe to regard granite as a rock of plutonic or deep-seated origin (pp. 65–7). Some of the coarsest intrusions, usually found near the margins of activity, are known as *pegmatites* (p. 68); crystals 12 m (40 ft) in length have been found in the Black Hills of South Dakota.

Where the intrusions are along cracks and lines of weakness in the country rock, the cooling is more rapid than in a large mass but slower than on the surface; this produces a very variable and intermediate category of *hypabyssal rocks*. Some, termed porphyries, consist of well-formed, conspicuous crystals (*phenocrysts*) of different minerals embedded in a ground-mass of glassy material.

Rocks resulting from the cooling of magma which has been poured out or erupted on to the surface are usually small crystalled (notably rhyolite, andesite and basalt) or even glassy (obsidian). They are known collectively as *volcanic rocks*.

There is, therefore, a two-fold classification, based (*a*) on chemical composition and (*b*) on their mode of origin. The table on page 9 classifies the more common igneous rocks on this dual basis.

Pyroclasts A small group of rocks consist of igneous materials, but are fragmental in character. They have been thrown out from a volcanic vent, and include fragments of solid lava, cinders, ash and dust, more or less consolidated; they are described on pp. 69–70 under the volcanic activity which has produced them.

Some important igneous rocks Many of the igneous rocks are of interest chiefly to the petrologist, but a few occur very widely and are responsible for striking relief forms. *Granite* (of which there are many varieties) is a massive acid plutonic rock; it is coarse grained and large crystalled, consisting mainly of quartz, felspar, hornblende and biotite mica; the quartz grains are clear and glassy, the felspar may vary from white to pink, the hornblende and mica are black, so that the appearance of the rock is coarsely speckled, forming either grey or pink granite. Where the dominant ferromagnesian (or *mafic*) minerals are mica, biotite- or muscovite-granite is formed; where they consist of hornblende, it forms hornblende-granite. This rock tends to occur in great masses or batholiths (p. 65 and figs. 29, 30, 41, 42), revealed by the denudation of softer overlying rocks. Examples include the South-west Peninsula of England (fig. 30), the Cheviots, the northern part of the Isle of Arran (fig. 29), the Cairngorms, south-eastern Skye, the Wicklow Hills and the Mourne Mountains in Ireland.

Gabbro is a plutonic rock, occurring like granite in masses, but

4

basic in composition. It forms a dark grey or even black rock, and appears in north-eastern Scotland, in south-western Skye (figs. 42, 43 and plate 25) and in Jersey in the Channel Islands.

Basalt is the result of the solidification of lava flows, and is extremely widespread, frequently forming great plateaus (fig. 34 and plate 2). It is a black or dark-coloured rock, with extremely fine crystals, and often contains iron; this accounts for the rust-like limonite which forms on the surface of weathered basalt.

Igneous jointing A feature of many igneous rocks is the presence of joints, set up through tensile stresses caused by cooling. This is well shown in basalt, where clear-cut vertical joints sometimes outline perfect hexagonal columns, as in the Giant's Causeway in Antrim (plate 3), at Fingal's Cave on the island of Staffa, and in the Devil's Post-pile in the eastern Sierra Nevada in California. The Devil's Tower in Wyoming (plate 20) consists of pentagonal columns of phonolite-porphyry. Granite too may have clean-cut vertical joints, as in the Chamonix *aiguilles* in the Mont Blanc massif, and in the 'Needles' of the Black Hills of South Dakota (plate 34). Sometimes granite has three sets of master-joints at right-angles, forming castellated piles of blocks (plate 27).

SEDIMENTARY ROCKS

When an igneous rock is exposed at the surface of the earth, it is subjected to the processes of earth-sculpture, carried on by the agency of the weather, rivers, ice-sheets, ocean waves, in fact all the processes grouped together under the term denudation (Chapter 4). The result is that an igneous rock may be broken down. This may take place either as an exceedingly complex *chemical disintegration* (thus various silicates may break down into 'clay minerals' and alkalis in solution, e.g. felspar into kaolin) or by *mechanical disintegration*. These products are transported elsewhere, then laid down or deposited in layers (*strata*) on dry land or in fresh or salt water, and consolidated or cemented in various ways, a process known as *lithification* or *diagenesis*. The plane of division between each stratum is a *bedding plane*, which usually indicates where one phase of deposition has ended and another begun. Where a stratum thins out in a horizontal plane and disappears, it is said to *pinch-out*, while a thick stratum is described as *massive*. The finest scale of stratification, sometimes defined as less than 1 cm in thickness, is known as a *lamina* (hence lamination); there may be a hundred or more layers in 2 cm of rock. The Green River Shales of Wyoming, which total 800 m (2600 ft) in thickness, were laid down in layers only 0·18 mm (0·007 in) thick; it is estimated that it took 6·5 million years for these rocks to be deposited, each lamina representing a thin layer of mud, which was compacted, and then the next flood added another layer.

False-bedding, or *current-bedding*, occurring especially in sandstone, reveals thin laminae inclined at varying angles to the general stratification. This pattern is the result of changing currents, either of water or wind, responsible for the deposition of the sand-grains.

Jointing may develop in sedimentary rocks during consolidation and compaction, frequently forming at right-angles to the bedding-planes, and so dividing the rock into massive blocks. The Checkerboard Mesa in Zion Canyon, Utah, is so called because the joints intersect the bedding-planes with a most striking rectilinearity.

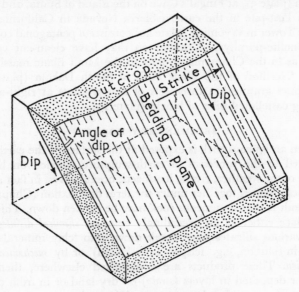

Fig. 1 The nomenclature of rock strata

Individual strata of sedimentary rocks, though usually laid down in a horizontal plane, rarely remain that way, since earth-movements may subsequently tilt or warp them. The angle of slope of a stratum is known as the *dip*, expressed in degrees between the bedding-plane of the stratum and the horizontal (fig. 1). It is important to remember that the dip is not necessarily the slope of the land surface; in a cuesta (fig. 61) there may be a distinct difference between the dip-slope and the back-slope. *True dip* is the maximum angle of stratum slope; its direction is measured in degrees clockwise from

6

2 The columnar basalt cliffs at the edge of the plateau of Antrim

This is a view of the 'Spanish Organ' and the projecting Benanouran Head, near the Giant's Causeway. Note (i) the level surface of the plateau; (ii) the different lava flows; (iii) the scree slopes flanking the base of the cliffs.

(*Aerofilms Ltd*)

3 The hexagonal basalt columns of the Giant's Causeway, Antrim

(*Mustograph*)

4 Quartz vein traversing pebbles of conglomerate in Old Red Sandstone, near Oban

(*R. Kay Gresswell*)

5 A Pre-Cambrian rock contorted by metamorphism at Holy Island, Anglesey

(*R. Kay Gresswell*)

true north. A line at right-angles to the true dip is known as the *strike*; any other line between the strike and the true dip is the *apparent dip*. The extent of the *outcrop* of a stratum on the surface depends partly on the dip; thus a gentle dip results in a broad out-crop, while the outcrop of a vertical stratum is no wider than the stratum itself. Thus the Eocene rocks on the northern side of the Hampshire Basin dip very gently south, so outcropping over a distance of 50 km (30 miles), but in the Isle of Wight they rise almost vertically, outcropping for only a few hundred metres.

Sedimentary rocks may be classified either according to their mode of origin or to their composition. Under the first classification, they may be described as (i) mechanically formed, (ii) organically formed and (iii) chemically formed. Under the second classification, they may be divided into groups according to their composition— whether they primarily consist of coarse-textured sands, or of fine clays, or of calcium carbonate, or of carbon, or of other substances. It is convenient to combine the two, as has been done in the table on p. 9.

(i) **Mechanically formed sedimentary rocks** This group includes a variety of either coarse- or fine-textured rocks, formed by the compaction, desiccation or cementation of detrital sediments such as sand, silt, clay and gravel, hence the term *clastic* (from the Greek word for 'broken') sometimes applied to this category. The materials may be consolidated by cementing solutions of siliceous, calcareous or ferruginous materials. The most common mineral in the sandy rocks is quartz, cemented in the case of yellow or red sandstones with iron compounds, in the case of a white sandstone with a siliceous or calcareous cement. These sandstones are wide-spread and of diverse age; thus very old sandstones occur in northern and southern Devon and in south-eastern Wales, while younger sandstones appear along the Cumberland coast forming St Bees Head, and in the Eden valley, and underlie much of south-western Lancashire and a large part of the western Midlands. Some of the oldest rocks in Britain, in the North-west Highlands of Scotland, are sandstones, the Torridonian, of Pre-Cambrian age.

A much coarser, well cemented rock, with both sand-grains and small pebbles, is known as a grit, such as the Millstone Grit of part of the Pennine moorlands, while a rock formed by the cemen-tation of large fragments may be either a *conglomerate* (plate 4), where the pebbles are rounded, or a *breccia*, where they are angular.

At the other end of the scale are rocks composed of deposits of fine mud, the constituents of which largely comprise minute particles of quartz and mica. Various clays and hardened mudstones are examples. The vales of south-eastern England are floored with clays; the greyish-blue Lias Clay lies to the west of the Cotswolds, and the dark greyish Kimmeridge Clay and the bluish-grey Oxford Clay extend diagonally across England from Dorset through the Oxford Vale into northern Yorkshire. The yellow or brownish Wealden Clay occurs between the Downs and the centre of the Weald. The bluish-grey London Clay of the London Basin weathers to a brown colour near the surface. There are also the extensive smothering layers of till laid down by ice-sheets (p. 239). Rocks composed of sand and grit are known as *arenaceous*, those of fine clay particles as *argillaceous*, and those of coarse particles as *rudaceous*.

(ii) **Organically formed rocks** These rocks are composed of the remains of once living organisms, the hard parts of which have accumulated over long periods of time. The most widespread and varied group is that of the limestones, composed of shells or skeletons, whole or comminuted, which consist to a large extent of calcium carbonate. These include shelly limestones, such as the 'Crag' of Norfolk, coral limestone built up in reefs, the soft white pure Chalk of the Downs and Wolds, the often fossiliferous Carboniferous Limestone of the Pennines, the oolitic limestones of the Cotswolds which consist of rounded grains like fish-roe, and calcareous clays known as marl, such as the chalk-marls of Cambridge.

Minute bacterial organisms, produced by decomposing vegetable matter, assist in the precipitation of hydrated iron oxide from the waters of lakes and marshes. This 'bog iron-ore' has long been valuable in Sweden and Finland. Blackband and Clayband ironstones, in the form of ferrous carbonates, were probably deposited in the Coal Measures in much the same way.

Siliceous rocks may be organically formed from the remains of animals such as sponges and radiolaria (a minute organism whose hard parts remain as a complex framework of silica), and of plants such as diatoms. Nodules of chert (brittle, minutely crystalline silica) may be found in some limestones and sandstones, and nodules of flint in chalk. Great beds of diatomite, or diatomaceous earth, the remains of countless diatoms, sometimes occupy dried-up lake-bottoms or former swamps, as in north-eastern Skye, in Kentmere in Westmorland (p. 205), and in California.

The carbonaceous rocks consist essentially of carbon combined with other elements. They are formed from plant accumulations, and peat, lignite and various types of coal occurring in seams in the Coal Measures represent stages in their gradual change under

CLASSIFICATION OF ROCKS BY ORIGIN AND COMPOSITION

Igneous rocks

	Acid	Intermediate	Basic	Ultrabasic
Silica (%)	65	65–55	55–45	Under 45
Basic oxides (%)	35	35–45	45–55	Over 55
Intrusive				
Plutonic	Granite	Diorite	Gabbro	Peridotite
Hypabyssal	Granophyre	Porphyries (various)	Dolerite	
Extrusive				
Volcanic	Rhyolite Obsidian	Andesite	Basalt	

Sedimentary rocks

Mechanically formed

    (i) Arenaceous—sand, sandstone, grit

    (ii) Argillaceous—mud, clay, mudstone, shale

    (iii) Rudaceous—breccia, conglomerate, tillite, gravel, scree, boulder-clay

Organically formed

    (i) Calcareous—chalk, various limestones (crinoidal, coral, oolitic, shelly)

    (ii) Ferruginous—ironstone

    (iii) Siliceous—diatomaceous earth

    (iv) Carbonaceous—peat, brown coal, lignite, cannel coal, bituminous coal, anthracite

Chemically formed

    (i) Carbonates—travertine, dolomite (calcium and magnesium carbonate)

    (ii) Sulphates—anhydrite, gypsum ⎱ *evaporites* formed by

    (iii) Chlorides—rock-salt   ⎰   desiccation

    (iv) Silicates—sinter, flint, chert

    (v) Ironstones—limonite, haematite, siderite

Metamorphic rocks

Thermal or contact metamorphism ⎧ producing compaction, cleavage,
Regional metamorphism   ⎨ foliation, complex mineralogical
                                 ⎩ changes

Of Sedimentary rocks

    (i) Arenaceous rocks—quartzite (from sandstone and grit)

    (ii) Argillaceous rocks—slate, schist (from clays, shales, mudstones)

    (iii) Calcareous rocks—marble (from calcium carbonate)

    (iv) Carbonaceous rocks—graphite (from organic matter)

Of Igneous rocks

Actual mineralogical changes—augite into hornblende
                                 —granite into gneiss

pressure. There is a wide range of hydrocarbons, solid, liquid and gaseous; pitch, asphalt and bituminous shales are the solid forms, crude mineral oils the liquid, and natural gas the gaseous form.

(iii) Chemically formed rocks For the most part, these rocks are precipitated or evaporated from solutions of salts. All water falling on to the earth as rain, and either running away over the surface, or sinking in and later reaching the surface again (Chapter 5), carries salts in solution. The salts may be precipitated by the direct evaporation of the water, or by chemical interaction, or by the release of pressure where underground water reaches the surface.

Deposited carbonates afford examples of these chemically formed rocks. The deposition of calcite on the beds of streams and in the form of stalactites and stalagmites in caves is a familiar feature in limestone country (p. 124). Where the calcareous substance is spongy, it is known as tufa. Sometimes this deposition takes place round hot springs, when the result is known as travertine, harder and more compact than tufa.

Dolomite is a chemically formed compound of calcium and magnesium carbonate; an example is the Magnesian Limestone. In England this outcrops from Nottingham northward to the coast of Durham; it forms a marked ridge, with a sharp escarpment to the west, and where it reaches the coast it has been eroded into striking yellowish cliffs.

Other rocks are formed by the precipitation of sulphates from solution; thus hydrated calcium sulphate, known as gypsum or in its granular form as alabaster, is formed by evaporation in inland drainage basins, such as the Dead Sea. Beds of sodium chloride (rock-salt) are widespread, both on the surface, as around the Great Salt Lake in the western U.S.A., and at depth, as near Droitwich in Worcestershire, near Northwich in Cheshire and near Middlesbrough. Silica may be deposited from hot springs (p. 80) round their vents to form siliceous sinter, as in Iceland and the Yellowstone National Park, Wyoming. Finally, many scattered deposits of salts are industrially of great importance, such as potash and nitrates.

Next to aluminium, iron is the most widely distributed metal, of which each ton of the earth's crust contains a hundredweight. Most iron ores have accumulated as a result of chemical precipitation within sediments, though some are directly due to igneous activity in the form of magmatic differentiation (pp. 3-4), as in the case of the magnetites of northern Sweden. Subsequently processes of oxidation and hydration have modified their composition and increased their concentration (p. 18).

METAMORPHIC ROCKS

Igneous and sedimentary rocks may undergo changes, both physical and chemical, which produce either new minerals or new structures within the rocks. These changes may be brought about by earth movements, subjecting the rocks to great pressure; these are usually on a large scale and the results are widespread, hence the names *dynamic* or *regional metamorphism*.

Other changes may be brought about by the influence of heat; a rise in temperature is most usually caused by the intrusion of a mass of igneous rock, so that the nearby rocks are affected. The effects result from direct contact between the magma and the country rock (hence the names *thermal* or *contact metamorphism*), and as the changes are usually on a more local scale than those produced by earth movements, they are sometimes described as *local metamorphism*.

The results of metamorphism are extremely complex. Thermal metamorphism can produce a fusion or a recrystallization, such as the change of coarse grained sandstone into quartzite, or of a limestone into a granular crystalline rock known as marble. Much more complex is the production of new minerals. This can be seen around a large mass of granite, where a zone of contact alteration (known as an *aureole*) can sometimes be distinguished (fig. 26), and a gradual transition in the character of the rocks can be traced as new minerals appear.

Regional metamorphism may produce complex mineralogical changes, but more often the alteration comprises a re-shaping of the rock particles, inducing a tendency to split along parallel planes which have no relationship to the bedding; this is known as *flow cleavage*, and is revealed most clearly in slate. Fine-grained sediments such as shale are turned into slates and schists; coarser grained or crystalline rocks form quartzites, gneisses and granulites. Many rocks show *foliation*, a sort of 'wavy-grain' or laminar structure; most schists, for example, are foliated (plate 5).

Metamorphic rocks are in general compact and resistant to erosion. They tend to form masses within areas involved in mountain building; much of the Highlands of Scotland, and large parts of the ancient blocks strung out across central Europe from Brittany to Bohemia, are composed of metamorphic rocks. The eastern part of the Belgian Ardennes, for example, consists mainly of slates and quartzites, worn into rounded eminences separated by swampy depressions and covered with thin, poor soils. A specially interesting British example is the Isle of Anglesey, much of which consists of ancient gneiss and schist, the remnants of a former upland area.

THE AGE OF THE ROCKS

So far the rocks have been considered and classified according to their composition and their mode of origin. Another important consideration is their age, and one branch of geology, known as *stratigraphy*, has the function of working out and arranging the record of the events in the earth's history as evidenced by a systematic study of the rocks which form the crust. In other words, a tabulation of rocks according to their age affords a convenient time-chart or chronology of the earth's history. If, for example, certain strata are affected by mountain-building movements, it is possible to date those movements and to compare the results in different parts of the world. Moreover, particular rocks are being produced at the present time under specific conditions by certain processes; if similar rocks have been formed during bygone geological time, it may be assumed that broadly similar conditions obtained then. Thus certain sandstones were laid down under desert conditions, and some limestones under periods of widespread inundation by the sea.

Unconformities Sometimes gaps occur in the continuity of the geological record, where the sediments have been deposited on a surface produced by a long period of denudation, so interrupting the normal order of succession and producing an unconformity. Three distinct types have been recognized. Where a series of folded or tilted strata have been eroded over a long period of time, and then covered by new layers of sedimentary rocks, an *angular unconformity* is formed; the overlying strata lie at a distinct angle to the underlying ones. Thus, for example, the basement of the Pennines consists of ancient, much metamorphosed rocks affected by the Caledonian folding (p. 58), the strata in places inclined at a high angle. The Carboniferous Limestone and younger rocks lie almost horizontally on the planed-down rocks (fig. 45), and there is a gap (representing about 70 million years) with no rocks of Devonian age. This can be seen at quarries in the Ribble valley near Horton, in Yorkshire.

A second type is a *disconformity*, which involves a non-sequence of beds, but both the older and newer strata are broadly parallel. Prolonged denudation removed certain strata, and then new material was deposited on them, but without any folding or other earth-movements involving angular change in either the under- or overlying rocks. The section of the walls of the Grand Canyon of the Colorado river (fig. 71) reveals two disconformities, one where the Redwall Limestone of Devonian age is deposited directly on the Muav Limestone of Cambrian age, representing an interruption of about 80 million years, and the second where Cambrian rocks rest on the eroded surface of the Pre-Cambrian basement rocks.

The third type is the *nonconformity*, where igneous rocks are eroded and then overlain by sediments. Near Cape Town, in South Africa, a dark sandstone nonconformably overlies pale-coloured granite.

Relative age As all beds of stratified sedimentary rocks were deposited in layers, it follows that in normal cases the oldest rock is at the bottom, and each overlying bed is successively younger; this is known as the *law of superposition*. (Complete inversion can be caused by earth movements, such as overfolds of strata.) A second obvious fact is that any rock is newer than the fragments of which it is composed. A third help is that fossils are found in many sedimentary rocks; these may be either the preserved hard parts of organisms or their replacement with great exactness by silica or calcium carbonate. In many cases each group of beds is characterized by a definite set of organic remains. To an analysis of the physical and chemical composition of a rock can be added information about the associated fossils, and thus the sedimentary rocks can be arranged in a definite order of date.

Absolute age The actual age of the rocks, that is, of the earth itself, has long exercised man's interest. In the seventeenth century Archbishop Usher of Ireland stated with pleasing exactitude that the world was created at 8 p.m. on 22 October 4004 B.C., worked out by complicated mathematics from the chronologies of the Old Testament. Gradually scientists realized that they needed an increasing span of time for the development of the earth's features and its age was pushed back and back, until now some rocks have been dated as more than 3400 million years old, and the earth's crust must be at least 4700 million years old.

It is only in recent years that methods of absolute dating have been devised, and a subzone of the sedimentary rocks can now be usually dated with confidence to within about 150,000 years. This may sound very approximate, but if the record back to the beginning of the Cambrian is represented by a 24-hour clock, this period of 150,000 years represents only 22 seconds; man, incidentally, would have appeared only a little over 2 minutes ago.

One of the earliest methods was the counting of laminae in shales, as in the case of the Green River Shales in Wyoming, where each sheet, only 0·18 mm thick, represented the annual accretion of mud. But most sedimentary deposition is too uneven to make this practicable. A special application of this principle, developed by the Swedish geologist De Geer, affords a means of calculating postglacial time with some degree of exactitude. It depends on counting distinctive banded layers, known as *varves*, deposited annually in melt-water lakes on the margins of ice-sheets. Each varve consists of two layers of sediment, one coarse, deposited during the summer thaw, the other fine, suspended material laid down as the surface freezes with the onset of winter and movement in the water ceases. The varves vary in thickness and can be correlated in different areas; from them the post-glacial history of Scandinavia has been deduced (pp. 260–1), and similar work has been done in North America.

The counting of tree rings is useful for establishing recent chronologies where trees of sufficient age are available, as in the south-west of the U.S.A. where the *Sequoia gigantea*, over 3000 years old, and the Douglas fir have been used to make a tree-ring index; each ring differs, corresponding to wet and dry years, and can be differentiated and counted. This process, known as *dendrochronology*, is mainly useful to archaeologists studying the prehistoric Indians in Arizona and Utah.

The most valuable method, which has been used to date the older rocks, is based on the progressive radioactive breakdown of certain unstable elements at a known rate. One of the most widespread elements of this nature is uranium $(_{92}U^{238})$, which in the form of uraninite is present in granite in the proportion of about four parts per million. It breaks down into a lead isotope, Pb^{206}; its half-life, that is, the time required for half the nuclei in the element to decay, is 4.51×10^9 (4510 million) years. Thus the age of a piece of rock containing a uraninite mineral can be obtained by finding the proportion of the uranium still present to the amount of the lead isotope, the so-called *lead-ratio*. Other ratios also used are thorium/lead, rubidium/strontium and potassium/argon. Sometimes it is possible to get two ratios from the same rock, thus affording a valuable check, as in the case of rocks containing mica minerals.

Another application of this principle to materials younger than about 60–70,000 years was developed in Chicago by Dr W. F. Libby in 1949, known as *radiocarbon dating*. This method depends on the fact that a radioactive isotope of carbon, known as carbon-14 $(_6C^{14})$, circulates in the atmosphere as a form of carbon dioxide $(C^{14}O_2)$, ultimately reaches the earth's surface, and is absorbed by living organisms. On death an organism not only ceases to assimilate it, but the content diminishes at a rate indicated by its half-life of 5570 years. Thus the age of a piece of wood in a peat-bog or of bone in a tomb can be established with a considerable degree of accuracy; this is especially useful in the study of post-glacial time.

From the point of view of the geological chronology, the most useful method is the dating of the igneous rocks of various ages, and transferring them to the sedimentary record of relative time. The *law of cross-cutting relationships* emphasizes the somewhat obvious fact that an igneous rock is younger than any other rock across which it cuts; thus a dyke (p. 61) is younger than the sedimentary strata across which it cuts, a sill (p. 63) is younger than the strata between which it lies. Occasionally a dyke can be seen cutting through another dyke, which obviously must be older.

The geological time-scale A scheme of age classification has been evolved for geological time extending back to about 600 million years ago, which is divided into three *eras*. These are named *Palaeozoic* (after the Greek words for 'ancient life'), *Mesozoic* ('middle life') and *Cainozoic* ('new life'). The first two eras are sometimes called Primary and Secondary respectively, while Cainozoic denotes Tertiary and by some authorities includes the Quaternary also. The Quaternary, which until recently was believed to have lasted for much less than a million years, more or less coincides with the appearance of man in the world, and its beginning marks the onset of the last great glaciation. (Some authorities now believe that the onset of this glaciation occurred from 1·8 to 2 million years ago (p. 247), and that the earliest form of man was much older still.)

Actually these 600 million years make up only a small proportion of geological time since the earth was formed. But evidence is so scanty that there is no generally accepted classification for the vast span of time before the Palaeozoic era. Some authorities have used the term Eozoic, from the Greek 'dawn of life', to cover the whole oi this time; others have divided it into three further eras, Eozoic (the oldest), Archaeozoic and Proterozoic. The simplest method is to refer to the whole of this time, and its rocks, as Pre-Cambrian.

The eras are subdivided into *periods*, then into *epochs* and finally into *ages*. The rocks laid down during these particular time divisions are themselves classified accordingly into *groups*, *systems*, *series* and *formations* respectively. Some of the names bestowed on these systems are derived from localities where the rocks were originally studied (such as *Cambrian* in Wales, *Jurassic* in the Jura Mountains), others from the type of rock (*Cretaceous* from the Latin *creta*, chalk).

The names of the groups and systems are now of world-wide application, though there are some modifications; in the U.S.A., for example, the Carboniferous is divided into two, the early part known as the *Mississippian*, the later as the *Pennsylvanian*. In that country, too (and increasingly in Britain), the Tertiary and Quaternary are regarded as periods of time and systems of rocks within the Cainozoic.

One of the commonest rocks is granite, which contains uranium elements and, fortunately, granites have been formed at different times in the earth's history, particularly in association with mountain-building periods, and their *relative* ages are known. It follows therefore that if their *absolute* ages can be obtained by means of the lead-ratio or other method, an absolute chronology can be ascribed to the time-chart of the rocks. There is general agreement that the Palaeozoic era lasted from 600 to 225 million years ago, the Mesozoic from 225 to 70 million years ago, the Cainozoic from 70 million years ago; the exact onset of the Quaternary is disputable (p. 247). The geological time-scale is summarized on p. 19.

THE ECONOMIC GEOLOGY OF THE ROCKS

Directly or indirectly, the rocks are of very great value economically. On the nature of the bedrock depends to a large extent the soil which is produced by weathering processes; generally the old hard rocks have a thin, poor soil, while the newer rocks tend to develop a deeper, more workable soil. Soils are considered more fully in Chapter 19.

The whole question of man's vital water supply is bound up with the nature of the rocks: the way in which they affect the circulation of ground-water (Chapter 5), the formation of springs, the sinking of wells to reach water-holding strata, the possibilities of natural or artificial reservoirs, and the amount of surface runoff in the form of rivers. Obviously, the rocks of the floor of a hollow to be used as a reservoir should be such that leakage due to permeability or to other causes is not possible. Hard, compact rocks in high moorlands, preferably in areas of non-calcareous rocks so that the water will be soft, are often chosen; numerous high valleys in the Millstone Grit of the Pennines are so utilized.

From the rocks many substances of great importance are derived. One category is that of fuel: peat, brown coal, lignite, bituminous and steam coals, anthracite and mineral oil. Another group of substances of economic value consists of building materials. Some limestones and sandstones afford easily worked 'freestones', of even grain and texture, such as the Portland and Purbeck Limestones and the 'Bath Stone' of the Jurassic system, and some of the Triassic sandstones. Granite is quarried especially at Peterhead and Aberdeen in Scotland, at Shap and in Cornwall. A few other ornamental stones, such as serpentine, porphyries and variously 'marked' Carboniferous Limestones are quarried. Roofing slate is obtained from fine grained metamorphosed rocks, which have a well defined cleavage so that the sheets can be readily split. The largest slate producing area in the world lies in North Wales between Llanberis and Bethesda, where slates of Cambrian age are quarried, while at Blaenau Ffestiniog, to the south of Snowdon, Ordovician slates are exploited. The Delabole quarry in Cornwall has worked Devonian slates since the reign of Queen Elizabeth I.

A wide range of rocks is used for road metal. These are mostly igneous rocks—diorite, gabbro and dolerite, while granite chippings are used for non-skid surfacing. Many of the old intrusive masses in the Midlands (such as at Mountsorrel in Leicestershire), the granites of Cumberland and Westmorland, and the igneous rocks of Wales are employed. Sedimentary rocks are seldom satisfactory, since they tend to crumble, but some Carboniferous Limestone is used.

Various limestones are worked to produce lime, for mortar and for agricultural use (plate 6). Portland cement is made by burning a mixture of two-thirds limestone or chalk and one-third clay, and grinding the result to fine powder. Along the Thames and the Medway, cement works use Medway mud and North Downs chalk, while Lower Lias limestones are used at Rugby, Chilterns chalk at Luton and Dunstable, Oolitic Limestone at Ketton in Rutland, and Magnesian Limestone at Ferryhill in Durham. Plaster of paris is obtained from gypsum mined in the Keuper Marls of the Trent valley near Newark and in northern Cumberland.

Bricks are made from many clays, the most useful of which is the Oxford Clay in the English Midlands, where brick works are situated near such centres as Peterborough, Bletchley and Bedford. Special firebricks are made from Coal Measure fire-clays. The North Staffordshire pottery industry uses china clay from Cornwall, 'ball clay' from Dorset, and local coarse grey clays, as well as imported raw materials. Gravel is quarried extensively to make concrete for use in the building industry. Sand, such as the red sands dug near Mansfield, is needed for moulding at the blast-furnaces. Only coarse glass-sands are found in Britain, and finer qualities are imported. Other useful materials obtained from the rocks include millstones and grindstones, mainly from the Millstone Grit, and whetstones from fine grained lavas or metamorphic rocks.

Salts which are obtained in Britain include rock-salt occurring in the Permian and Keuper Marls in the Droitwich, Northwich, Middlesbrough and other districts. There are many large deposits of salts in various parts of the world, such as phosphates in Tunisia and Morocco, and potash at Stassfurt in East Germany, in Alsace near Mulhouse and in Saskatchewan, Canada. In many parts of the world masses of salt are forced upwards by great pressures as domes or plugs, uplifting the overlying strata, as in Texas, Louisiana and southwestern Persia. These structures are commonly associated with oil and natural gas, sometimes with gypsum and anhydrite.

Metallic ores Metals are found usually in the form of a metalliferous mineral known as an *ore*. Many mineral ores are associated with former igneous activity, so that they occur as veins occupying fissures and cavities in the rocks. The formation of these veins is extremely complex; the ore may be deposited directly from the molten magma, or it may have been produced through the activity of very hot gases reacting with both the magma and the country rock, or again it may be due to ascending solutions of mineral-impregnated hot water. Another group of ores comprises those which occur in beds, such as some iron ores. Bauxite, the chief source of aluminium, also occurs in beds, in the form of a clay-like mass of hydrated oxide of aluminium, formed by the partial decomposition of various rocks which originally contained a high proportion of aluminium silicate.

Most metals occur in combination with other elements, forming sulphides, oxides and carbonates. Thus lead, zinc and silver occur commonly as sulphides, known respectively as galena, blende and argentite. Manganese and tin are found mainly as oxides, the latter being known as cassiterite. Copper occurs in a wide range of minerals, in fact in about 360, but most usefully as a sulphide at Katanga in the Republic of the Congo, in Zambia, Canada, in Utah and in Nevada. Iron is a widespread constituent of the earth's crust, found most commonly in oxide forms such as (i) *limonite* (hydrated ferric oxide), worked in Lorraine; (ii) *haematite* (red ferric oxide) in West Cumberland, in the 'iron ranges' near Lake Superior, and in a slightly different form at Krivoi Rog in the Ukraine; (iii) *magnetite* (the black oxide of iron), found at Gällivare and Kiruna in Sweden; and (iv) *siderite* (ferrous carbonate), found in the Coal Measures and the Jurassic limestones of England (plate 7).

Sometimes the various agents of erosion, working on the parent deposits, may wash out and deposit elsewhere either the metallic ores or even the parent metals. Owing to their high density, stream action sometimes sorts and concentrates the metals in beds of gravel or in alluvial flats, known as *placers*; gold, tin and platinum may occur in this way.

THE CLASSIFICATION OF THE ROCKS BY AGE

In the following table, the rocks are arranged in their normal order, with the newest at the top and the oldest at the bottom. The names are those in use in the British Isles, but the systems (in italics) are of world-wide application.

Holocene or Recent
Alluvium, peat
Pleistocene
Till, glacial sands
Quaternary
Pliocene
Crag (e.g. shelly sands and gravels)
Miocene
(Absent from Britain)
Oligocene
Clays, marls, sands, limestones (e.g. Bembridge Limestone), lignite
Eocene[1]
Sands (e.g. Thanet Sands), clays (e.g. London Clay), pebbles (e.g. Black-heath Pebble-beds)
Cainozoic (Tertiary)[2]
Cretaceous
Chalk
Upper Greensand
Gault (Clay)
Lower Greensand
Wealden (e.g. Weald Clay and Hastings Sands)
Jurassic
Purbeck
Portland
Kimmeridge Clay
Corallian (shelly limestone and clays)
Oxford Clay
Great Oolite (limestone)
Inferior Oolite (limestones, clays and sands)
Lias (clay, sand, marl and limestone)
Triassic
Rhætic[3] (marls, shales, limestones)
Keuper (marls and sandstones)
Bunter (sandstones and pebble-beds)
Mesozoic

Permian
New Red Sandstone, Magnesian Limestone
Carboniferous[4]
Coal Measures (coal, sandstone, limestone, shale)
Millstone Grit
Yoredale Beds (shales, sandstones limestones)
Carboniferous Limestone
Devonian
Devonian slates and limestones, Old Red Sandstone
Silurian
Shales, slates, limestones, sandstones (e.g. Ludlow Shales, Wenlock Limestone, Llandovery Sandstones)
Ordovician
Slates (e.g. Skiddaw Slates), limestones (e.g. Bala Limestone), volcanic rocks (e.g. Borrowdale Volcanics)
Cambrian
Slates (Llanberis Slates), grits (Harlech Grits), quartzites and flags
Palaeozoic

Ancient igneous, sedimentary and metamorphic rocks (e.g. Torridonian Sandstone, Lewisian Gneiss)
Pre-Cambrian

[1] In the U.S.A., and increasingly in Britain, the earliest part of the Cainozoic is known as the Palaeocene, lasting from 70 to 60 million years ago, and succeeded by the Eocene.
[2] In a revised time-scale, introduced by A. Holmes, the Cainozoic is divided into two periods, the *Tertiary* and the *Quaternary*; the former includes from the Palaeocene to the Pliocene epochs (from 70 to 2 million years ago), the latter the Pleistocene and Holocene epochs.
[3] Some geologists regard the Rhætic series as a transition between the Triassic and Jurassic systems.
[4] In the U.S.A. the Carboniferous is divided into two systems: the Mississippian (Lower) and the Pennsylvanian(Upper).

CHAPTER 2

The Structure of the Earth

A STUDY of land-forms necessitates some knowledge of the processes which have operated in the past, and are operating in the present, in order to understand in some measure how the landforms have come to be what they are. A description of the relief of any area, to be adequate, must also be explanatory. The features of the land surface are the product of two sets of forces. On the one hand, the 'internal' forces (contracting, expanding, uplifting, depressing, distorting, disrupting, intruding and outpouring) have directly affected the crust on both a major and minor scale. On the other hand, the 'external' forces of denudation (or gradation) can wear away the loftiest mountains uplifted by earth movements and deposit elsewhere the material so derived; the description of these processes occupies Chapters 4 to 9.

THE EARTH'S INTERIOR

The study of the interior of the earth belongs to the geophysicist. Nevertheless, certain facts are relevant to the geographer in order to help the understanding of surface features.

From the evidence of the velocity of earthquake waves, which can be very exactly recorded (p. 30), and more recently from seismic waves generated by man-made explosions, it seems that the surface layer of the continents is composed mainly of granitic rocks, with a density of about 2·65 to 2·70. Because these rocks contain a large proportion of silica and alumina (p. 3), they are often referred to collectively by the group name of *sial*. This sialic layer is very variable in thickness, and is wholly absent over much of the ocean basins, especially in the Pacific. On the surface of the sialic layer lie sediments, several miles thick in basins where deposition has long continued, though absent in some ancient upland areas.

Beneath the sial occurs a layer of denser rocks. These, consisting of silica (but in less proportion than in the sial) and of minerals rich in magnesium and iron, have a density of about 2·9; they sometimes appear on the surface as basalt lava. This layer of basic rocks is termed *crustal sima* or the *mafic crust*, and is grouped with the sialic layer to form the *crust*. The overall thickness of the crust ranges from 16–50 km (10–30 miles).

Beneath the crust, the semi-plastic *mantle* extends downward for about 2900 km (1800 miles), within which the rocks become denser (3·0 to 3·3), not gradually but in a series of concentric shells, consisting chiefly of the pale green mineral olivine (ferromagnesian

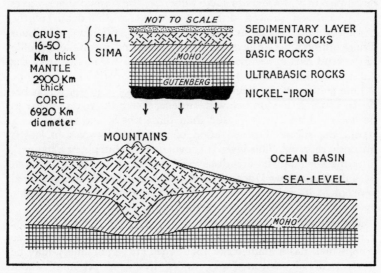

Fig. 2 Diagrammatic representation of the earth's interior

silicate), in the form of an ultrabasic rock in a state of high rigidity, known as *dunite*, a variety of peridotite. The surface of discontinuity between the crust and the mantle is known popularly as the *Moho*, after the scientist A. Mohorovičić who discovered it in 1909; here the speed of propagation of earthquake waves suddenly accelerates from about 5·0 to 8·1 km per second. American scientists planned a programme of deep drilling into the crust in order to reach the Mohorovičić Discontinuity and penetrate into the mantle. The pilot drilling was carried out in the eastern Pacific between the coast of the peninsula of Lower California and the island of Guadalupe, in 1961. This went well, and 'cores' of the sediments covering the ocean floor were retrieved. The main Upper Mantle Project (inevitably known as 'Operation Mohole') involved the drilling of a much deeper boring at a site 270 km (170 miles) off Honolulu in the Hawaiian Islands, but this was abandoned in 1966 because the U.S. Congress was unable to sanction the expenditure. The Russians made similar drillings off the Kuril Islands in the northern Pacific; they penetrated the superficial sediments and entered a layer of basalt.

An international Upper Mantle Project was however initiated by Russian scientists in 1960, with whom American, Canadian and other geophysicists have cooperated fruitfully. During the Great Lakes Seismic Experiment, when underwater explosions were detonated in the waters of Lake Superior, the seismic shock-waves when analysed showed distinct variations in the depth of the boundary of the crust and the mantle. The still more ambitious Transcontinental Geophysical Survey produced a complete profile of the crust and upper mantle right across North America from the Atlantic to the Pacific.

One feature of the mantle is the presence of a thin layer at between 100 and 200 km below its surface, where the rocks appear to be less rigid and more plastic than those above and below. As a result, the speed of propagation of earthquake waves (p. 32) is abruptly reduced. This layer is known as the 'Gutenberg Channel', after the scientist who first discovered it. (This is not to be confused with the 'Gutenberg Discontinuity' discussed below.) The theoretical concept of its existence was confirmed by detailed seismographic evidence during the Chilean earthquake of 1960.

The Gutenberg Discontinuity demarcates the mantle from the core; here the temperature is estimated to be about 3700°C. The core, which has a diameter of about 6920 km (4320 miles), may be a mass of nickel-iron *(nife)* under immense pressure (about $3·88 \times 10^6$ kg per sq. cm or 24,500 tons per sq. in), with an average density of about 10·5; it was thought that this core was wholly liquid, but from the detailed study of the behaviour of earthquake waves in recent years it is now believed that a central core, with a density of 16 to 17 and a diameter of about 2600–2700 km (1600–1700 miles), is a solid mass. This would account for the fact that the average density of the earth as a whole is about 5·517, though that of the crust and mantle are much less than this. The total mass of the earth has been calculated to be $5·945 \times 10^{24}$ kg. ($5·882 \times 10^{21}$ tons).

Isostasy The interesting fact has emerged from the study of earthquake waves (p. 30), and from gravity observations by means of pendulums, that the layer of sial is largely missing from the Pacific floor, and is present only in thin patches beneath the waters of the other oceans. The sial seems to be the major constituent of the continental crust down to about 13 km (8 miles), while the sima not only underlies this continental crust but mainly forms the floor of the oceans, except for a thin veneer of sediments. This is exemplified

6 Quarry in Carboniferous Limestone, near Pateley Bridge
Note the remarkable stratification, with slight foldings.
(*Eric Kay*)

7 Open-cast iron-ore mining, near Scunthorpe
Thick beds of iron-stone in the Lower Lias (Jurassic) formations are exposed by
the removal of the overburden of clays and shales, and can then easily be worked
by large-scale open-cast methods.
(*Eric Kay*)

8 Road fissured by an earthquake, near Kyoto, Japan
(Paul Popper)

9 Multiple normal faults in Bunter Sandstone, Bromborough, Wirral, Cheshire
The small throw enables the individual strata to be traced across the faults.
(R. Kay Gresswell)

by what is called the *Andesite Line* (fig. 3), a petrological boundary in the Pacific Ocean from Alaska, Japan, the Marianas, the Bismarck Archipelago, Fiji and Tonga to the east of New Zealand. To the west of the Line the rocks are intermediate in type, mainly andesite, dacite and rhyolite; to the east they are basic, including basalt, trachyte and olivine. In other words, the Andesite Line forms the real eastern margin of the Asiatic continental land-mass, outside which is the true ocean basin floored with simatic rocks.

It is tempting to think of the lighter outer crust as 'rafts floating' in the denser sima; where masses of sial project higher (thus forming the mountains), the penetration of the sial downward into the sima is greater to compensate for the excess mass. Conversely, under the oceans, where the sial layer is thin or absent, the sima comes nearer to or actually reaches the surface of the solid earth (fig. 2). That is why the

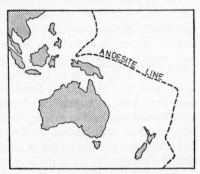

Fig. 3 The Andesite Line

site for Project Mohole was in the ocean off Hawaii, where the surface of the mantle is only about 5 km (3 miles) down, instead of on land where it might be 30–50 km (20–30 miles).

This concept is formulated as the theory of *Isostasy*, a term derived from two Greek words meaning 'equal standing', and signifying a state of equilibrium or balance in the earth's crust, with equal mass underlying equal surface area. This must involve two movements of adjustment. The first is movement in a vertical direction. For example, as great ice-caps formed, the weight of the ice may have caused sinking or sagging of the crust, and when the ice melted, so recovery gradually took place, producing considerable changes in the level of the land relative to that of the sea (pp. 259, 304). The continental shelf around Antarctica is covered with water to a depth of about 750 m (2500 ft), compared with 180 m (600 ft) around other continents; this may be the result of the weight of the present ice-sheet (p. 218). Similarly, but for the enormous weight of the Greenland ice-cap the surface of the underlying plateau would be 1100 m (3500 ft) higher. The uplift (or isostatic recovery) of northern Europe since the ice-sheets largely disappeared is still in progress; numerous former beaches occur round the coast of Scandinavia high above the present level of the Baltic, and the land at the northern end of the Gulf of Bothnia is rising at the rate of about 1 m per century. In other words, Scandinavia is still out of equilibrium. Another result of this isostatic equilibrium is that where a mountain range is gradually worn down and the resultant material is deposited on the sea-bed, as, for example, on a delta (p. 169) or on the floor of a geosyncline (p. 51), that portion of the crust underlying the area of deposition is depressed and the part underlying the area of erosion is uplifted.

The second movement of adjustment involves some sort of slow horizontal flow (or 'convection current') in the simatic layer, specifically in the Gutenberg Channel of maximum plasticity. One major problem is to account for the energy involved. Continuous radioactive disintegration within the rocks (p. 14) liberates heat, which may accumulate in the simatic layer, prevented by the overlying more solid sialic layer from escaping at the earth's surface by radiation. This, it is postulated, might make the sima layer more mobile, and also lead to sinking of the sialic continents, which could be a possible explanation of the periodic oceanic transgressions. Then the accumulated heat is dissipated (perhaps through widespread vulcanicity) at a peak of activity, the sima becomes less fluid, and the continents begin to rise, so producing a regression of the sea and the exposure of the former sea-floor on which sediments had accumulated in the interim. These major transgressions and regressions have occurred at intervals in the earth's history.

This theory of heat accumulation appears to have been observationally confirmed to some extent in 1963–4 by the discovery with the use of magnetometers of a number of 'hot spots' underneath the Andes in Peru and Bolivia. These are associated with crustal movements, and, in the words of the report of the Carnegie Institution sponsoring the research, 'appear to be repeating in our own time the vast spectacle of mountain building'.

The whole concept of isostasy, formulated and named as long ago as 1889, is much more complex than this explanation would seem to imply, but the principle appears to be well established through the researches of geophysicists and geodesists.

Continental drift One resultant possibility is that the continental masses have changed their positions relative to each other, perhaps because large blocks of sial have broken up and moved apart; this is the fascinating hypothesis of 'continental drift', a concept even mentioned by Francis Bacon in 1620, and developed by F. B. Taylor and A. Wegener. The earliest theories were based on the apparent geological similarity of coastlines along each side of the Atlantic, especially the close fit of South America into Africa, and the fact that all four southern continents reveal signs of a period of large-scale glaciation in Carboniferous-Permian times (pp. 246–7), some 250 million years ago, when they were closer together and nearer the South Pole. In brief, it was suggested by Wegener that there was an original single sialic land-mass (which he called *Pangaea*), surrounded by a primeval sima-floored ocean (*Panthalassa*). The land-mass broke up in late Pre-Cambrian times to form a northern continent (*Laurasia*), and a southern (*Gondwanaland*), separated by a long narrow ocean known as *Tethys* (p. 52). Probably the former land-mass then lay across the Equator, the latter across the South Pole. During the 150 million years between the mid-Permian and the Cretaceous periods, Laurasia broke up into two fragments (the Laurentian Shield and Fennoscandia), while Gondwanaland formed a whole series of 'ancient nuclei' described on p. 26; all these moved north relative to the present position of the Poles.

In recent years the concept of continental drift has received new support from the study of the magnetism of rocks, known as *palaeomagnetism* when very old rocks are considered. This is based on the fact that igneous rocks when cooled retain a certain magnetization, the result of the presence of grains of substances containing iron which were aligned and permanently magnetized in accordance with the direction of the earth's magnetic field at that time. This 'fossil magnetism' thus provides a record by which the position of the magnetic poles at various times may be located, and information derived from rocks of the same age indicates that the continents must have moved relatively to each other in order to accommodate the direction of magnetization in the differing areas.

Here again the suggestion has been made that radioactive heat is responsible for the energy required, which sets up convection currents moving (at an estimated rate of 1 cm per annum) in a series of 'cells' throughout the upper part of the mantle and the simatic layer. These currents rise under the oceans, then move out horizontally in either direction, taking with them the 'continental rafts' of the lighter sialic material in opposite directions, and causing them to drift apart at present rates of about 2·5 cm a year. This would result in tensional rifts or cracks in the ocean floor midway between the continents, forming a world-wide Y-shaped pattern, in length some 64,000 km (40,000 miles). The most clearly defined section is the Mid-Atlantic Ridge (pp. 337–8 (fig. 162)), which seems to be a double ridge separated by a gaping crack. It can also be traced in the eastern Pacific Ocean, and trending south-eastward from the Arabian Sea towards south-western Australia and the Antarctic.

Other rifts on a world scale include the line of the Red Sea and the East African Rift-valley (p. 45 and fig. 15), and along the west coast of North America, as indicated by the San Andreas Fault (pp. 38–9) and the Gulf of California. Along these lines are indications of crustal activity, volcanic islands and frequent earthquakes; at points along the Mid-Atlantic Ridge the heat-flow has been measured and shown to be about eight times the normal. Where the heat-flow is descending to complete the movement in the 'cell', possibly where two flows in opposite directions converge, the simatic layer and mantle may be dragged down, forming the great troughs or deeps (p. 331). It must be stressed, however, that this is a theory, admittedly supported in part both by observational evidence and mathematical calculations, but it is still a matter of controversy among geophysicists.

Another theory, based on the concept of an expanding earth (possibly caused by the diminishing of the force of gravity with the increasing age of the universe), postulates that the earth was once covered with a complete layer of sialic rocks. The expansion caused the sialic shell to break up and move apart, which would account for the oceanic ridges and rift-valleys.

THE CONTINENTS

The land-surface occupies 29 per cent of the surface of the globe (p. 328). In point of fact, the continents and the ocean basins are only slightly higher and slightly lower irregularities in the earth's surface. The difference in height between the summit of Mount Everest (8848 m, 29,028 ft above sea-level) and the bottom of the Mariana Trench (11,033 m, 36,198 ft below), amounts only to about 19·9 km (12·4 miles), which is small in relation to the earth's diameter. The globe has a slight 'flattening' at each pole (the result of its daily rotation); as a result its equatorial diameter is 12,755 km (7926 miles), its polar diameter 12,714 km (7900 miles). It can therefore be called a *spheroid*, or even a *geoid* (meaning 'an earth-shaped body').

Various hypotheses seek to account for certain broad patterns which can be discerned; both oceans and continents are roughly triangular, land predominates in the northern hemisphere and water in the southern, land seems to form a broken ring round the Arctic region, and the Arctic Ocean is counterbalanced in the south by the Antarctic land-mass (Chapter 12). There seems, in fact, to be a broad generalization that land-masses are antipodal to water areas. These distributions are facts; the hypotheses seek to explain them.

The ancient nuclei It seems that in the continents certain

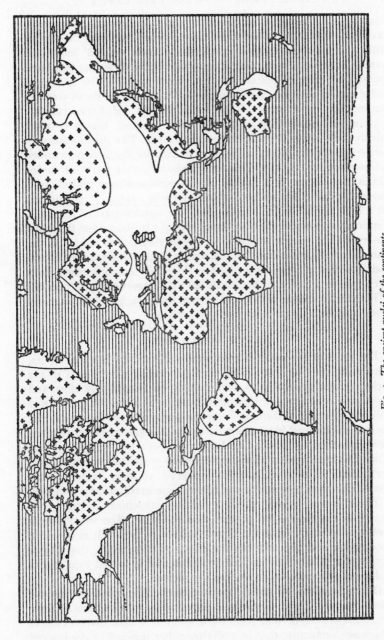

Fig. 4 The ancient nuclei of the continents
Based on a map in *The University Atlas*, edited by G. Goodall and H. C. Darby.

'nuclei', the fragments of Laurasia and Gondwanaland (p. 25), have remained more or less stable since an early period in the earth's history (fig. 4). On to these nuclei have been 'welded' mountain systems of various ages, and around and upon their margins have grown the sheets of sediments which cover much of the lowlands. These nuclei are the 'rigid masses' or 'shields'. In north-western Europe the rigid mass of the Baltic Shield (or Fennoscandia) appears on the surface in Sweden and Finland, but is covered thickly with newer deposits in western Russia. The shield-masses of Asia consist of the Siberian Platform, buried for the most part under sedimentary rocks, the Chinese Platform, small parts of Indo-China, the Deccan and Arabia. Africa consists almost entirely of a single vast shield, the rocks of which have lain relatively undisturbed for a long period of geological time. The western part of Australia comprises an ancient plateau, some 180–460 m (600–1500 ft) above sea-level. The northern part of North America consists of the Canadian (or Laurentian) Shield, and in South America the Guiana and Brazilian plateaus are tilted so that their Atlantic edges form steep ranges. Finally, in the south lies the Antarctic shield (p. 218).

The shields consist of very old metamorphic rocks, which in many areas contain mineral veins and ore-bodies.

EARTH MOVEMENTS

The crust of the earth is continually affected by internal forces, the results ranging from rapid and discernible movements of the rocks, known generally as earthquakes, to exceedingly slow and large-scale continent- or mountain-building movements, perhaps of millions of years' duration. Some movements have fractured the ancient shields and have caused areas to subside, while other areas have remained upstanding or have even been bodily uplifted and tilted. Other movements have squeezed up great thicknesses of rock, particularly of new sedimentary material, to form compressional arcs or chains of mountains and islands, bordered by deep trenches.

These forces, which produce various structural features, are known as *diastrophic*, the process as *diastrophism*. The word *tectonic* (noun *tectogenesis*), from the Greek 'tekton' (a builder), is of wider implication and also includes the effects of vulcanicity, the subject of Chapter 3, whereby molten rock-material is forced into cracks and zones of weakness, or reaches the surface.

Classification of earth movements The most convenient classification of these diastrophic movements, on the basis of their structural results, divides them into two groups. One involves forces which act vertically, that is, radially, either upward or downward along a radius from the centre of the earth to the surface. They are usually on a large scale, and hence are called *epeirogenic*, after the Greek word 'epeiros' (meaning a continent). They may involve *en bloc* vertical uplift or subsidence, which is possibly the result of isostatic readjustment, or warping and tilting.

These movements may be on a local scale, and are especially discernible around the coast, thus affecting the relative level of land and sea and therefore the nature of the coastline (p. 280). Care must be taken, however, not to confuse these actual movements of the land with changes in the level of the sea itself, known as *eustatic* movements (p. 304). In either case, a transgression of the sea over the land or the emergence of land from the sea may occur.

A fascinating example of the oscillation of the land itself in historic times is afforded by the town of Pozzuoli, the records of which are preserved in three marble pillars, 12 m (40 ft) high, of the Roman temple of Serapis, situated in an unstable part of the crust, as evidenced by earthquakes and volcanic activity, on the shores of the Bay of Naples. From about the tenth to the fifteenth centuries the town rose 8 m (25 ft), but then began to sink again, which until recently continued at an average of about 2·5 cm a year, though occasionally sudden drops occurred, as in 1913. This made it necessary to raise the streets by 1·4 m (4·5 ft), filling in the ground-floors of houses. However, in 1969 this sinking of the land not only ceased, but it began to rise rapidly; estimates indicate that in six months Pozzuoli has risen by over 75 cm (30 in), causing much structural damage and testifying to the instability of the whole area.

The second group of forces comprises those which operate horizontally, that is, at a tangent to the surface of the earth, involving both compression and tension in the crust, with resulting strains and stresses in the rocks themselves. These are the forces which have produced the great folded ranges of mountains, and hence the term *orogenic*, from the Greek word 'oros' (a mountain), has been applied; similarly, a period of mountain building is known as an *orogeny*. For reasons not wholly understood, the mountain-building movements seem to have been cyclic, that is, they have recurred periodically throughout geological time. Many of the orogenies are so ancient that it requires patient investigation to piece together and interpret the data and to produce some conception of their original nature.

These two broad types of movement afford a basis for analysing the resulting land-forms. But it must be remembered that their results cannot be arbitrarily separated and distinguished. When tangential forces come into play to form fold mountains, there may be vertical forces operating within the area affected. When the fold

29

movements took place which produced the Alpine ranges in Europe, other parts of the continent were involved in vertical movements of uplift, depression and tilting.

EARTHQUAKES

It has already been emphasized that much knowledge concerning the interior of the earth is derived from the recording and analysis of earthquake waves. The very 'shallow' earthquakes are caused by sudden diastrophic movements resulting from strain in the rocks of the crust. But the majority originate in the mantle, 'triggered off' by localized energy increases of an explosive nature; the cause of these is not yet understood, but they may be due to changes in the crystalline structure of the silicate minerals composing the mantle. The resulting vibrations or waves in the rocks travel outward from the centre of disturbance as one main shock, with 'fore-' and 'after-shocks'. The point at which the shock occurs is known as the *origin* or the *focus*; a large proportion occur in the upper 20 km (12 miles) of the crust, though the deepest yet recorded is 700 km (435 miles). A point on the surface of the earth vertically above the origin is known as the *epicentre*. It is possible to plot places suffering from equal degrees of damage, according to a recognized scale; *iso-seismal lines* are drawn through these places to show successive degrees of intensity (fig. 5). This scale used to be based on the observation of actual results; thus on fig. 5, intensity VI indicates 'oscillation of chandeliers, stopping of clocks', etc., VIII is 'cracks in buildings', and X is 'total destruction of buildings and disaster generally'. The modern Richter scale is an indication of the magnitude of shock based on instrumental records, rather than on results; a magnitude of 7·0 is a major earthquake, and the greatest so far recorded is 8·9. Delicate instruments, known as seismographs, can record an earthquake, the origin of which was thousands of miles away, and are based on the principle of a pendulum which when disturbed transmits this motion to a needle tracing a continuous record on a moving drum. The most modern types are photographic, whereby a mirror on the pendulum reflects a beam of light on to sensitized paper, or electro-magnetic, in which a small coil within the pendulum generates an electric current fed to a galvanometer; in each type friction is eliminated. It is necessary to have separate tracings of movement in all three dimensions. Another very delicate instrument, used to detect earthquakes, is the *seismometer*.

When the seismograph record is examined of an earthquake which occurred more than 1100 km (700 miles) from the instrument, three distinct types of vibrations in the strata can be distinguished; if

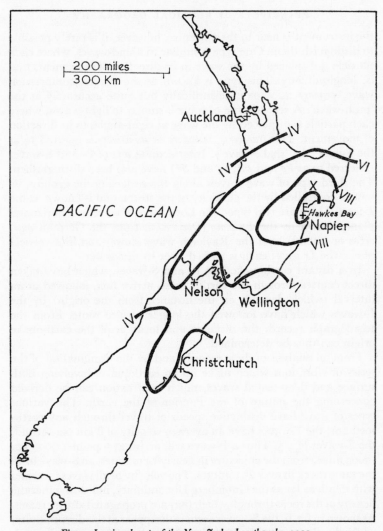

Fig. 5 Isoseismal map of the New Zealand earthquake, 1931

Based on a map in J. Milne, *Earthquakes and other Earth Movements* (London, 1939).
The isoseismal lines show the areas affected, and how severely, by the earthquake,
the epicentre (**E**) of which was near the shores of Hawkes Bay. The earthquake
caused an uplift of 1 m (about 3 ft), adding 13 sq. km. (5 sq. miles) to N.Z. territory. Unfortunately, 256 people were killed.

X. Widespread destruction; **VIII.** damage to buildings; **VI.** hanging objects
set in motion; and **IV.** doors and windows rattle. Intensities less than **IV** are not
shown on this map. The lowest intensity (**I**) would be scarcely noticeable by an
ordinary observer. These figures refer to the Rossi-Forel Scale, based on observed
effects. It has since been replaced, first by the Modified Mercalli Scale, then in
1935 by the Richter Scale, based simply on instrumental records.

the instrument is near to the epicentre, however, it is rarely possible to distinguish them. One type is similar to sound-waves, where each particle is displaced by the wave in its direction of movement, that is, longitudinally. This type is known as the *P-wave*, compression wave, *primary wave*, or unscientifically but quite accurately as the 'push-wave'. A second type of wave is similar to light-waves, where each particle is displaced by the wave at right-angles to its direction of movement. This *transverse, secondary* or *shear-wave* is referred to as the *S-wave* (the 'shake-wave'). Intermediate sets of P- and S-waves (referred to as *Pg* and *P**, *Sg* and *S**) have also been distinguished. The third type of wave travels along the surface of the ground, its nature controlled by the elasticity of the strata, and is known as the *L-wave*. There are two types, the Love wave, a wave in a horizontal plane, known as the *Lq wave* (*Querwellen*, from the German *quer*, cross or lateral), and the Rayleigh wave, slower and in a vertical plane, the *Lr wave*; each is named after its discoverer.

In a distant earthquake, the P- and S-waves, which have taken direct courses through the earth's mass, arrive first, followed at an interval (which depends on the distance from the origin) by the L-waves which have followed the longer surface route. From the seismograph record, the distance and location of the earthquake origin can thus be determined.

From an analysis of these records, and by the examination of the series of vibrations which make up an earthquake, involving both surface and deep-seated waves, much information can be derived concerning the nature of the interior of the earth. The various types of wave have distinctive speeds of travel through any earth-medium; the P-waves have an average velocity of 8 km per second, the S-waves of 4·5. Thus a P-wave will arrive at a point 11,000 km (7000 miles) from the epicentre in nearly 14 minutes, an S-wave from the same shock in over 25 minutes. The velocity of all waves increases with depth as far as the Gutenberg Discontinuity, indicating that the density of the rocks through which they are propagated also increases. As shown in fig. 6, the S-waves seem to disappear at this discontinuity, while the P-waves are refracted and travel on at a reduced speed. As a result, on the opposite side of the world to the epicentre there is a complete 'shadow zone' around the globe where no waves at all are recorded, enclosing an antipodal area which receives the weakened P-waves only. S-waves can travel only through solids, so here is direct evidence for the liquid nature of the outer core (p. 22).

The effects of earthquakes If shocks are experienced in densely populated and closely built-up areas the result can be disastrous. Some of the worst include Lisbon (1755), San Francisco

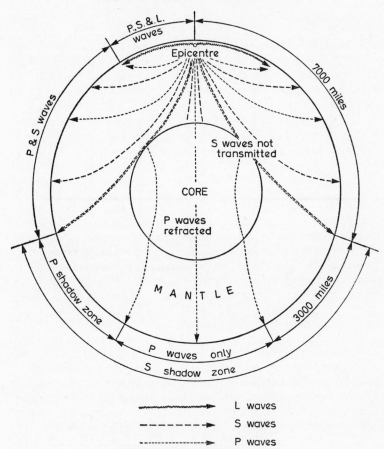

Fig. 6 *The paths of earthquake waves*
Based on various geophysical memoirs.

(1906), Kansu in China (1920), Tokyo–Yokohama (1923) which killed 143,000 and destroyed over 500,000 houses, Assam (1933), Agadir in Morocco (1960), Chile (1960), Skopje in Jugoslavia (1963), Anchorage in Alaska (1964), Bogotá, Colombia (1967), and Persia (1962, 1968). The surface waves, though often small in amplitude, can do great damage to buildings; it is said that a surface wave with a vertical range of a cm would cause most houses to collapse. On the other hand, the Alaskan earthquake of 1899 involved a vertical displacement of 14·3 m (47 ft), though exceeded by that of the Alaskan earthquake of 1964; the uplift of 15 m (50 ft)

33

of part of the sea-floor in the Gulf of Alaska is the maximum ever recorded. In most areas liable to earthquakes, modern buildings are specially constructed so that they transmit the shock and vibrate; steel and prestressed concrete are used, and the building may be erected on a thick 'raft' of concrete sunk beneath the surface. Sometimes gaping cracks or subsidences are formed, landslides occur, railways and main-pipes are cut, bridges collapse, and a wide degree of structural damage, and sometimes loss of life due to this damage and the accompanying fires, may result (plate 8). The earthquake of 17 August 1959, in the north-western U.S.A., for example, caused a huge rock-slide which blocked the Madison River with a barrier of rocks 60–90 m (200–300 ft) high (p. 197) and cracked the Hebgen Dam across the river. Farther away from the origin, the effects of the tremors become progressively milder. Where an earthquake affects the ocean floor, great waves, known as *seismic waves* or by the Japanese name of *tsunamis*, may spread outwards across the ocean at speeds of 500–800 km (300–500 miles) per hour, and sometimes cause great damage to coastal areas. These waves are frequently not very marked at sea; fishermen out at sea during the submarine earthquake of 1896 off Japan did not notice them, but the waves surged over the land, drowning 27,000 people and destroying nearly 11,000 houses.

Earthquakes are very frequent in the earth's crust, though the majority are extremely slight. Between 1913 and 1930, for example, research workers tabulated 6738 shocks pronounced enough for the epicentres to be determined. During the great periods of mountain building, major earthquakes must have been much more frequent than at present; as L. D. Stamp has said: 'The earthquakes of today are like the final murmurs of a great storm which has passed.'

No place on the earth's surface is wholly free from earthquake shocks, but in many parts the crust is so stable that the chance of a discernible tremor is negligible. Even in Britain minor shocks can be experienced, as on 11 January 1956, when tremors were widely felt over the north of England, causing pictures and crockery to be broken, in the Midlands on 10 February 1957, and on 25 October 1963, when shocks were felt near Portsmouth. Britain's worst recorded earthquake is said to be that experienced in the Colchester area in 1884, which destroyed or damaged 1200 buildings over an area of 400 sq km (150 sq miles), completely ruining Langenhoe church to the south of Colchester. Such events are extremely rare in many apparently stable parts of the earth.

If the epicentres of major shocks are plotted, it will be seen that some areas are much more liable than others, and these are shown in a simplified form on fig. 7. There is one great belt around the Pacific, and another stretching through the Mediterranean eastward through Asia Minor, the Middle East and northern India. The association of these areas with recent fold-mountain building, and with lines of island arcs and neighbouring linear 'deeps', is not surprising; these are the regions of crustal instability. Many areas, for a similar reason, coincide with active volcanoes, but a comparison with fig. 38 will demonstrate that other parts liable to earthquakes, such as northern India, have none.

THE EFFECTS OF EARTH MOVEMENTS ON THE CRUST

Rocks may be exposed to stress, causing deformation of their volume and shape. If two forces act away from each other the resulting strain is a tension, which involves an extension of the surface ('tensional splitting'), and results in the production of *joints* and *normal faults*. When the two forces act towards each other it is a compression, which involves a contraction of the surface and results in the formation of *folds* of varying degrees of amplitude, *reversed faults* and *thrust-faults*. If the two forces act parallel to each other, though in opposite directions, it is a *shear*, and it is also usually associated with jointing and faulting; as the two adjacent portions of the rock slide past each other their character is modified, sometimes by shattering and crushing. Should the stress be very great, as in the case of rocks at immense depths subject to extreme pressure and heat, the result may be the plastic deformation of the rocks, causing actual *viscous flow* rather than rupture. In the study of the flow of a solid, the relationship between its resistance to viscous flow and to elastic deformation is known as *rheidity*. This relationship obviously involves the time-factor. A substance undergoing deformation in this way is therefore known as a *rheid*.

On a world scale, the whole mid-ocean rift system (p. 25) represents a tensional zone, the lines of fold mountains and arcuate islands (p. 51) compressional zones.

Joints Where there is little or no actual movement or displacement of the rocks, a local tensional stress or shearing may produce cracks known as *joints*. Sometimes there are two sets of joints, one set parallel to the dip of the strata, the other at right angles to it, i.e. parallel to the strike. Joints are of course formed by other than diastrophic causes (p. 51).

FAULTS AND RELIEF FORMS PRODUCED BY FAULTING

When a fracture or rupture takes place and the rocks are displaced on either side of it relative to one another, the result is

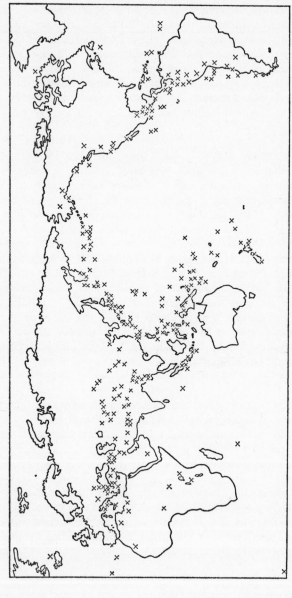

Fig. 7 Epicentres of major earthquakes during the last century

Based on a map in *The Great Soviet World Atlas* (Moscow, 1938), with the addition of recent earthquakes. Only 'catastrophic' and 'destructive' earthquakes are located. In some areas, notably in Japan and Indonesia, several earthquakes have occurred in virtually the same place.

known as a *fault* (plate 9). The total movement is known as the *shift*, which involves both *slip* (the relative movement along the fault), and *throw*, the vertical change of level of the strata. Where the fault is inclined there must be some lateral displacement, known as the *heave*. The angle of inclination of the fault-plane from the vertical is the *hade* (fig. 9); the rock-face on the upper side of the fault is the *hanging-wall*, that on the lower side is the *foot-wall*.

A distinction is made between a *normal fault* due to tension, when the inclination of the fault-plane and the direction of the down-throw are both to the left or both to the right, and a *reversed fault* caused by compression, where the beds on one side of the fault-plane are thrust over those on the other side. Over-

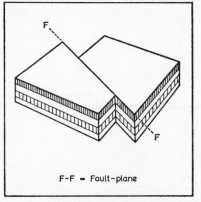

Fig. 8 Block-diagram of a tear-fault

thrusting, where a reversed fault of a very low angle is involved, occurs| when compressional earth movements have operated very powerfully.

A *tear-, strike-slip-, wrench-* or *transcurrent fault* is a vertical fracture, but instead of a vertical movement of one side relative to the other, the displacement is horizontal along the line of the fault. The effects are sometimes seen in earthquakes, when, for example, railway lines are cut across, so that the broken ends are several yards apart, although still on the same level. The San Andreas fault (fig. 10) is of this nature; in the 1906 earthquake, the land to the west of the fault-line was moved 6 m (20 ft) to the north, while the vertical displacement was only about 1 m (3 ft). One of the largest tear-faults in Britain is that of the Great Glen of Scotland, where the horizontal displacement of the strata is about 100 km (65 miles). In many instances, major tear-faults define the true margins of the ocean basins, notably the Pacific, and the boundaries of the ocean trenches (p. 332).

A *monocline* is a tensional feature in which the strata are bent or flexed; it must not be confused with the monoclinal fold which is really an asymmetrical fold with one limb markedly steeper than the other. A monocline is closely related to a normal fault, and may turn into one at depth or further along the line of movement.

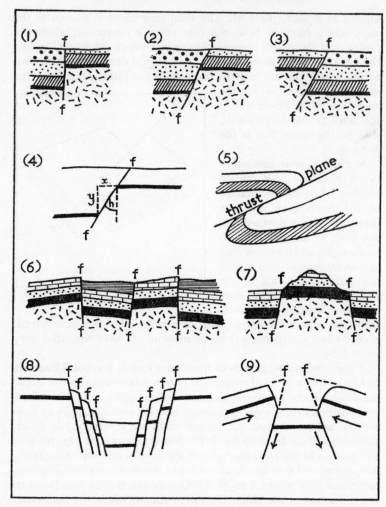

Fig. 9 Types of faults and rift-valleys

ff indicates the line of the fault.

1. Normal fault, vertical **2.** normal fault, inclined; **3.** reversed fault; **4.** terminology of a fault; **h** = hade (in degrees); **y** = throw (in m); **x** = heave (in m); **5.** thrust-fault in a recumbent fold (see p. 49); **6.** 'basin-and-range' faulting; **7.** a horst; **8.** step-faulted rift-valley; **9.** compressional rift-valley. In **2, 3** and **4,** the hanging-wall is to the left, the foot-wall to the right.

Faults rarely occur singly, but a number may be parallel or *en echelon* to each other. One of the longest single faults is that in western North America, the San Andreas already mentioned, which

10　The rift-valley of the Jordan, south of the Sea of Galilee
(*Paul Popper*)

11 Folded strata in Purbeck Limestone, Stair Hole, Lulworth, Dorset
(*R. Kay Gresswell*)

12 Folding in the Jura Mountains
An example of a small anticline and syncline in Jurassic limestone near Delémont, Switzerland.

(*Eric Kay*)

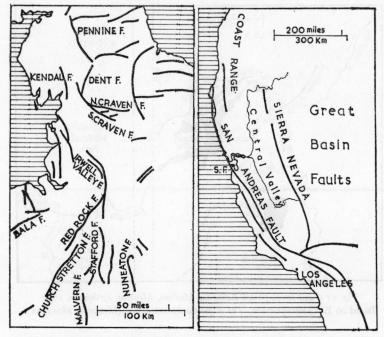

Fig. 10 *Fault-lines in northern England* (left) *and in western North America* (right)

The left-hand map is based on the Geological Survey.

The right-hand map, much simplified, is after R. A. Daly. It should be noted that there is an immense number of both major and minor faults east of the Sierra Nevada. **S.F.** = San Francisco.

can be traced on the surface for at least 1000 km (600 miles) almost through San Francisco, more or less parallel to the coast (fig. 10). This fault is still active, and the Los Angeles aqueduct has been strengthened where it crosses the line. The movement is of the order of about 5 cm a year, which is probably better than the storing up of the elastic energy without any relief, which might produce another major earthquake. On a different scale, a whole series of faults has been traced in northern England (fig. 10). Sometimes sets of faults intersect at or near right-angles, a phenomenon known as *cross-faulting*; one result is a distinct angularity of pattern in the drainage which develops, since streams tend to follow the easily eroded shatter-belts. This is shown on a small scale on the Wasdale face of Scafell Pike (p. 153), and again in the Highlands of Scotland. The dissected plateaus in East and South Africa reveal the effects of cross-faulting on an enormous scale.

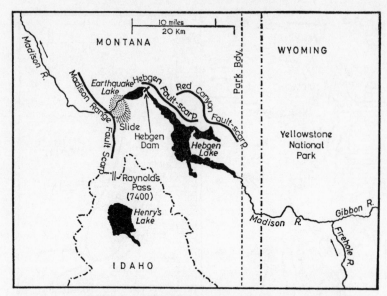

Fig. 11 *Fault-scarps formed during the Madison, Montana, earthquake, 1959*
Based on E. Christopherson, *The Night the Mountain Fell* (Missoula, Montana).

Sometimes the rocks forming the surface-plane of a fault may be highly polished, and occasionally they show fine striations or fluting. These surfaces, known as *slickensides*, are caused by friction, commonly resulting in fusion by heat generated through differential movement along the divisional planes of the fault. They sometimes give an indication of the direction of movement on either side of the fault. Some excellent examples in sandstone are shown in the Frodsham area in Cheshire. The chalk cliffs along the Dorset coast (plate 85) west of Lulworth Cove have been so severely affected by earth-movements that the Upper Chalk has been inverted; the packed vertical strata exhibit clear slickensiding.

Fault-scarps and fault-line scarps The vertical movement of the rocks along the line of a fault will produce a fault-scarp, the face of the foot-wall (fig. 12). This is an initial land-form, the direct result of crustal movement, and is so rapidly affected by denudation that it may soon be destroyed; where it exists it is the result of a recent earthquake. The Madison (Montana) earthquake of 17 August 1959 created two fault-scarps (fig. 11), with down-throws of about 6 m (20 ft). These faults continue vertically for an unknown depth, and can be traced across country for several

FAULT – SCARP

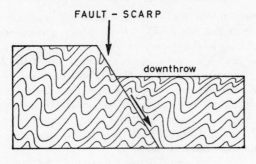

OBSEQUENT
FAULT-LINE SCARP

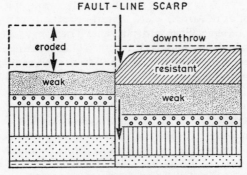

Fig. 12 *A fault-scarp and a fault-line scarp*
The lower diagram is of an obsequent fault-line scarp.

km, like terraces on the hillside. Weathering has attacked the face of the scarps which, as they are composed of boulder-clay, will soon be smoothed out and will ultimately disappear as surface features.

It usually happens that faulting brings rocks of differing resistance into close juxtaposition, so that differential denudation will emphasize the line of the fault, forming a *fault-line scarp*. Thus to the north of the Mid-Craven Fault in the Malham district of Yorkshire is an extensive plateau of Carboniferous Limestone. To the south this has been down-faulted about 1000 m, so that the surface rocks consist of less resistant Bowland Shales. Denudation has caused the line of the fault to stand out as a series of 'scars'—Giggleswick and Attermire Scars, Malham Cove and Goredale Scar.

Occasionally denudation may progress so far that the resistant strata on the higher side of a fault-line may be removed, exposing the less resistant underlying strata. The strata on the down-throw

41

side may now be the more resistant and will gradually stand out, so that a fault-line scarp will develop along the line of the same fault, but facing in the opposite direction. This is sometimes called an *obsequent fault-line scarp* (fig. 12); an example is the Mere Fault on the north-western side of the Vale of Wardour in Wiltshire.

If denudation still continues, this obsequent fault-line scarp may also be obliterated, as great thicknesses of strata are removed. Far below the original surface a fault-line scarp may develop yet again, facing in the original direction; this is termed a *resequent fault-line scarp*.

Plateaus and basins Movements of uplift or subsidence are responsible for some marked relief features. The crust may warp on a large scale, producing a broad arch or a shallow basin, or the strata may be fractured and faulted to form sharply defined blocks of upland, basins and troughs. These movements may affect the shield areas, as in Africa, which consists essentially of a great series of basins, faulted valleys and plateaus. They may also affect the fold-mountain belts, and very complex movements of uplift and depression, accompanied by faulting on both a large and small scale, may diversify fold-mountain ranges. The ranges of the North American Cordillera enclose a series of plateaus and basins (fig. 13), while much of central Asia consists of fold ranges of different ages enclosing plateaus on an enormous scale—the Tsaidam and Tarim Basins, the Red Basin of Szechwan, the plateau of Mongolia and many others. The vertical movements involved have been enormous; thus the Tarim Basin lies at a height of 1800 m (6000 ft) above sea-level between the Altyn Tagh and the Tien Shan, but in the north-east of the basin is the deep down-faulted Turfan Basin, the floor of which is 210 m (700 ft) below sea-level. These plateaus enclosed by higher ranges are known as *intermont*. The name *table-land* is used where the edges of the plateau are sharply defined, dropping steeply on each side, as in South Africa and Arabia.

The term *basin* is widely employed: it can denote an intermontane plateau, like the Great Basin of Nevada and Utah, a sediment-filled 'sag' in the crust, such as the Congo or Chad Basins in Africa, and a sea-filled depression, usually lying among a fold-mountain system, e.g. the basins of the Mediterranean, the Aegean and the Black Seas.

Fault-blocks Areas of the crust may be divided into individual elevated or subsided masses by faults, known as *block-faulting*. Sometimes the blocks have been tilted, hence the name *tilt-blocks*. One of the most striking examples of these tilted blocks is in the west of the U.S.A. What seems to have been an original folded arch has been strongly faulted, leaving the bounding tilt-block ranges of the Sierra Nevada on the west and the Wasatch Mountains on

Fig. 13 Intermont plateaus in the Cordillera of western North America
The main trends of the ranges are shown diagrammatically.

the east, each with steep in-facing walls. The Sierra Nevada is a
block 650 km (400 miles) long and 130 km (80 miles) wide, which
has been uplifted and tilted so that it is markedly asymmetrical. Its
eastern face rises 2400 m (8000 ft) along a major fault-line scarp at
an average slope of about 23°, deeply dissected to form peaks
culminating in Mount Whitney (4420 m, 14,495 ft). The western

43

slope on the other hand descends at an angle of only 2° to 4° towards the Great Valley of California. The fault-line scarp is not the product of a single master fracture but of a multitude of minor ones which have occurred at different times since *en bloc* uplift started in the Eocene. The effects of faulting in the early Pleistocene are shown by the fact that some moraines are actually cut by faults, and in one place a small volcanic cone is sliced through. This movement continues, as shown by frequent small earthquakes; in 1872 a larger shock in the Owens Valley area caused some damage and loss of life, leaving a clear fault-scarp which can still be seen. Another example is the range of the Grand Tetons in Wyoming, to the south of Yellowstone National Park; this is the dissected eastern edge, rising to 4190 m (13,766 ft), of a tilted block along the line of the Teton Fault; the block was uplifted about 2000 m (7000 ft) in Cainozoic times.

Between the Sierra Nevada and the Wasatch numerous faulted blocks are tilted usually with a steep eastern face and a more gentle western one. These blocks have been much eroded, the material filling the intervening basins to great depths. This is the Basin-and-Range country, and so such a series of tilted blocks is sometimes called *basin-and-range structure* (fig. 9).

The steep outward-facing edges of tilted blocks form very striking relief features. The Cévennes are the south-eastern rim of the Central Massif of France, the Sierra Morena are the south-eastern edge of the Spanish Meseta, the Erzgebirge and Riesengebirge form the northern edge of Bohemia in Czechoslovakia, and the Western and Eastern Ghats are the edges of the Deccan. An example of a gently tilted block is provided by the North Pennines (fig. 14). This, sometimes known as the Alston Block, is bounded on the west by the Pennine Fault, which has its down-throw on the west. Thus the long, steep edge of the block, the highest point of which is Cross Fell (882 m, 2893 ft), falls steeply for more than 500 m (1600 ft) as a fault-line scarp to the Triassic floor of the Eden valley, while the surface of the block slopes gently eastward toward the North Sea.

Horsts When an individual fault-block is sharply defined, it is given the name of *horst*. It may be left upstanding by differential movement, either the sinking of the crust on either side of a pair of faults, or the bodily uplift of a mass between them. Many of the Hercynian masses of central Europe, the Vosges, the Black Forest and the Harz, are horsts, also Sinai, between the Suez and Akaba faults, and Korea, between the Yellow Sea and the Sea of Japan.

Rift-valleys In a sense, a rift-valley is the reverse of a horst, because it consists of a fault-trough, let down between parallel

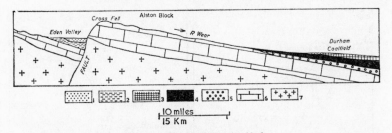

Fig. 14 *The North Pennine block*
The section is drawn along a west–east line approximately through Cross Fell.
1. Triassic sandstones; 2. Permian sandstones; 3. Magnesian Limestone; 4. Coal
Measures; 5. Millstone Grit; 6. Carboniferous Limestone; 7. Lower Palaeozoic
rocks. *Note:* The Great Whin Sill is omitted to avoid complexity.

faults, with throws in opposite directions. There may be a series of
faults on either side of the trough, producing *step-faulting*, or the
sides of the rift-valley may be clean-cut due to the immense down-
throw of a single great fault on either side. Exactly how a rift-valley
has been formed is not clear; the chief problem is to account for
the mass of the subsided block. Tension in the crust may have
pulled the two sides apart, leaving the centre to subside in the form
of the classic metaphor, 'the collapsed keystone of an arch' (fig. 9).
Another hypothesis is that the movements were of deep-seated
compression from the sides, so that the masses on either side of the
faults were thrust up higher than the central block, which was forced
down between the two to form the valley. Again, during a movement
of general uplift a long line of country between two faults may have
risen more slowly than did the masses on either side. Yet another
theory is that the rift-valley is the result of the gentle up-bending
of somewhat rigid strata, so that a gaping large-scale crack developed
along the crest of the swell. Many rift-valleys are associated with
igneous activity along their margins.

Whatever the mechanics of their formation, rift-valleys form
striking features, especially where denudation has not appreciably
modified their outlines. The steep walls of the Jordan rift-valley,
with the wilderness of Judaea on one side and the bare mountains
of Jordan on the other, enclose the Dead Sea in a clean-cut trough
(plate 10). This is the northern end of a very long rift-valley system
(fig. 15). It does not consist of a single trough, rather a series of
connected curving branches, but it can be traced for a distance of
more than 5000 km (3000 miles) from northern Syria, through the
Jordan Valley, the Gulf of Akaba, the Red Sea, Abyssinia and East
Africa (where it bifurcates into eastern and western branches), to

45

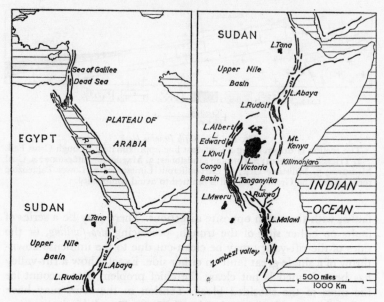

Fig. 15 *The East African rift-valley*

The faults are shown diagrammatically by heavy lines. The right-hand diagram forms the southerly continuation of that shown in the left-hand diagram.

the Zambezi. The down-throw was on a considerable scale; the floor of the Dead Sea is more than 800 m (2600 ft) below the surface of the Mediterranean Sea.

Other rift-valleys include that of the middle Rhine between the two horsts of the Vosges and the Black Forest (fig. 16), and the Midland Valley of Scotland between the Highlands and the Southern Uplands. The last example, however, is less clearly defined; although it undoubtedly lies between two major boundary faults, it has been subject to so much denudation and volcanic activity that its rift-valley form is to a large extent obscured.

The German word *Graben* is often used synonymously, though incorrectly, for rift-valley. The latter is a relief feature, while a graben is a structure enclosed by parallel faults which may or may not be a valley.

Various terms have been used in recent years for other types of depression formed by crustal tension, compression or shearing. These include the *rhombochasm*, a parallel-sided rift-valley (the Dead Sea), and the *sphenochasm*, a triangular depression between continental blocks, such as the Arabian Sea.

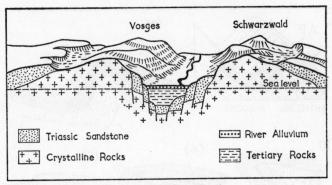

Fig. 16 *The Rhine rift-valley*

Fault-valleys A single fault may determine the line of a valley. The Eden Valley is bounded on the east by the long Pennine Fault (figs. 10, 14). The Vale of Andalusia (the Guadalquivir valley) lies in a valley demarcated by the faulted edge of the Sierra Morena. The Great Glen of Scotland (Glen More) cuts right across Scotland from the Firth of Lorn and Loch Linnhe in the south-west to the Moray Firth in the north-east, 160 km (100 miles) in length. The floor of this glen is never higher than 30 m (100 ft) above sea-level, and parts of the floors of the lochs lying along the line of the fault are 240 m (800 ft) below sea-level. Of course, the present form of the glen is largely due to glaciation, but it does reveal how a fault may provide an initial line of weakness. The Glen More fault follows the 'grain' of the country, and the map of Scotland shows many more glens and lochs with the same orientation.

FOLDING

Folds (fig. 17) Compressive forces in the earth's crust can cause widespread crumpling or folding (plates 11, 12), usually along well-marked zones which indicate lines of weakness. These folding movements may produce small wrinkles only a few cm in size, yet they have been responsible for most of the world's great mountain systems. The simplest form of folding is symmetrical, when the strata are bent into a simple upfold (*anticline*) and into a corresponding downfold (*syncline*). The central line of either fold is known as the *axis*, while the two sides are the *limbs*. If one side of the fold is steeper than the other, it is said to be *asymmetrical*. If the axis of a fold dips, the fold is said to *pitch* in the direction of dip. A *drag fold* is a minor fold, formed either subsidiary to a main fold, or along the sides of a fault, where the vertical displacement has made

47

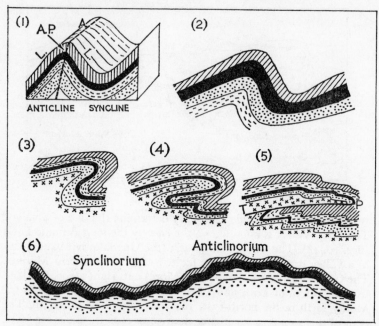

Fig. 17 *Types of folds*

1. Simple folding (**A** = axis; **AP** = axial plane; **L** = limb); **2.** asymmetrical folding; **3.** overfold; **4.** recumbent anticline; **5.** overthrust fold or nappe (**TP** = thrust-plane); **6.** an anticlinorium (an anticlinal complex of folds) and a synclinorium (a synclinal complex of folds).

flexures and puckers in the rocks on either side. The crest of a fold is usually structurally weak, and igneous material or masses of salt may be forced through the cracks and fissures; these intrusions are known as *diapirs* and the folds as *diapiric* or *piercement* folds.

A *pericline* (or *centrocline*) is a form of anticline, usually of small dimensions, which pitches along its axis in each direction from a central high point. It is in a sense an elongated dome. Examples are common in the chalk country of southern England; they include the Kingsclere and Shalbourne periclines to the west of Basingstoke. Most periclines, from their nature, are particularly vulnerable to breaching along their axes (p. 181–2). The Bighorn Mountains of Wyoming afford an example on a large scale; a folded wrinkle in the crust, 160 km (100 miles) in length from north to south and about 50 km (30 miles) across, rises abruptly from the plain. It has been so eroded that the granite core of the centre of the fold has been exposed, culminating in Cloud Peak (4011 m, 13,165 ft).

48

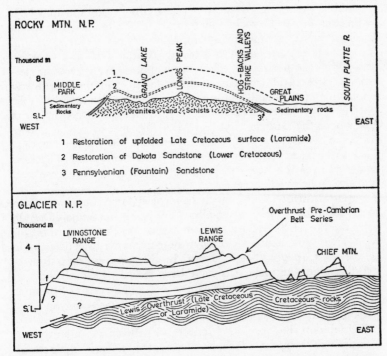

Fig. 18 *Sections across the Rockies in Rocky Mountain National Park, Colorado* (above) *and Glacier National Park, Montana* (below)

Based on diagrams appearing in pamphlets on the respective National Parks published by the National Park Service, U.S. Department of the Interior.

In some fold systems, the main anticlines and synclines appear to have numerous minor folds superimposed upon them. Such an anticlinal complex forms much of the Southern Uplands of Scotland and another the Weald; it is given the name of an *anticlinorium*, while the synclinal complex is a *synclinorium*.

The nature of the fold depends on the intensity of the forces involved. In the ranges of the Rockies in the neighbourhood of Rocky Mountain National Park in Colorado, the folding was quite gentle and regular (fig. 18); denudation has now removed the central part of the folds, but the overlying strata can be seen dipping away on either flank.

If the folding movements are very intense, the asymmetrical anticline is pushed right over, and *overfolds* or even *recumbent folds* may be formed, while a whole series of parallel overfolds is known

49

as *isoclinal folding*. If the pressure exerted on a recumbent fold is sufficiently great, the rocks may fracture, and a mass will be thrust bodily forward, often for many km, along the plane of the fracture, known as the *thrust-plane*; this is really a kind of fault at a very low angle from the horizontal. These thrust-planes are very characteristic features in all areas of complex folding; in the North-west Highlands of Scotland four main ones can be distinguished (the Glencoul, Ben More, Moine and Sole); in the northern Ardennes of Belgium is the Grande Faille du Midi (fig. 25), over which rocks from the south have been driven north; and in Glacier National Park, Montana, the Pre-Cambrian 'Belt' Series have been forced eastward over the Lewis thrust-plane (fig. 18), so that these very old rocks now overlie Cretaceous strata.

The overthrust masses which have been forced far away from their 'roots' over these thrust-planes are given the name of *nappes*, after the French word for a cover or sheet. They are responsible for some of the most complex mountain systems, as the brief description of the Alps on pp. 54–4 indicates. In the most extreme form of overthrusting, the rocks may have been forced over each other in slices, as has happened in the North-west Highlands of Scotland; this is known as *imbricate structure*; the strata lie like a steeply inclined pack of cards (fig. 19).

Denudation has profound effects on these nappe-structures, especially as the rocks, having been subjected to considerable strain,

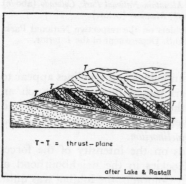

T–T = thrust–plane

after Lake & Rastall

Fig. 19 *Imbricate structure*

are structurally weak. The development of a river system, for example, in a folded region is a story of great complexity (pp. 179–83). Two structural features resulting from denudation are the *Klippe* and the *Fenster* (or in French, *fenêtre*). The klippe, or nappe-outlier, is a surviving portion of a nappe almost destroyed by denudation, and as a result a mass of older rocks may rest on younger; this is well exemplified in Glacier National Park, Montana (fig. 18), where Chief Mountain is a superb example of a klippe. By contrast, denudation may wear a hole in a nappe, thus exposing the younger rocks beneath as a 'window', which is very helpful to the geomorphologist since it affords him information about the underlying structures. In the eastern Alps in Switzerland and Austria, the underlying nappes are exposed in three 'windows': in the Lower Engadine, Semmering and Hohe Tauern districts. The Fenêtre de Theux in the Belgian Ardennes is a mass of Devonian rocks revealed by the removal of the overthrust Cambrian-Silurian rocks.

Warping Where only a gentle deformation of the crust has taken place over a considerable area and time, the result is known as *warping*. Though it may involve a vertical movement of only a few metres, this warping may have important results along coastal margins, or on the surface of an uplifted peneplain (p. 94), or on near-horizontal sedimentary rocks. The Lake Superior basin, for example, is still being warped in such a way that the northern shore is being slowly raised and the southern shore slowly lowered; thus on the north raised beaches can be seen (p. 312), on the south drowned river mouths and submerged forests (p. 310).

FOLD-MOUNTAIN SYSTEMS

If a physical map of the world, or of a continent, is examined, the great mountain ranges appear to be arranged in long lines and interlinked arcs. The more recent fold-mountain systems are often called 'younger folds' because the majority of the movements took place in mid-Tertiary times. 'Younger' is of course used in a relative or geological sense; this folding was at its maximum about 35 million years ago in the Miocene. The main results comprise (i) an *Alpine-Himalayan system* extending from the Straits of Gibraltar, through the Mediterranean Basin and Asia Minor, then across the north of India to south-eastern Asia, and (ii) a *Circum-Pacific system*, comprising the Andes, the North American Cordillera and the island arcs (which represent partially submerged folds) off eastern Asia. The two systems meet in the complex Indonesia area to the north of Australia. From these main systems various loops swing away, such as round the Caribbean Sea, and another running through New Guinea into New Zealand. There is still much doubt about the exact relationship of the individual fold-lines making up these systems.

Rarely do the mountain systems consist of a single upfolded ridge, but more usually of either a double arc separated by a trench or of a whole series of folds of all degrees of complexity, though more or less parallel to the main trend, and sometimes involving upstanding blocks of ancient rocks and basin-like depressions.

Geosynclines One of the problems concerned with the understanding of fold mountains is the explanation of the vast thicknesses of sedimentary rocks involved. At least 7500 m (25,000 ft) of sedimentary strata were included in the building of the Appalachians, for example. The only feasible explanation of how a thickness of 8–10 km (5–6 miles) of sediments could accumulate is to postulate a linear depression of the sea-floor, to which the name *geosyncline* has been applied, that is, a down-fold on an 'earth scale'. This subsided at more or less the same rate as sediments were deposited in it over long periods of geological time, derived from the wearing away of

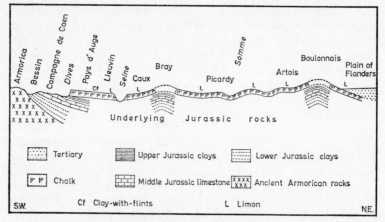

Fig. 20 The minor fold flexures of the northern Paris Basin

Based on A. Cholley, *La France* (n.d.), p. 77. The length of the section is about 300 km (190 miles).

The chalk strata reveal the 'ripples' of the outer margins of the main Alpine folding. Two of the anticlines (Bray and Boulonnais) have been 'unroofed' (p. 184), exposing the underlying Jurassic rocks.

land-masses on either side of this depression. They are classified into various types, notably the *miogeosyncline*, where little or no vulcanicity (p. 61) has occurred, and the *eugeosyncline*, where vulcanicity was a striking associated feature of the infilling of a geosyncline.

Stretching across what is now Eurasia was a geosyncline, to which the name 'Tethys' is given; to the north lay a land-mass consisting of the various shields and other old mountain systems (known as *Laurasia*), to the south another land-mass of which Africa, Arabia, the Deccan and western Australia are perhaps remnants (known as *Gondwanaland*) (p. 25).

Forelands At first it was thought that earth movements caused one side of the trough to move towards the other, thus squeezing up the strata at the sides and causing them to buckle up and splay

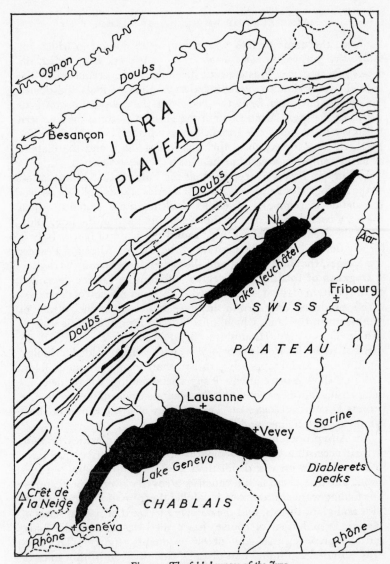

Fig. 21 The folded ranges of the Jura

The arc of the Jura Mountains extends across the frontier between France and Switzerland, which is shown as a pecked line. Most of the French Jura consist of a faulted plateau, known as the *Jura tabulaire*, descending from east to west in three main steps. The folded Jura (Jura *plissé*) consist of a series of anticlines and synclines; the anticlines are here shown only diagrammatically as heavy lines, since there are about 160 individuals in the central part. The anticlines here mostly form ridges, the synclines valleys (known as a *val*). The rivers in places break through the anticlines in right-angled bends, forming steep-sided gorges or *cluses*. The highest point, the Crêt de la Neige, attains 1723 m (5653 ft). **N** = Neuchâtel.

forward; the name *foreland* was given to the side on to which the folds were pushed, *backland* or *hinterland* to the side from which the motive force came. This concept has largely been replaced by the idea of two forelands moving together, so that the folds and thrusts are forced outwards from the centre of the geosyncline towards them. In the great Asiatic system the Himalayas form the southern fold overthrust on to the Indian Foreland, the Kwen Lun are the northern folds forced on to the Asiatic Foreland, and the plateau of Tibet comprises a vast 'median mass' between the two.

In Europe, however, it seems that the bulk of the folding was on to the northern foreland, consisting of a series of stable blocks, largely the denuded remnants of earlier fold-mountain systems, which played a considerable part in the exact alignment of the later folds. Thus the great curve of the Alps round the plain of North Italy is largely determined by the position of the Central Massif of France, the Vosges, the Black Forest and the Bohemian Plateau (fig. 23).

Degrees of folding The degree of folding in Europe depended on the position of the strata involved. The 'outer ripples' of the movements produced only gentle flexures, as shown in the Weald of south-eastern England, which in its simplest form is a gentle upfold, and in Dorset and Hampshire (fig. 95) and in the Paris Basin (fig. 20) equally gentle downfolds with slight folded 'ripples' pitching-out and reappearing rather than long continuous folds. The Jura show a more intense if superficial folding, where the overlying strata moved easily over a basement-surface of low friction, a process known as *décollement*. They consist of a succession of short symmetrical anticlines and synclines (fig. 21).

The Alps proper, however, reveal very great complexities. Because of their accessibility to the geologists of many countries, they have been studied in greater detail than any other range, and yet it is significant that there is still much controversy over some features. The folding was very complex, involving the formation of recumbent folds and giant thrusts, and the carrying forward of several series of nappes. Denudation, of course, has etched deeply into the folds, and the present peaks are merely remnants (fig. 22). The Pre-Alps (situated south and east of Lake Geneva) consist of a mass of klippen whose origin is still much of a mystery. South of these come the nappes of the High Calcareous Alps, which make up much of the Bernese Oberland (with the Diablerets, Jungfrau, Wildhorn and many more fine mountains). The Pennine nappes lie still farther south beyond the Rhône valley, a series of six forced out of the main geosyncline, now forming peaks such as Monte Rosa and the Matterhorn. The latter (figs. 22, 112 and plate 13) is a fragment of

13 The summit of the Matterhorn
An unusual view of this fine rock-pyramid; the more familiar triangular profile visible from Zermatt is of the face on the left (the east face). The ridge in the left-centre is the Hörnli, Edward Whymper's route on the first ascent in 1865, the easiest way to the summit. On the extreme left is the Fürggen ridge, on the extreme right the Zmutt ridge. The fourth ridge of this pyramid is on the far side, the Italian ridge.

(*Swissair-Photo A.G.*)

14 Cascade Mountain in the Canadian Rockies, near Banff, Alberta
A remarkable example of a rock peak carved from a synclinal fold. In the foreground are sections of the Canadian Pacific Railway and the Trans-Canada Highway.

(*The Photographic Survey Corporation Ltd, Toronto, Canada*)

15 Granitic dyke on the Saddle, Arran

A well jointed granite dyke, situated in the Saddle (a col between Goat Fell and Cìr Mhòr) stands out because of the removal of less resistant rocks.

(L. J. Monkhouse)

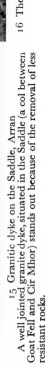

16 The Great Whin Sill at Bamburgh Castle, Northumberland, overlying sandstone of the Carboniferous system

(R. Kay Gresswell)

17, 18 Dykes on the coast of Portugal
The left-hand dyke is more resistant than the country rock and stands out as a
wall; the right-hand dyke is less resistant and has been eroded by the waves to
form a trench.

(*F. J. Monkhouse*)

19 Salisbury Crags and Arthur's Seat, Edinburgh
Salisbury Crags are part of a sill, Arthur's Seat a much eroded composite
volcanic neck, probably the remnants of a volcano with two vents.

(*Eric Kay*)

20 The Devil's Tower, Wyoming

(F. J. Monkhouse

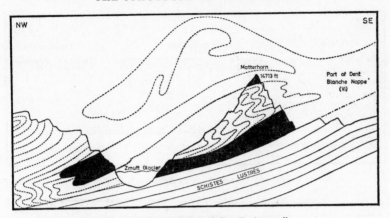

Fig. 22 The Matterhorn (after E. Argand)

This magnificent rock pyramid (see also plate 13), rising to 4505 m (14,713 ft), has been carved out of the core of a great inverted fold of Hercynian rocks, part of Nappe VI, pushed over the younger *schistes lustrés*. The Valpelline Series (in black), which form the summit cone and part of the base of the peak, consist of dark-coloured, much altered, igneous and metamorphic rocks—granite, gneiss, marble and gabbro. The tremendous cliffs below are composed of the Arolla Series (pecked lines) of lighter coloured granite and gneiss. The underlying *schistes lustrés*, so called because of their gleaming appearance, are derived from deep-sea sediments deposited in the Tethys geosyncline.

the Dent Blanche nappe (number VI in the accepted Alpine series), consisting of crystalline rocks (schist and gneiss) of Hercynian age. Involved in these folds are several blocks of ancient granitic rock, which probably represent either portions of the 'splintered edge' of the foreland overwhelmed by and caught up in the folding or crystalline rocks from the floor of the geosyncline forming the 'cores' of the folds, or even may be batholiths (p. 65). The granite mass of Mont Blanc itself and neighbouring peaks such as the Grandes Jorasses from one of these remnants, and the Aar massif (out of which is carved the Finsteraarhorn) is another. These crystalline massifs are exposed at intervals because the plane of the nappes, i.e. a line at right-angles to the direction of the folding movements, seems to undulate; the higher parts are called *culminations*, the lower parts *depressions* (fig. 23). The culminations carried upwards the higher parts of the nappes so that they have been more exposed to denudation and therefore largely removed, thus exposing the crystalline basement rocks.

Synclinal valleys A valley formed by a syncline would seem to be an obvious result of folding. Broad, shallow synclines do indeed often produce gently sloping basins, such as the London Basin

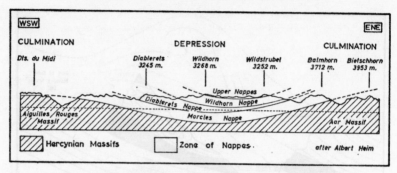

Fig. 23 Section across the Bernese Oberland

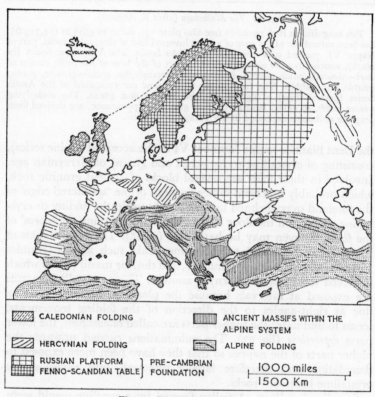

Fig. 24 Structural map of Europe
Based on E. de Martonne.
The general trends of the Hercynian and Alpine foldings are indicated by lines.

between the Chilterns and the North Downs, the Hampshire Basin and the Paris Basin. The longitudinal valleys of the Jura, each known as a *val*, are mainly the original synclines (fig. 21). Other examples of synclinal valleys are the upper (Vorder) Rhine, reveal-

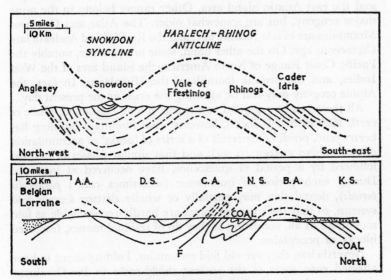

Fig. 25 Old fold mountains

The upper diagram is based on a section by E. Greenly, in *The Mountains of Snowdonia*, edited by H. R. C. Carr and G. A. Lister (London, 1948). It reconstructs two great Caledonian folds, an upfold known as the Harlech-Rhinog anticline, and a downfold over the area where Snowdon now stands; the folding was so complex that these two folds are an anticlinorium and a synclinorium respectively (p. 49). It was the pressure of this folding which formed the fine slates of North Wales. Long ages of denudation have largely destroyed this folded form, but it can be traced on the face of Snowdon overlooking Glaslyn, on the cliff of Clogwyn Du'r-arddu, and in the Devil's Kitchen cliffs above Llyn Idwal. But at least 6000 m (20,000 ft) of rock has been removed from the former anticline.

The lower diagram (based on M. Gignoux and other Belgian geologists) reconstructs the series of Hercynian folds: the Ardennes anticline (**A.A.**), the Dinant syncline (**D.S.**), the Condroz anticline (**C.A.**), with its thrust-plane (**F.F.**), the Namur syncline (**N.S.**) enclosing the southern Belgian coalfield, the Brabant anticline (**B.A.**), and the Kempen syncline (**K.S.**) containing the Kempen coalfield concealed under a thick layer of Cainozoic sediments (stippled).

ing a trend-line continued eastward by the Engadine, and the valleys between the main chains of the Atlas Mountains, containing numerous salt-lakes or *shotts*. True synclinal valleys are, however, much less common than might be expected (pp. 179–80).

THE OLDER FOLD MOUNTAINS

Fold-mountain ranges of approximately the same age as the Alps (Lower and Middle Tertiary) are to be found in many parts of the world, including the Pyrenees, the Carpathians, the Himalayas and the east Asiatic island arcs. Other ranges belong to the same major orogeny, but are somewhat older. The Atlas and Caucasus Mountains are of late Jurassic age, and the Rockies and Andes of late Cretaceous age. On the other hand, some are younger, notably the Pacific Coast Range of North America, the island arcs of the West Indies, and the Siwalik foothills of the Himalayas. In fact, the Alpine orogeny may still be affecting the crust at the present day.

All these can be regarded as a single, if long continued, phase of earth movements. It seems, however, that mountain building has been cyclic, possibly the result of a series of 'peaks' of accumulation of radioactive energy (p. 24), and that similar phases of folding, followed by a period of quiescence, have occurred at intervals. During each period of quiescence (sometimes called *pediocratic periods*), denudation may partially or wholly destroy former fold systems, so that their present forms are hardly recognizable as fold mountains at all, consisting as they do of residual masses, fractured blocks, or peneplains.

Nevertheless, they are old fold mountains. Folding seems to have affected even some of the ancient shield-rocks in Pre-Cambrian times; American geologists claim to have discovered traces of three orogenies recorded in the rocks in the neighbourhood of the Great Lakes (p. 60), and some geomorphologists consider the Wrekin, Charnwood Forest and the Malvern Hills to be the remnants of a Pre-Cambrian 'Charnian' orogeny. The vestiges of at least four Pre-Cambrian orogenies can be traced in the Baltic Shield in Finland, the worn-down 'roots' of mountains trending from north-west to south-east.

Caledonian folding The earliest of the post-Cambrian fold-systems is known as the Caledonian, so called from its development in Scotland (after the Roman name *Caledonia*). This orogeny occurred in late Silurian and early Devonian times, and the mountains produced may once have been as lofty and as extensive as the Himalayas. Now only the worn-down, broken relics remain, forming the uplands of north-western Ireland, the Highlands of Scotland, and western Scandinavia, where the folding came up against the Baltic Shield, which acted as the stable foreland.

Folds of the same age are to be found in the Southern Uplands of Scotland, the English Lake District and North Wales (fig. 25). In the Lake District, for example, the rocks were uplifted into an elongated dome with an axis running approximately east-north-east to west-south-west through Skiddaw, where the oldest rocks, the Skiddaw Slates (figs. 41, 94), crop out.

This folding must have been extremely intense, as shown by the great overthrusts in the North-west Highlands of Scotland. The ancient folds were worn down by long continued denudation, then newer sediments were deposited, next the whole eroded mass was uplifted without folding (although with much faulting), and finally the present surface, with its deep glens and uniform summit-levels, was carved out. The Highlands of Scotland are thus a dissected plateau, and folding accounts only for their far-distant origin.

Hercynian folding Then followed long-continued denudation during a period of crustal quiescence until Upper Carboniferous times, when another orogeny occurred, extending into Permian times. This is given various names: sometimes the whole European system is called the Altaid mountains, sometimes 'Hercynian' is applied both to the orogeny and to the resulting mountains, after the Roman name, *Hercynia Silva*, for the Harz Mountains. The term *Armorican* is usually given to the more westerly mountains in Brittany and south-western Ireland, *Variscan* to the rest.

The results of the same orogenic phase, of broad contemporaneity, can be traced in most continents, and include the Urals, ranges in central Asia such as the Tien Shan and Nan Shan, the eastern highlands of Australia (in part at least), the Appalachians of North America and the foothills of the Andes.

The Hercynian ranges of central Europe survive as isolated massifs, worn down, covered with sediments, uplifted and partially exposed, and now standing out from the newer sedimentary rocks around them. They are shown on fig. 24, where some indication is given of the trend-lines of the original folds, and they form substantial elements in the physical make-up of Europe. The Hercynian ranges acted as part of the foreland in the Alpine orogeny, helped to control the trends of the Alpine folding, and then were partly overwhelmed by it, as already discussed. Thus the present form of Europe has developed through the 'welding on' to the old rigid mass of the Baltic Shield of the products of three orogenies: the Caledonian, the Hercynian and the Alpine. Similarly, Asia includes the ancient masses of the Siberian Shield, the Chinese Platform, the Deccan and Arabia, on to which are 'welded' some Caledonian, many Hercynian and a series of Alpine folds. The North American continent has the Canadian Shield, the old folds of the Appalachians corresponding in part to the Caledonian but mainly to the Hercynian orogenies, and the Western Cordillera formed during the Alpine orogeny.

The North American orogenies American geomorphologists distinguish eight main orogenic phases which have affected the North American continent. Three of these occurred in Pre-Cambrian times, and their remnants have been traced mainly in the rocks of the Canadian Shield; they are known as the *Laurentian*, *Algomanian* and *Killarnean* orogenies. During late Ordovician-Silurian times, the *Taconian* orogeny was responsible for folding movements in Nevada and Utah, and the *Acadian* in the Devonian and Lower Carboniferous for folding in New England. Although the Appalachians are of complex origin, the orogeny chiefly responsible seems to have been the *Appalachian* during the Permian. The main ranges of the Rockies were created by the *Laramide* orogeny, corresponding broadly to the Alpine-Himalayan in Eurasia, but occurring rather earlier, as the movements seem to have been in progress during the Cretaceous. Finally, the *Cascadian* comprised a diverse series of earth-movements at the end of the Tertiary, when the Cascade Mountains were uplifted and the fault-block of the Sierra Nevada (p. 43) was bodily uplifted and tilted. There was then a period of quiescence in the early Quaternary, followed by widespread volcanic activity when the present Cascade peaks (p. 85) were built up.

CHAPTER 3

Vulcanicity

THE term vulcanicity includes in its widest sense all the processes by which solid, liquid or gaseous materials are forced into the earth's crust or escape on to the surface. In either case, this activity may have a profound effect on surface features. The materials injected into the crust, known as *intrusive rocks* (p. 3), may later be exposed by denudation, while the molten rock and other material which reaches the surface, there solidifying as *extrusive rocks*, can build up relief forms ranging from tiny cones to widespread sheets of lava. Magma, the molten rock material beneath the earth's crust (p. 3), is normally maintained by overlying pressure in a state of apparent solidity, but when the pressure is locally relieved by earth movements, various phenomena of vulcanicity may result. It follows that this activity is associated both with periods of earth movement and also with the main areas of crustal instability. Even in Great Britain, where there are now no active volcanoes, the results of past vulcanicity are to be found in western England and Wales and much of Scotland.

INTRUSIVE FORMS OF VULCANICITY

The results of the forcing into the earth's crust of molten magma depend on firstly its degree of fluidity, and secondly the character of the planes of weakness, such as joints and faults, or cracks and fissures in the crests of anticlines (*diapiric intrusions*) along which it can penetrate. A mobile magma flows farther, as a thin sheet, while a more viscous magma solidifies rapidly in a dome-shaped or lenticular mass. The main intrusive forms are shown in fig. 26.

(i) **Dykes** These are formed when the magma has risen through near-vertical fissures, solidifying to form 'walls' of rock cutting discordantly across the bedding planes of the country rock. Sometimes dykes occur in large 'swarms'; hundreds of parallel dykes can be traced in north-western Scotland, particularly in the islands of Mull and Arran (plate 15). In southern Arran the dykes penetrate the Triassic sandstone country rock, running out from the cliffs across the beaches. Where affected by denudation, the dyke may either stand up as a wall (where the dyke-rock is harder than the

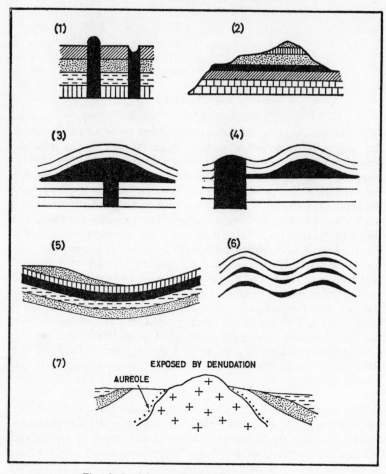

Fig. 26 *Land-forms produced by igneous intrusion*

  1. Dykes; 2. sill; 3. laccolith; 4. bysmalith and laccolith; 5. lopolith; 6. phacoliths; 7. batholith.

country rock, as in the case of the Cleveland Dyke of north-eastern Yorkshire and some of the dolerite and granite dykes in Arran, or it may be worn away to form a long narrow ditch-like depression (plates 17, 18). Two remarkable dykes radiate due west and south from the fantastic Ship Rock Peak in New Mexico, forming rugged walls 3 and 8 km (2 and 5 miles) in length respectively, and rising 60–90 m (200–300 ft) sheer above the dissected plateau around.

Sometimes a zone of dykes may surround a circular or dome-shaped intrusion in more or less arcuate form; these are known as *ring-dykes*. The magma forming the intrusion seems to have exerted pressure upwards and outwards, so forming fractures which then filled with magma (a process known as *magmatic stoping*). If the fractures are vertical in section, ring-dykes are formed, but if they are inclined inwards towards the top of the intrusion they are known as *cone-sheets*. These are often commonly associated, and occur in Mull, Skye, Arran and Northern Ireland.

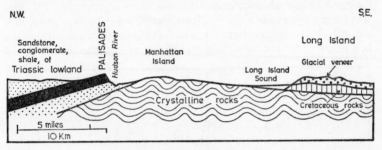

Fig. 27 The Palisades Sill, New Jersey

Based on N. M. Fenneman, *Physiography of the Eastern United States* (New York 1938).

(ii) **Sills** Horizontal sheets of rock solidify from magma which has been injected concordantly between bedding planes; they may be of any thickness and extend for many square km. The most striking example in Great Britain is the Great Whin Sill, consisting of dolerite, and varying in thickness from 1 to 70 m (2–3 to 240 ft). It extends as a series of lens-shaped masses over about 4000 sq km (1500 sq. miles) from the Farne Islands and the coast of northern Northumberland (plate 16) diagonally across northern England to the western edge of the Pennines overlooking the Eden valley. The only site where both the top and the bottom of the Whin Sill are seen together is at Ritton White House quarry in Northumberland. A sharp edge is in places exposed by denudation, as where the crest of the north-facing Whin Sill in Northumberland is followed for some distance by Hadrian's Wall. Other examples include the Salisbury Crags near Edinburgh (plate 19), and the cliffs of Cader Idris overlooking the Barmouth estuary. Sills are numerous in western Scotland and in the Western Isles, sometimes as much as 30 m (100 ft) in thickness. In north-western Skye, overlooking Loch Bracadale, some near-horizontal sills form a series of tabular hills rising to 450–490 m (1500–1600 ft), the Macleod's

Tables. Again, Garbh Eilean and Eilean an Tighe, two of the Shiant Islands off the south-eastern coast of the Island of Lewis, and now united by a shingle beach, are portions of a thick sill of dolerite, syenite and other igneous rocks. In North America the Palisades extend for over 80 km (50 miles) along the western bank of the Hudson River near New York; this sill is probably 300 m (1000 ft) thick (fig. 27).

(iii) Laccoliths These features, the name of which is derived from two Greek words meaning 'rock-cistern', are produced where tongue-like lateral intrusions of viscous magma, probably fed sideways from a bysmalith (p. 65), have forced the overlying strata into a dome. In its simplest form the magma solidifies as a cake-like mass, but often there are subsidiary laccoliths around the main one, or

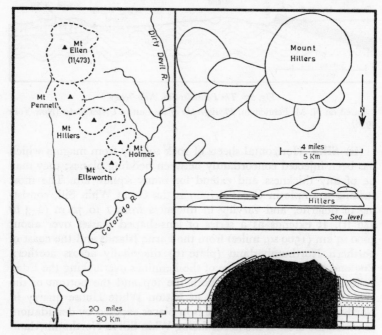

Fig. 28 The Henry Mountains, Utah

Based on diagrams and maps in G. K. Gilbert, *Report on the Geology of the Henry Mountains* (1877).

The left-hand map locates the five main groups of laccoliths. The top-right diagram gives the pattern of the Mount Hillers cluster, the middle-right section its plan. The upper horizontal line marks the base of the Cretaceous, the middle one the base of the Jurassic, the rocks into which the laccoliths were intruded.

The bottom right diagram is a reconstruction in section of Mount Hillers, a bysmalith.

several laccoliths may be formed one above another, resulting in a 'cedar-tree' laccolith. The classic example of laccoliths occurs in the Henry Mountains in southern Utah, to the west of the Colorado river, the subject of close study in 1875-6 by G. K. Gilbert, who produced a memoir in 1877. He distinguished five groups of laccoliths (fig. 28), masses of trachyte injected among sandstones ranging in age from the Carboniferous to the Cretaceous. They vary from relatively simple forms, such as Mount Ellsworth in the south, which is a single large laccolith, to Mount Ellen in the north consisting of a cluster of about thirty, forming the highest mountain in the group (505 m, 11,473 ft), about 1500 m (5000 ft) above the surrounding plateau surface. Mount Hillers consists of one large and eight small laccoliths. The hills have been subject to much denudation, and the laccoliths, with their associated dykes and sills, reveal various stages of exhumation and destruction; some are still covered with arched sedimentary strata, others are part-bared, and some like Mounts Ellen and Hillers are wholly exposed and carved into rugged peaks. Another very striking example of an unroofed laccolith is the Navajo Peak (3177 m, 10,416 ft) in Utah. On a smaller scale, Traprain Law forms a striking hill in East Lothian, and the cake-like masses of gabbro which make up the Black Cuillins of Skye are regarded as laccoliths.

A distinction is sometimes made between a *phacolith* and a laccolith. While the latter has 'blistered-up' horizontal strata, the former occurs either near the crest of an anticline or the base of a syncline in folded strata. It is believed that Corndon Hill in Shropshire is a phacolith.

Another variety of this form of intrusion is a *lopolith*, a saucer-shaped mass forming a great shallow basin. Examples include sheets in the Transvaal (the Bushveld Complex), the rhyolite plateau of Yellowstone National Park in Wyoming, U.S.A. (fig. 37), and the Duluth gabbro mass south-west of Lake Superior.

Where the intrusion is of very great vertical extent, the term *plutonic plug* or better *bysmalith* is applied. The overlying strata are arched, but the sides of the mass plunge steeply for an unknown depth, unlike the laccolith proper which is relatively 'shallow'. Mount Hillers in the Henry Mountains is of this nature. Another striking example is Mount Holmes (3155 m, 10,350 ft) in Yellowstone National Park. Here the intrusion is so vast that the overlying strata have not only been arched but faulted around the circumference.

(iv) Batholiths Large masses of rock occur in the heart of mountain ranges, formed by deep-seated movements on an enormous

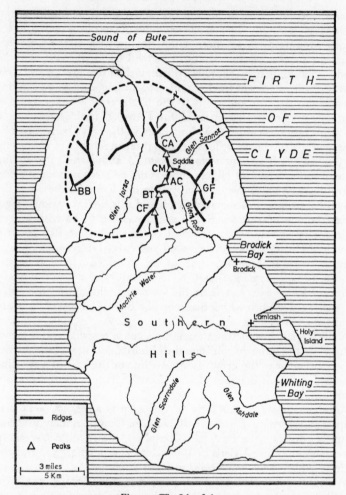

Fig. 29 *The Isle of Arran*

The extent of the main granite mass is shown by a bold pecked line, although it is deeply dissected by glacial valleys which are thickly drift-covered. The smaller 'central complex' further south is not shown.

The abbreviations for the peaks are as follows: **AC** = A'Chir; **BB** = Beinn Bharrain; **BT** = Beinn Tarsuinn; **CA** = Caisteal Abhail; **CF** = Caisteal an Fhinn; **CM** = Cir Mhor; **GF** = Goat Fell (874 m, 2866 ft).

scale, so that the masses of magma cooled slowly to form large-crystalled rocks such as granite, later exposed by prolonged denudation as massive upland areas. The edges of a batholith descend steeply to unknown depths, while the country rocks with which the intrusions have come into contact are often metamorphosed by thermal contact (p. 11). There are many examples of batholiths—the granite masses of the South-west Peninsula and the Isles of Scilly, the Wicklow Mountains in Ireland, and the uplands of Brittany form striking relief features (fig. 30). The last are elongated roughly from west to east along the trend-lines of the old fold mountains, and have been exposed by denudation to form flat-

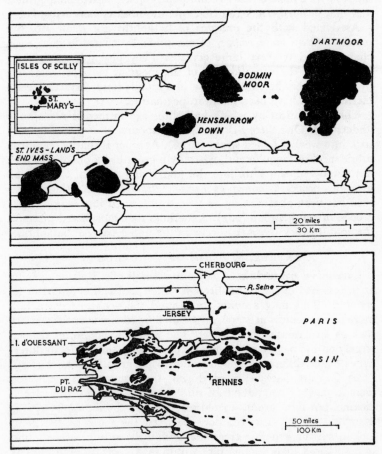

Fig. 30 *Granite batholiths in the South-west Peninsula of England* (above) *and in Brittany* (below)

The granite masses are shown in black. Note the alignment of the Breton batholiths along the trend-lines of the Armorican folding.

topped moorlands that drop sharply to the lower areas of ancient sedimentary rocks. The largest known batholith was thought to be in British Columbia, but it is now realized that this mass comprises a series of individuals emplaced during several distinct periods ranging from the Lower Jurassic into the Cainozoic, sometimes even intruding into each other. They form elongated masses mainly of quartz-diorite, and are known as the Coast Intrusions.

Smaller but similar intrusions are *bosses* or *stocks*: bosses are masses with a more or less circular plan, e.g. the area of Shap granite and that in northern Arran (fig. 29); the stock is more irregular.

Associated with the margins of batholiths are small intrusive masses, known as *pegmatites*, of exceptionally coarse igneous rock, characterized by very large crystals. They probably represent a late stage in the consolidation of the magma, perhaps where the presence of very hot water kept it in a mobile state. Crystals several metres across have been found in pegmatites.

The mechanism by which batholiths are formed is not wholly understood. One theory is the upward forcing of *diapirs* of igneous rock into weaker folded strata (p. 48). Another is that of 'cauldron subsidence' or 'engulfment', whereby a great block of country rock has foundered and sunk into the underlying magma. On a small scale this process is known as *magmatic stoping*. Possibly the country rock has been absorbed by or even transformed into magma by a process of granitization (p. 4). Sometimes metamorphosed pieces of the country rock (*xenoliths*) occur in the upper parts of batholiths.

EXTRUSIVE FORMS OF VULCANICITY

Extrusive materials A volcano consists of a vent or opening at the surface of the crust through which material is forced in an eruption. This may accumulate around the vent to form a hill, more or less conical in shape, or it may flow widely over the country rock as an extensive level sheet. The resultant landforms depend largely on the nature of the material ejected, which varies in different cases and different stages in an individual eruption (plates 21-3).

(*a*) *Gaseous* Gases emitted during the course of an eruption include gaseous compounds of sulphur and hydrogen with carbon dioxide, but most are dissipated directly into the atmosphere. It is thought that the inter-reactions between the gases contained in molten magma generate great heat within the ascending lavas, and so maintain a high temperature within lava pools in a crater. The eruptions of Mont Pelée and Soufrière in the West Indies in 1902 were accompanied by vast outpourings of gases, as well as incandescent fine lava-fragments (a *nuée ardente*), which rolled down

the mountain slopes and caused great loss of life. The same phenomenon happened in May 1915 at Mount Lassen in northern
California; the 'Great Hot Blast', as it is referred to, swept down
over an area of forest, snapping off thousands of trunks and forming
a 'Devastated Area' which is still clearly visible today.

Steam is possibly derived from surface water such as in a crater-lake
or from the sea in the case of a volcano near the coast, but most
probably originates from the water to be found in magma. The
expansive force of this steam, as well as that of included gases, may
well contribute largely to the mechanism of a volcano, particularly
where the eruption is of an explosive or paroxysmal nature. Frequently the masses of steam condense to form clouds and then
torrential rain, causing mud-flows.

(b) *Solid* When an eruption is accompanied by a series of
explosions, solid materials are ejected, known generally as pyroclasts
(p. 4). These may include fragments of the country rock disrupted when the pipe was blown through the crust, angular fragments of solidified lava from a previous eruption that had cooled in
the pipe, and finer material such as scoria, pumice, cinders (*lapilli*),
dust and 'ash'—a misnomer, since it is not the product of com-
bustion, but consists of finely comminuted particles of lava.

A dramatic example of the emission of ash was afforded in recent
years by the volcano Irazu (3432 m, 11,260 ft) in Costa Rica,
which had been dormant for a long time until March 1963. It
produced no lava, but it poured out continuously dry acid dust
over an area of 650 sq. km (250 sq. miles), including the capital, San
José. The ash lay everywhere 'like a deep fall of black snow, which
in several places had drifted into dunes', in the words of a correspondent. Agriculture was ruined over a large area, and conditions
in the city were comparable to those in London during a bad smog.

The coarser fragments form breccia, while the finer materials
may be loosely cemented as tuffs. Sometimes small amounts of
liquid magma are thrown into the air and solidify before reaching
the ground in globular masses known as 'volcanic bombs'. Other
masses remain liquid until they hit the ground, when they spatter
and congeal, forming small *spatter-cones* 3–6 m (10–20 ft) high, as
around Sunset Crater in Arizona.

Occasionally the torrential rains which often accompany eruptions cause the fine dust to flow down the slopes as a stream of mud;
Herculaneum was buried under such a stream during the eruption
of Vesuvius in A.D. 79. The Mount Lassen eruption (referred to
above) also produced an enormous mud-flow; the water, provided
by the snow melted through the heat of the eruption, mixed with

fine ash to make a paste, which picked up cinders and large lava blocks, and flowed as a stream 6 m (20 ft) thick for about 30 km (18 miles), its impetus causing it to cross a rise of 45 m (150 ft) *en route*.

(c) *Liquid* Usually the most important product of an eruption is lava, the molten magma which reaches the surface. The form of a volcanic cone depends to a large extent upon the nature of this lava, and so to some extent does the nature of the eruption. Some lavas contain much silica, i.e. are said to be acid lavas (p. 3), with a high melting-point; they are very viscous, solidify rapidly and so do not flow far. Acid lavas build high, steep-sided cones, and, moreover, they may solidify in the vent and cause recurrent explosive eruptions (pp. 75–7). Where, on the other hand, the lava is relatively poor in silica and rich in iron and magnesium minerals (i.e. a basic lava), it has a lower melting-point and will flow readily for a considerable distance before solidifying. Such a lava tends to produce a much flatter cone of great diameter, and, moreover, as the flow from the vent is unchecked and widespread, the eruption is quiet, without much explosive activity. The mobility of the lava also varies with the amount of gas dissolved within its mass; in Hawaii lavas containing gas have remained mobile at a temperature of 850° C, while others without much gas content have solidified at 1200° C. The process by which hot gases mix with fine-grained lavas, stimulating their flow, is known as *fluidization*. This has the further result of causing the creation and enlargement of cracks by both physical and chemical means, thus forming volcanic *pipes*, as in the Kimberley area of South Africa and in the Swabian Jura in West Germany.

When the lava solidifies it may take a number of forms, some of which have been given Hawaiian names as the result of the vulcanological research carried out there. The *aa* (pronounced *ah-ah*) type has solidified into irregular block-like masses of a jagged, angular clinkerous appearance, the result of gases escaping violently from within the lava and the effects of the drag of the still molten material under the hardening crust. The *pahoehoe* type has a wrinkled, 'ropy' or 'corded' surface, formed where the lava is at a high temperature, but the gases escape quietly and the flow congeals smoothly. *Pillow* lava has solidified in the form of a pile of pillows, probably under water, particularly in the case of such basic lavas as those which form basalt and andesite. Some pillow lavas in the Canadian Shield have been dated as 2800 million years old, the oldest known lavas. Basalt commonly hardens in a very striking columnar form, the result of the creation of hexagonal joints (p. 5).

21 Volcano de Fuego, Guatemala

Note the crater-rim and the sides furrowed by rain-wash. This volcano lies 30 miles south-west of Guatemala City in Central America.

(Fairchild Aerial Surveys, Inc)

22 The active inner cone of Vesuvius, within the caldera-rim of Monte Somma

(Aerofilms Ltd

23 Crater Lake, Ruapehu (9175 feet), Tongariro National Park, North Island,
New Zealand
(High Commissioner for New Zealand)

24 Jewel Geyser, Yellowstone National Park, Wyoming. This geyser erupts
regularly at intervals of 5 to 6 minutes.
(F. J. Monkhouse)

Types of volcanoes (fig. 31) The two major eruptive forms are *central*, that is, when the activity proceeds from a single vent or a group of closely related vents, and *linear*, when lava wells up along a line of weakness, either from its whole length simultaneously or at intervals along it. The nature of the resultant relief in both cases

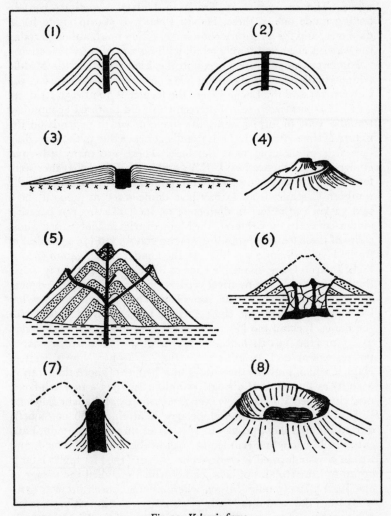

Fig. 31 Volcanic forms

1. Ash-cone; **2.** acid lava dome; **3.** basic lava 'shield'; **4.** secondary cone within old crater-ring; **5.** composite volcano; **6.** caldera; **7.** volcanic plug or neck, exposed by denudation; **8.** crater lake.

depends on the character of the eruption together with the variety of the erupted material, but later denudation may considerably modify the original form of the volcano.

(*i*) In an *explosion vent* a small hole has been blown through the country rock and is surrounded by a low crater-ring of rock fragments. The crater-pits at Krafla in Iceland sometimes contain boiling mud; one of these, Helviti ('hell'), is 360 m (1200 ft) in diameter, and for a century contained boiling mud, although today the water is clear and barely tepid, with no sign of volcanic activity.

Numerous small craters occur in the Eifel region of the Middle Rhine Highlands. Some contain lakes (*Maaren*) (plate 66), e.g. the Pulvermaar and the Lachermaar; the former is 76 m (250 ft) deep.

(*ii*) *Ash- and cinder-cones* Fragments of solid material accumulate round a vent to form a cone, the shape of which depends on the nature of the material, but it is usually concave due to the spreading outwards of material near the base. There are many ash- and cinder-cones in the western U.S.A., and it is believed that they were formed at a late stage of activity. Thus Mount Newberry in the southern Cascades is an extinct lava dome, 1200 m (4000 ft) high and 32 km (20 miles) in diameter; on its flanks are 150 parasitic cinder-cones. In the Craters of the Moon district in Idaho are a whole series of these features; standing among 200 sq. km (75 sq. miles) of frothy lava and cindery rubble are Big Cinder Butte, 240 m (800ft) high, Inferno Cone, the triple cone of Big Craters, and many more. Sunset Crater is a symmetrical cinder-cone near Flagstaff, Arizona, about 300 m high, to the top of which the writer somewhat laboriously plodded; the cinders at the top are tinted pink, hence the name. Iceland too has many examples; ninety such ash-cones, 35–45 m (120–150 ft) high, are at Ravoholar, near Reykjavik. A perfect example of an ash-cone is that of Monte Nuovo, west of Naples, which grew in three days to a height of more than 140 m (450 ft) as the result of a single eruption. In 1937 a new ash-cone, now christened Vulcan, grew even more rapidly at Blanche Bay near Rabaul in the Bismarck Archipelago; it attained 180 m (600 ft) during the first day and 225 m (740 ft) in three days. Again, Paricutín, in Mexico, first erupted in 1943 in the middle of a field, and within a year formed a cinder-cone 450 m (1500 ft) high; later it began to issue streams of lava, one of which engulfed the village of San Juan 5 km (3 miles) away, where now a church projects desolately from the solid lava; activity abruptly ceased nine years later. A few major cones seem to be made entirely of ash; Volcano de Fuego in Guatemala is a symmetrical example, 3350 m (11,000 ft) in height (plate 21).

(*iii*) *Lava cones* Where explosive activity is absent, so too is fragmentary solid material, and the lava flows placidly from the vent, building a volcanic form which varies according to the nature of the lava. A viscous acid lava does not flow far, and so it produces a steep dome, mainly of rhyolite, dacite or trachyte, with convex sides and usually without a visible crater. The trachyte lavas of Auvergne, in the Central Massif of France, form such domes (fig. 44); an example is Puy de Dôme, the highest (1465 m, 4806 ft) of the 'puys'. The symmetrical Puy de Grand Sarcoui, and some of the other puys, were probably developed by pressure from within. The surface layers of acid lava hardened, and further uprisings of lava, unable to reach the surface, forced these layers outwards. Thus the dome-shape was accentuated by internal expansion. This is sometime known as a *cumulo-dome* or (specifically in the island of Réunion in the Indian Ocean) as a *mamélon*. If such a dome forms in an existing crater, it is known as a *tholoid*, e.g. Bezymianny, which developed in Kamchatka in 1956.

Lassen Peak affords an example of a dome volcano of a plug-like character, made of a stiff acid lava which hardened into dacite and rhyolite as it was forced up vertically about 5000 years ago; it is likely that the plug rose to a height of 760 m (2500 ft) in five years or so. Its summit formed steep spines of dacite, which rapidly crumbled, so masking the slopes with rock debris that the plug can only be seen in a few places. There is, however, one unusual feature of Lassen, the summit of which reaches 3186 m (10,453 ft): a plug-dome usually does not have a crater, but in 1916 a series of eruptions took place and it now has four distinct craters.

Some of the more highly viscous lavas may be squeezed out slowly to form great 'spines'. A very striking example of a spine was that of Mont Pelée in Martinique; a mass of viscous lava was forced out as a result of the 1902 eruption to a height of over 210 m (700 ft) above the former summit, but was largely destroyed by contraction during cooling and by rapid weathering during the succeeding year. The Santiagito plug in Guatemala was formed between 1922–4, during which it rose to a height of 460 m (1500 ft).

The Devil's Tower in Wyoming consists of a mass of phonolite-porphyry, which is believed to have been squeezed upward through Jurassic and Cretaceous sandstones and limestones, solidifying in remarkable pentagonal columns (fig. 32, plate 20). It now forms a most striking feature, rising abruptly for over 180 m (600 ft), its summit at 1560 m (5117 ft) being a level surface an acre in extent.

Mobile basic lava can flow for very considerable distances. Thus the great volcanoes of Hawaii have cones with a small angle of slope. Mauna Loa has a broad, shallow crater, 16 km (10 miles) in circumference, from the rim of which the angle of slope is only 6°, but it is about 110 km (70 miles) in diameter at sea-level. The summit is 4168 m (13,675 ft) above the sea, while the total height from its

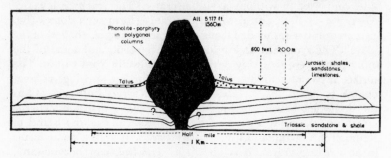

Fig. 32 Cross-section of the Devil's Tower, Wyoming

Based on C. S. Robinson, *Geology* of *Devil's Tower National Monument, Wyoming: Geological Survey Bulletin,* 1021–11 (1956). What happens below the surface is not known, hence the query marks, but it is possible that the mass thins off into a narrow pipe up which moved the molten magma (cf. plate 20).

base on the ocean floor is about 9750 m (32,000 ft); the base must be at least 500 km (300 miles) in diameter. Mauna Loa is still active, and large-scale lava-flows have issued on half a dozen occasions during the last fifty years. In 1926 a large flow reached the south-western coast, destroying a village. Mauna Kea, 50 km (30 miles) to the north, the highest peak in Hawaii (4213 m, 13,825 ft), is extinct. These great basic-lava cones are often known as *shield volcanoes.*

An interesting feature of Mauna Loa is that on its south-eastern flank, 32 km (20 miles) from the summit, is another cone, Kilauea, the summit of which just exceeds 1200 m (4000 ft). Its steep-sided crater is 5 km (3 miles) across, and its floor is studded with pools of boiling lava.

(*iv*) *Composite cones* Probably the most common and typical volcano is the composite cone (called in America a *strato-volcano*), built up over a long period of time as the result of a number of eruptions; this category includes most of the highest volcanoes in the world. The main cone consists of layers of ash and lava fed from the main pipe, which culminates in a crater. Sometimes a later explosion may blow off the top of the cone, and so form a much larger crater, within which a secondary cone may develop (plate 22), and perhaps parasitic cones grow on the flanks. Etna, in Sicily, has some hundreds of secondary cones on its slopes. Some volcanoes have several major cones and may therefore be termed 'multiple volcanoes'. The neighbouring volcanoes of Ruapehu and Tongariro in New Zealand are of this nature.

One of the best examples of a composite volcano in Europe is Stromboli, in the Lipari Islands to the north of the 'toe' of Italy. Its eruptions are frequent and gentle, often at intervals of an hour or less, while the glow of the hot lava on the clouds of smoke and condensed moisture above the crater have caused it to be named 'the lighthouse of the Mediterranean'.

By contrast, many volcanoes have suffered repeated eruptions of a paroxysmal nature. After a long period of quiescence, a violent eruption takes place, due probably to the accumulation of a solid plug of lava below which magma has become heavily charged with gases, with an enormous build-up of pressure. The eruption of Vesuvius in A.D. 79 is an example; a large part of the crater was blown away and the town of Pompeii was buried in debris. The broken crater-ring is known as Monte Somma, and a new cone and crater with a circumference of about a mile have since accumulated within it (plate 22). The last large eruption was in March 1944, when an 8 km wide lava stream flowed down the slopes towards the coastal plain, destroying the little town of San Sebastiano. In the words of *The Times* correspondent, this lava stream now lies as 'a monstrous black snake, stricken, silent and still, between banks of trees'. After this eruption the well-known 'umbrella-plume' of smoke disappeared.

The name *caldera* or *basal wreck* is given to the great shallow cavity which remains when a paroxysmal eruption removes the top of a former cone. In 1883 occurred probably the most stupendous eruption of recent times, when a group of islands, representing the caldera of an ancient volcano in the Sunda Straits between Java and Sumatra, was involved in a tremendous explosion; most of the main island of Krakatoa disappeared. The finest European example of a caldera is Askja, in central Iceland, within which lies the large lake of Öskjuvatn (fig. 33); this erupted most recently in 1961. Crater Lake (fig. 104), in the state of Oregon, fills a caldera nearly 10 km (6 miles) across, the remains of a peak at least 1800 m (6000 ft) higher, while a small more recent cone projects from the water as Wizard Island; the maximum depth of the lake is about 600 m (2000 ft). The town of Crater, in Aden (now Southern Yemen), is built within a long-extinct caldera. Two large calderas, Aniakchak and Galtes, are situated in Alaska; the former is 11 km (7 miles) across and 900 m (3000 ft) deep. The world's largest is said to be

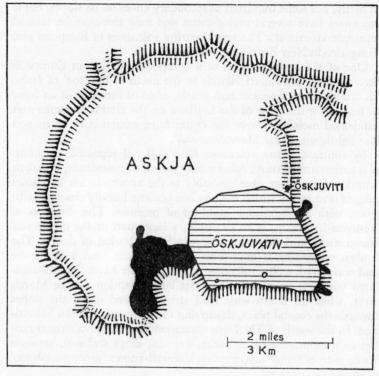

Fig. 33 The Askja caldera, Iceland
After N. H. van Doorninck and B. B. Roberts.
The Askja caldera, to the north of Vatnajökull, consists of a lava plain surrounded by a ring of tuff mountains rising to about 700 m (2300 ft) above the surrounding country and about 350 m (1150 ft) above the floor of the crater. In the south-east is the crater-lake of Öskjuvatn, 168 m (550 ft) deep. Öskjuviti is a crater-pit in the floor of the caldera. It erupted in 1875, pouring out a tremendous amount of ash and pumice and devastating many farms in Iceland; some ash fell in Stockholm twenty-four hours after the eruption. There have been several later lava-flows from small fissures and vents; these are shown in black. The large flow in the south-west of the caldera took place in 1922. Askja again erupted in October 1961.

Aso in Japan, which is 112 km (71 miles) in circumference and 27 km (17 miles) in maximum width.

A large caldera in south-western Alaska was formed in 1912, when a violent eruption blew off the summit-cone of Mount Katmai. What had been a glacier-swathed peak rising to 2285 m (7496 ft) above sea-level became a caldera 4 km (2·5 miles) across, with a jagged summit-rim in places attaining only 1400 m (4500 ft). The

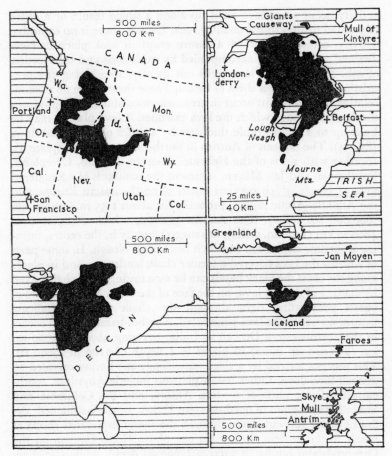

Fig. 34 Basalt plateaus
The basalt-flows are shown in black.

floor of the crater lies from 460–1070 m (1500–3500 ft) below the rim, and a lake nearly 150 m (500 ft) deep has accumulated.

It must be noted, however, that some calderas were formed not by a paroxysmal eruption, but by the process of 'cauldron subsidence' mentioned in connection with batholiths (p. 68). In the case of Crater Lake, Oregon (fig. 104), a vast amount of material must have been removed for which it is otherwise difficult to account, for there is little or no fragmental material around the present rim.

(*v*) *Fissure eruptions* Some of the largest-scale examples of volcanic

77

activity occur when lava wells up along a single fissure or a series of parallel fissures in the country rock. Usually there is no explosive activity, although in 1886 a fissure eruption took place in New Zealand at Tarawera, accompanied by explosions and the ejection of ash. Normally the lava wells out quietly, smothering the pre-existing relief under a sheet of basalt. These sheets have no obvious relation to craters, but occur in areas where crustal tension produces the fissures through which the lava extrudes. A lava plateau may be built up to a considerable thickness by a series of individual flows (fig. 34). The plateau of Antrim in north-eastern Ireland (plate 2), together with parts of the Hebrides, Faeroes, Iceland, Greenland, Spitsbergen and Jan Mayen, represent the remnants of a foundered area of basalt which covered much of north-western Europe and the Atlantic–Arctic region, probably for about 1·55 million sq. km (600,000 sq. miles). The Antrim plateau, which occupies over 4000 sq km (1500 sq. miles), has sagged slightly in the centre, forming a shallow depression in which lies Lough Neagh. In some parts of Antrim the lava covered a former chalk landscape, and in places along the coast the black basalt can be seen resting upon chalk baked by contact of the hot lava, on the face of the cliffs. Vertical dykes of basalt penetrating right through the chalk probably indicate fissures up which the lava penetrated. Another basalt plateau occurs in the north-western Deccan in India, covering about 650,000 sq. km (250,000 sq. miles) in places to a depth of 2100 m (7000 ft). This was not formed by a single upwelling; one boring revealed 29 distinct lava-flows. Other basalt plateaus include Abyssinia, south-eastern Africa, parts of western Arabia, and the Columbia and Snake River regions of north-western U.S.A. In the last-named, the Snake River has cut down through the many individual sheets of basalt to the ancient surface (in one place, Hell's Canyon on the Oregon–Idaho border, is 2400 m (7900 ft) deep), while elsewhere peaks project through the basalt like nunataks (p. 219) through an ice-sheet. A lava plateau may end sharply and form a steep edge, as in the eastern slopes of the Drakensbergs in South Africa, and on a smaller scale along the coasts of Skye and Rhum.

The fissure eruptions in Iceland are of a rather different character. The most recent was in 1783, when, in addition to outpourings of lava from many points along a fissure 30 km (20 miles) long, a string of ash-cones was built up along the line of the fissure (fig. 35).

(*vi*) *Minor volcanic forms* A variety of minor volcanic forms can be distinguished, usually though not necessarily associated with volcanoes approaching extinction. A *solfatara* is a volcano which only emits steam and gas; it has been given this name from a volcano

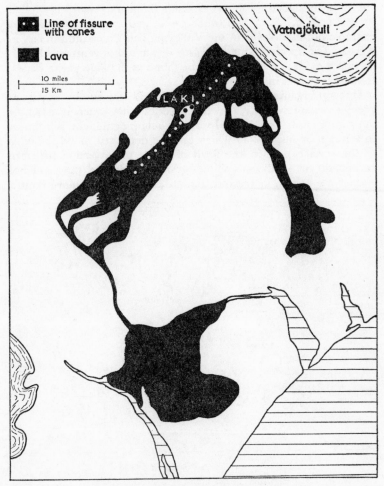

Fig. 35 *The Laki fissure-eruption, Iceland* (1783)
After P. Thoroddsen and B. B. Roberts.
The ice-cap of Vatnajökull is indicated. For its location see fig. 36.

in the Phlegraean Fields (Campi Flegrei) to the west of Naples,
which has not had a major eruption since the end of the twelfth
century. The floor of this crater is comparatively cool, and one can
walk about on it, although here and there are pools of boiling water,
while jets of water-vapour or puffs of sulphurous gas are emitted
periodically. The term solfatara is commonly limited to cases where

79

the gases emitted are sulphurous, while the name *fumarole* is applied to emissions of steam and other gases. Perhaps the most famous example of fumaroles is in the Valley of Ten Thousand Smokes in Alaska. The special name of *mofette* distinguishes vents which emit carbon dioxide; examples are to be found in the Phlegraean Fields, in Auvergne and in Java.

Hot springs and geysers In some areas associated with past or present volcanic activity, the chief product is hot water. It may flow out continuously as a thermal spring, containing mineral substances in solution or suspension, which may be deposited round the edge of the surface pool in the form of crusts of travertine (calcium carbonate) or of siliceous sinter (geyserite). Several thousand hot springs are known in Iceland (fig. 36); one area has more than a

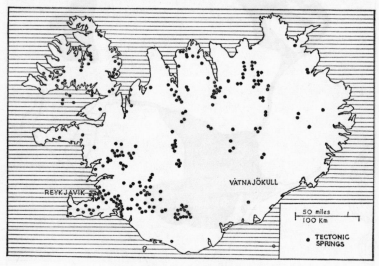

Fig. 36 Thermal springs in Iceland
After T. Barth and B. B. Roberts.
The dots indicate the important centres and groups of springs.

thousand in less than 2 sq km. Some are piped and the water led into Reykjavik, where it is used for central heating and for supplying swimming baths. Hot springs, such as the waters at Bath, do occur in non-volcanic areas, however.

Where hot water is ejected with considerable force accompanied by steam, an intermittent paroxysmal fountain occurs, known as a *geyser*, as in Iceland, Yellowstone National Park (fig. 36, plate 24), and the North Island of New Zealand. Some erupt periodically at

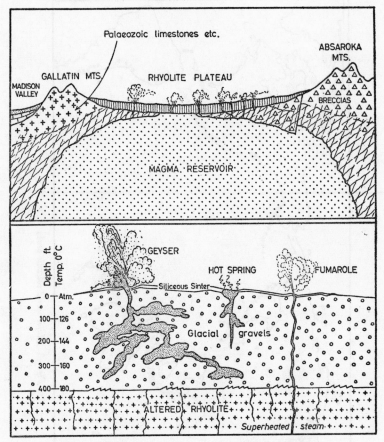

Fig. 37 The geysers of Yellowstone National Park, Wyoming
Based on W. A. Fischer, *Yellowstone's Living Geology* (Wyoming, 1960).

regular intervals, others more spasmodically. The cause is complex, but it is due to superheating far down in the pipe of the geyser; the temperature of the water at depth increases, but convection cannot easily take place up the column of water because of its length. The water at depth is heated beyond 100° C because of the pressure of the column above, and its ultimate sudden conversion into super-heated steam causes the water in the upper part of the pipe to be violently emitted. Cooler water flows into the pipe, and the heat increase begins again. One of the most famous geysers is 'Old Faithful', in Yellowstone, which has an average eruption interval of

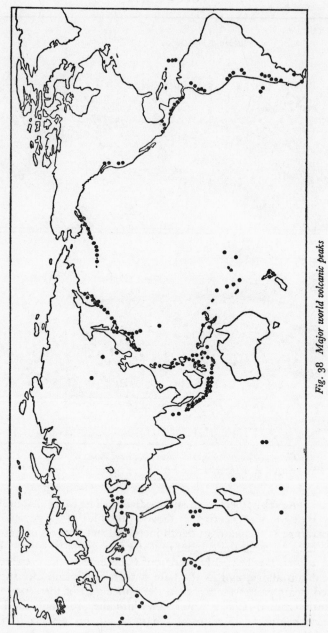

Fig. 38 Major world volcanic peaks

about 65 minutes, ranging between 33 and 95 minutes. It throws 50–100 cubic m (10–20,000 gallons) of near-boiling water, with steam, 40–60 m (120–180 ft) into the air, lasting from 2 to 5 minutes.

Where the ejected water contains fine material, a *mud volcano* is formed. A string of these occurs along the line of a fissure near Paterno in eastern Sicily, and others are at Krafla in Iceland and in the North Island of New Zealand.

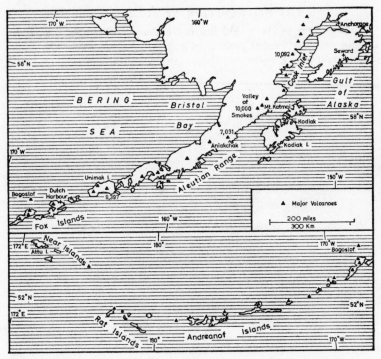

Fig. 39 The volcanoes of the Alaska Peninsula and the Aleutian Islands

DORMANT AND EXTINCT VOLCANOES

An active volcano is one which is definitely known to have erupted periodically in historic times. Some volcanoes, however, may be described as dormant when a renewal of eruptive activity is possible. Vesuvius had been dormant so long before its eruption of A.D. 79 that it was thought to be extinct. Other volcanoes may definitely be regarded as extinct, since they were formed in long-past geological times, and are situated in areas with no sign of any volcanic activity. They may retain their original forms, as the miniature cones of

Auvergne, but for the most part many ancient cones have been destroyed beyond recognition through long-continued denudation. In the British Isles little remains of volcanic cones beyond the hard lava plug which formerly filled the vent; the softer surrounding ashes have been removed. Arthur's Seat, near Edinburgh, is a volcanic plug of a complex nature with two vents (plate 19), while Castle Rock, also in Edinburgh, forms a steep-sided mass.

THE DISTRIBUTION OF VOLCANOES

A large part of the world has not experienced volcanic activity during the most recent periods of geological time; there are some 520 known active volcanoes, but many thousands of extinct ones.

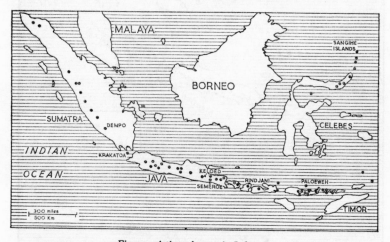

Fig. 40 Active volcanoes in Indonesia

There are estimated to be 167 volcanoes in Indonesia, of which 77 are active; some lie east of the area shown on this map, in the Banda Sea and on Halmahera.

Occasionally a completely new volcano may appear, such as Paricutín (p. 72), and a volcano, now named Little Surtsey, which began to rise out of the sea off the south coast of Iceland in November 1963, where after the first steam-cloud (which rose to more than 6000 m (20,000 ft)), lava was poured out at an estimated rate of 5×10^8 kg (500,000 tons) an hour; within three weeks the crater was 120 m (400 ft) high and 0·8 km (0·5 miles) across. Volcanic activity has continued along a ridge in the sea-floor; a second volcano appeared in 1965 and was given the name of Surtling, reaching 170 m (560 ft), but was rapidly destroyed by marine erosion; a third appeared in 1966.

The main volcanic peaks are indicated on fig. 38; it is evident that there is a close relationship between their location and the major lines of weakness in the earth's crust. The most striking development is around the basin of the Pacific Ocean—the graphically named 'Pacific Ring of Fire', where two-thirds of the world's volcanoes occur. Most are associated with fold-mountain ranges or faulted blocks. One group, for example, occurs in the Cascade Mountains of western U.S.A., an uplifted and peneplaned (p. 93) fault-block 90–130 km (60–80 miles) wide, from which rise several volcanic peaks, notably Mounts Rainier (4391 m, 14,408 ft) (fig. 110), Hood, Lassen and Shasta.

A chain of volcanoes nearly 3200 km (2000 miles) in length can be traced through southern Alaska, the Alaska Peninsula and the Aleutian Islands, one of the most volcanically active parts of the world (fig. 39). There are 80 active volcanoes, ten calderas over 1·6 km (1 mile) in diameter, and since 1760 225 major eruptions have been recorded. One interesting island-peak is Bogoslof, which appears to emerge periodically; actually it rises 1500 m (5000 ft) from the sea-floor, where its base is 13 km (8 miles) in diameter, and its summit appears by accumulation after each eruption, soon to be worn down again below sea-level by the waves.

In South America most of the highest peaks are volcanoes rising from the folded ranges, including Aconcagua (7021 m, 23,035 ft), which, though extinct, is the world's loftiest volcano, and Guayatiri (6060 m, 19,882 ft), which erupted in 1959. On the Asiatic margins of the Pacific, volcanoes are strung along the island arcs; the most famous is snow-capped Fujiyama, rising to 3776 m (12,390 ft) within 24 km (15 miles) of the sea. Another line runs through the East Indies (fig. 40) towards New Zealand; the island arcs of the western Pacific, which are often on the edge of submarine trenches, have hundreds of volcanoes, while many of the more remote islands (Hawaii, Tonga, Samoa) are volcanic cones rising from the ocean floor (pp. 334–5). In Africa volcanoes are to be found along the line of the East African rift-valley (p. 45); Mount Kenya (5194 m, 17,040 ft) and Kilimanjaro (5889 m, 19,320 ft) are probably only recently extinct. The Alpine-Himalayan belt of folding is not associated to the same extent with active volcanoes, except in the central Mediterranean, where they are probably related to geologically recent movements of subsidence. Farther east, in Asia Minor and to the south of the Caspian, are many extinct and a few solfataric cones. The Himalayas form the most striking exception to the general occurrence of volcanoes along zones of recent folding, for none is to be found there. Conversely, volcanic activity is wide-

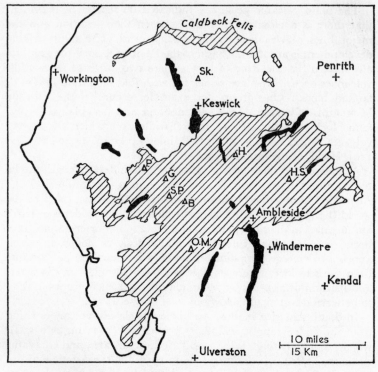

Fig. 41 The Borrowdale Volcanic Series, English Lake District

The outcrop of the Borrowdale Volcanic Series is shown by diagonal shading, the lakes in solid black.

B = Bowfell; **G** = Great Gable; **H** = Helvellyn; **HS** = High Street; **OM**= Old Man of Coniston; **P** = Pillar; **Sk** = Skiddaw; **SP** = Scafell Pike.

spread in Iceland, where there has been no recent folding of the crust. Several of the Atlantic islands, notably along the line of the Mid-Atlantic Ridge, have volcanoes; a peak on Tristan da Cunha erupted in October 1961, and the Azores and Canaries have experienced eruptions within historic times. The West Indian arcs have many indications of past vulcanicity, as well as a few active cones.

VULCANIC LANDSCAPES

The evolution of a vulcanic landscape is the result of the normal processes of denudation upon the forms produced by vulcanicity. The present-day features represent the stage reached by these processes, depending upon how recent was the activity. The effects

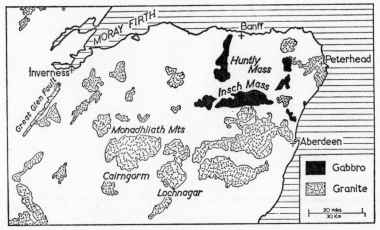

Fig. 42 *Granite and gabbro massifs in north-eastern Scotland*

Intrusive masses of gabbro (and kindred ultrabasic rocks) and of granite are a major feature of north-eastern Scotland. They are believed to be of late-Silurian age, possibly related to the Caledonian folding. The gabbro occurs in large sill-like masses; the Insch Mass is about 180 sq. km. (70 sq. miles) in extent. The granite masses vary in size and form; some may be batholiths, while it has been suggested that the Cairngorms may be a laccolith, the surface of which has been exposed by denudation. All these masses are very complex and are surrounded by aureoles (p. 11) as a result of the thermal metamorphism.

of differential denudation are exceedingly well marked, since the products of vulcanicity vary from the hardest of lavas to the softest of ashes, often in close juxtaposition. Thus the central part of the English Lake District, which consists largely of the Borrowdale Volcanic Series (fig. 41), presents a most rugged aspect. These rocks, poured out in early Palaeozoic (Ordovician) times, were covered by other rocks now removed by long-continued denudation, and the crags of hard lava, the joy of the rock-climber, now stand out from the gentler slopes worn in the ashes, consolidated into beds but much less resistant. Again, in north-eastern Scotland intrusions of granite and gabbro have been exposed by denudation (fig. 42).

Fig. 43 *The Cuillins of Skye*

The upper map shows the generalized relief (the black lines indicate ridges, the hachured areas the rounded hills).

The lower map shows the simplified solid geology.

The main peaks on the upper map are indicated by letters, as follows: *Black Cuillins:* **A** = Sgurr Alasdair (the highest peak in Skye, 991 m (3251 ft); **Ba** = Sgurr na Banachdich; **BD** = Bidein Druim nan Ramh; **Bl** = Blaven; **F** = Bruach na Frithe; **Ga** = Gars-bheinn; **Gi** = Sgurr nan Gillean. *Red Cuillins:* **BC** = Beinn an Caillich; **BDM** = Beinn Dearg Mhor; **GM** = Glas Bheinn Mhor; **Gl** = Glamaig; **M** = Marsco.

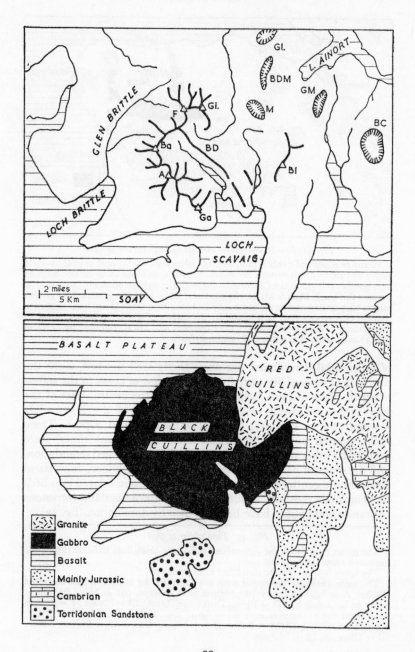

Two detailed examples must suffice to summarize the varied landscapes which are directly or indirectly due to vulcanicity— the Isle of Skye and Auvergne.

(i) **The Isle of Skye** Most of Skye (fig. 43) is composed of igneous rocks of early Cainozoic (Eocene) age, one of the notable periods of vulcanicity in the North Atlantic area. There are four distinct features resulting from this activity. The first is the out-pouring of sheets of lava, which consolidated on the surface as basalt, and now cover most of northern and central Skye, although the fact that the sheets terminate seaward as steep cliffs shows that their former extent was far greater. They have been abruptly cut off, partly by marine erosion, for the most part by faulting. In western Skye, between Lochs Bracadale and Brittle, the thickness of the basalt is between 150–275 m (500–900 ft), made up of a number of individual flows; at Dunvegan Head farther north about twenty-five distinct sheets have a total thickness exceeding 300 m (1000 ft). The second is the injection within the basalt of material which cooled at depth into a cake-like mass of coarsely crystalline gabbro; (some regard this as a laccolith (p. 64)). The overlying basalt has largely been removed. The third is the similar injection of material of very different chemical composition, which cooled at depth into a granite mass. Finally, at a much later date immense numbers of minor intrusions—dykes and sills—of basalt and dolerite were intruded among both the basalt and gabbro.

The varied rocks are due to several forms of activity—in part to intrusion, in part to fissure eruptions, in part to central eruptions, although very little trace of the vents can be detected; some theories put the main cone responsible in the adjoining island of Rhum. The three dominant rocks produce the main elements in the relief: the basalt forms the moorland plateaus in the centre and north, the gabbro forms the Black Cuillin hills of the south-west, and the granite the neighbouring Red Cuillins. The later intrusions produce terracing due to horizontal sills of hard rock, and cause deep gullying due to the removal of dykes of more rapidly eroded rock.

The total thickness of the igneous rocks is difficult to estimate, for since early Tertiary times denudation has removed thousands of metres. The main ridge of the Black Cuillins (fig. 43) is largely the result of ice action, which has carved glens, cut back corries (p. 230 and fig. 114) into the gabbro, and left a long, sinuous ridge,

Fig. 44 The Puy country of Auvergne

The hachured map (based on the official French map on a scale of 1:80,000) emphasizes the variety of the volcanic forms. The Puy de Dôme (1465 m, 4806 ft) lies 16 km west of Clermont-Ferrand.

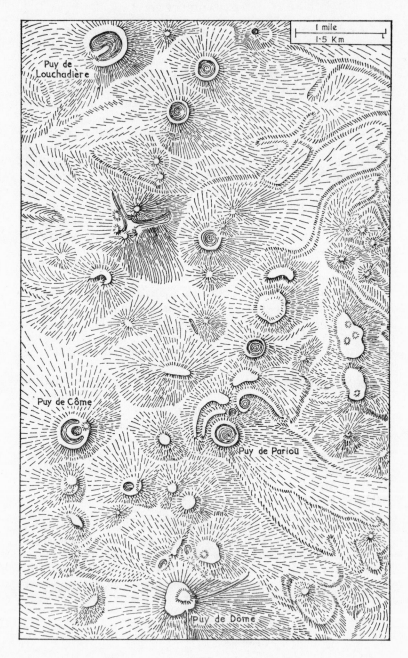

90

24 km (15 miles) in length, at its lowest dropping only to 765 m (2510 ft), and with a score of peaks over 900 m (3000 ft).

The Red Cuillins, of more uniform granite, have been worn into smooth, rounded hills, with slopes rising gently from the glens, but with none of the stark profile of the Black Cuillins. The basalt hills of the north have been planed into a gently undulating plateau, 350–450 m (1200–1500 ft) above sea-level, although the edges of the lava sheets have sometimes been worn into pillars and obelisks.

(ii) **Auvergne** Parts of Auvergne, in the Central Massif of France, present a fascinating example of a miniature vulcanic landscape, in which many of the different results of vulcanicity are exhibited in various stages of destruction by denudation. The region extends for 140 km (90 miles) from St Geniez to northward of Clermont-Ferrand. In the south the lava plateau of Aubrac is a level sheet of basalt lying upon the crystalline basement, with its western edges dissected by the head-streams of the river Lot. Then farther north, across the river Truyère, the volcanic cone of the Plomb du Cantal rises to a height of 1858 (6096 ft) above sea-level from a basal circumference of more than 140 km (90 miles). Farther north again the more complex cone of Mont Dore culminates in the Puy de Sancy (1886 m, 6188 ft)). These two cones have been severely attacked by denudation; radiating streams have trenched their sides, and these, emphasized by glaciation in Quaternary times, now consist of triangular lava slopes known as *planèzes*, which narrow upward to culminate in bold projections as a broken rim of peaks around the old crater-rim.

Beyond Mont Dore, extending northward for about 30 km (20 miles), is the curious country of the puys (fig. 44), with about seventy cones; some are of ash, some of viscous acid lava are dome-shaped, some have double cones, others an outer crater-rim. They rise only a few hundred metres above the general level of the plateau; the highest, the Puy de Dôme itself (1463 m, 4806 ft), is a rounded, craterless dome.

Near the town of Le Puy in the valley of the upper Loire are several hard lava plugs, from around which the softer ashes have been removed; the Rocher d'Aiguille, 85 m (279 ft) high, is crowned with the Church of St Michel, while the Rocher de Corneille, 130 m (427 ft) high, has a great statue of Notre Dame de France on its summit and a cathedral at its foot. Yet another contrast is shown near the village of Polignac, which clusters round the foot of a flat-topped basalt mass crowned by a fourteenth-century tower.

CHAPTER 4

The Sculpturing of the Earth's Surface

Land-form development The basic features of the earth's surface are determined by large-scale movements of and within the crust which have occurred at intervals during geological time. But these features undergo constant modification—ultimately destruction—by the combined agencies of earth sculpture, which operate as soon as one part of the earth's surface begins to rise relative to another. For purposes of land-form description, it is convenient to regard the work carried out by these agencies—like that of mountain building —as time-progressive in nature; in such a hypothesis, a landscape may be considered to suffer a series of changes, whereby the initial features are slowly modified. As S. W. Wooldridge wrote, "Land forms . . . are seen in their true significance only if it be remembered that they have developed, and are developing"; in the words of W. M. Davis, ". . . landscape is a function of structure, process and stage". The rocks, and the forms in which they have been arranged by earth movements, are conveniently described as 'structure'. The 'process' is the collective work of the various agents involved; the 'stage' is the extent to which these agents have operated and the resultant alteration of relief so far achieved.

The use of the term 'cycle' is well established in geographical literature to describe the modification of the physical landscape as a result of natural agencies in an orderly progressive sequence. The full hypothetical cycle ranges from the uplift of the land as a result of orogenic forces into an upland to an ultimate low, almost featureless plain. Though this concept of the cycle was first used in 1894, it was developed and formalized by W. M. Davis in 1902. However, in recent years exception has been taken by some to the use of 'cycle' in this context, since it implies and involves an infinite repetition of a whole series of changes which return to a point of origin, as does, for example, the hydrological cycle (pp. 110, 427). On these grounds, some prefer to use the word *sequence*. Even so, many authorities maintain that the concept of a cycle of erosion in landform development is perfectly logical and exact, since almost all the earth's surface (or at any rate, any mountain range folded before mid-Cainozoic times) has been planed at least once, and has therefore passed through at least one cycle of erosion.

It must not be assumed that a cycle always runs uninterruptedly its full course, nor does it begin just when the earth movement is completed. Uplift by earth movement and down-cutting will proceed simultaneously; some of the great gorges, such as those of the Brahmaputra (p. 155) or the Colorado, are due to rapid uplift and equally rapid simultaneous down-cutting by the rivers, so that the latter have been able to maintain their original levels. Moreover, further uplift may take place before the denudation cycle is completed, so that a new cycle will be initiated producing fresh features which are superimposed on a landscape still bearing the impress of some of the old characters, producing what may be called *polycyclic relief*. Hutton, the distinguished eighteenth-century geologist, summed up this 'endlessness' of the sequences of changes in the earth's surface in the solemn words: 'No trace of a beginning, no prospect of an end'.

The idea of a cycle of development is helpful, if only because it implies, and indeed stresses, gradual change in the evolution of slopes towards a level surface. The implication of age is furthered by the use of terms such as 'youth', 'adolescence', 'maturity' and 'old age'; one may thus refer to 'a maturely dissected landscape'. Generally 'youth' implies a stage in which the structure of a fold-mountain system or of an uplifted plateau is the dominant feature; gradients are irregular and processes rapid. 'Maturity' occurs when the last of the original land surface is on the point of being removed, about the middle of the gradual evolutionary process. 'Old age' indicates great areas of lowland, with scarcely perceptible undulations; this landscape is called a *peneplain* ('almost a plain') though rarely does it attain this ultimate perfection. Usually masses of more resistant rock stand out, forming residual hills or 'erosional survivals', to which the name *monadnock* (fig. 45) has been given after

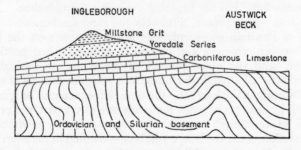

Fig. 45 Ingleborough

Ingleborough is an erosional remnant standing out prominently in the Craven District of the West Riding of Yorkshire (see also fig. 56).

the mountain of that name in the Appalachians, rising to 965 m (3165 ft) above sea-level.

But should uplift intervene at any stage of the cycle, the forces of down-cutting are revived, and another term in the age-cycle, 'rejuvenation', can be logically used. If the cycle had attained the peneplain stage, the uniform summit-levels remaining between the new valleys are preserved as eloquent relics. To such an area of relief the term *dissected peneplain* can be applied.

The cycle concept was first postulated with reference to river systems in mid-latitudes; the name 'cycle of normal erosion', or better 'water erosion', is therefore usually applied (p. 134). The same idea can be used, however, to describe desert landscapes ('the cycle of arid erosion'), glacial landscapes ('the cycle of glacial erosion') and coastlines ('the cycle of marine erosion'), with obvious modifications in each case. It must be remembered in studying polycyclic relief that a large-scale change of climate (an 'ice age' or the onset of desert conditions) may result in the superimposition of a glacial or arid sequence upon the features produced by a 'normal cycle', leaving its legacy of landscape details should a return occur to a renewed 'normal' sequence.

There has been much recent criticism of the use of these somewhat evocative age-terms in the descriptive analysis of land-forms, a belief that using the human life-cycle as a simile can be carried too far. It cannot be assumed, for example, that the stage of youth is necessarily the most active in terms of denudation, nor that old-age streams are the most sluggish. Some geomorphologists indeed are wont to dismiss all Davisian concepts as traditional and conservative, if not outworn and outmoded; others seek to take 'the best of what is new and the most tested and trusted of what is old', relating new concepts in climato-morphology to the traditional approach, stressing 'process-form' but not ignoring evolution through time.

Dynamic equilibrium Probably the most fundamental modification of Davisian geomorphology is the emphasis on the interaction between structure and process in explaining any particular land-form, and the abandonment of 'stage' in the limited and restricted sense of a particular phase in a progressive cycle of evolution. Obviously in the long term the time element cannot be ignored; there will be changes between the landscape of the past, the present and the future, though not with the rigorous 'cyclic inevitability'. This evolution of the landscape will be interrupted in the short term by periods of 'steady state', balance or dynamic equilibrium. These conditions of equilibrium can develop, for example, between the rate of rock weathering and the rate of removal of the weathered pro-

25 Scree-slopes and ice-worn gabbro outcrops beside Lochan Laggan in the
southern Black Cuillins, Isle of Skye
(*F. J. Monkhouse*)

26 Widening of rectangular joints by frost-action in Glen Orchy, Scotland
(*R. Kay Gresswell*)

27 Granite block-piles on the north-western ridge of Goat Fell, Arran
The north-western ridge of Goat Fell consists of a continuous series of these massive strongly jointed granite outcrops, forming tor-like projections along the ridge.
(*L. J. Monkhouse*)

28 Granite weathering in Arran
Chemical weathering can rapidly decay granite. Some types of granite are more prone to disintegration, and form mushroom-like blocks.
(*L. J. Monkhouse*)

29 A frost-riven rock, Aletschwald, Switzerland
(*Eric Kay*)

30 The shattering of granite by pressure release, Yosemite National Park, California, U.S.A.
(*F. J. Monkhouse*)

31 Infilled solution pipes revealed in a chalk quarry, near Woodhead, eastern Devon
(*Eric Kay*)

32 Earth pillars developed in a mass of boulder-clay, cut through by a road
tunnel in the Val d'Hérens, Switzerland
(*Eric Kay*)

ducts on a slope. The existence of such an equilibrium may be indicated by constant angles of slope within a small region of uniform relief, rock-type, climate and vegetation. If this equilibrium is disturbed, as by a climatic change affecting the rates of weathering and removal, or by actual uplift of the land, it will be restored by an adjustment of the angle of slope to meet the requirements of the new situation and conditions. Although some modern authorities are wont to refer to this concept of dynamic equilibrium as 'the new geomorphology', it was actually postulated as long ago as 1877 by G. K. Gilbert.

THE DEFINITION OF DENUDATION

The term denudation is used widely to cover all the agencies by which parts of the earth's surface are undergoing destruction, wastage and loss; the material thus removed is deposited elsewhere, to form the sedimentary rocks (pp. 7–10). The loosening, decaying and breaking up of the rocks, the formation of a mantle of more or less disintegrated waste (the *regolith*), is largely due to various agents of the weather, and so is known as *weathering*. Rock-waste moves inevitably downhill as creeps, slides, falls, slumps and flows, on various scales and under various conditions; this is known as *mass-movement*, its results as *mass-wasting*.

Apart from the effects of weather and gravity, earth sculpture is carried out by agents which attack the surface and at the same time remove the materials resulting from their attack—by running water, i.e. rain-wash, streams and rivers (what is sometimes described as 'linear erosion'), by moving ice in the form of glaciers and ice-sheets, by wind, and on the margins of the land by the waves, tides and currents in the sea. The sculpturing work of these agents is termed *erosion*. They transport vast quantities of weathered rock, usually downhill, towards the area which is the final scene of deposition in the current sequence; hence *transportation* is an essential phase of earth sculpture. But during the various transporting processes, the load itself acts as an abrasive tool in wearing away the rock over which the transporting medium passes. The joint work of the transporting agent (which supplies the energy or momentum) and of the load (which provides the abrasive medium) is known as *corrasion*, which leads to the *abrasion* of the surface. The load itself undergoes constant wear and tear during transport; this is termed *attrition*.

There is a good deal of confusion in the use of the words denudation, erosion and corrasion; frequently, indeed, they are used indiscriminately and synonymously. Denudation is a wide term covering all the processes of earth sculpture, including weathering. Erosion comprises those processes of earth sculpture that involve transport of material, that is, excluding weathering, which is essentially carried out *in situ*. Corrasion is the result of the abrasive work done by the load actually carried, and is really mechanical erosion; it excludes erosion by solution (*corrosion*).

The term *gradation* is used as synonymous with denudation, particularly in America. Its collective processes ultimately reduce outstanding land-forms to a low uniform surface. These processes involve *degradation* (cutting-down and removal), and *aggradation* (accumulation or building-up).

The processes of weathering and mass-movement are described in this chapter. It is, however, more convenient to examine the work of erosion under the heading of each particular agent—rivers (Chapter 6), glaciers and ice-sheets (Chapter 8), the wind (Chapter 9) and the sea (Chapter 10).

WEATHERING

Weathering involves the disintegration and decay of solid rock. It depends on both the nature of the climatic elements involved (sun, rain, frost and temperature change, though not wind, which implies transport and corrasion), and on the character of the rock—its chemical composition, its hardness, its texture, whether it has clearly defined joints (plate 34) and bedding planes, and its permeability (which will allow the penetration of water into the mass of the rock). Where rocks of markedly different properties occur close together, differential weathering may strongly affect the surface. In many areas, beds of shale, limestone and sandstone, sometimes only a few centimetres thick, may rapidly succeed each other. The variety of the Jurassic beds emphasizes this well—the Purbeck and Portland Limestones, Kimmeridge Clay, Corallian Limestone, Oxford Clay, the Great Oolite Limestone, the Bridport Sandstone and the blue-grey Lias Clays. These closely associated rocks have differing resistance to weathering, as well as to the other denuding forces.

Weathering is carried out in two ways, though usually working in conjunction: by *mechanical* (*physical*) means, i.e. disintegration without chemical change, and by *chemical* means, i.e. a decomposition of some or all of the mineral constituents in a rock causing ultimately a collapse within the rock itself, sometimes known as *rotting*. Mechanical weathering tends to break down the rock into progressively smaller fragments, while chemical weathering forms residual materials; the joint result is the production of a loose layer which can readily be removed by the agents of transportation. The influence of plants and animals, although not strictly part of weathering, may be considered as a biological agency which directly contributes to the layer of rock waste.

Mechanical weathering This is carried out in deserts by rapid changes of temperature or in mountains by the action of frost. In dry, cloudless regions a very marked diurnal range of temperature is due to direct sun-heating by day and rapid radiation by night (pp. 382, 482–3). The rocks successively expand and contract and so tend to enlarge joints, and the masses will ultimately break into smaller blocks, a process known as *block disintegration*. In some rocks the surface may become so intensely heated that a 'shell' pulls away and splits off from the interior which maintains a constantly lower temperature, thus causing the surface rock to 'peel off' (*exfoliation*). An isolated mass of rock may be rounded-off to form an *exfoliation dome*. These processes are particularly effective on the face of a steep crag which receives and loses the sun's rays rapidly, and so emphasizes the effect of sudden heating and cooling. A well-known example of disintegration into blocks along the joints, followed by the rounding of the blocks themselves by exfoliation, is seen in the neighbourhood of Rhodes's grave, in the Matopo Hills of Rhodesia. Temperature changes, moreover, will open up fissures into which water can penetrate, rather than flowing away over the surface, thus stimulating chemical weathering.

If the rock is heterogeneous, i.e. composed of various minerals, probably with differing coefficients of expansion, complex strains are set up inside the rock, and sudden fractures occur. During the half-hour after sunset, when one is encamped in the desert, reports like pistol-shots are commonly heard as the rock shatters; this is especially marked in coarse-grained or large-crystalled rocks. An occasional shower of rain, not unknown in the driest desert, may cause a sudden chilling of the rock and its shivering into fragments. The result is to produce a mantle of angular fragments swathing the base of the harsh, sharp-cut desert ranges.

It must be admitted that considerable doubt has been cast on the efficacy of alternate heating and cooling as an agent of mechanical weathering. Cases have been quoted where weathering was more rapid on rocks in total shade (therefore probably with more moisture present) than the same outcrops in alternating sunshine and shade; and laboratory experiments have shown that rocks can stand long periods of heating and cooling at temperature ranges far greater than would be experienced in the hot deserts. Some authorities maintain that the presence of water is essential, that, in fact, some of this weathering is really chemical rather than mechanical.

Frost There can be no doubt, however, that block disintegration is effectively carried out, both in mid-latitudes in winter and in high mountains, by frost action, a process known as *congelifraction*. When water fills the interstices of a rock, it may freeze at night; as its volume thus increases by about 10 per cent, it exerts great pressure, and the rock tends to shatter (plates 26, 29). A surface covered with a spread of angular blocks broken off in this way is known as a *Felsenmeer*, *block-spread* or *boulder-field*, as on the Great End–Scafell Pike ridge in the English Lake District, and on the top of the Glyders in North Wales. Slopes of rock fragments, known as *scree* (see p. 103, plate 37), fall away below the buttresses in the British mountains; in fact, frost is the most potent sculpturing agent of the pyramidal peaks or fretted ridges of a glaciated mountainous area, such as the Black Cuillins of Skye (plate 25), and of the rock peaks of the Alps and Himalayas, projecting above their swathing snowfields. Frost can loosen any surface exposed to it, hence the ploughing of heavy clayland by farmers, in order that winter frosts will break it down into a more friable mass. This process, particularly potent in porous rocks which can absorb large quantities of water, is known as *granular disintegration*; it can be seen on some of the high but gently rounded British mountains, such as the Carneddau and Moel Famau in North Wales, Grasmoor and High Street in the English Lake District, and many Scottish summits, such as the Cairngorms. The surface freezes and then crumbles when thawing occurs (pp. 255-7). This process of frost-heaving and -churning is known as *congeliturbation*.

Rain hardly enters into mechanical weathering except as a source of water for frost action, for it involves transport, either as rain-wash or in the form of river erosion. But a whole slope may become unstable after heavy rain. Moreover, some clay-rich sedimentary rocks are subject to a process called *slaking*. When a drought succeeds a rainy period, the rocks give up moisture previously absorbed; as a result, a rock such as shale may crumble into small elongated fragments. When a mass of clay dries out, it shrinks and its surface becomes seamed with cracks, facilitating its break-up and subsequent removal, especially on a slope.

'Unloading' One process of physical disintegration of rocks is known as 'unloading' or 'pressure release'. If overlying layers of rock are removed by denudation, the release of this weight-caused pressure may allow the newly exposed rock to expand and form new curvilinear joints, causing curved rock-shells to pull away from the mass, a

process known as *sheeting* (plate 30). Granite appears to be particularly prone to this; evidence seems to be very convincing in the Cairngorm Mountains and on the Etive Slabs on the slopes of Ben Trileachan overlooking Loch Etive. The enormous domes of the Yosemite Valley (plate 35) in California are probably the result of this outward expanding force of pressure release on little jointed granite. This has caused a succession of curved rock-shells, from 1 cm to 6 m in thickness, to pull away.

Chemical weathering Various changes in the chemical composition of a rock, with resulting changes in the coherence of its cemented particles, may result from long-coninned exposure to weathering agents. Some minerals, such as quartz, are virtually unaffected, others such as olivine, augite, hornblende, biotite, orthoclase and muscovite are very susceptible, and a few such as rock-salt can be completely removed in *solution*. The first essential for chemical weathering is the presence of water, which is plentiful in a humid climate. But even in a hot desert occasional heavy showers occur, and despite the rapid evaporation some water penetrates the rocks. Sometimes the relative humidity in a desert is high, resulting in a heavy dew (p. 435), and possibly water can penetrate a pervious rock in the form of vapour.

Where percolating water contains carbon dioxide derived from its passage through the atmosphere, it acts as a dilute acid upon calcareous rocks such as limestone and chalk, dissolving and removing them in the form of calcium bicarbonate, except for a thin indissoluble residue (p. 487); this is known as *carbonation*. In the Pennines bare limestone 'pavements' (plate 38) are seamed with solution grooves separated by sharp ridges, known as *grikes* and *clints* respectively. Felspars break down when attacked by rainwater, a process known as *hydrolysis*; the chemical equation is complex, but in effect the crystalline felspathic rocks crumble and form clays, while colloidal silica is removed in solution. Thus granite may break down to form a friable clay mass mixed with resistant quartz crystals and mica flakes. The two photographs (plates 27, 28), taken on the Goat Fell ridge on the Isle of Arran, illustrate the weathering of strongly jointed granite. The results of *oxidation* are most readily shown when the rocks affected contain iron, a very common element (p. 18). The weathered surface of many rocks reveals a yellow or brown crust; the ferrous state in which iron commonly occurs changes into the oxidized ferric state, and this crust readily crumbles. In deserts, a 'varnish', 'rind', or film of iron or manganese oxide is formed on rocks by solutions drawn to the surface through intense evaporation. Sometimes, incidentally, such a hard unbroken varnish may be an actual protection against other weathering or erosion, yet inside the rock the cementing material may have collapsed; once the rind is ultimately broken, the whole rock may be removed by eroding forces such as the wind. Such hollow 'shells' are known as *tafoni*, after a Corsican patois word. Finally, certain minerals have the property of taking up water and thus expanding, so stimulating the disintegration of the rock containing them; this is *hydration*.

Thus five main types of chemical weathering may be distinguished: (i) solution; (ii) carbonation; (iii) hydrolysis; (iv) oxidation; and (v) hydration. The result is the conversion of the original minerals into secondary minerals, usually more readily removable. The consolidation or adhesion of unaffected minerals in the parent rock is weakened, for cements are washed out and the rock tends increasingly to crumble. Mechanical weathering and erosional agents can then act with greater potency.

One result of chemical activity akin to physical weathering is known as *spheroidal weathering*, and its effects are very similar to exfoliation. The outer shell of such rocks as basalt is affected by penetrating water, and the chemical reactions cause this to swell or expand, and so pull away from the solid core, presenting a fresh surface to the atmosphere. A well-jointed rock allows this to go on readily, so that the blocks become more and more rounded as each shell of decayed rock breaks away.

Tors Some recent work indicates that the familiar tors of Cornwall may be the result of the sub-surface (rather than sub-aerial) rotting of granite through the action of acidulated rain-water penetrating along joints into the body of the granitic mass. Thus the pattern of the tor is controlled by the joints, which will leave between them broadly rectangular 'core-stones'; this may well have taken place in pre-glacial or interglacial times. Where the jointing is widely spaced, massive core-stones remain; where the jointing is close, there is more shattering and rapid removal of debris, thus forming depressions between the tors. Then followed a post-glacial period of 'exhumation', when the overlying weathered material

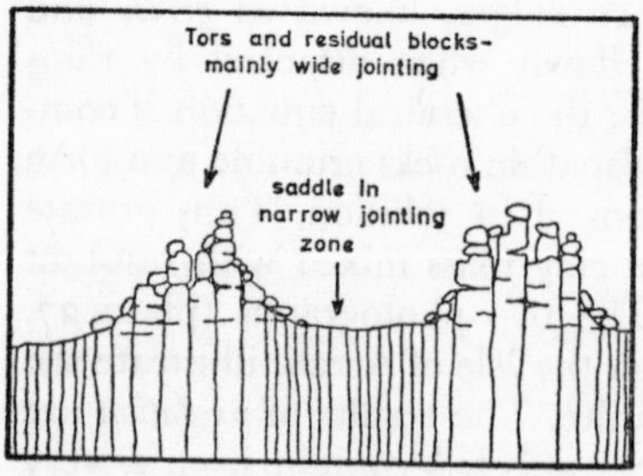

Fig. 46 The development of tors in jointed rocks

and the fine-grained products of rock-decay (known as *growan* on Dartmoor) were removed by solifluction or by melt-water, thus revealing the tor. This may be modified by sub-aerial weathering, but the 'block-pile' character of the tor remains. It should be mentioned, however, that some authorities believe that periglacial processes are largely responsible for tors.

Biological or organic weathering Plants assist in surface weathering by both chemical and mechanical means. Algae, mosses, lichens and other vegetation retain water on the surface of the rock, and various organic acids help to decay the rock beneath, so that a tuft of moss may lie in a small and growing hollow in the rock. The presence of vegetation increases the acid content of the soil-water (p. 488), which will be effective in the chemical disintegration of calcareous rocks. It seems too that water containing bacteria can assist the decomposition of some rocks, particularly limestones. The mechanical disintegrating effect of vegetation is mainly due to the penetrating and expanding power of roots, which exert considerable force as they grow and help to widen cracks and crevices, thus allowing water and air to enter.

It must be remembered, however, that a close mat of vegetation may actually prevent disintegration, since it binds the surface layer, hinders its removal, holds up water, and prevents the exposure of a surface of fresh rock to the elements. The injudicious removal by man of vegetation is, in fact, one of the major causes of soil erosion (pp. 506–8).

Various forms of animal life, such as worms, rabbits and moles, may have a contributory effect. Worms bring large quantities of fine material to the surface in the form of casts, while burrowing animals in some measure help to loosen the surface material.

Weathering in various climatic regions The effects of the weather upon rocks vary according to the potency of the different climatic elements. In equatorial latitudes, where both humidity and temperatures are consistently high, chemical weathering is continuously active, and as it is generally much more rapid and effective than the transport and removal of the weathered material, a thick detrital layer will accumulate. Similarly, in tropical regions, with a marked dry season during which evaporation is potent, but with a wet season which allows leaching, great thicknesses of laterite may be formed (p. 499), a reddish clay-like substance consisting mainly of hydrated oxides of aluminium and iron. In desert areas there is little weathering by ordinary leaching, but considerable mechanical weathering, while chemical weathering takes place by the drawing of strong solutions to the surface by capillarity. Sometimes this process leads to a concentration near the surface of various minerals, which form a hard compact layer known as *duricrust*. This is especially common in semi-arid areas with a brief rainy season. In mid-latitudes frost is by far the most powerful agent, while solution, particularly in limestone areas, exerts great effect. Under polar conditions great areas of permanent snow prevent any ordinary weathering, but where nunataks (p. 218) project from the ice-sheets, frost action is rampant. Chemical and organic agencies here seem to be negligible in their effects, though recently it has been suggested that chemical weathering is more active than was realized. Carbon dioxide is more soluble at low temperatures than at high, and as the melt-water has therefore a higher carbonic acid content, chemical weathering may be quite active under a glacier or at the edge of an ice-sheet.

Weathering therefore produces a mantle of rock-waste *in situ*, the *regolith*, but no movement is involved; this is the first stage in the creation of slopes.

MASS-MOVEMENT

The movement of debris, the loose material derived from the weathering of bedrock, down a slope is known as *mass-movement* if no actual transporting agency, such as running water, is involved; its results are *mass-wasting*, causing effects on the landscape sometimes graphically known as 'slope collapse' or 'slope failure'. This takes a variety of forms, some of which are slow, almost imperceptible, and continuous over a long period of time, while others act suddenly, rapidly and sometimes catastrophically. The movement is basically gravitational; the steeper the slope the more rapid it will be. The presence of water is a contributory factor; if the debris is dry, it will slide or fall as a loose mass, but if the material is impregnated with water, the movement will be of the nature of a flow. Water is not a lubricant in the strict sense of the word, though the term is often used. But if water fills the pore spaces between individual grains in a deposit, the surface tension that assists in maintaining its cohesion will be diminished. Water also adds to the sheer bulk and weight of the mass on the slope, which will accentuate the outward and downward motion, and springs have an undermining effect. Engineers endeavour to remove excessive amounts of water from vulnerable slopes to reduce the danger of slope-failure.

The more rapid movements may be 'triggered off' by some external influence, natural or artificial, such as a sudden intensive rainstorm, a concentrated snow-melt, an earth-tremor or earthquake. Incautious quarrying, the collapse of a dam, the clear-felling

33 A 'fossil forest' exposed in the steeply dipping Purbeck Beds to the east of
Lulworth Cove, Dorset
(*Eric Kay*)

34 The 'Needle's Eye', Black Hills of South Dakota, U.S.A.
An example of weathering concentrated along a major joint in granite just off
the Needle's Highway, in the Black Hills.
(*F. J. Monkhouse*)

35 The 'domes', Yosemite National Park, California, U.S.A.
The deeply cut Yosemite Valley lies in the foreground; the rounded granite
domes rise to heights of about 7500 feet.
(*F. J. Monkhouse*)

36 Solifluction near Brook, Isle of Wight
Wealden Clays are slumping down from the cliffs and flowing across the darker sands.

(Eric Kay)

37 Scree slopes in the Pass of Brander, Argyllshire, Scotland
(Eric Kay)

[103

of trees from a hillside, ploughing down a slope, the burrowing of animals, the vibrational shock of a passing train, and the passage of grazing stock or human beings across a slope may all start some form of movement. Other erosional agencies may cut into the base of a slope, such as a river, a glacier or the waves at the foot of a sea-cliff, so steepening its angle. In critical areas, such as above a rail-way or a mountain road, it is possible to reduce the danger of falls by various engineering measures, such as revetment, terracing and the insertion of drains to prevent the accumulation of water within the mass of material on the slope. Further precautions are the maintenance and stimulation of a vegetation cover, contour ploughing, in fact all the measures used to combat soil erosion (pp. 506–8), the removal of the thin precious surface layer of the soil.

The processes comprise various kinds of falling, sliding, slipping, flowing and creeping, and can be broadly divided into two groups, involving slow and rapid movements.

Slow movements The most common type is known as *creep*, which may be almost indiscernible, especially when it proceeds under a turf mat. Gradually, however, it becomes apparent as posts and fences are first tilted and then displaced downhill; even growing trees may be moved; a ribbed or stepped pattern develops across the slope in the form of *terracettes*, about the mechanics of whose origin little is known; the turf bulges and even rolls up; and overhanging banks develop above roads and rivers. Where the strata are inclined inwards from the slope, their outcrops may be 'cambered' downhill (fig. 47). Angular rock debris will creep down a mountain side, forming scree (p. 98), sometimes, especially in the U.S.A., known as talus[1]; this normally lies at an angle of rest, which may be disturbed by heavy rain, by the addition of frag-ments through frost-weathering of the crags above, and even by the passage of animals or an incautious climber across the slope. Occasionally the fragments form a tongue down the slope, known as a 'rock-glacier'. One very important category of slow movement is *flow*, where especially clays containing an excessive amount of water behave as a plastic mass moving as a thickly viscous fluid. A potent type of flow is *solifluction* or *sludging*; at the present time this is largely the result of the periglacial processes (p. 253–9) active in high latitudes and at high altitudes, which affected much greater areas during the Quaternary glaciation than it does at the present.

[1] There is some uncertainty about the use of the terms scree and talus. Some regard them as synonymous; some refer to scree as the ingredient, talus as the slope; others define scree as all loose material lying on a hillside, while talus accumulates specifically at the base of a crag or cliff.

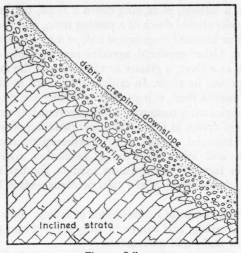

Fig. 47 Soil-creep

Rapid movements The various types of rapid movement, generally referred to as landslides, include slumping, mudflows and earthflows, and rock-slides. *Slumping* (or *slipping*) involves an actual shearing of the rocks, a tearing away of a mass of material (a *slump-block*) from the face of an escarpment or a sea-cliff. There is often a distinct rotational movement on a curved concave-up plane, tilting back the block as it moves and leaving a fresh scar on the hillside. This is particularly common when massive, well-jointed rocks, such as limestone, overlie clay or a weak shale, which may readily rupture. It can be seen along the south coast of the Isle of Wight; along the Kentish coast between Folkstone and Dover where the Chalk overlies Gault Clay; along the Kentish coast east of Herne Bay where the London Clay overlies sandstone; along the Devon coast between Exmouth and Lyme Regis where the Gault, Upper Greensand and Chalk rest unconformably on Jurassic shales and clays; and on the escarpment of the Cotswolds between Gloucester and Cheltenham, where Jurassic limestones slump down over the clay. An outward dip of the beds is a contributory factor in most cases. The first sign of this slumping along sea-cliffs is usually a curving crack a few yards inland from and parallel to the top of the cliff, down which rain-water will penetrate.

Mudflows and *earthflows* occur where material contains a good deal of water. The former can flow as a slurry down quite gentle slopes, the latter, consisting of drier material, down steeper slopes. These movements are intermittent; they are especially common in semi-arid lands where sudden torrential rain sweeps away the unconsolidated weathered debris, at first moving very rapidly, then becoming increasingly pasty until they come to rest, where they may splay out as tongues at the foot of the slope (plate 36).

A *rock-slide*, or on a smaller scale a *debris-slide*, may involve quantities of rock moving at great speed, sometimes with devastating, even catastrophic results. For example, in 1881 the Swiss village of Elm was largely destroyed when a mass of rock slid from the Tschingelberg to the south-east, which had been weakened by quarrying and affected by several weeks of rain; over a hundred people lost their lives. More recently the Madison, Montana, earthquake of 1959 (p. 40) triggered off one of the largest slides on record. The side of the Madison river canyon consisted of a dolomite stratum behind which was a mass of unstable schist; the dolomite retaining wall was fractured, and some 80 million tons of rock crashed down to the valley floor, surging up the other side, and creating a natural dam behind which Earthquake Lake (p. 197) was ponded up. In October 1963 a landslide from the side of the 1800 m (6000 ft) Toc peak overlooking the Piave valley in northern Italy fell into the reservoir behind the Vaiont dam, causing a wave which spilled over and swept away five villages, with considerable loss of life. As a final example, in April 1964 earth tremors cracked the side of Darnovoz, a mountain in Uzbekistan in Soviet Asia, a huge slide blocked the valley of the river Zeravshan, and ponded up a lake; the collapse of the dam would have caused great damage to the city of Samarkand, and so engineers cut down through the debris and relieved the pressure.

A distinction is sometimes made between a rock-slide, an *en masse* movement of material down a slope, and a *rock-fall*, a free fall (or rock-avalanche) of boulders down a steep mountain-side.

In these various ways the downhill movement of material contributes to the denudation of the land, in collaboration with weathering, which helps to furnish the material and with the agents of transport —running water, moving ice, the wind, and waves and currents, which continue the work.

SLOPES AND SURFACES

Only a very small part of the land surface is either truly horizontal or vertical. Plains are rarely plane surfaces, though the term 'planation' is used often enough in describing the processes which may ultimately form a more or less level area. Vertical features are limited to cliffs along the coast and in some parts of the mountains.

There are indeed vertical or even overhanging sections in which the exponents of modern tension climbing delight but few mountains are really vertical; some of the great Dolomite faces, perhaps, and parts of the Chamonix *aiguilles*. The average gradient of an Alpine face or ridge is about 20° to 30°.

It follows therefore that most of the land surface is made up of slopes, extending from the summits of mountains and the crests of watersheds to the valley-floors and to the level of the sea, ultimately to the bottom of the greatest deeps.

Slopes The analysis of the form and evolution of slopes forms an integral part of geomorphology, and can be studied in two ways.

The earliest approach to slope study, of which W. M. Davis was a master, is by means of general descriptive, explanatory and inductive methods applied to actual examples studied in the field, sometimes illuminated by flashes of inspiration.

The second, more modern, approach involves the application of geometrical methods, both inductive and deductive, which attempt to erect a hypothetical sequence of development of a given slope under carefully specified conditions. Some of these methods are essentially qualitative, discussing only the general trend of change, while others are strictly quantitative and establish precise mathematical models. An example of the former method is afforded by A. Wood's concept of a vertical or very steep rock-face, which he called the *free face* (fig. 48). As this weathers, the debris falls to its foot and builds up a heap of scree which gradually accumulates to form the *constant slope* (so called because it maintains a constant angle of rest), which extends upwards to the rock-cut slope of the free face; the area of the latter decreases as the accumulation of scree grows. Above the free face is the convex curve of the hill crest, which Wood termed the *waxing slope.* Below the constant slope another slope of fine material becomes progressively less steep and merges into the valley floor; it is called the *waning slope*, but as it is derived from the material washed down as the scree of the constant slope is weathered and comminuted, some authorities prefer to call it the *wash* (or

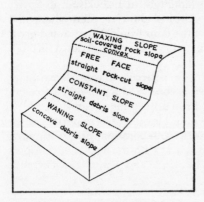

Fig. 48 The development of a slope

debris) slope. These low-angle wash slopes may become more and more dominant and will gradually coalesce to form a *depositional pediment* (cf. p. 276).

The exponents of quantitative methods claim that some degree of objectivity may be given by the examination of statistical information concerning the slopes of a particular area, derived either from the morphometric analysis of a topographical map or from actual instrumental surveys.

Some controversy arises as to whether slopes are the product of wearing down or wearing back, of declining or retreating, though it must be appreciated that both can proceed simultaneously. The former involves the gradual down-cutting of the land surface, the creation of progressively gentler slopes of lower angle as the stage of peneplanation is approached, in other words, conforming to Davis's concept of the 'normal' cycle of erosion in humid mid-latitude areas (fig. 49, right). At any time the upper part of the slope will be convex in profile, the lower part concave, and there may be a straight section in between, particularly in the later stages. Some theories regard the upper convexity as the result of creep and 'wash' by rain-water not in defined or concentrated channels, while the lower concavity is the result of the more concentrated action of small rills as they develop into larger streams. This is of course a generalization, and varies both with the nature of the rock and with the climate. Thus a resistant stratum or capping near the summit may cause a vertical step in the profile; highly permeable rocks such as chalk and limestone or conversely a dry climate will tend to produce convexity; and impermeable rocks or a very wet climate will cause much surface runoff and consequent concavity.

The other school of thought, largely derived from the work of Walther Penck and L. C. King, postulates the concept of a slope retreating parallel to its original angle, a process known as the *parallel retreat of slopes* (fig. 49, left). King accepts Wood's fourfold slope elements in a hillside, and suggests that where the free face and constant slope are being actively weathered, with removal of material from the scree slope, the whole slope will move back, or retreat, without having its gradient altered to any great extent. This concept was first hinted at in humid Britain in the 1850s and was precisely stated for northern Europe by Penck. Its application to arid and semi-arid areas came much later, and does much to explain the existence of inselbergen (p. 276) and pediments (p. 274–6). What has aroused controversy is whether this is possible and practicable in a humid mid-latitude landscape. One problem is whether the removal of debris from the base of the slope can be effected sufficiently rapidly. Possibly a river flowing at the foot of the slope or powerful springs bursting out may be able to do this; perhaps under periglacial conditions (p. 255) solifluction may have been adequate. But some authorities claim that in a humid mid-latitude climate parallel retreat is not an easy hypothesis to maintain.

Surfaces The term *erosion surface* has been widely used to indicate some fundamentally level area of the landscape; perhaps *planation surface* is better, since much of the land area is a surface of erosion, whereas the term is intended to refer only to near-level sections.

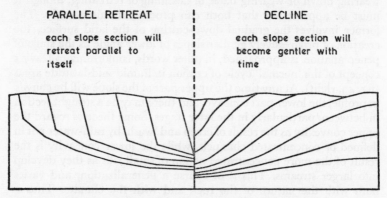

PARALLEL RETREAT | DECLINE

each slope section will retreat parallel to itself | each slope section will become gentler with time

Fig. 49 The development of a slope: parallel retreat and decline

The peneplain has already been mentioned as the ultimate stage in Davis's concept of the cycle of normal erosion, an area of gentle relief, though not absolutely flat, as a certain minimum gradient is necessary for water to be able to flow to the sea.

Peneplains do not necessarily remain low-lying, as uplift may occur at any stage in a cycle of erosion. If the uplifted surface is sufficiently extensive, it may be called a 'hill-top plain' or 'upland plain'. In some districts several of these 'erosion surfaces' can be distinguished, resulting in a *polycyclic relief*, each the result of a prolonged period of denudation corresponding to a particular sealevel. The distinguishing of these features, both by field-work and by cartographic analysis, is an important part of geomorphology. Frequently the dissected remnants are so fragmentary that their interpretation produces most stimulating dispute. There is no doubt however that some of the younger surfaces, mostly of late Tertiary age, can be well authenticated. While possible Scottish peneplains are now represented only by accordant summits at various heights (1200 m, 900 m and 600 m, 4000, 3000 and 2000 ft) are all suggested), there is no doubt about the 55 m (180 ft) surface in the Gower and Tenby districts, the 120 m (400 ft) surface of the London Basin, and the 170–200 m (550–650 ft) surface in the southern Weald. Apart from the problem of reconstructing dissected peneplains, uplift is by no means uniform, involving sometimes both tilting and warping.

It is possible for plane surfaces to originate in ways other than peneplanation. The term *panplain* was coined in 1933 by C. H. Crickmay to denote a plain formed by the coalescence of several adjacent flood-plains, each created by the lateral erosion of its respective river (p. 156), in contrast with the peneplain, which is formed by the degradation of divides between adjacent rivers. In connection with parallel retreat, a *pediplain* is a multi-concave surface resulting from the coalescence of several adjacent pediments (p. 276). Widespread surfaces in Africa are thought to have been formed by pediplanation; the 'Gondwana pediplain' at 1200 m (4000 ft) probably dates from early Cretaceous times, and the 'Africa pediplain' at 600 m (2000 ft) is possibly of late Cretaceous to early Tertiary age.

Mention should also be made of marine planation, the work of the sea, as exemplified by the present development of the Norwegian *Strandflat* (p. 288). A long period of marine transgression in the past may have worn such a surface, now uplifted by earth-movements. Some maintain that the production of an extensive surface in this way would be impossible because of the weakening of wave action in shallow water after quite a narrow bench has been cut; they would limit marine activity to a mere trimming of an existing sub-aerial peneplain which has been later submerged by a transgression. There is, however, some agreement that the Pliocene platform of the southern Weald at 170–200 m (550–650 ft) was caused by marine planation during a long period of widespread marine transgression.

CHAPTER 5

Underground Water

THE movement of water below the surface of the earth is a subject
of much interest and complexity. Part of our drinking-water is
obtained by sinking wells to tap this sub-surface water, which is
filtered naturally during its passage through the rocks, while many
villages owe their siting to the reappearance of underground water
on the surface as springs. Moreover, the erosive activity of water
as it works its way down through the rocks can not only affect
'underground scenery', which in limestone regions is often spectacular, but it can produce results on the surface in the shape of
sink-holes, subsidence depressions and collapsed caverns.

THE HYDROLOGICAL CYCLE

At a later stage (p. 427) it is necessary to discuss the endless
interchange of water between the sea, the air and the land, since
this is a vitally important climatic feature. In the words of *Ecclesiastes*,
i, 7, "All the rivers run into the sea; yet the sea is not full; unto the
place from whence the rivers come, thither they return again."

A small amount of subterranean water may have been retained
in sedimentary rocks since the time of their formation, known as
connate water, and a certain amount is liberated during igneous
activity, usually hot and mineralized, known as *juvenile* or *magmatic
water*. Further, in coastal areas some *oceanic water* may percolate
inland through the rocks. But these sources are relatively small
compared with *meteoric water*, which is directly derived from rainfall
or snow-melt.

When water reaches the surface of the earth in the form of rain
or snow-melt, part runs off down the slopes (hence *runoff*) to join
streams through which it ultimately reaches the sea, part is directly
evaporated (p. 432), and part is absorbed by vegetation and then
mostly returned to the atmosphere (p. 514). The rest percolates
through the surface soil into the bedrock to form the *ground-water* or
phreatic water. Various factors determine to what extent each of these
categories accounts for the total water received from the atmosphere.
On some steeply inclined surfaces, runoff will represent almost 100
per cent of precipitation. Runoff in hot deserts is short-lived, for the

rain quickly sinks into the sand, or should it fall on level rocky land it rapidly evaporates. On a rock such as sandstone, rain sinks in throughout the whole mass of the rock, while on a jointed rock such as chalk or limestone it disappears rapidly down joints, the lines of least resistance.

THE PERMEABILITY OF ROCKS

Classification of degree of permeability Rocks may be classified into the *permeable* ones, which will allow the downward passage of water, and the *impermeable* ones, which do not allow any appreciable passage. Permeable rocks owe this quality either to their *porosity* (their open texture, coarse-grained constituents and loose cementation, with pores of a certain minimum size), notably sand, sandstone, gravel and oolitic limestones, or to their being *pervious* (that is, traversed by joints, cracks and fissures through which water can flow), such as Carboniferous Limestone, chalk, quartzite and jointed granite. All impermeable rocks, such as slate, shale and gabbro, are also impervious. Some rocks which are impermeable are, however, porous; clay consists of extremely fine particles separated by minute pore spaces, but when wet the pores are filled with water held by surface tension, so sealing the rock against the downward passage of water. Clay can therefore hold water, but will scarcely permit it to flow through. 'Puddled clay' is almost completely impermeable; dew-ponds on chalk hills are usually lined with clay, and many streams which persist over chalk country with undiminished volume often do so because they have lined their own beds with fine-grained alluvium.

A bed which allows water to pass through it will if it is underlain by an impermeable stratum (*aquiclude*), become water-holding, and is known as an *aquifer*.

The water-table Except where there is standing water on the surface in the form of lakes or swamps, the layer immediately below the surface is known as the *zone of non-saturation*, through which penetrates most water not used by plants. Below that a layer may contain ground-water after long continued rain but will dry out after a period of drought, known as the *zone of intermittent saturation*. Below that again the *zone of permanent saturation* extends as far down as the impermeable layer forming the limit to downward percolation. The upper surface of saturation is known either as the *water-table* (not a good term, as it indicates a horizontal surface which in fact rarely exists) or the *level of saturation*. If this water-table is plotted on a section (as in fig. 50), it will be seen that

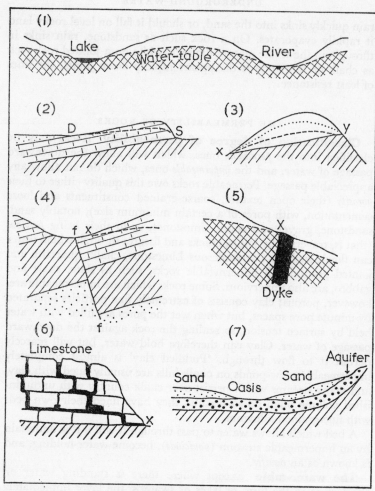

Fig. 50 The water-table and the occurrence of springs

Impermeable rocks are indicated by diagonal shading. In (1) and (5) saturated rocks are shown by cross-hatching. The permanent water-table is shown by a pecked line.

1. The relationship of the water-table to the surface profile. **2.** The water-table in relation to an escarpment; S indicates the position of scarp-foot springs. **D** of dip-slope springs. **3.** The dotted line indicates the maximum height of the water-table, the pecked line the permanent water-table; **Y** is therefore a temporary or intermittent spring, **X** a permanent spring. **4.** A fault-spring where a permeable stratum is brought up against an impermeable stratum along a fault, producing a spring at **X**. **5.** A dyke-spring, where a dyke cuts the surface, acting as an impermeable layer. **6.** A Vauclusian spring or resurgence. **7.** The occurrence of an oasis, where a surface depression is deep enough to reach the ground-water contained in the underlying aquifer.

it roughly follows the surface profile, with its gradients somewhat flattened out. The water-table tends to sink as it approaches a valley, since water more readily escapes at the surface on the hillsides above the valley, and so drains off more rapidly. Water tends to flow through beds of saturated permeable rock towards areas where the water-table is lower; the rate of flow is much slower than

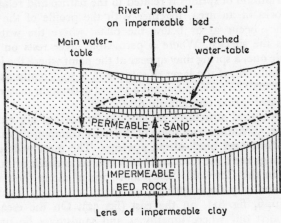

Fig. 51 *A perched water-table*

on the surface, since capillarity and friction with the grains through which it passes tend to arrest movement. The term *vadose* is applied to this 'wandering' water found beneath the surface but above the water-table.

Seasonal changes in the height of the water-table are of great importance in connection with water-supply; in Britain it is lower in summer, higher in winter. As a result, but for the case of an exceptionally wet period, the rainfall experienced during the summer has little effect on the water-table; it moistens the dry upper layer, but most is speedily evaporated or used by growing plants. The autumn rains first remoisten the upper layer, and then the winter rains percolate through to the water-table, which, lowered by summer consumption during which it has not been replenished, begins to rise again.

Occasionally an independent area of ground-water, above the water-table proper and separated from it by unsaturated rocks, may form a *perched water-table* (fig. 51); the usual cause is an isolated lens of impermeable clay.

SPRINGS

A spring is a natural outflow of water from the surface of the ground; it may flow strongly and even gush out with considerable force, or it may just ooze or seep out. Where a line of springs appears, the term '*spring-line*' is used; this is often indicated by a string of villages originally dependent on the springs for their water-supply. The distribution of springs is related to the nature and relationship of the rocks of an area, together with the profile of the surface relief; they occur at or below the plane where the water-table intersects the surface. Where a permeable layer rests on an impermeable one, a spring may appear at the point where the junction cuts the hill-side; if the layer of saturation is sufficiently extensive, the spring may be *permanent*, if not a period of drought may result in an *intermittent* spring temporarily drying up.

The geological structures responsible for springs are extremely varied; fig. 50 summarizes some of the more common types.

Scarp-foot and back-slope springs The most widely distributed springs are those associated with cuestas in chalk or limestone country; in the Cotswolds 'scarp-foot springs' along their western margins and 'dip-slope springs', or more accurately 'back-slope' springs (p. 6, fig. 61), on the east (fig. 52). On the escarpment springs burst out where limestones and sandstones lie upon the impermeable Lias Clay, so there is a string of 'spring-line villages'. On the back-slope the oolitic rocks dip gently to the east, and the springs break out here and there on the slope; thus 6 km (4 miles) south-south-east of Cheltenham is one of the reputed sources of the Thames, Seven Springs, which is only 0·8 km from the crest of the cuesta, while the river Dikler (a head-stream of the Windrush) issues strongly near Stow-on-the-Wold, 11 km from the crest.

Fault-, dyke- and joint-springs Springs may be thrown out along a fault where a permeable bed such as sandstone is brought up against an impermeable rock such as shale, or where a system of jointing in a rock such as granite or limestone reaches the surface, or where a dyke or sill cuts the surface. In Skye several 'joint-springs' occur high up among the Black Cuillins; the water penetrates through joints in the gabbro and emerges along the line of a basalt or dolerite sill. Some of these springs are as high as 750 m (2500 ft), but as their catchment areas are so limited they are very intermittent and a week's drought will cause them to disappear. On Great Gable, in the English Lake District, there is a spring which flows from a joint in a rectangular rock corner near Kern Knotts above Sty Head, which fails only in rare prolonged droughts.

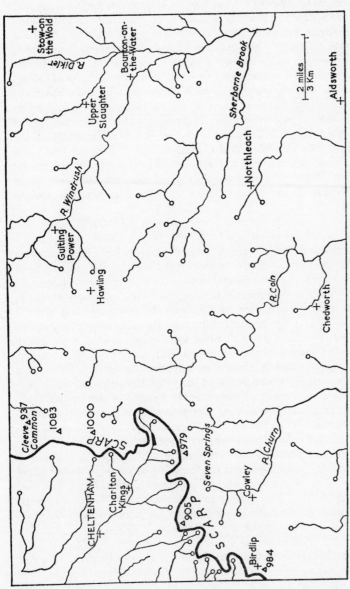

Fig. 52 Scarp-foot and back-slope springs in the Cotswolds

The springs are shown by open circles, the approximate position of the escarpment edge by a heavy line. The Seven Springs occur close together in a small hollow, claimed to be the main source of the Thames. In some cases the back-slope streams disappear beneath the surface after a short distance. Heights are given in feet.

Vauclusian springs One type of spring is sometimes referred to as Vauclusian, after the Fontaine de Vaucluse in the Rhône valley, where the river Sorgue rises under a 'horseshoe' of white limestone cliffs nearly 300 m (1000 ft) high. These occur in limestone country, where water has worn subterranean ramifications, finally issuing from the limestone near its base. Sometimes a series of springs bursts out, as at Kilnsey Crag in Wharfedale, where the basement of impermeable Silurian slates is not far below. More commonly a single stream emerges spectacularly from a large cave, such as the river Axe, which flows strongly out of the Wookey Hole in the Mendips, the Aire from under Malham Cove, and the Peakshole Water from Peak Cavern at Castleton in Derbyshire. These will be considered later, since they are the exits of underground rivers where water reappears at the surface (i.e. *resurgences*) rather than springs (pp. 121–6).

WELLS

The position of the water-table is of vital concern in the sinking of wells. In most wells this water-level rises and falls in the same way that springs flow strongly or weakly or even cease altogether. Shallow wells, which merely tap surface water, easily dry up, and, moreover, the water they contain may be contaminated, since it has not undergone natural filtration through any considerable thickness of rock. For a well to be satisfactory, it should be drilled to a depth considerably below the lowest possible point of the water-table (fig. 53), and it should be supplied by an aquifer with a steady source of water received directly through overlying permeable rock or by way of some surface outcrop of the aquifer itself. It must be remembered, too, that pumping will cause the water-table near a well to fall locally, so forming a 'cone of exhaustion' (fig. 53); thus a large well may produce such a considerable cone round it that nearby shallower wells gradually dry up. This has happened in East Anglia and the Weald, when a big country house installed a modern pump and so cottagers' wells ran dry and had to be dug deeper. The many pumps raising part of London's water-supply have lowered the water-table in the London Basin by more than 30 m (100 ft) during the last century.

The chief water-bearing strata in south-eastern England are the Chalk and the Lower Greensand. In the London Basin the former provides the aquifer, and about 900 wells have been sunk to 180–210 m (600–700 ft) below the surface; these now, however, supply only about one-eighth of the vast amount distributed by the Metropolitan Water Board. Practically all the water used in the Weald, by

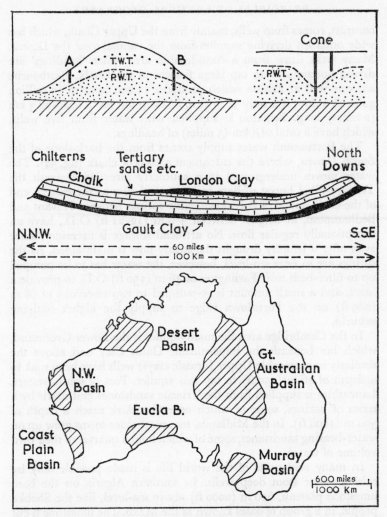

Fig. 53 *Temporary, permanent and artesian wells*

The top diagram (left) shows the temporary (**TWT**) and permanent (**PWT**) water-tables, while the impermeable layer is shaded diagonally. **A** indicates a temporary well, liable to dry up in the summer, **B** a permanent well sunk below the lowest possible point of the water-table. The top diagram (right) shows how the 'cone of exhaustion', caused by the deep well lowering locally the water-table, results in the depletion of neighbouring shallow wells.

The middle diagram (simplified from the Geological Survey Memoir, *The Water Supply of the County of London*) indicates the position of the chalk aquifer, with catchment areas on the Chilterns and North Downs, between the impermeable London and Gault Clays.

The bottom diagram locates the main Australian artesian basins.

contrast, comes from wells, mainly from the Upper Chalk, which has wide outcrops deriving supplies from the rainfall over the Downs. Many wells draw from a considerable area, since 'headings' are driven horizontally to tap large fissures in the Chalk. Eastbourne has one well at Friston which produces 14,000 cubic m (3 million gallons) a day; it is only 34 m (110 ft) deep but has a heading 3 km (2 miles) long. Brighton is supplied with water from five wells, which have a total of 8 km (5 miles) of headings.

The Portsmouth water supply comes from the back-slope of the South Downs, where the catchment area is of chalk (fig. 54). The water moves underground in a southerly direction beneath the Tertiary-filled Forest of Bere syncline, and rises at the eastern end of the chalk upfold of Portsdown. Nine springs between Havant and Bedhampton, at heights between 3–6 m (12–21 ft) O.D., have an exceptionally regular flow. No artificial storage is necessary, since even after a lengthy drought the steady flow is maintained. As the springs are so near sea-level, however, the water has to be pumped up to filter-beds near Farlington at 46 m (150 ft) O.D. to provide a head, and a small amount is re-pumped to two reservoirs at 88 m (289 ft) on the Portsdown ridge to supply the higher outlying suburbs.

In the Cambridge area the main aquifer is the Lower Greensand, which lies beneath the impermeable Gault Clay and above the similarly impermeable Upper Jurassic clays; wells have been sunk to a depth of 46 m (150 ft) to tap this aquifer. Part of south-western Lancashire is supplied from the Triassic sandstones and marls by a series of borings, some of which near Ormskirk reach a depth of 300 m (1000 ft). In the Midlands, too, most older towns grew up on water-bearing sandstones, some of which hold a quarter of their own volume of water.

In many arid parts of the world life is made possible only by raising water from deep wells. In southern Algeria on the bare limestone plateau, 600 m (2000 ft) above sea-level, live the Shebka people, in a group of oases known as the Mzab. The limestone is cut up by dry valleys, so wells are sunk from their floors to reach the water-table, which may be 60 m (200 ft) below. There are estimated to be 3300 wells in these oases, and water has to be raised ceaselessly by rope and pulley to supply the gardens.

Artesian wells Where the strata are arranged in a gently inclined downfold, forming a shallow basin with the aquifer out-cropping on the margins, conditions are particularly suitable for the supply of well-water. It sometimes happens that the aquifer is enclosed above and below by impermeable beds, while its rim is

38 Pen-y-Ghent, in the Pennines

Pen-y-Ghent (2273 feet), a monadnock of Millstone Grit (the level summit cap) and Yoredale Series (note the outcrops of the resistant bands on the face of the south ridge to the left), stands prominently above the surface of the Carboniferous Limestone platform at about 1250 feet. Note the limestone pavement in the foreground, eroded by solution along the well-marked joints into rectangular blocks.

(Eric Kay)

39 A stream-sink near Malham Tarn, in the Craven District of Yorkshire

This small stream, deriving its water from the drift-covered gritstone uplands, crosses on to the Carboniferous Limestone and takes the first opportunity to disappear down a sink-hole. The water reappears at the base of Malham Cove (Fig. 51).

(Eric Kay)

40 The Winnats Gorge, near Castleton, Derbyshire
(*Paul Popper*)

41 The Devil's Dyke, in the South Downs, near Brighton
A fine example of a chalkland dry-valley. See also Fig. 52.
(*R. J. Small*)

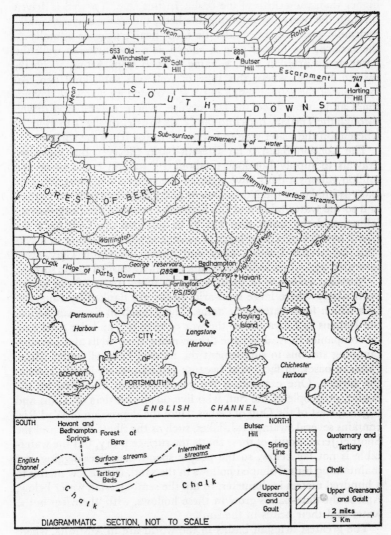

Fig. 54 *The water supply of the city of Portsmouth*

Based on a map after D. H. Thomson, in the Portsmouth Water Company publication, *Description of the Principal Works of the Company* (1937). The needs of the city have grown so considerably that some 25,000 cubic m (5 million gallons) of water is also taken daily from bore-holes at Walderton in West Sussex. (Heights in feet.)

higher than the centre of the basin. In this case, if a well is driven through the overlying impermeable beds, the water may rise under hydraulic pressure, i.e. under its own 'head', near or actually to the surface; the former case is known as a semi-artesian well, the latter as an artesian well, a name derived from Artois, in north-eastern France, where this type of well was sunk long ago. An example of the semi-artesian effect is provided by a well at Bovington Camp, near Wool in Dorsetshire, where a bore was driven through the Tertiary beds into the Chalk to a depth of 221 m (726 ft), whereupon the water rose to within 28 m (93 ft) of the surface.

The London Basin is an artesian basin (fig. 53), but the withdrawals to supply Londoners (about 1·3 million cubic m or 260 million gallons a day) have reduced the supply. Consumption now exceeds intake on the North Downs and Chilterns catchment areas, and so the water-table and the hydraulic pressure are progressively falling, and the water now has to be pumped up. Originally the Chalk and the overlying layer of Eocene sandstones, lying in a syncline below the London Clay and above the Gault Clay, were saturated and when wells were sunk the water gushed to the surface.

Artesian wells are of great value in many parts of the world, particularly where large semi-arid basins are flanked by ranges of hills which form catchment areas. The French in Saharan Africa, the Italians in Libya before the war of 1939–45, and the American oil companies in Arabia have driven many deep wells and installed pumping stations to supplement the natural head of water. Many Saharan oases, in fact, exist because artesian water locally reaches the surface. One example is the group of oases inhabited by the Soafas, or people of the Suf, who live on the borders of Tunis and Algeria, west of the Gulf of Gabes. Here the depression of Wad Rir' contains several swampy salt-lakes, such as the Shott Melrir and the Shott el Jerid. South of these shotts the surface is dry, but the water-table is not far below, so the people spend their lives digging and maintaining funnel-shaped hollows in the sand, from a few metres to a hundred metres in diameter, to tap the semi-artesian water below. And so the date-palms grow in these hollows, with 'their feet in the water and their heads in the sun'.

The greatest artesian basins in the world are in Australia, where the aquifers (Jurassic sandstones) underlie more than 1·3 million sq. km (500,000 sq. miles), deriving their intake from rain falling on the Eastern Highlands (fig. 53). Many of the 9000 wells are extremely deep, descending to a km or more, but unfortunately much of the water is slightly saline and so is used for watering stock rather than for irrigation. A constant Australian worry is that the artesian wells may run dry, for some experts consider that the amount withdrawn each year far exceeds the annual intake from rainfall; in other words, part of the water is derived from long accumulation in the past which is not being replaced, part possibly from juvenile water.

UNDERGROUND STREAMS AND CAVERNS

Chalk streams The drainage of country with markedly permeable surface rocks is to a large extent sub-surface. In such a rock as the Chalk, which is highly permeable and has numerous close joints, percolating rain-water rapidly makes its way down, dissolving the calcium carbonate (p. 99). It is estimated that each square km of chalk country in England loses 35,000 kg (35 tons) of matter by carbonation each year. The surface of the Chalk shows many shallow depressions. Fissures may be enlarged by solution; mention has been made of 'headings' driven horizontally from wells in the Weald to strike these fissures, which may give considerable water supplies. Many hollows develop at the edges of patches of superficial deposits of Clay-with-flints, or of isolated areas of Tertiary clays, particularly on the back-slope of the North Downs. They are fairly shallow and are usually filled with sand and gravel; such are the 'pipes' often seen in chalk-pits (plate 31). Some rivers, which flow across chalk country only because the water-table is actually at the surface for most of the year, may during a dry spell disappear down a hole in the bed. Examples are the river Mole, north of Dorking, along whose course a whole series of conical swallow-holes has developed, probably where large joints cross in the river's bed, and the Mymmshall brook, a head-stream of the Colne, the waters of which, except in wet weather, disappear down a group of large swallow-holes (fig. 55). In other parts of the chalk country large depressions are formed by the coalescence of several solution hollows.

Limestone streams While these solution hollows are quite common in chalk, in Mountain Limestone, with its marked jointing, they are characteristic features. This rock is hard and coherent and the joints are clean-cut, so that surface water works its way down, now vertically, now horizontally, towards the base of the limestone, where it may emerge as a Vauclusian spring (p. 116). Thus a veritable labyrinth of caverns and interconnecting galleries, underground rivers, waterfalls and lakes may be formed. In some interglacial and immediately post-glacial periods, the underground streams carried much more water than at present; many enormous caverns were excavated, now left normally dry with the lowering of the water-table.

The water-table within the limestone fluctuates appreciably, since

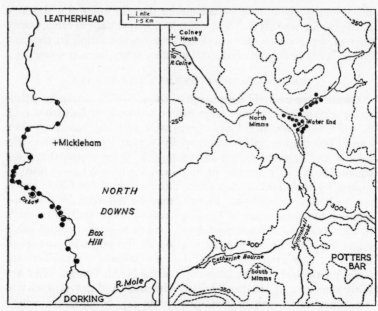

Fig. 55 *Swallow-holes in the Mole valley, Surrey* (left), *and in the neighbourhood of North Mimms in Hertfordshire* (right)

The swallow-holes are indicated by black discs.

The left-hand map (based on a map in *The Wealden District*, British Regional Geology, H.M.S.O.) shows the river Mole, which flows across the Chalk from Dorking to Leatherhead. Normally the river flows on the surface, but during low-water part of the stream disappears through conical swallow-holes in its bed.

The right-hand map shows the Mymmshall brook in Hertfordshire, which flows northward near South Mimms, and for most of the year disappears down a group of swallow-holes in the bed of a large depression known as Water End, near North Mimms. Colouring tests have revealed that this water reappears at four springs in the Lea valley to the east: at Woolmers Park (after a passage of 60 hours), at Chadwell Spring (90 hours), at a spring between Chadwell and Hoddesdon (92 hours), and at Lynch Mill (79 hours). During heavy rain in winter, the depression may form a lake, and the excess water then flows off westward through the small surface stream in the north-west which joins the river Colne.

the runoff is so rapid; many cavern systems can only be explored after a considerable drought, and a sudden rainstorm may cause such a rapid rise that cave explorers have to beat a hasty retreat.

Where the limestone is capped by a more impermeable rock, a surface stream of some size may form on the latter, but when it flows down on to the limestone it takes the first opportunity of plunging down a joint, which it rapidly enlarges to form a swallow-hole or sink (plate 39).

The Craven district of the West Riding of Yorkshire presents

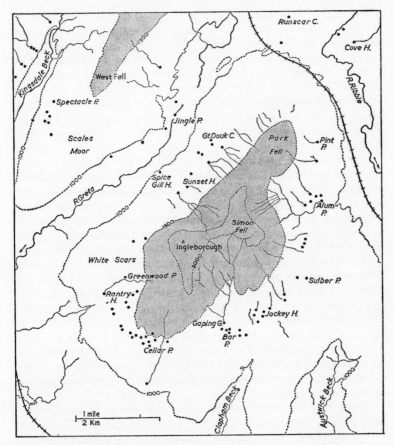

Fig. 56 Sink-holes near Ingleborough in the Craven District of the Pennines

Each sink-hole, indicated by a black disc, is listed and described by N. Thornber, *Pennine Underground* (Clapham, 1947).

The main railway line between Leeds and Carlisle is shown in the north-east, the Ingleton and Clapham branch-line (now closed) in the south-west. The land over 460 m (1500 ft) is stippled.

numerous examples of these phenomena. From the gritstone and Yoredale cappings of Ingleborough (figs. 45, 56), Pen-y-Ghent and Fountains Fell, streams flow down in all directions; the lower limestone slopes are pitted with holes often of great depth. Ramifications of interconnecting caverns, many of which have been explored and mapped by caving clubs, extend for long distances. Gaping Gill, on the slopes of Ingleborough, which the writer has descended on a wearisome rope ladder, is the most spectacular of them all. From the surface of the moorland to the floor of the main cavern is a vertical hole, 111 m (365 ft) deep, down which falls a stream, Fell Beck. At the bottom is the Main Chamber, 150 m (500 ft) long, 34 m (110 ft) high and 27 m (90 ft) wide, from which open off 5·6 km (3·5 miles) of surveyed passages and caverns. The water which enters the main hole reappears 1·6 km away at Beck Head, in Clapdale, as Clapham beck. There are hundreds of named swallow-holes in Craven, and many have been proved to be inter-connected.

Two caverns in the Gaping Gill system are known respectively as Stalactite and Stalagmite Chambers. Calcium carbonate is deposited on cave ceilings and on points vertically below, from drops of water seeping through crevices and joints, either because the water evaporates, or more probably because some of the carbon dioxide in the water escapes and so part of its dissolved bicarbonate becomes once more insoluble. Long *stalactites* grow downwards from the roofs, while *stalagmites* grow upwards from the floors. Many caverns in limestone districts are encrusted with these deposits; 'organ pipes', 'fluted screens', 'hanging curtains', 'portcullises' and other fanciful descriptive names are bestowed on them. In the Ingleborough Cave, near Clapham, the rate of growth of stalactites was once measured and found to be 7·493 mm (0·295 in) per annum, or about 76 cm (30 in) per century, much more rapid than might be expected. They probably grew still more rapidly in the past, when the water-table was higher, and more mineral-charged water was passing through the fissures.

There are also *helictites*, crystalline formations of most varied shapes; they may be as thin as threads, arranged in spirals or loops, or festooned with tentacles and occasionally they rejoin the ceiling of the cave. Where stalactites and stalagmites meet, they may fuse to form a column joining the ceiling to the floor. Some are of enormous size; in the Aven d'Armand, in the department of Lozère in the Central Massif of France, these pillars are 23–24 m (75–80 ft) high, fancifully called 'The Virgin Forest'. In some caverns, notably in the Appalachians, the calcite is deposited on their roofs in delicate, flower-like formations known as *anthodites*.

One of the most interesting examples of complicated subterranean drainage is in the neighbourhood of Malham Tarn (fig. 57). There are two sinks, at point A (plate 39) and south of the Tarn at B; the water from the former reappears at the foot of Malham Cove where the latter ought apparently to emerge, but the water from sink B actually reappears much farther down the valley at

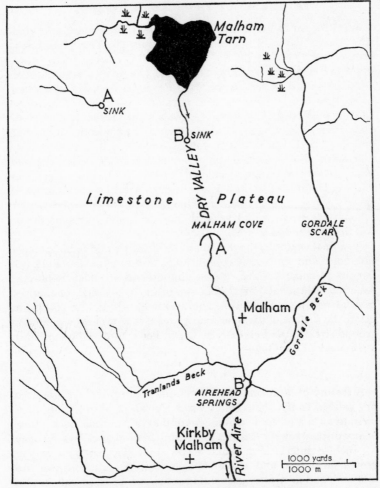

Fig. 57 *The drainage system of the upper Aire, Yorkshire*
For explanation see p. 124. The sink at **A** is shown on plate 39.

Airehead Springs. Thus these two streams actually cross underground, but maintain their identity, as coloration of the water with
fluorescein has shown.

There are many more examples of water-sinks and underground
drainage in the limestone country of the Pennines and the Mendips.
The deepest cave-system so far discovered in the British Isles is the
Pen-y-Ghent Pot, in the Craven District of the Pennines. These
features are also to be found in the Causses of the south of the

Central Massif of France, in the Jura, the Pyrenees and the Fore-Alps. In the Pyrenees the remarkable Gouffre de Pierre St Martin, on the crest-line near the Spanish frontier, is entered by a vertical shaft no less than 346 m (1135 ft) deep. The series of deepest caves yet descended in France is the Puits Berger in the Isère *département*, where speleologists have attained 1134 m (3720 ft) from the surface. In Italian Venetia the deepest cave system known is the Abisso della Preta, 637 m (2090 ft) deep, which was descended by Italian speleologists. In the Karst region of Jugoslavia sink-holes (known as a *ponor*, plur. *ponore*), shallow, funnel-shaped depressions (*dolina*, plur. *dolinas*) and large enclosed basins (*polje*, plur. *polja*), all solution features of various kinds, occur on an enormous scale. A cavern-system in north-western Greece is estimated to be 1370 m (4500 ft) deep; its initial vertical shaft, of 400 m (1300 ft) depth, was first descended in 1968.

As a final example, the plateau of Kentucky is said to have 60,000 sink-holes and numerous complicated cave-systems, including the famous Mammoth Cave, the neighbourhood of which is now an American National Park. About 240 km (150 miles) of mapped caverns occur on five levels, the lowest of which, 110 m (360 ft) below the surface of the limestone plateau, is occupied by an underground stream, the Echo River, which flows into the Green River, a tributary of the Ohio.

DRY VALLEYS

A feature of both chalk and limestone country is the existence of dry valleys. In the chalk country the dry valleys of back-slope areas seem to form a pattern reminiscent of that of a normal river system. Many display similar features to stream-containing valleys (Chapter 6), such as interlocking spurs, accordant valley junctions and ingrown meanders, and almost always their floors are strewn with alluvium and dry valley-gravel. However, not all chalk valleys show such 'normal' characteristics. Those which dissect the escarpments are often cut to an unusual depth, and have steep walls and abrupt heads; in plan they frequently follow curious courses, and zig-zag patterns are not uncommon. An excellent example of such a 'coombe' is the Devil's Dyke, near Brighton (fig. 58 and plate 41).

It is hardly surprising that widely divergent views have been expressed concerning the origin of these dry valleys; perhaps the only certain thing about them is that they have been cut under conditions of drainage and runoff which no longer exist. Some authorities regard them as in no way unusual features, and have explained their desiccation in terms of a gradual lowering of the

42 Typical dale country in the Carboniferous Limestone of
northern England
This is Littondale, through which flows the Skirfare, one of the two main headstreams of the River Wharfe in Yorkshire.

(*G. Douglas Bolton*)

43 The Cinque Dita (Fünffingerspitze) (main summit, 9830 feet) in the Sasso
Lungo (Langkofel) group in the Dolomites, east of Bolzano, Italy
Note the magnificent buttresses and ridges of dolomite, rising sheer from the
scree-slopes.

(*Paul Popper*)

44 Bryce Canyon, Utah
The rapid weathering and erosion of Eocene strata with well-defined bedding
and jointing has produced this remarkable surface relief.
(*Paul Popper*)

45 The River Rhine deeply entrenched between the Hunsrück and the
Taunus, West Germany
Note the remarkably uniform surface of the peneplaned plateau.
(*Eric Kay*)

46, 47 Two aspects of the 'Badlands' of South Dakota

The Badlands, so called by the early settlers who found them so hard to cross, cover a vast area of South Dakota (now a National Monument). The differential erosion of Oligocene sandstones and clays has produced a chaotic relief of buttes, ravines and basins.

(*Ewing Galloway, N.Y., Paul Popper*)

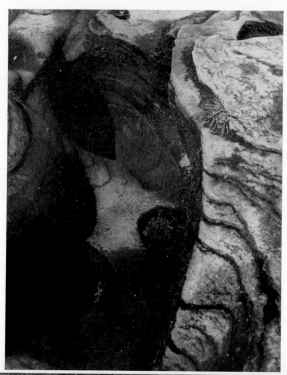

48 Multiple pot-holes in the bed of the River Wharfe at the Strid above
Bolton Abbey, Yorkshire
(R. Kay Gresswell)

49 'Spring-sapping' in Borrowdale, English Lake District
(R. Kay Gresswell)

Fig. 58 *Dry valleys in the South Downs near Brighton*

This map shows an escarpment dry valley, the Devil's Dyke (see also plate 41), and several back-slope dry valleys. The streams, originating as springs near the foot of the escarpment, ultimately join the Adur.

water-table with which the streams in the smaller, higher valleys were unable to keep pace. An important factor to be borne in mind here is that the water-table itself is controlled to a large degree by the height of the springs. Those occurring in scarp-foot positions are 'thrown out' by impermeable beds such as the Gault and Kimmeridge Clay, both of which are, by comparison with the Chalk, easily erodible rocks, and thus allow the formation of scarp-foot vales. As the latter have been etched out, the escarpments have been driven back, and the spring-line has been progressively lowered.

With the consequent change in the level of saturation in the Chalk, many of the valleys have slowly dried from their heads downwards. There can be no doubt that this process has been a major influence in the development of the dry valleys; only its relative importance is open to question.

Another widely held view is that the valleys have been cut at least in part under periglacial conditions (p. 255). During the glacial periods of the Pleistocene, runoff on the chalk surfaces may have been nourished during the summer by brief rains or more probably by the melting of small snow-caps accumulated under winter conditions. Possibly torrents of great erosive power were released at certain times, and the deep escarpment coombes, which often give the appearance of having been cut by large bodies of rapidly flowing water, may thus have been formed. A more recent view is that the process of 'spring-sapping' (p. 137) has played a major rôle in the excavation of this type of valley. It is certainly noticeable that many are still occupied in their lower reaches by feeble but definitely eroding springs. Furthermore, this hypothesis offers the best explanation both of the steep valley-heads, which seem to be indicative of some process of undermining, and also of the sharp angular courses, which may be due to the spring working back along such well-defined lines of weakness as intersecting major joints.

Finally, periods of much higher rainfall during the Pleistocene, perhaps under interglacial conditions, must have rendered some dry valleys 'wet'. But it would be unwise to attribute these valleys in their entirety to such abnormal climatic influences, for in many of the higher chalk areas the floors of the valleys are at present 100 m or more above the zone of saturation. The increase of rainfall in the past necessary for the formation of these particular valleys would be far beyond the bounds of reasonable expectation.

It is, however, not surprising to find that today some dry valleys may be occupied in their lower parts by temporary streams after heavy and sustained rainfall. These are known in various chalkland areas as 'bournes', 'winterbournes', 'nailbournes', 'gypseys' and 'lavants', and the term 'bourne' or 'borne' is a common place-name element. The bourne breaks out in a normally dry valley or some distance above the usual source of a chalkland river; this is due to a considerable if temporary rise in the water-table. Some bournes flow every winter, others only two or three times in a decade. A number of these, such as the Croydon Bourne and the Hertfordshire Bourne, have been studied for many years and their flows related to rainfall records. There is a considerable time-lag between the period of heavy rain and the bourne-flow; the water percolates slowly through the closely jointed chalk towards the saturated layers.

Limestone valleys Dry valleys, often gorge-like in character, are also common in limestone country. Some of these may be due to surface erosion, possibly during and just after the Quaternary glaciation, when the joints were filled with either ice or boulder-clay,

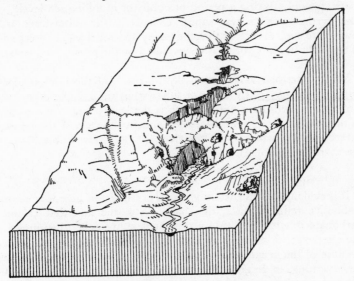

Fig. 59 *The transformation of a limestone cavern into a gorge*
Based on a diagram by J. Cvijić.
The collapse of the roof of an underground cavern may leave a steep-sided gorge. Sometimes portions of the former roof survive as arches; this can be seen at Gordale Scar, near Malham.

and so the river flowed on the surface. The dry valley above Malham Cove exhibits many features of stream erosion, and there is even a 'dry waterfall'. There are several records of periods of exceptionally heavy rain, following which a stream flowed again down this valley, terminating in what must have been a magnificent fall over the lip of Malham Cove. Many dry valleys are around the slopes of Ingleborough and Pen-y-Ghent, with swallow-holes strung out along their courses; the present streams plunge down the highest of these, but the water rushes down the dry valleys if, after heavy rain, the upper holes are unable to cope with the amount. There are examples of streams flowing down on to a limestone area, with their volume gradually decreasing as one follows the course downhill. Sometimes they finally disappear completely; occasionally, as in the case of the

Gordale beck at Malham, they just survive, with progressively reduced volume, until they flow on to impermeable rocks again.

Many dry valleys in limestone areas have such steep craggy walls that they may be called gorges; notable examples are Cheddar in the Mendips and Winnats in Derbyshire (plate 40). While the origin of these features is not certain, it seems that they were cut by former rivers, possibly during a period of uplift (or of falling sea-level), so that vertical erosion was rapid. Ultimately the water-table was lowered to such an extent that the valley became dry. The steepness of the walls in some gorges is accentuated by the presence of vertical joint-planes.

Some limestone valleys may be the result of the erosion of an underground cavern by a stream, followed by the collapse of the roof (fig. 59). Sometimes a natural arch is the last remnant of the roof, as at Gordale Scar, near Malham, and in the case of the famous 'Marble Arch', which spans the river Cladagh near Enniskillen, in Northern Ireland.

CHALK AND LIMESTONE LANDSCAPES—A SUMMARY

It is evident that the extensive areas of chalk and limestone reveal distinctive relief features, due basically to the fact that calcium carbonate is soluble in rain-water. It is convenient to summarize the characteristic features of their typical landscapes at this point. Because of the great permeability of these calcareous rocks due to their jointing, underground water has played a major part in sub-terranean sculpture itself and so in producing, directly or indirectly, distinctive surface features.

Chalk This rock often forms rolling hills, with open expanses of downlands. Where the strata are tilted, a cuesta (fig. 61) may be formed (p. 173), with a steep escarpment and a gentle back-slope.

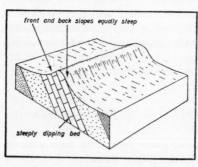

Fig. 60 *Block-diagram of a hogback*

Where the dip of the strata is very steep, as in the Hog's Back, west of Guildford, the chalk outcrop is narrow, since on both sides the slopes fall steeply away. This feature has been given the name of *hogback* (fig. 60). Sometimes the Chalk forms more extensive gently un-dulating uplands, such as the expanse of Salisbury Plain. Dry valleys are fre-

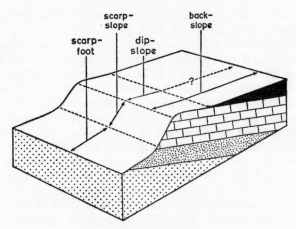

Fig. 61 *The terminology of a cuesta*
Many authorities prefer the term escarpment to scarp-slope.

quent and everywhere surface drainage is slight, except where rivers rising outside the chalk area actually cross it, cutting striking gaps and so dividing the hills into blocks; thus both the North and South Downs are cut through completely by several river gaps, and the Goring Gap of the river Thames separates the Berkshire Downs from the Chilterns. The surface of the Chalk is sometimes diversified with a superficial cover; Clay-with-flints, probably derived from a former Tertiary cover, forms cappings in many parts of the southern chalklands of England, usually carrying clumps of beech, oak, hornbeam or pines. The Chalk of the Lincoln Wolds is more uniformly covered with glacial deposits—boulder-clay, sands and gravels—while in East Anglia the gently inclined chalk surface is for the most part deeply buried. Where the Chalk reaches the coast it forms impressive vertical cliffs and headlands, resisting weathering by its very permeability, but constantly attacked at the base by marine erosion (p. 283). Finally, man in the past, with his stone circles, tumuli or burial mounds, ridgeways, flint-works, 'white horses' and ancient hill-top villages and camps, and in the present, with the white scars of quarries, dew-ponds, farm buildings and race-horse stables, ploughed arable lands on the back-slope and spring-line villages below the escarpment, contributes materially here, as everywhere, to the landscape.

Limestone The landscapes of the limestone country are almost as varied as the types of limestone. There is the long line of the Cotswolds (and its continuation north-eastward into northern

Yorkshire) of bedded Jurassic limestone with its yellow-brown tints, stone walls, residual brown soils, beech-woods and dry valleys. Then there are the former prosperous medieval 'wool-towns' of the Cotswolds, and the modern open-cast workings of the Jurassic iron-ores in Northamptonshire, Lincolnshire and northern Yorkshire. The older crystalline limestones, the chief of which are of Carboniferous age, form the Mendip Hills, part of South Wales and considerable blocks of the Pennines, mainly uplands separated by fairly steep-sided valleys known as 'dales' (plate 42). The uplands are covered with dry, sweet turf growing on the thin residual soil, or, where the surface is bare rock, with clint 'pavements' (p. 99 and plate 38), especially where near-horizontal bedding-planes are exposed. Absence of surface drainage, numerous 'swallow-holes', depressions caused by the coalescence of several of these holes, intricate cave systems, dry valleys and gorges (plate 40) are all characteristic.

The Carboniferous Limestone in the Pennines includes both massive beds of limestone, with marked vertical joints, and thinner beds of shale. The latter are much less permeable, so that commonly seepages of water emerge along their outcrops, but they are also much less resistant than the limestone, so that they are more easily broken up and removed; both sapping and frost action are potent. Thus the bedding-plane of the limestone below is revealed as a horizontal step or platform as the shale is stripped off, while the stratum of limestone above forms a rock-wall or 'scar'. These scars are particularly pronounced where faulting has brought less resistant rocks (such as the Bowland Shales) into close juxtaposition with the limestone, so that differential denudation has full scope; this in fact produces fault-line scarps (p. 41). A striking example is the series of scars (including Malham Cove shown in fig. 57). along the line of the Mid-Craven Fault. Many scars are flanked along the base by scree-slopes; these are especially well developed where the limestone is closely jointed and readily subject to water penetration and therefore to frost disintegration. As elsewhere, this scree formation was more potent under periglacial (p. 253) conditions; some screes are now inactive and even turf-covered.

These features are also to be seen in the Central Massif of France (the Causses) and in the Karst of Jugoslavia; the long aridity of summer and the concentrated winter precipitation have emphasized and exaggerated the phenomena described in the Pennines. In these areas of long summer drought, the surface is either of bare limestone or is covered with 'garigue', a thin vegetation of spiny aromatic shrubs (p. 532).

As a final example, the Magnesian Limestone of the Dolomites, in northern Italy, reveals the shattering effects of frost at high altitudes on their fantastic pinnacled summits (plate 43), guided by enormous vertical joints and the shatter-belts of faults. Here are many exceedingly steep rock-faces, notably those of the Tre Cime di Lavaredo near Misurina. Possibly the hardest climbing route on the Cima Grande (some say in the whole Dolomites) is the well-named '*Direttissima*'; when it was first climbed by four young Germans, they had to use hundreds of steel pegs. They hauled their loads up from the base of the cliff on 300 m of rope, which was possible because the face is so steep that the loads nowhere touched the rock.

CHAPTER 6

Rivers and River Systems

RUNNING water is the most effective agent by which the surface of the earth is sculptured, except, of course, in the arid or frozen lands, where it infrequently exists. A river is a mass of water flowing over the land surface, from its source in a spring, a marsh, at the end of a glacier or as the collected surface runoff of rain-water, until it reaches its mouth, which usually opens into the sea. Occasionally it may end in an inland lake, as the Volga in the Caspian and the Amu and Syr Darya in the Aral Sea, or in a salt swamp, as the Tarim River disappears in the Lop Nor and the Humboldt River of Nevada vanishes in a great saline depression (Carson Sink).

As each river flows down the land-slope to the sea it receives tributaries, and so gradually evolves a system which occupies a basin or catchment area surrounded by a main watershed or divide. Sometimes a watershed may separate headstreams whose waters ultimately reach the sea far apart. Fig. 62 represents part of Switzerland, where lies what is virtually the drainage centre of Europe; within a few km rise the Rhine and its tributaries the Aar and the Reuss (ultimately reaching the North Sea), the Rhône (flowing to the Mediterranean Sea) and the Ticino (flowing to the Po and the Adriatic Sea).

As rivers flow they perform the triple function of erosion, transport and deposition, and in so doing they modify the surface features of their basins. They wear away the surface of the land, eroding valleys with distinctive features and leaving residual hills and ridges, or *interfluves*, between them; the original landscape is therefore slowly 'dissected'. As time passes, the endless removal of material causes the gradual lowering of the original surface until it reaches the stage of a peneplain (p. 93). It has been estimated that the surface of the Mississippi basin is being lowered by erosion at the rate of 0·3 m (1 ft) every 4000 years and of the U.S.A. as a whole at the same amount in 9000 years, though not taking into account isostatic compensation (pp. 22–4).

The work of a river goes on steadily and gradually, and its basin reveals similar changes; the concept of a cycle, or an evolutionary development, which has already been discussed (pp. 92–5), is

50 Rock débris transported by the Waters of Nevis in flood, near
Fort William, Scotland
(*Eric Kay*)

51 Interlocking spurs, Sail Beck, near Buttermere, English Lake District
(*F. J. Monkhouse*)

52 The gorge of the Wharfe at the Strid, cut by the perennial stream
within the winter bed of the river
(*R. Kay Gresswell*)

53 The 'stepped-fall' of the River Ure at Aysgarth, Wensleydale
(*R. Kay Gresswell*)

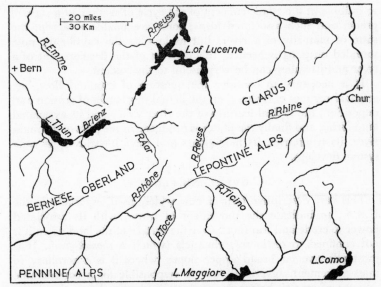

Fig. 62 *The drainage dispersal of Central Europe*

The Rhine (receiving also in due course the Aar (*Aare*) and the Reuss), flows northward to the North Sea; the Rhône flows south-westward to the Mediterranean Sea; and the Ticino flows (by way of the Po) to the Adriatic Sea.

therefore very apt, though perhaps the term *fluvial sequence* is preferable. Each stage of development has its characteristic features—the steepness of its course, the shape of its bed and containing valley, the results of the relative importance and potency of erosion of bed and of banks, of the weathering of the valley sides and the removal of this weathered material, and of the balance between erosion and deposition.

Any interruption of the sequence may cause revived erosive activity, so that the river becomes rejuvenated, and therefore begins a renewed stage, superimposing these new features on a landscape already modified by previous river work (i.e. producing polycyclic relief). Rejuvenation may result from a change of base-level due to a relative rise or fall of sea-level, or from tectonic causes (folding, faulting, tilting, uplift or depression); this is known as *dynamic rejuvenation*. Should, however, erosive activity be re-stimulated by an increase in the river's volume, either because of a rise of precipitation over its basin as a result of climatic change or due to river capture (p. 176), this is known as *static rejuvenation*.

It must be stressed that some modern workers find much at which to quibble in the whole concept of the cycle of erosion, of a state of grade, and such associated ideas; some are unable to accept the state of maturity of a river. But as long as the cautionary note sounded on p. 94 is heeded, this concept of the development of a river and its valley can be very useful to the student.

It is necessary to consider the question of overland flow, the seasonal regimes of rivers, then the actual mechanics of erosion and deposition, the typical features of the river valley and its associated land forms, and finally the patterns produced in the plan of the main river, its tributaries, its neighbours and their interfluves, as each system develops.

OVERLAND FLOW

This very descriptive term is now widely used, especially in the U.S.A., to indicate the movement of water, with its associated power to erode and transport, over the surface of the land when it is not confined to distinctive channels (which is *channel flow*). It is most effective on broad upper slopes, where it is a corollary to mass-movement. Overland flow is responsible for a considerable amount of erosion before streams attain identity lower down the slope.

The first contribution is the mechanical work of rain-drops on impact with the surface, what has been called *splash erosion*. Torrential rain with large individual drops can on the one hand move particles of soil, and on the other can cause compaction of the surface, thus accelerating runoff. Splash erosion is most effective in semi-arid regions, where precipitation is sporadic and intensive, where there is a loose friable surface and no protective vegetation cover, with no trees to break the direct fall of rain, nor turf to provide a shielding and absorbent mat. The next stage is when the individual effects of splash erosion are widened and coalesced to cover a considerable area, resulting in *sheet erosion*. When the water begins to concentrate in a number of tiny superficial streamlets, though still insufficiently defined to occupy channels, this is *rill erosion*. Both processes may lead to severe soil erosion, depending on the intensity of the precipitation, the angle of slope, the nature of the surface soil and underlying bedrock, and the absence or presence of vegetation. If rill erosion is not checked it will lead to the formation of more localized and concentrated runoff, ripping a deep gash into the land, a process known as *gully erosion*, which however comes into the category of channel flow rather than overland flow.

Near the source of a stream, therefore, erosive work is carried out by rain itself. It enables headward erosion to cut back into the slope above the point at which the headstream really begins to flow. In other words, the source of the stream slowly recedes and gradually cuts into the ridge which forms the divide (p. 174).

Spring-sapping Another form of erosion which may occur before channel flow commences is involved when the source of a stream consists of a strongly-flowing spring (p. 114). When the water issues, it is at once able to carry a certain load, and may acquire a considerable proportion of this from its immediate surroundings. In this way the material around the spring (especially where this is either unconsolidated or soluble) is gradually removed or 'sapped' away. This process is known as 'spring-sapping', and the net result is the formation of a small amphitheatre-like hollow, the head-wall of which is cut farther and farther back into the slope. Plate 49 shows an incipient stage in this process.

RIVER REGIMES

The regime of rivers, that is, their seasonal variation in volume, has received much attention in recent years in connection with problems and schemes of flood control and hydro-electricity production. The seasonal precipitation (both rainfall and snowfall), the presence of snow-fields and glaciers, the steepness of slopes in the catchment area, the nature of the rocks (particularly their permeability), and the character of the vegetation cover, are all salient factors. If a stream is fed wholly by melting snow and ice, low water is experienced during the winter-freeze (this is one of the problems of Alpine hydro-electric generating stations), but extensive floods occur during the early summer melting. Where summer rains also occur, as in the case of most Alpine rivers, maximum high water is in June–July, and lowest water in late autumn. On the other hand, streams in mid-latitudes fed almost wholly by rain (such as the Seine and the Saône) tend to have low water in summer, when evaporation and the demands of vegetation are both at their maximum.

In tropical latitudes, where temperatures and evaporation are constantly high, floods closely follow the rainfall regime. A short river, such as the Mahaweli-Ganga in Ceylon, often almost disappears in May, but by June, after the summer monsoon has broken, it may flood wide areas, sweeping away bridges and even villages. The larger rivers of south-east Asia, such as the Irrawaddy, the Mekong and the Yangtse (fig. 63), have floods in summer and low water in winter, corresponding to the rainy summer monsoon and the dry winter monsoon.

The equatorial rivers, such as the Amazon and the Congo,

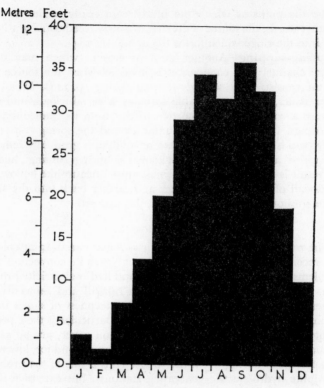

Fig. 63 *The regime of the Yangtse at Ichang*

The graph (based on J. Sion) shows the average depth of the river for each month at the Ichang Gorges. It reveals the variation in the regime of the river between the summer high level (caused by the heavy precipitation brought by the South-east Monsoon) and the winter low level.

maintain a considerable volume throughout the year, but with maxima as a result of the periods of maximum rainfall following the Equinoxes (p. 459).

The Nile floods, on which the agricultural prosperity of Egypt and the Sudan largely depend, are mainly caused by the great volume of water brought down by the Blue Nile as the result of monsoonal rains on the Abyssinian Plateau. The White Nile rises on the equatorial plateau of East Africa, and its headstreams emerge from lakes. The flow of this river is therefore remarkably consistent, but its volume is reduced by evaporation, particularly where the river spreads out in shallow lakes in the Sudd area of the southern Sudan. Were it not for the contribution of the Blue

Nile, and to a less extent that of the Sobat and Atbara, the Nile might not reach the Mediterranean.

It is interesting to compare the regimes of the neighbouring rivers Tigris and Euphrates, each of which rises in the mountains of Armenia. The Euphrates is in many ways similar to the Nile, since it receives no tributaries during its long course through Iraq, though its current is swift because of its considerable gradient to the sea. Its minimum flow is in autumn, after the long hot dry summer. The river begins to rise in December during winter precipitation in Asia Minor, and reaches its maximum in May, when the effects of melting snows on the higher parts of the basin are at their greatest. The Tigris, by contrast, flowing nearer the Zagros Mountains, receives many torrential tributaries with very steep gradients, and sudden disastrous floods are common. It has low water in autumn, but its maximum is in April, earlier than on the Euphrates, because of the more speedy runoff. W. B. Fisher quotes some revealing figures. The Euphrates at Hit discharges 8800 *cusecs* (p. 140) during September low water, but rises 3·4 m (11 ft) higher in May, with a discharge of 64,300 *cusecs*. The Tigris discharges 11,900 *cusecs* at Baghdad in September, but rises 5·5 m (18 ft) with a discharge of 106,650 *cusecs* in April.

In the light of these various factors, the French scientist M. Pardé, in his work *Fleuves et Rivières*, has divided the regimes of rivers into three classes. The first is the *simple regime*, where each year includes one period of high water and one of low (the Yangtse, Volga). The second is the *regime of the first degree of complexity* (or more simply a *double regime*), with two distinct periods of high water, the result of early summer snow-melt and autumn–winter rain(Garonne) or of two rainfall maxima (Amazon, Congo). The third is the *regime of the second degree of complexity* (or more simply a *complex regime*), a feature of many of the world's largest rivers with extensive basins covering various climatic regions, and receiving numerous tributaries each perhaps with a different regime. Thus this is in effect a *composite regime* (Rhine, Danube, Mississippi).

CHANNEL FLOW

The channel A stream is confined within a definite channel incised into the surface of the land, with a distinctive pattern and cross-section which varies from source to mouth and changes its form as the drainage system develops. In order to introduce some definitiveness, a distinctive if somewhat obvious terminology has been developed, mainly by civil engineers. The depth and width of the water measured from bank to bank (the margin of the channel) are not always easy to specify, since unless the stream has been regularized for flood control, navigation or power production the cross-section is rarely a distinct rectangle; a mean depth is commonly used (below). Their relationship is expressed as the *form ratio*, an expression derived from depth/width, as 1 : 50 where the width is 50 m (150 ft), the mean depth 1 m (3 ft). Sometimes the banks are clear-cut and well defined, at other places uneven beds of sand and shingle may slope gradually from the water's edge, which itself fluctuates. The term *wetted perimeter* denotes the length from bank to bank of the section in contact with the stream, and the *cross-sectional area* is, as its name would imply, the area of a cross-section of the water at any particular point. Another relationship, important because it is a guide to the amount of friction between the water and the channel, and therefore an indication of energy loss, is the *hydraulic radius*, the cross-sectional area divided by the wetted perimeter. A low hydraulic radius, where a shallow stream occupies a very wide channel, is the least efficient as far as its energy is concerned.

The river at lowest water (*base flow*) may not occupy the entire channel. After rain or snow-melt the depth will increase; when it just occupies the entire channel, it is at *bankfull stage*, beyond which it may rise to the *flood* (or *overbank*) *stage* when it overspills the banks. There is a lag after heavy rain over the basin, depending on the nature of its surface and constituent rocks, the overall gradient, and the number of tributaries, before the flood wave passes down the main river, its maximum being the *crest*. In countries such as the U.S.A., where flooding is frequent, systems of warning and forecasts are operated; the U.S. Weather Bureau operates a River and Flood Forecasting Service, and in many districts exact information about heights of rivers is given periodically over local radio stations.

River energy The energy of a river depends essentially upon (i) its volume and (ii) its velocity, which together are summed up by the term *discharge*; this is usually expressed in *cusecs*, that is, the number of cubic feet per second passing a particular section of river or, in metric measure, cubic metres per second (m^3/sec). A *cusec* is equivalent to a flow of 538,000 gallons per 24 hours (fig. 64). The number of *cusecs* is obtained by multiplying the rate of flow in feet per second by the cross-section of the river at the particular point in square feet. The rate of flow may be obtained by carefully timing a float down a measured reach of the river, or by means of a current-metre, an instrument in which rotating vanes linked to a calibrated dial are propelled by the water. The cross-section is obtained by sounding depths at close intervals along the line of section across the river. In practice, river discharge is measured at a weir or flume, where there is an artificial rectangular cross-section, often by an automatic recorder. In Britain the Ministry of Housing and Local Government maintains a Surface Water Survey Centre, which publishes statistics of rainfall and runoff at 205 gauging-stations, and in the U.S.A. are about 6000 gauging-stations maintained by the Geological Survey.

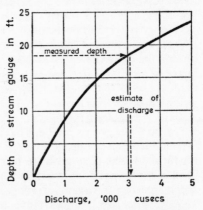

Evidence provided by the records of gauging-stations emphasizes a fact which is not always appreciated: that the average velocity of a river[1] increases from its head-waters down its valley, even though the gradient of its bed may steadily decrease. The cause of this apparent paradox is that as the stream and its channel both enlarge down-valley, the loss of velo-city through the friction of the running water on both

Fig. 64 *A rating curve showing the discharge of a river*

If a number of measured discharges are plotted and a smooth curve is drawn through them, discharge can then be estimated from a measurement of depth.

bed and banks becomes proportionately less, which more than compensates for the gentler gradient, and increased velocity is necessary to cope with the larger discharge in its lower sections.

The actual mechanics of the hydraulics of flowing water are almost as complex as those of moving air. Just as an aero-engineer may have recourse to a wind-tunnel, much of the information about water flow is derived from thousands of observations at gauging-stations. Basically, the movement of water confined to a channel is *laminar flow*, a stream-lined transference of its mass in a downhill direction, or it would be if the channel were straight and smooth; this is rarely the case, except in an artificial flume, sluice or mill-race, and any obstacle, major or minor, in its bed or projecting from its banks will cause eddies and turbulence, or *turbulent flow*. Turbulence actually reduces the overall speed of flow, because of

---

[1] Average velocity implies 'the long-term average of velocity through the cross-section' (G. H. Dury), at the stage of average discharge when the river is lower than bankfull. At bankfull or overbank stages the velocity in the main channel is constant along the whole length of the stream.

frictional eddies within the mass of water and on the bed of the river, but it is a major eroding and transporting agent. Sometimes in a winding river a 'corkscrew' motion will develop in the current within a bend, known as *helicoidal flow* (fig. 70). Finally, where the body of water spills near-vertically over a sudden change of gradient, forming rapids and waterfalls, the water will achieve *plunge flow*, with very powerful effects.

The transported load The energy of a river is expended partly in friction with its sides and bed, partly in friction within the mass of moving water due to turbulence, partly in transporting its *load*, the term used to denote the conveyed matter derived from the products of weathering and mass-movement down the sides of the valley, from glacier melt-water, from the contribution of the tributaries, and the river's own erosive activity. When the velocity of a stream increases in time of flood (p. 137), its carrying capacity is also vastly increased. The measure of the ability of a river to carry a load is known as its *competence*, indicated by the weight of the largest fragment that a stream of given velocity can transport. As the velocity increases, so does the maximum weight of the particles, though not in direct ratio. The 'sixth power law', postulated by W. Hopkins as long ago as 1842, suggests that the weight of the largest fragment that can be carried increases with the sixth power of the stream velocity. For example, if the velocity becomes twice as great, then the maximum particle weight will be increased 64 times. Thus for each stream velocity there is a corresponding maximum weight of particle that can be carried. Likewise, for a given particle size there is a critical water velocity which must be attained before that particle can be picked up. However, once in motion the particle may be transported by a current of much lower velocity. The difference between 'pick up' and 'carry' velocities is particularly marked with very small grain sizes; it is difficult for slow-moving water to erode a mud-bank, though mud particles can be carried by an extremely low velocity current. Another concept postulates that the diameter of a particle which a stream can carry varies with the square of its velocity. Load-carrying ability may also be indicated by *capacity*, that is, the total load. A large slowly flowing stream may have a high capacity, but a low competence; its load consists of a large quantity of fine material in suspension. Various tables and graphs of particle size related to capacity have been devised; as a rough indication, a stream flowing at 0·5 km ph (0·3 mph) can carry sand, at 1·2 km ph (0·75 mph) gravel, at 5 km ph (3 mph) small stones 5–8 cm (2–3 in) in diameter, at 10 km ph (6 mph) large stones, and at 32 km ph (20 mph) boulders.

The load is carried in a number of different ways. Small particles are swirled along in suspension; this is the *suspended load*. Larger particles proceed in a series of hops (the result of a 'hydraulic lift' which is caused by variations in turbulence), touching the bed at intervals, a process termed *saltation*; pebbles are rolled along by gravity and by the pushing power of the water; and big stones and boulders are moved by rivers in flood. All this material carried along the bed is the *bed-load* or the *traction load*.

When most of the material consists of particles of more or less the same size, there is a definite limit to the total load; in other words, if any more material is added to the load some must be dropped. A period of low water, a marked change in the slope of the bed, the presence of a lake through which the river passes—any of these may cause its carrying capacity to be checked and some material to be deposited, the largest particles first. A river therefore acts as a sorting agent; the coarsest material, derived from steep upper slopes where erosion is rampant, is found in the upper part, while the finest material is laid down in the flood-plain or swept out to sea. Much of the load in the lowest part of a great river is so fine and uniform that the water often has a consistently dark-brown colour.

It has been calculated that 100 million tons of sediment are carried past Wadi Halfa each year by the Nile, of which 30 millions are of fine sand, the same amount of clay and the rest of silt. Much of this is derived from the denudation of volcanic rocks in Abyssinia and is rich in minerals which help to fertilize the land when spread out by the flood-waters. A resulting problem is silting at the barrages, so that the flood peak is allowed to pass with much of the load before the sluices are closed to preserve water for perennial irrigation.

It is when rivers are in flood that the most impressive movement of material occurs (plate 50). Lingmell Gill flows into Wastwater (English Lake District), normally a clear stream flowing over hard rock. During the August Bank-holiday week-end of 1938, 23 cm (9 in) of rain fell in thirty-six hours; the stream became a swollen torrent, brown with earth from the gashed and scoured slopes, and with stones crashing along its bed. When it subsided, two fields at the head of the lake had been covered with stones and small boulders to a depth of 0·3 m or more, the load of the stream in flood. Within forty-eight hours the stream returned to normal, having accomplished more erosive activity in two days than in the previous ten years. Devastating floods poured down the valleys of the East and West Lyn rivers in northern Devonshire, following heavy rain on Exmoor on 18 August, 1952. The discharge of water passing down the two rivers was estimated to be over 18,000 cusecs for a few hours, nearly as great as has ever been recorded on the lower Thames. Many thousands of tons of boulders were swept into the streets of Lynmouth, 23 people lost their lives, and over a thousand were rendered homeless.

Solution provides another means of transport; this is especially potent in the case of rivers such as the Shannon, which flow over limestone country (fig. 101). American scientists have calculated that the Mississippi carries to the sea each year 136 million tons of matter in solution, compared with 340 million tons in suspension and 40 million by saltation. It is estimated that 20 tons of material in solution and 120 tons of solid matter in suspension are removed annually from each square kilometre of the earth's surface.

River erosion It may seem illogical to have discussed the transporting work of rivers before that of erosion, but the load is an important agent in erosion. The erosive work of a river consists of four interacting processes: (i) the hydraulic action, (ii) corrasion, (iii) attrition and (iv) solution.

(i) *Hydraulic action* is caused by the force of moving water, which can sweep out loose material, and by surging into cracks can help to break up solid rock. Turbulence and eddying have a powerful effect. One result is the undermining of the banks on a curve, a process known as *bank-caving*. Erosion is also carried out in rapidly flowing water by *cavitation*, when bubbles of air or water-vapour collapse and form shock-waves against the banks. Poorly consolidated sands, clays and gravels, and weathered material generally, are particularly vulnerable. It must be remembered that even without a load, running water has an eroding potential. Indeed, while the load may be regarded as a 'grinding tool', fluvial mechanical activity may be actually reduced if a load is carried, because energy is used up in transport. A striking example is the river Colorado below the Hoover Dam, completed in 1936; deprived of its load, which was deposited in Lake Mead behind the dam, the river below the dam cut down its bed by more than 3 m (10 ft) in less than a decade.

(ii) *Corrasion* is the wearing away of the bed and banks, using the load as the 'grinding tool'. This can be seen most effectively where pebbles are whirled round by eddies in hollows in the bed, so cutting pot-holes (plate 48), a process known as *evorsion*. In mountain streams large deep pools are similarly worn. The result of corrasion is to scour and excavate the bed, adding the material so derived to the load, which in turn is itself used as a corrasive agent. There is a limit to this, since when the river attains a full load any vertical corrasion must be balanced by deposition. This means that vertical corrasion tends to stop, and lateral corrasion of the outside curve of a bank will be balanced by deposition on the inside curve in the slack. Between the stages of a stream with no load and one with a full load, there is a point at which vertical corrasion is at its greatest.

(iii) *Attrition* consists of the wearing down of the load itself, as the fragments are in constant collision with each other and with the bed. This is part of the progressive reduction in the size of the load materials as they move downstream, which thus become easier to transport.

(iv) *Solution*, as already mentioned, is the solvent work of water as it flows over such rocks as limestone.

THE LONG-PROFILE OF A RIVER

The classic concept of the development of the long-profile of a river was promulgated as an integral part of geomorphology by W. M. Davis. He maintained that a river's activity is devoted to the attainment of a slope from source to mouth which will result in such a velocity that erosion and deposition are exactly balanced.

When a river enters a reach without a full load for that part of its course, it erodes its bed, adding the material so derived to the load; this is known as *degrading* the bed. Erosion in the lower part of the reach, however, decreases as the load increases, and thus gradually the slope is lessened. As a result, the erosive power itself decreases, and ultimately a velocity is attained throughout the reach which just allows erosion to balance deposition.

On the other hand, if a river enters a reach which has such a slope that the load is too great for the velocity, deposition takes place in the upper part of the reach (*aggrading*), so steepening the slope and thereby increasing the velocity. Ultimately, the velocity is just sufficient to enable the river to carry its load (fig. 65).

In either case, when such a profile has developed by degrading or aggrading or both, so that the stream has just sufficient velocity to move its load, it is termed a *graded profile* or a *slope of equilibrium*.

Proceeding from this instance of a single reach, the same principles can be applied theoretically to the whole of the profile of a river, from source to mouth. If the velocity and power of erosion were uniform throughout, this graded slope would be a straight line. But erosion near the source is less than average because of the smaller volume and load, and in the lower course it is again less, probably because it is heavily laden. The result is that maximum erosion is in the middle course, and so a concave curve parabolic in

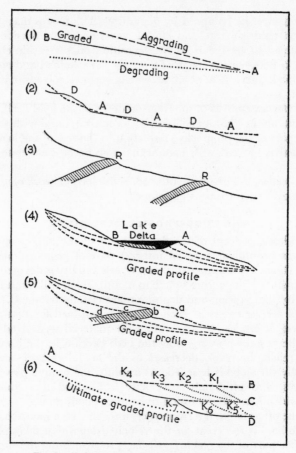

Fig. 65 *A theoretical concept of the long-profile of a river*

1. Simplified profile of a theoretical single reach. **2.** The original slope is a solid line, the ultimate graded profile is pecked; **D** = reaches in which degradation predominates, **A** = reaches in which aggradation predominates. **3.** The effect of resistant strata (**R**) upon the development of the graded profile. **4.** The elimination of a lake, **A** by lowering the level by down-cutting of the outlet, **B** by filling up the head with a lacustrine delta. **5.** The elimination of a waterfall: **a** graded to fall caused by hard band; **b** recession of falls, graded to new position; **c** falls cut back into a series of rapids; **d** final graded profile (4 and 5 after A. Holmes). **6.** The effects of successive rejuvenations: **AB** original profile, graded to sea-level at **B**; **AK₄C** profile attained after fall of sea-level to **C**, causing renewed erosion, with successive knickpoints K_1 to K_4; **AK₄K₇D** profile attained after renewed fall of sea-level to **D** with successive knickpoints K_5, K_6, K_7.

appearance is produced (though admittedly idealistic in concept), with its lower end at sea-level, the base level of erosion.

Even when this graded profile has been attained (and many of the rivers of the English Midlands, for example, flowing to the North Sea, would seem virtually to have done this), erosion still goes on very slowly, for even the most sluggish stream bears a burden of fine material to the sea, so that the basin is still suffering a net loss. This is derived mainly from the upper part of the course, and hence the curve of river erosion is ever being slowly flattened.

In recent years considerable criticism has been made of the whole concept of grade, whether indeed there is such a feature as a profile of equilibrium, and whether this nice adjustment of the river to this profile is ever attained. Some workers maintain that too much effort is devoted to working out an exact relationship between the profile, the velocity and the eroding power of the river. Some claim that a river may still be in grade without a smooth concave profile, that a change in grade of the long-profile may be balanced by changes in the cross-profile. Confusion has also arisen between the use of the word grade, both for a balanced condition of an 'old' valley, and for the slope of the bed itself. Still, it is quite clear that irregularities do gradually disappear from the course of a river, and it does seem to be striving to eliminate such features. If one is careful to realize that a smooth long-profile of a graded stream is an ultimate ideal which may never be reached, that any balance between competence, capacity and load is only a very long-term average, that in the short term a river is rarely in grade, the concept is still useful in helping to understand how a river develops its valley. As J. H. Mackin put it, grade refers to the 'climate', not the 'weather' (p. 372) of a stream. Probably it is advisable, to use the term 'provisional profile of equilibrium', which will allow for temporary changes and minor oscillations on either side of a mean state.

It must be further emphasized that it has been assumed so far, for the sake of simplicity, that the slope down which the river flows consists of rocks of uniform resistance to erosion, but this is rarely the case. A band of hard rock may outcrop transversely across the valley, and prove more resistant to erosion than other beds above and below (plate 53). The river will therefore grade itself above each hard band and then concentrate its attack upon the bands, so that each becomes the scene of waterfalls and rapids. The current is more rapid at each of these points and has additional erosive power, so that ultimately the waterfall disappears.

Another irregularity in the profile of a river is caused by a lake lying in its course, which acts for a time as a local base-level.

Sediment brought down by the river slowly fills it as a lacustrine delta is developed (p. 162), and as water flowing out at the lower end cuts down the outlet so the level is lowered, and ultimately the lake is drained. This has happened in certain English Lake District valleys (p. 203). The river then grades itself through both the lacustrine sediments and the underlying rock floor.

The effects of rejuvenation on the long-profile The most serious interruption to the development of a valley long-profile is a change of base-level, such as a fall in sea-level or local land movements of uplift. This may cause a steeper slope and greater velocity, and therefore renewed down-cutting. An increase in precipitation over the higher parts of the basin, and therefore of river volume, produces the same effect. The river cuts into its former flood-plain, leaving terraces on either side (pp. 161–4). The result on the long-profile is the development of a marked break of slope, known as a *knickpoint* or 'rejuvenation head', sometimes marked by rapids. The river begins to cut upstream from its mouth, so producing a new curve of water erosion, which interesects with the old curve at the knickpoint (fig. 65). The knickpoint recedes upstream at a rate depending upon the resistance of the rocks; it may therefore linger at a hard outcrop of rock, and it is sometimes difficult to distinguish between rapids due to knickpoints and those due simply to resistant bars. Some rivers and their tributaries may show several knickpoints, indicating several stages of rejuvenation.

Waterfalls It is useful at this point to summarize the causes of waterfalls, since these spectacular features are mainly, although not entirely, the result of river erosion (fig. 66), and form important interruptions to the long-profile.

(i) The first and commonest reason for a waterfall's existence has been mentioned—the presence of a bar of rock lying transversely across the valley of a river, thus interrupting its progress towards a graded profile (plate 53). If the bar is dipping gently downstream, it will cause a series of rapids, with much broken water, as the Nile in southern Egypt (fig. 67). If the resistant mass is horizontal or only slightly inclined (i.e. forming a cap-rock), and the underlying strata are less resistant, a vertical fall results. The scouring action of the water at the base of the fall undercuts into the soft rock, so that undermining causes pieces of the hard rock to break off and the fall recedes upstream, leaving a gorge. A large-scale example is Niagara, where a stratum of hard dolomite limestone overlies softer shales and sandstones; the fall is 51 m (167 ft) high on the American side, and the gorge about 11 km (7 miles) long. The estimates of the rate at which the gorge is receding vary

(1)

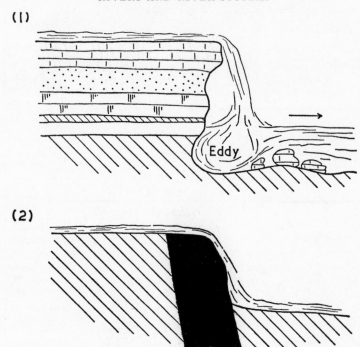

(2)

Fig. 66 The formation of waterfalls

1. The effect of horizontal beds of resistant rock; undercutting produces a steep overhanging face, while the less resistant beds are cut back, thus causing a rapid upstream recession of the fall.

2. The effect of a vertical or steeply inclined resistant stratum (shown in black).

from 0·3 to 2·0 m per annum (plate 54). The Kaietur Falls in Guyana are 251 m (822 ft) in height, due to the Potaro river flowing over a ledge of resistant conglomerate overlying softer sandstones and shales.

(ii) Waterfalls are found where there is a sharp, well-defined edge to a plateau. Many African rivers are hampered for navigation because of the rim which bounds much of the continent (plate 55); the Congo descends 270 m (900 ft) in a series of thirty-two rapids, known as the Livingstone Falls, and the Orange plunges over the 140 m (460 ft) high Aughrabies Falls at a point 80 km (50 miles) below Upington. Numerous smaller falls and rapids occur on the American rivers where they descend the 'fall-line' between the ancient rocks of the Piedmont Plateau and the newer rocks of the Atlantic coastal plain (fig. 68). As a minor example, most Pennine

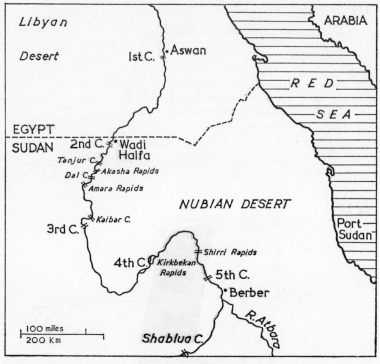

Fig. 67 The Nile cataracts

The map shows the location of the five main cataracts (numbered 1 to 5) and of some of the minor named cataracts.

The Nile has cut its way vertically through Nubian Sandstone until in places it reached the ancient crystalline rocks below. These harder rocks offered greater resistance to erosion, and so delayed the grading upstream, forming a short reach with a steep slope, interrupting longer reaches of more gentle gradient. The hard rocks therefore cause complicated rapids and broken water, divided channels and many picturesque rocky islets. Most of these crystalline rocks consist of gneiss. In places, however, bands of still more resistant granite among the gneiss cross the channel at right-angles, thus forming ridges and therefore more striking rapids within the general broken water of the cataracts.

walkers will be familiar with Kinder Downfall, where the Kinder (which drains the peat bogs of the Kinder Scout plateau) falls over the vertical gritstone edge.

(iii) Some waterfalls are due to faulting. The commonest type is where a fault-line scarp brings a less resistant rock on the downstream side against a more resistant rock. Gordale Scar, where the Gordale Beck falls through an arch and over a curtain of tufa (p. 10), has cut back a gorge (possibly when it was a subterranean

54 Niagara Falls
(*The Photographic Survey Corporation Ltd, Toronto, Canada*)

55 Howick Falls, Natal
(*The Aircraft Operating Co of Africa Ltd*)

56 The Rainbow Falls section of the Victoria Falls on the River Zambezi
(Paul Popper)

57 The Yosemite Falls, California
(Paul Popper)

stream) into the edge of the Carboniferous Limestone plateau. The Mid-Craven Fault brought the limestone against the softer Bowland Shales on the down-throw (south) side, and the latter have suffered great erosion. Similarly, the Folly Dolly Fall near Meltham (south-west of Huddersfield), some 9 m (30 ft) high, is due to a stream falling over a fault in horizontally bedded gritstone; on the downstream side the grits have been down-faulted, bringing soft shales, which have now been removed by the stream, against the gritstone face.

It may be mentioned that the Victoria Falls on the Zambezi, 110 m (360 ft) high, owe their development in part to faulting (plate 56). The river crosses a basalt plateau which is traversed on its eastern margins by a series of fractures, forming shatter-belts or lines of weakness. The gorge below the falls follows a zigzag

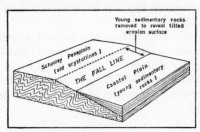

Fig. 68 The Fall-line of North America

course, due to the intersection of these faults almost at right angles, and its recession for more than 100 km (60 miles) has been due to the rapid erosion of the faulted and weakened basalts by the river.

(iv) Waterfalls are commonly found in glaciated districts, where over-deepening of the main valley (p. 232) leaves hanging valleys and cirques high above the main floor. Snowdonia, the English Lake District and Scotland abound in examples; we may mention Sour Milk Ghyll, plunging in a series of white cascades down the dark fell-side from Bleaberry Tarn into the outlet of Buttermere, and the well known Lodore Falls above Derwentwater. Numerous magnificent falls can be seen in the Yosemite Valley, California; plate 57 shows the Upper Falls descending 436 m (1430 ft) from a hanging valley, the Lower Falls a further 98 m (320 ft); with the intermediate cascades the total descent is 780 m (2565 ft). The first 20 m (70 ft) of the Upper Falls are in a deep chute worn in the solid granite, below which the water makes a free parabolic leap one and a half times the height of the Eiffel Tower in Paris.

(v) Falls sometimes occur where streams flow over the edge of a steep cliff into the sea. At Osmington Mills, east of Weymouth, a small stream cascades over a horizontal stratum of relatively resist-ant Nothe Grits (one of the Corallian beds of the Jurassic system). On the night of 18–19 July 1955 torrential rain fell in southern Dorset; about 18 cm (7 in) fell on the 8 sq. km of country which constitutes the catchment area of this stream. A vast volume of water poured down to the sea, deepening the bed of the stream by 1–1·5 m (4–5 ft), and cutting into the hard stratum of the terminal cascade. Much more striking are the delightful falls along the Devon coast near Hartland; Litter Water has a fall of 23 m (75 ft) from a valley well above sea-level, and has cut a small notch in the cliff top. These valleys are the result of rapid marine erosion, so that the cliffs have receded landward. Milford Water descends in a series of five separate falls, interrupted by two sections at right-angles. Again, numerous torrents fall over the steep basalt cliffs around the coasts of Skye.

Thus waterfalls are formed whenever the efforts of a river to produce a smooth profile have been temporarily interrupted, either by the differential resistance of rocks to erosion, or by various earth movements which have produced locally steep slopes. Ultimately, however, even the highest fall will be cut back to form part of the graded valley.

THE DEVELOPMENT OF A RIVER VALLEY

As a river gradually develops its long-profile, so the cross-profile and the general features of its valley also develop, as a result of down- and lateral-cutting by the river, combined with the weathering of the valley sides, involving sheet-movement and slope collapse. The actual form is determined (i) by the nature of the rocks, their resistance to erosion and their structural arrangement; (ii) by the erosive power of the river and of its tributaries, together with the effects of weathering and mass-movement on the valley sides; and (iii) by the stage to which the processes mentioned in (ii) have attained in their effects on (i). We can therefore describe in turn the characteristic forms of the valley of a river in its upper or mountain, middle, and lower or flood-plain tracts (fig. 69). Rejuvenation can initiate a new cycle of erosion, and so develop fresh features.

(i) **The upper valley** In its upper course a torrent is chiefly concerned with vertical corrasion, mainly by pot-holing, and it cuts a steep-sided V-shaped valley, the angle of whose sides depends on the resistance of the rocks both to erosion and to the weathering of the containing slopes. An upland stream follows a winding course, since it tends to flow round obstacles formed by more resistant rocks. These bends are gradually emphasized, since the current tends to be strongest on the outside of a bend, and so spurs alternating on either side of the river 'interlock', 'overlap' or 'interdigitate' with each other (fig. 69, plate 51). The bed of the stream consists of pools, pot-holes, boulders, and small rapids or falls due to locally resistant rocks.

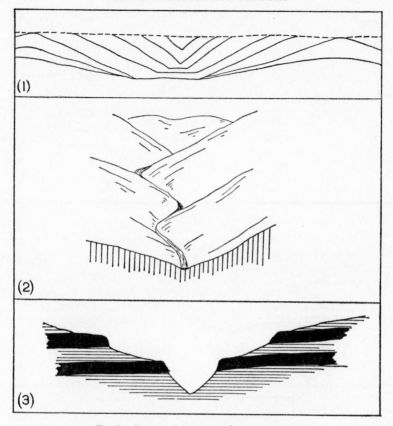

Fig. 69 *Features of the cross-profile of a river valley*

1. Successive stages in the evolution of a cross-profile.
2. Interlocking spurs (see also plate 51).
3. The effects of almost horizontal resistant strata (in black) upon the form of the cross-profile.

River gorges and canyons The term gorge is used to refer to an elongated steep-sided hollow, deep in proportion to its width. A river gorge is found where fluvial erosion cuts down more rapidly than the forces of weathering can wear back and 'open up' the sides. Most mountain torrents in North Wales, the English Lake District and Scotland have their upper courses contained more or less in a gorge, especially where they follow a line of weakness. Piers Ghyll flows down the face of the Scafell range into Wasdale in a series of right-angled gorges, in places 24 m (80 ft) high, with falls

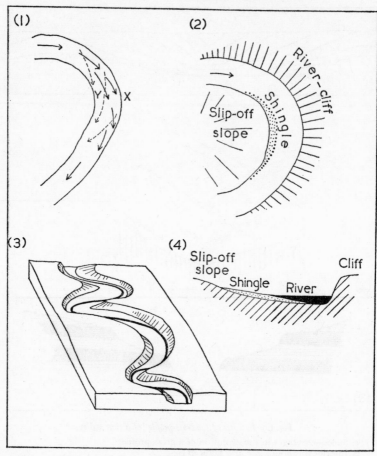

Fig. 70 *Meanders*

1. The direction of the main current is shown by solid arrows; as it swings round the bend, it impinges on the outside of it, thus eroding a steep river-cliff or bluff at **X**. As the water heaps up on the outside of the bend by centrifugal force, a return bottom-current (as shown by the pecked arrows) flows towards the inside bank (helicoidal flow), depositing a shingle- or sand-bank at **Y**.

2. The developing meander.

3. Block-diagram of meander development; each bend is enlarged, both sideways and downstream. The valley floor is therefore gradually widened.

4. Section of meander.   (See fig. 102 for ultimate development of ox-bows.)

alternating with deep rock pools; the stream is following the shatter-belts of several clearly defined faults.

A gorge may also occur either in resistant rock, so that the sides stand up steeply, or where rainfall is so slight that the weathering back of the sides can only proceed slowly. The first has already been mentioned in connection with waterfalls, which form a gorge as they recede. The famous Aar (Aare) Gorge, near Meiringen in Switzerland, occurs where the rapidly flowing Aar has cut through the Kirchet, a hill of hard limestone separating the wide Ober- and Unter-Haslithäler, while on a smaller scale the Wharfe above Bolton Abbey in Yorkshire flows through 'The Strid' (plate 52).

There are some other reasons for gorges which can be mentioned here for completeness. They may be formed as the result of a powerful river cutting down at the same rate as earth movements are uplifting the surrounding land; the gorges of the Indus in Kashmir, of the Brahmaputra where it crosses from Tibet into Assam, and of the Ganges and its headstreams, exceed the stupendous depth of 5 km (3 miles). On a smaller scale is the Rhine Gorge from Bingen to Bonn, where the river cuts diagonally across the Middle Rhine Highlands, and with its tributaries (such as the Moselle) has divided this plateau into a series of blocks (plate 45). The river occupied more or less its present course before the uplift of the plateau in Tertiary times. This is sometimes known as *antecedent drainage* (p. 186). Other special reasons for gorges include the cutting of fluvioglacial overflow channels (p. 207) and the collapse of the roofs of long caverns in limestone (p. 130 and fig. 59).

A gorge is formed readily in arid climates, where a river deriving its volume mainly from snow-melt on mountains beyond the desert is able to maintain its flow and its erosive power. Such gorges, known as *canyons* in America, are found on the Colorado, Snake, Yellowstone and many other rivers, especially emphasized where uplift has occurred. The Grand Canyon of the Colorado River (plate 58, fig. 71) is nearly 500 km (300 miles) long, and has a maximum depth of 1900 m (6250 ft), cut through near-horizontally bedded limestones, sandstones and shales, the youngest of which on the plateau surface are of Permian age, into the Pre-Cambrian crystalline rocks below; the North and South Rims are 19 km (12 miles) apart.

(ii) **The middle valley** In the middle part of its course a river develops mature features, and the valley becomes wider through the weathering of its sides. As the water flows round a bend the river tends to accentuate the curve (fig. 70), since the current impinges most strongly on the concave side or outside of the curve, and maximum erosion, even undercutting, takes place there. There is little erosion, and even some deposition, in the slack of the current on the inside of the bend (plate 61). Thus an initial 'swing' in the river may be transformed into a *meander*, with

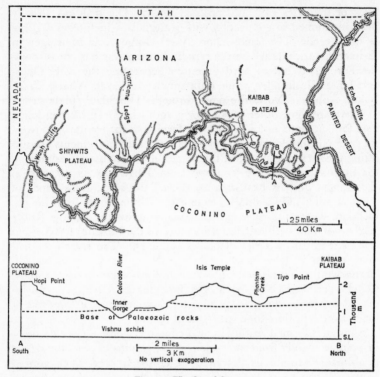

Fig. 71 The Grand Canyon

Based on the United States Department of the Interior, Geological Survey
1:48,000, Bright Angel Quadrangle.

a *river cliff* overhanging the under-cut bank and a sloping spur,
known as a *slip-off slope*, projecting from the opposite side.

While this may explain the development of meanders, it does not
account for their initiation. Some of the most striking examples are
found in virtually homogeneous and almost level alluvial deposits,
as in the lower Mississippi. The size (or *wavelength*) of the meanders
and the width of the meander-belt seem to be related to such factors
as discharge and bed-load of the river, and to varying depths and
bottom friction as the stream crosses an uneven surface.

Gradually the river extends its lateral erosion, widening the
swing of each meander (and so its overall valley) by continued
undercutting and the removal of weathered material. The lateral
shifting of a river's course is referred to as a *divagation*. In addition,
the current cuts into the upstream sides of spurs, so that each

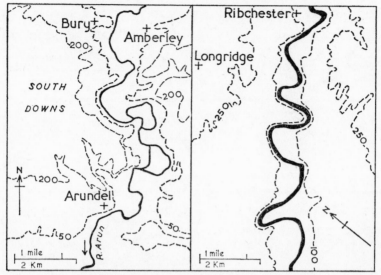

Fig. 72 The meanders of the rivers Arun (left) and Ribble (right)

On the right-hand map, Preston lies 3 km (2 miles) downstream. The contour heights are in feet. The two direct cuts on the Arun are artificial.

meander begins to move downstream (fig. 70), planing an increasingly wide valley, Ultimately each spur is removed, leaving a cusp or bluff overlooking the floor of the valley. Thus the river is beginning to create a broad, nearly level valley, bounded by the low walls of the bluffs (the *bluffline*).

(iii) **The lower valley** In its lowest stage the river wanders or 'shuffles' in a series of sweeping meanders over a broad almost level valley, bounded by low bluffs which recede through weathering, reducing the interfluves and so gradually forming *panplains* (p. 109) where coalescence of adjoining floodplains takes place. Ultimately the floor of the valley will be considerably wider than the actual meander belt (fig. 72). A convenient distinction between maturity and old age is, in fact, that while in the former case the meander belt fills the valley, in the latter case the valley is considerably wider. Relics of former meanders are left in the form of 'ox-bow lakes' or 'cut-offs' (p. 198, figs. 73, 102, plate 60).

Deposition In this lowest section of a river's course deposition becomes of major importance, and vertical corrasion ceases except for the incising of the channel in the floor of the flood-plain as the meanders move downstream.

The coarser river gravel is laid down in the bed, to be left there

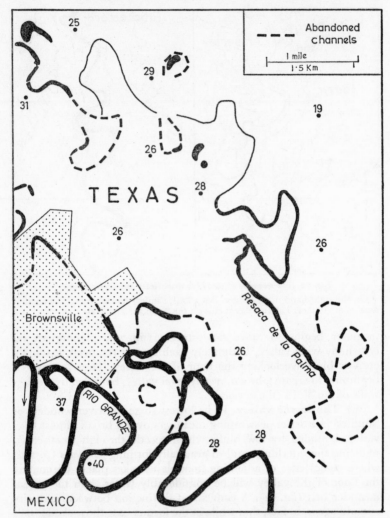

Fig. 73 *Meanders and ox-bows on the Rio Grande, between Texas and Mexico*

Based on the United States Department of the Interior, Geological Survey, 1:31,680, East Brownsville Quadrangle. Heights are in feet.

The course of the main river (in the south-west), which marks the border between Mexico and the U.S.A., has been stabilized by means of embankments. Numerous abandoned ox-bows and sections of stagnant channel cover the flood-plain.

as each meander moves on. When the volume of the river for some reason is swollen, this deposition of gravel takes place on a much larger scale. The river Meuse in immediately post-glacial times

58 The Grand Canyon of the River Colorado, in Arizona
(Fairchild Aerial Surveys, Inc)

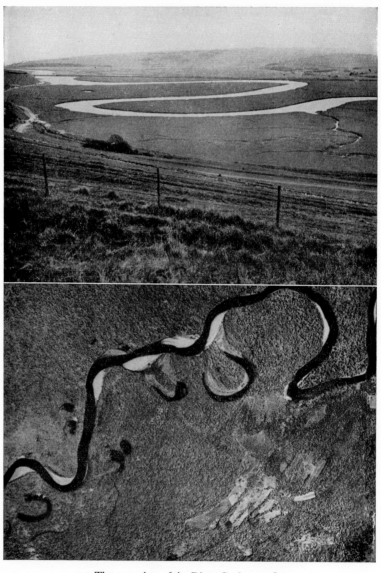

59 The meanders of the River Cuckmere, Sussex
(*R. Kay Gresswell*)

60 Ox-bows on the River Arkansas
This vertical infra-red photograph was taken for forestry conservation purposes.
(*Fairchild Aerial Surveys, Inc*)

61 An incised meander of the River Rheidol in central Wales
This shows the level surface of the meander-core.
(*R. Kay Gresswell*)

62 Deeply dissected landscape in central Iran
(*Aerofilms Ltd*)

63 The lacustrine delta of the River Kander where it enters Lake Thun, Switzerland

(Swissair-Photo A.G.)

was heavily laden with coarse sand and gravel as it left the Ardennes, where its gradient was abruptly checked. The Meuse laid down a sheet of this material, varying in thickness from 5–15 m (15–50 ft), over the Kempenland of north-eastern Belgium (now covered with heathlands), as the river gradually altered its course in a series of slowly moving curves. Many large, heavily laden rivers, especially when they flow from an extensive mountainous area, spread out sheets of material which may split the stream into complicated channels; such rivers are said to be 'braided' (figs. 78, 81). This braiding is encouraged where the banks of the main stream consist of easily erodible sands and gravels.

The flood-plain Most rivers, certainly in their lower parts, transport a load consisting of finely divided silt. This may be deposited to a great thickness; in the lower Nile Valley, for example, no boring has ever reached the rock floor. In times of flood a river may spread a thin veneer of alluvium over its whole flood-plain, a fact of major importance to the farmers of the Nile, the Tigris–Euphrates, and the Asiatic valleys where teeming millions depend on rice cultivation.

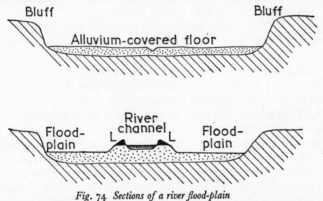

Fig. 74 Sections of a river flood-plain

The upper diagram shows the broad smooth floor, cut by lateral erosion as the meander belt swings downstream, with bluffs bounding the flood-plain, and with a veneer of sediment.

The lower diagram indicates how a large river which floods periodically builds up its bed above the level of the surrounding country; natural or artificial banks (*levees*) (**L**) are but a temporary safeguard.

Most mud is, however, deposited at the edge of the channel, where the current slackens, so that when the floods subside natural embankments remain. Aggradation continues in the bed, which thus tends to be raised above the flood-plain (fig. 74); this raising is partly caused by the slumping of the embankments into the bed itself. The danger of inundation to the surrounding plain during time of flood becomes progressively greater, since once the flood-waters breach the embankments (known as *levees* along the Mississippi), they can spread widely over the lower-lying lands. Moreover, when drainage of the surrounding lands is effected, their level is appreciably lowered because of settling and compaction, especially in an area of peat-fen (p. 538). Along many rivers, such as those in the Fen District of eastern England, the Po in northern Italy (which flooded disastrously during the winter of 1951–2) and the Mississippi, efforts have been made to strengthen and enlarge the natural embankments. This often makes matters worse, since the river also continues to raise its bed, and so becomes progressively higher and more threatening. One of the most dangerous rivers is the Hwang Ho, known as 'China's Sorrow' from the devastation and loss of life it has repeatedly caused. In 1852 the river burst its banks and temporarily shifted its mouth 500 km (300 miles) to the south of the Shantung Peninsula; over a million people were estimated to have been drowned. In 1938 during hostilities with the Japanese, the river was deliberately shifted south for strategic purposes, and was not returned to its northern course until 1947.

Levees are not a successful long-term method of coping with the menace of floods. The control of water in the catchment by planting forests on steep slopes to retard runoff, by using the upper valleys as reservoirs to hold flood-water, and by cutting through meanders, so providing a straight, more direct channel and therefore a steeper gradient, are more effective measures.

The Dutch have had to face great problems of water-control, for their country largely consists of the combined flood-plain of three rivers—the Rhine with its several distributaries, the Maas (Meuse) and the Scheldt (Escaut). There are three contributory factors to the problems: the swinging meanders, the low level of the surrounding country (in places below sea-level), and the periods of flood due to heavy rain and snow-melt in the uplands of central Europe. The safeguarding of their lands against sea and river floods forms a large part of Dutch history. The rivers have been straightened and separated, new outlets to the sea have been cut, reservoirs have been made to store temporarily some of the flood-water, and massive dykes or embankments, set back sometimes as much as half a mile from the main channel, have been constructed.

Where a sluggish river has built up its bed, creating natural levees which have to be artificially raised and strengthened, it is obviously difficult or impossible for tributaries coming across the low-lying plain to join the main river. On a major scale the Yazoo enters the

Mississippi bottom-lands and flows parallel to the main river for 280 km (175 miles) before it can effect an entrance. This is known as a *deferred junction*.

A river flood-plain is therefore an area of gentle or even imperceptible slopes, across which wanders the heavily mud-laden river, bounded by levees, beyond which lie marshland, ox-bows and stagnant creeks. And so the river passes indistinguishably into either its delta or its tidal estuary.

(iv) **A rejuvenated river valley** The main results of rejuvenation (p. 135) on the cross-profile of the river are (i) the production of terraces on either side of the channel by both vertical and lateral erosion, and (ii) the deep incising of the channel into the surface, while maintaining the plan of the original meanders.

River terraces When a river renews its down-cutting it sinks its new channel into the former flood-plain, leaving the latter well above present river level, so that the remnants form terraces on either side. Gradually the new valley is widened, and the terraces are cut back by lateral erosion. If, however, renewed rejuvenation takes place the process is repeated, and a second pair of terraces, at a lower level than the first, will remain. These terraces can frequently be correlated with corresponding knickpoints in the long-profile (fig. 75).

The original material deposited consists of gravel and alluvium; the latter is, however, more easily removed when renewed erosion takes place, and so most of the older terraces are covered with gravel-sheets, the 'terrace-gravels' which are given various names and assist in identification and correlation.

A very nice example of terraces is shown in fig. 75, where the Ill and the Rhine in eastern France flow in almost parallel courses. The present flood-plain is the *Ried*, consisting of marshland and backwaters, covered with clumps of willows and poplars and with damp pasture, flooded in spring. The terraces, however, are higher and drier, and on them are roads, railways and villages, with arable lands and vineyards.

The terrace patterns of the Thames have been studied in detail, both in the London Basin (fig. 75) and in the Oxford Clay vale. In the latter the terraces formed by several stages of rejuvenation have been named after villages situated on them. The Hanborough Terrace is the highest, at about 30 m (100 ft) above the present river-level. A period of erosion was succeeded by the deposition of gravels now forming the Wolvercote Terrace, which in turn was cut into by the river. Renewed deposition of gravels formed the Summer-town–Radley or Twenty-foot Terrace (on which the city of Oxford

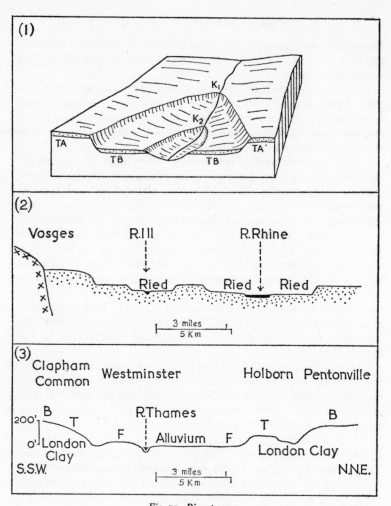

Fig. 75 River terraces

1. (After S. W. Wooldridge.) This shows how successive rejuvenations (with knickpoints K_1 and K_2) can be related to successive pairs of terraces (**TA, TB**).

2. (After A. Cholley.) This illustrates the terraces of the Rhine valley in Alsace, complicated because the Rhine and its tributary the Ill flow almost parallel to each other, leaving the remnant of a terrace between them. The *Ried* is the present marshy flood-plain.

3. (After the Geological Survey.) This shows diagrammatically the Thames terraces: **B** = Boyn Hill Terrace; **T** = Taplow Terrace; **F** = Flood-plain Terrace.

The Taplow Terrace lies from 15–30 m (50–100 ft) above the level of the Thames; gravel patches on this terrace can be traced in a number of places, for example in Hyde Park and Holborn. The Boyn Hill Terrace lies 15 m (50 ft) higher.

stands). Further rejuvenation then caused the Thames to cut a bed to a depth of 9 m (30 ft) below its present level; this channel is buried under the alluvium of the present flood-plain.

Incised meanders If down-cutting is potent the channel may be deeply incised into both the alluvium and the solid rock, forming 'incised' meanders. A distinction is sometimes made between *intrenched meanders*, when the valley sides are steep and symmetrical, and *ingrown meanders*, when there is some lateral cutting as well, producing a more open valley with slip-off spurs. A fine series of incised meanders is provided by the Wye, which winds and twists through a gorge-like valley, in places forming almost complete loops (fig. 76). The river has cut vertically through the Coal Measures, Carboniferous Limestone and Old Red Sandstone.

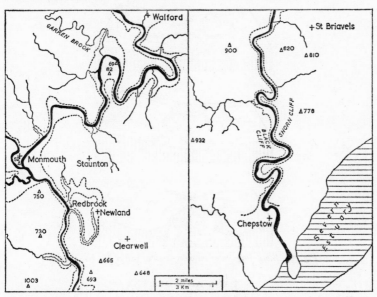

Fig. 76 The Wye valley

The right-hand map lies south of the one on the left. The pecked line indicates the approximate edge of the valley floor. Heights are in feet.

Sometimes an incised meander may cut right through the 'neck' of the meander, and so the river abandons it, but continues to erode the 'short-cut'. This is clearly seen in the Wye valley near Redbrook (fig. 76); the old meander, though occupied by two small tributaries, is now 120 m (400 ft) above the bed of the main river. The Dee north of Llantisilio in North Wales is also deeply incised, and large abandoned meanders can be seen to the south-east of Llangollen. These meanders are plugged with boulder-clay, probably a contributory cause towards their abandonment by the river.

Incised meander-terraces Series of terraces, not regularly paired like those mentioned above, are formed in the lower part of a flood-plain, when the meanders are swinging freely but where the river is still to some extent down-cutting. This is especially common in alluvium or in glacial drift, and also where masses of hard rock are exposed in the valley floor, thus checking lateral erosion and preserving the higher terraces. As the meanders move downstream, swinging from side to side, they erode part of a higher level and leave it as a terrace (fig. 77). This is well seen in the middle Rheidol valley, east of Aberystwyth (plate 61).

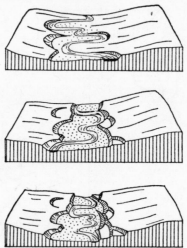

Fig. 77 Block-diagram of meander-terraces (after W. M. Davis)

FANS AND DELTAS

Deposition of material takes place more or less gradually over the whole lower course of a river, but it is more concentrated where the speed of the current is abruptly checked for some reason. This may occur (i) by a sudden change in gradient; (ii) where its constricting narrow valley suddenly opens out into the broad gentle floor of the main valley, so forming a fan or cone; (iii) by the entry of the river into the still waters of a lake, so forming submarine turbidity currents, and building up a lacustrine delta; and (iv) by its entering the sea, where the material is deposited to form either estuarine banks or a delta.

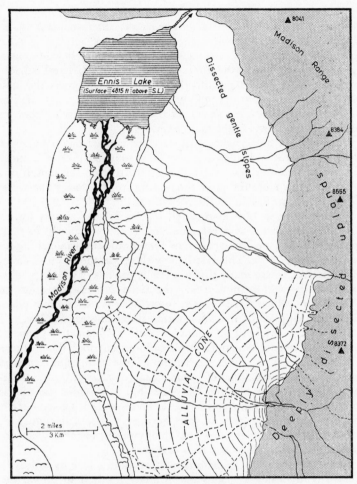

Fig. 78 Ennis Lake in southern Montana, U.S.A.

Based on the United States Department of the Interior, Geological Survey,
1:62,500, Ennis Quadrangle.

The Madison river flows northward from its source in the Yellowstone National
Park to join the Missouri. It flows into Ennis Lake, which it is steadily filling
up; the area of swamp and alluvium to the south, crossed by the braided river,
represents the gradual northward extension of the lacustrine delta. Note the huge
alluvial cone in the south-east, built up by torrential streams deeply dissecting the
mountains.

Land over 6000 ft (1800 m) is stippled. Heights are in feet.

Alluvial fans and cones A fan-shaped mass of material is deposited by a mountain torrent where it enters a main valley. This is found especially where the torrent occupies a hanging valley and then reaches the flat trough-floor of a glaciated valley, as in Lauterbrunnen and in the upper Rhône valley in Switzerland. The stream usually divides into distributaries before it enters the main river. Some fans, consisting of almost level alluvium, form the sites of villages in the Rhône valley, at the foot of the steep slopes but above the flood-level of the main river.

In semi-arid areas, where deposits are carried down by short-lived torrents of great power, the form may be that of a much thicker and steeper cone extending outward from the mouth of a wadi (p. 273). Note the huge cone on fig. 78, bordering the valley of the Madison river in southern Montana.

Where numbers of near-parallel streams descend steeply from a mountain range on to a marginal lowland, particularly in these semi-arid areas, neighbouring fans may gradually coalesce into a *piedmont alluvial plain*. This is strikingly shown in the Central Valley of California, where many streams, fed by the heavy winter precipitation on the west-facing slopes of the Sierra Nevada, have formed alluvial fans which have merged to create a gently sloping plain. This area, supplied by irrigation water, is one of great fertility.

Many fans were built up in the past during more active phases of erosion and deposition. An example can be seen along the northern flanks of the central Pyrenees, where a large composite coalesced fan, known as the plateau of Lannemezan, is crossed by the tributaries of the Adour and the Garonne (fig. 79). While material is still being deposited during the winter floods and spring snow-melt, this fan was mainly built in late Tertiary times, and consists of coarse Pliocene pebbles worn from the Pyrenees, forming what the French geologists call *anciens cônes de déjection*.

Lacustrine deltas A delta is built out into a lake by a heavily laden stream (plate 63), and the ultimate result is to fill up the lake as part of the aggrading progress towards a graded profile (fig. 65). Many of the English Lakes contain fine examples of developing deltas (p. 203 and fig. 98).

One striking example may be mentioned here (fig. 80). Below Martigny in western Switzerland, the Rhône flows transversely across the folded ranges of the western Oberland, cutting the 18 km (12 miles) gorge of St Maurice through the limestone ridges. The river is strongly flowing and heavily laden at this point. At Bex the valley opens out and the gradient is abruptly reduced. It is clear that Lake Geneva once extended as far as Bex, but deposition for long

Fig. 79 *The drainage system of the Tertiary gravel-fans on the northern flanks of the Pyrenees*

The map shows the extraordinary pattern of streams flowing northward over the gravel plateau of Bigorre, situated about 30 km west of Lannemezan. Long narrow ridges separate the valleys, their sides furrowed with short rivulets.

periods of time has filled up its head; the former lake bed is marshy, and although the river has been regularized, there is much braiding, with backwaters and cut-offs. The delta is still growing into the lake. There is a remarkable contrast between the milky-grey colour of the river at its entry and the clear water which flows out at the western end; from the air the underwater banks of the extending delta, built up by these submarine *turbidity currents*, stand out from the clear waters of the rest of the lake.

The biggest lacustrine deltas are those which are being built

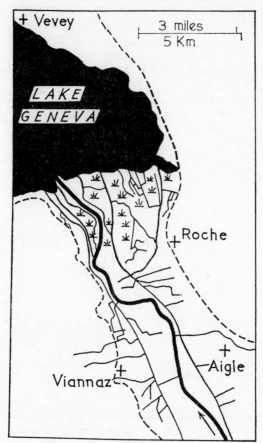

Fig. 80 *The Rhône delta in Lake Geneva*
The approximate edge of the upland is indicated by a pecked line.

out into the Caspian Sea by the Volga, the Ural, the Kura and several other rivers.

Estuarine banks The tidal portions of rivers are features of the coastline, described in Chapter 10. However, estuaries may be mentioned briefly, since they are the scene of extensive river deposition. Much of the river's load is deposited here, although the finest material may be swept many km out to sea, as in the case of the Amazon and the Congo. The extent and nature of the deposits depends on (i) the relative strength of river current and tidal stream; (ii) the configuration of the estuary, whether bottle-necked and therefore flushed by strong tides (Mersey) or wide and open (Dee); and (iii) artificial causes, such as dredging and the construction of training walls.

Marine deltas A delta is formed at the mouth of a river where the deposition of sediment comprising its load exceeds the rate of removal. Strong tidal currents, wave action and longshore drift often combine in open seas to prevent or limit delta formation, but it is not necessary to postulate near tideless seas. Several large deltas have developed in the Mediterranean Sea, which has only a small tidal range, at the mouths of the Nile, Po and Rhône, but they also occur in markedly tidal seas: the Colorado in the Gulf of California, the Amazon in the Atlantic, and the Ganges and Irrawaddy in the Indian Ocean. At the mouth of each of these rivers the tidal range exceeds that of, say, Southampton Water. But in each case a load of sediment, far greater than can be removed, is being brought down. The suggestion has been made that deltas are less obvious features than they might be because many are concealed either by the post-glacial eustatic rise of sea-level (p. 304) or by isostatic sinking due to the loading of sediment (p. 24) or both.

The materials deposited in a delta are classified into three categories. The fine materials carried out to sea and deposited in advance of the main delta are known as the *bottomset beds*. Over these, inclined layers of material are gradually built out, each one above and in front of the previous ones, so that the delta advances seaward; these are the *foreset beds*. Finally, on the landward margins of the delta fine material is laid down in a sheet continuous with the river's alluvial plain, forming the *topset beds*.

In a partially or wholly enclosed sea, such as the Gulf of Mexico, the Mediterranean and the Caspian, a broad, sloping bank of sediment is built up very rapidly. Where the coast is slowly subsiding for tectonic reasons, as is happening at the Mississippi delta, the sediment may be of immense thickness. But the sinking is not as great as the thickness of the added layer, so there is a net gain in level.

Not only is the river current checked where it enters the sea, but it is a physico-chemical fact that fine particles of clay rapidly coagulate (*flocculate*) and settle when they mix with salt water. As the bank of sediment grows seaward, so the water becomes shallower than it was, and the main channel may split into several streams, known as *distributaries*. Most sediment is deposited at the margins of a river current, so that elongated mud-banks are built up along the edge of the channel, to be breached at flood-time when bifurcating channels are formed. Meanwhile sedimentation proceeds all over the delta, especially at flood-time, thus slowly raising its general

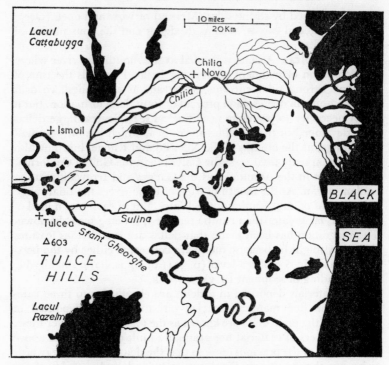

Fig. 81 The Danube delta

The Danube delta encloses a lowland area covered with marsh and shallow meres (*lacul*), between its two main distributaries, the Chilia (which is itself building out a subsidiary delta) and the Sfant Gheorghe to the south. In the south-west the Tulce Hills rise quite steeply to over 180 m (600 ft), contrasting with the flat delta-lands. The two main Danube ports, Galati and Braila, lie about 60 km (40 miles) upstream; most traffic uses the regularized Sulina channel. Note the lagoons, meanders, ox-bows and braided streams.

level. Lagoons and marshes, separated by mud-banks, replace the open sea as the delta grows seaward, aided by marsh vegetation and blown sand. These delta-lakes are discussed on pp. 198–9 (see also figs. 81, 102).

It is possible to distinguish varieties of delta, but the most important are the *arcuate* type (the Nile, Po, Hwang Ho and Rhône deltas), the *cuspate* type (the Tiber) (fig. 82), and the more extended *bird's foot* or *finger* type (the Mississippi). The '*bird's foot*' type is caused by a river bringing down enormous amounts of fine silt, deposited for a long distance along the edge of the channels.

The Rhône delta The Rhône divides into two distributaries, the

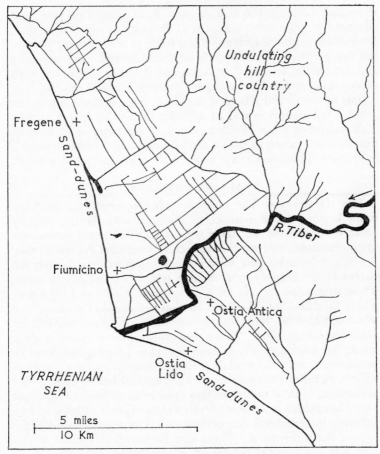

Fig. 82 The cuspate delta of the Tiber

Most of the Tiber delta is now reclaimed, as shown by the number of drainage channels. This reclaimed area is known as the '*bonifica di macares*'.

Grand and Petit Rhône (fig. 102), which enclose its present delta. At Arles, where the river is 150 m (500 ft) wide, it is 50 km (30 miles) from the sea but only 2 m (6 ft) above sea-level. Between the two distributaries lies the Camargue, a solitude of wind-swept marsh, covered with shallow meres (*étangs*) cut off from the sea by sandbars and dunes; the extensive Etang de Vaccarès is little more than a metre in depth. Low sandy ridges (*lônes*), winding across the delta, indicate abandoned distributaries. Some idea of the immense deposition is given by the fact that the river carries down about

17 million cubic m (600 million cubic ft) of sediment a year, and advances its delta south-eastward at about 150 m (500 ft) a year. Most of the southern Camargue is a waste of reed-covered swamp, but the northern part has been largely reclaimed, both naturally and artificially, so that there are great stretches of pasture, with groves of cypress and tamarisk, and some arable land growing rice.

East of the Grand Rhône is the Crau, a triangular plain in the angle of the main river and its tributary the Durance. It is covered with a sheet of gravel, brought down by the latter in Pleistocene times when it joined its parent farther south and together formed a joint delta. Thus the Crau is an example of an ancient 'dry delta'.

DRAINAGE PATTERNS

The tributaries The theoretical development of a drainage basin begins with a number of main streams flowing directly down a slope to the sea; these are the result of or are consequent upon that slope, and are therefore called *consequent streams*. As they develop, tributaries flow towards these main valleys, joining the parent river obliquely, and in turn minor tributaries join them. These tributaries are referred to as *insequent streams*, and the points where they flow into the main streams are *accordant junctions*. The somewhat obvious 'law of accordant junctions' was enunciated by John Playfair in 1802: "Every river appears to consist of a main trunk, fed from a variety of branches, each running into a valley proportional to its size, and all of them together forming a system of valleys, communicating with one another, and having such a nice adjustment of their declivities that none of them join the principal valley at too high or too low a level; a circumstance which would be infinitely improbable if each of these valleys were not the work of the stream that flows in it." There are, of course, exceptions to this principle, the result of special sets of conditions, such as the entry of a stream from a hanging valley (p. 232), which are of a temporary nature. If the rocks do not vary in resistance over the basin, each consequent river will become the centre of a converging stream pattern, to which is given the name *dendritic drainage* (fig. 83 and plate 62), after the Greek *dendron*, a tree.

Should, however, the basin of each consequent stream consist of rocks of varying degrees of resistance, tributaries will develop which have a definite relationship to structure. Where these flow along the strike they are known as *subsequent streams*, which develop by headward extension from the consequent streams. If outcrops of alternately more and less resistant rock occur at right angles to the consequent slope, the subsequent streams develop along the strike

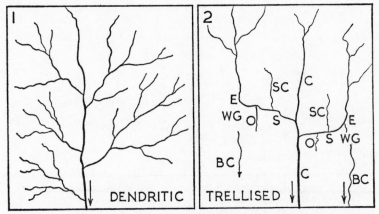

Fig. 83 Dendritic (left) *and trellised* (right) *drainage patterns*

In **2**, the following abbreviations are used: **C** = consequent stream; **S** = subsequent stream; **O** = obsequent stream; **BC** = beheaded consequent; **SC** = secondary consequent; **E** = elbow of river-capture; **WG** = wind-gap.

of the less resistant rocks, entering the consequent stream at right-angles (fig. 83). The more resistant bands of rock become elongated hills more or less parallel to the subsequent streams, through which the consequent streams, continuing to deepen their initial courses down the slope, cut steep-sided gaps. Tributaries to the subsequent streams then develop, flowing from the upstanding ridges of more resistant rock; those flowing down the initial slope (i.e. seaward) are *secondary consequent streams*, those flowing down the opposite slope are *obsequent streams* (fig. 83). However, there is now some ambiguity in the term obsequent. Many American and French geomorphologists still use it in the original sense of W. M. Davis as simply 'anti-consequent', but some English authorities reserve it for an 'anti-dip stream', that is, one flowing in the opposite direction to the dip of the strata, which has not always the same meaning.

The drainage system so developed is much more rectilinear than the dendritic, and is known as a *trellised pattern*. A most striking development is seen in south-eastern England (fig. 84), where less resistant beds (the Oxford, Kimmeridge, Gault and Wealden Clays) alternate with more resistant ones (Oolitic Limestone, Corallian Limestone and Chalk), all dipping gently south-eastward. The result is the evolution of scarplands, with complicated drainage patterns developed in the clay vales and the more resistant rocks forming cuestas. The steep escarpment faces north-west, the gentle back-slope south-east.

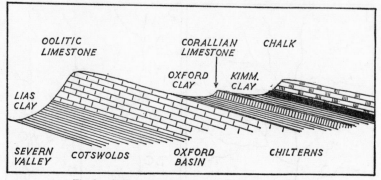

Fig. 84 Section across the cuestas of south-eastern England

The section shows in simplified form (with a considerable vertical exaggeration) the succession of cuestas and valleys. The solid black band represents Gault Clay and Greensand. **Kimm.** = Kimmeridge.

A remarkable pattern of cuestas and vales can be seen in the Paris Basin, where the Seine and its convergent tributaries have developed upon the gentle synclinal structure (fig. 85).

A pattern in which large numbers of rivers converge from all directions on one main stream is known as *centripetal drainage*. In the valley of Katmandu, Nepal, for example, many streams converge upon the Bagmati river, which drains southwards through a gorge cutting across the surrounding mountains.

Divides Between the catchment areas of two individual streams an area of higher land separates the surface waters which find their way into each; these may be major divides, between two complete river basins, minor divides between tributaries, and continental divides separating the great waters of a continent flowing to different oceans (fig. 62). The word *watershed* is commonly used in Britain for a divide, but this term in the U.S.A. signifies the whole gathering-ground of a single river system, equivalent to a drainage basin. In areas of strong relief the divides may be sharply defined, such as the Andean ridge-line, or, on a smaller scale but even more sharply cut, the ridge of the Black Cuillins in Skye (fig. 43). In Glacier National Park, Montana, the prominent and well named Triple Divide Peak separates water flowing to the Pacific Ocean, to the Gulf of Mexico, and to Hudson Bay. The water-divide of central Europe has already been mentioned (p. 134 and fig. 62).

By contrast, many divides are vague and indeterminate. For example, the heathland plateau of north-eastern Belgium has a very uniform surface, interrupted only by faint swelling eminences rising but a metre or two above the general level, and alternating with

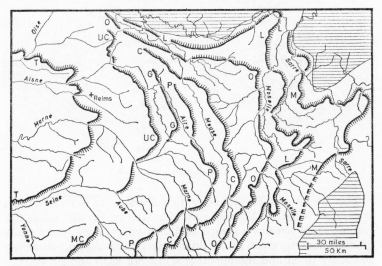

Fig. 85 The scarplands of the eastern Paris Basin and Lorraine

Based on the *Carte géologique de la France*, 1:1,000,000 and on F. Machatschek, *Das Relief der Erde* (1938), p. 54.

The escarpments are shown in generalized form by heavy lines with hachures on the scarp-slopes, and are lettered as follows: **T** = Tertiary (Falaise de l'Ile de France); **UC** = Upper Chalk; **MC** = Middle Chalk; **G** = Upper Greensand ; **P** = Portland Limestone; **C** = Corallian Limestone; **O** = Oolitic Limestone; **L** = Lias; **M** = Muschelkalk (a shelly limestone of Triassic age). The Palaeozoic rocks are indicated by horizontal ruling. The area west of the Upper Chalk escarpment is 'Dry Champagne'. Between the escarpments are the successive clay-vales.

shallow marshy depressions. This forms the watershed (called by its Flemish name of 'Waterschei') between tributaries of the Meuse and the Scheldt rising only a metre between them. Rather incredibly, Lake Isa, situated in Yellowstone National Park (Wyoming) precisely on the Continental Divide at 2518 m (8260 ft) above sea-level, drains to both the Pacific and Atlantic Oceans. In an area such as the Lake Plateau of Finland (fig. 100), where there is a chaos of lakes and moraines, divides are equally chaotic and, incidentally, are very shortlived. As a final example, from the undulating surface of the Columbia Icefield in the Canadian Rockies in Jasper National Park, the melt-waters form the headstreams of three major rivers: the 1230 km (765 miles) Athabasca river, a tributary of the Mackenzie which flows to the Arctic, the 1940 km (1205 miles) Saskatchewan river which flows into Lake Winnipeg and via the Nelson river into Hudson Bay, and the Columbia river which flows for 1945 km (1210 miles) to the Pacific Ocean.

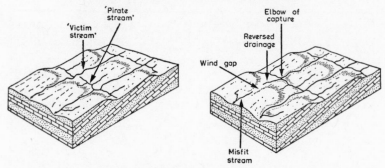

Fig. 86 Block-diagram of river-capture

This introduces a further point, that divides tend to change their position or 'migrate'. Rainfall may be heavier on one side, hence the volume of the streams is greater and their valleys are lowered more quickly, and regression of the divide takes place. Again, where the strata are tilted and the opposing slopes are steep and gentle, the former erode more rapidly; this is sometimes known as 'the law oᵢ unequal slopes'. The river on each side of a divide cuts back towards the crest, ultimately producing a notch in the skyline which may develop into a col.

The process of river-capture The development of contiguous river systems must lead to one river becoming more powerful than its neighbours; slowly it becomes the 'master-stream' of the area. It accomplishes this partly by watershed regression, partly by diverting parts of a neighbouring system into its own basin, known as *river-capture*. The tributaries of the more powerful river push back their minor divides by headward erosion, and as their base-level is lower (because the main stream is more powerful and therefore more deeply entrenched), the head-waters of the adjoining system are captured, making the capturing river still more powerful. At the point of diversion there is usually a marked bend, known as the *elbow of capture*. The river which has lost its headwaters will be much reduced in volume, and therefore too small for its existing valley, hence the term *misfit* or *underfit*. It may be so much smaller that its source will now be some distance below the point of capture, leaving a *dry-gap* or *wind-gap* to mark its former valley (figs. 83, 86).

River-capture takes place readily in scarpland areas, where subsequent streams can cut back along belts of less resistant rock (such as clay) at right-angles to the consequent streams, so capturing neighbouring head-waters and producing a trellised pattern of great complexity. A study of a large-scale map of the Weald will illustrate

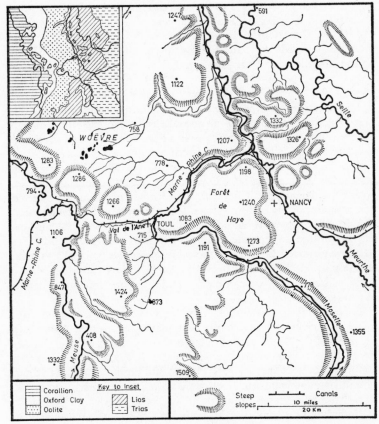

Fig. 87 The Toul gap

Based on *Carte de France et des frontières au 200,000e*, sheets 18,27.

Heights are given in feet.

Although the diversion of the former upper Meuse (now the upper Moselle) cannot be disputed, it seems from recent work by J. F.-L. Tricart that this was not just a simple case of river-capture by an active subsequent which cut back westward into the Oxford Clay vale. He adduces that during the Quaternary glaciation (p. 247) much rock-waste sludged down into the valley of the present Moselle above where Toul now stands. This filled the valley to such an extent that the Moselle was forced to spill over eastward and join the present Meurthe-Moselle. In so doing the river cut itself a new course which it has maintained, leaving the Val de l'Ane as the 'Toul gap', occupied by small misfit streams and by the Marne–Rhine Canal.

this. But river-capture does not need special structural conditions to occur; as S. W. Wooldridge has said: 'It is a normal incident in a veritable struggle for existence between rivers. . . . There are few

regions in Britain which do not bear the impress of river-capture.'
Examples in Europe include the capture of the upper Doubs from
the Rhine, the remarkable series of captures in the scarplands of
the Paris Basin (fig. 85), and the capture of the former upper Meuse
by the Moselle near Toul (fig. 87).

RIVER BASINS AND STRUCTURE

This description of the development of drainage patterns has been
based on the assumption of an initial surface sloping uniformly to
the sea, upon which a river system has developed. It was convenient
to assume relatively undisturbed sedimentary strata, horizontal or
with but a slight dip towards the coast, that is, of *uniclinal* structure.

Actually drainage systems develop on a variety of initial surfaces,
both lithological and structural. The former need not be stressed
here; the importance of limestone, for example, in the development
of a drainage system is considered in Chapter 5. Structural surfaces
are the product of tectonic activity of various kinds, responsible for
what may indeed be termed the *initial landforms* (p. 322). As soon
as one portion of the crust begins to rise relative to another, all the
forces of denudation come into play, and a sequence of erosion begins
to run its course (pp. 134–6).

These initial structural surfaces imply two controls in the development of a river system. In the first place, the height, shape and
steepness of the land-form will determine to some extent the character of the drainage pattern: whether it is a lofty fold-mountain
region, an area of smaller scale but possibly quite complex folding,
a fault-bounded plateau, a volcanic cone, an extensive horizontal
lava-sheet, or a dome-shaped uplift caused by periclinal folding or
by underlying intrusion. In the second place, complicated structures
usually involve rocks of varying degrees of resistance to denudation
lying in close juxtaposition. This differential denudation is responsible for many details of a valley, particularly in the river's youthful
stage, though ultimately it may produce a graded course unrelated
to the underlying rocks.

Changes of relative sea-level (pp. 303–4) may produce extensive
modifications, not only in the form of the cross- and long-profiles
of the river, but in the drainage pattern. The formation of the North
Sea and English Channel profoundly modified the river systems of
western Europe, which comprised essentially a major proto-Rhine
river to which the Thames was tributary. Again, in the Hampshire
Basin in Pleistocene times a 'Solent river' flowed from west to east
along the line of the river Frome in the Isle of Purbeck, and out
through the Solent and Spithead. It received tributaries both from

the north (the Avon and the stream now represented by Southampton Water, with its tributaries Test and Itchen) and from the south (Medina and Yar in the Isle of Wight and the Corfe river in the Isle of Purbeck). But the whole middle and lower part of this system was drowned, the last stages of submergence occurring in Neolithic times. The rivers were disrupted into a series of separate systems, their lower portions forming broad shallow bays (Poole Harbour and Christchurch Bay) and estuary-inlets (Lymington, Beaulieu, Southampton Water, Hamble, Medina).

Accordant and discordant drainage Where a systematic relationship is apparent between rock-type and structure on the one hand and the surface drainage pattern on the other, this is referred to as accordant drainage, sometimes as consequent drainage (though this is apt to lead to some confusion with consequent streams) (p. 172). Where, by contrast, the drainage cuts across the general lines of the structure or 'grain' of the country, and reveals no systematic relationship with this structure, it is referred to as a discordant (or inconsequent) drainage system (fig. 88). Examples of accordant drainage which have developed upon folded, domed and faulted structures will first be described, followed by examples of antecedent and superimposed drainage, which are usually discordant.

Folded structures It was mentioned on p. 57 that synclinal valleys are less common than might be expected. It frequently happens that an original *longitudinal consequent* stream flowing along a syn-

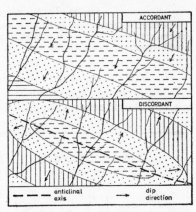

Fig. 88 An example of accordant and discordant drainage

cline will develop tributaries flowing at right-angles down the anticlinal slopes on either side (hence called *transverse consequent* streams). These powerful tributaries cut down rapidly, and may ultimately breach the crests of the anticlines. This is helped by the fact that the axis of an anticline is often structurally weaker, having been subjected to considerable tension, than the trough of a syncline. Ultimately the transverse consequent streams develop tributaries (*longitudinal subsequent* streams) parallel to the original main river in the syncline. These tributaries cut deeply along the crests of the anticlines, forming in-facing escarpments, which gradually recede outward, so reducing the area of the synclinal valley and thus the activity of the original longitudinal consequent river contained therein. In due course the anticlinal valley lies at a level below the synclinal one, the synclinal stream may disappear, and the remains of the syncline survive as a ridge or peak (fig. 89 and plate 14).

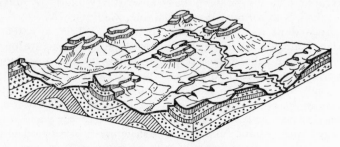

Fig. 89 The formation of synclinal mountains

The block diagram (after E. de Martonne) shows how the development of a river system, attacking the less resistant rocks forming the cores of the anticlines, can produce synclinal mountains (see also fig. 25 and plate 14).

This is by no means the end of the story, as denudation continues, striving to destroy the synclinal ridges and to attain a peneplaned surface. Should renewed uplift take place, a common enough event, the main rivers of the area incise themselves along the lines of the former anticlines. If resistant rocks lie beneath these anticlines, a river may be compelled to shift its course uniclinally (i.e. sideways) (fig. 90), so destroying completely the remnants of the synclinal ridges, and ultimately it may once again occupy a position more or less along the line of the longitudinal consequent stream in the original syncline. This is known as a *resequent stream,* and is a common feature in areas of ancient folding. In the Hampshire Basin the river Ebble, at present occupying a syncline in the Chalk, has at some time migrated from its original and now regained synclinal position

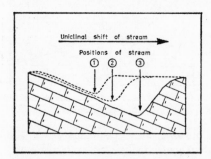

Fig. 90 The uniclinal shifting of a river in its valley

to follow a course along either the neighbouring Bower Chalke anticline or the Vale of Wardour anticline.

Numerous examples of resequent drainage can be seen in Hercynian Europe. In south-western Ireland the rivers occupy the lines of the original synclines, their valleys eroded in weak Carboniferous rocks. Between them anticlinal ridges of resistant Old Red Sandstone rocks project seaward as cliff-bordered promontories (p. 283). Similarly in western Brittany the main river is the Aulne, flowing towards the Rade de Brest in a valley (the Châteaulin basin) carved in the weak shales and slates of a former syncline. To the north and south respectively are the west–east ridges of the Mont d'Arée and the Montagnes Noires. They represent, as it were, the reincarnation of the ancient anticlines formed of tough resistant sandstones and quartzites, further emphasized by the granite batholiths forced into the cores of the anticlines and now revealed (fig. 30).

One of the most remarkable examples of the erosion of a folded area can be seen in the Appalachian Mountains of eastern North America. The 'Ridge and Valley Region' lies between the dissected Allegheny–Cumberland plateau on the west, and the rounded crystalline summits of the Blue Ridge to the east. This is an area of complex Hercynian folding, with large numbers of anticlines, parallel or *en échelon*, successively pitching-out and reappearing. The drainage systems have adjusted themselves to these ancient structural lines by eroding valleys along the anticlinal crests, thus lowering the general level, and then developing other valleys during a subsequent cycle of erosion along the long narrow bands of weak shales and limestones. Between them survive remarkably long, straight ridge-crests of much more resistant sandstones and quartzite-conglomerates, though occasionally broken through by transverse valleys, both wind- and water-gaps (fig. 91). An example of the development of anticlinal valleys on the edge of the Allegheny Plateau is given in fig. 92.

Some most interesting drainage patterns have developed on anticlinoria, anticlines and periclinal folds (p. 48), that is, on elongated domes of various dimensions. The systems developed on these structures are of considerable interest, since they involve the breaching and 'de-roofing' of the domes, the removal of the overlying rocks in the centre and the exposure of the underlying older ones. The Vale of Wardour, for example, was formed by the removal of the Chalk along an anticline, thus exposing older sandstones, clays and limestones, and the Vale is now drained by the Nadder flowing eastward to join the Avon at Salisbury. The breached and excavated Shalbourne and Kingsclere periclines in northern Hampshire

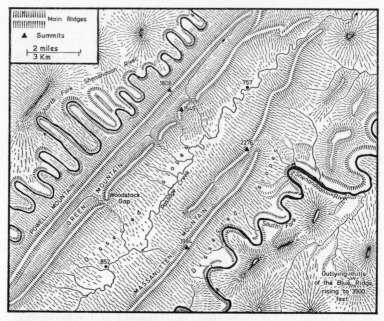

Fig. 91 *A section of the 'Ridge and Valley' region in Virginia in the Appalachian
Mountains*

Based on the United States Department of the Interior, Geological Survey,
1:62,500 Strasbourg Quadrangle. Heights are in feet.

The North and South Forks of the Shenandoah river converge and flow north-
eastward to join the Potomac near Harper's Ferry. Note (i) the deeply incised
meanders, (ii) the abandoned meander in the north-west, (iii) the incredibly
straight and continuous ridge-crests, and (iv) the water-gaps through other ridges.

provide further examples. The Vale of Pewsey is an excavated
anticline, bounded by infacing chalk escarpments; but it contains
no continuous strike stream along its axis, merely the Avon flowing
eastward in its western section, then turning abruptly south through
the chalk rim. In northern France is the Pays de Bray, an anticline
trending from north-west to south-east from which the overlying
Chalk has been removed, and which is now drained by the rivers
Béthune and Epte, flowing along the axis of the anticline.

The classic example of a denuded anticlinorium is the Weald, a
complex elongated dome with a number of subsidiary fold-axes
orientated more or less from west to east. Formerly it was believed
that consequent streams flowing north and south originated on the
initial chalk-covered flanks of the dome; subsequent streams developed along the strike-vales eroded in the softer clays and sandstones,

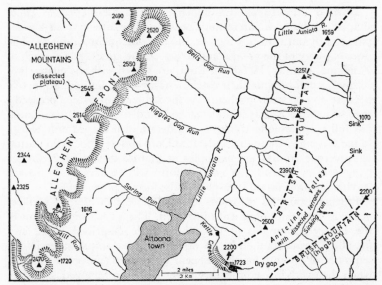

Fig. 92 *The development of anticlinal valleys, hogbacks and a dissected plateau-edge in Pennsylvania*

Based on the United States Department of the Interior, Geological Survey, 1:62,500, Altoona Quadrangle. Heights are in feet.

Sinking Run is a stream along the axis of an anticlinal valley, the Brush Mountain Ridges are the hogbacks on either side of the denuded anticline. The dry gap was formed when Kettle Creek beheaded the Sinking Run, which disappears down a sink-hole.

and the Chalk was removed except for the escarpments on the margins. River-capture has been frequent, wind- and water-gaps are common, and a trellised pattern has thus developed. Considered in detail, however, the problems are much more complex, mainly owing to the fact that the effects of more than one cycle of erosion have been superimposed. During an early cycle a series of west–east streams planed down the anticlines, and may even have formed a peneplain by the end of Miocene times. Then followed a period of marine transgression during the Pliocene, further eroding and smoothing the peneplain, except in the centre and west of the area (the so-called 'Wealden island'). At the end of the Pliocene came renewed uplift, with a further doming, and new rivers developed flowing at right-angles to the planed-down structural lines, northward into the London Basin and southward to the English Channel. At the same time, their strike-tributaries developed west–east vales, the escarpments became pronounced, divides migrated, and numerous river-captures took place.

Domed structures Land-forms of a dome-shaped or cone-shaped profile are interesting because they are responsible for patterns of outward-flowing, or *radial*, consequent drainage. The sides of many of the great composite volcanic cones (p. 74), such as Fujiyama, Rainier, Hood and Etna, are scored with the channels of streams, small because their catchment areas are limited, but active because of their steep gradients. As the sequence of erosion proceeds, the cone shape is drastically attacked, and differential denudation has full scope among the ashes, cinders and lavas. Mount Shasta in the Cascade Mountains and Mont Dore and the Plomb du Cantal in Auvergne (p. 91) are examples of mature dissection. If the sequence is continued, the conical form will be largely destroyed, leaving only a resistant lava-plug in the former pipe (fig. 31).

Further examples of drainage development on domed surfaces occur as the result of igneous intrusion. The laccolithic patterns of the Henry Mountains of Utah have been described (p. 64 and fig. 28). Granite batholiths do not provide an initial stage in the sequence of erosion, for they appear only when 'unroofed', that is, when the overlying sedimentary cover has been removed. But once they begin to stand out as massive granite uplands above the surrounding country, they will sustain a radial drainage pattern. Slopes are gentle though runoff is rapid, the bed-rock is resistant, and the streams are but shallowly incised until they reach the margins, where an increase of gradient may cause greater down-cutting and valley incision. Dartmoor in the South-west Peninsula of England and Limousin in the north-west of the Central Massif of France afford examples.

The Black Hills (fig. 93) of South Dakota and Wyoming were originally an elongated dome, with a central granitic and flanking schistose mass covered with limestone and sandstone strata. The Cheyenne and its tributaries have deeply dissected the dome, exposing its crystalline centre, with the sedimentary cover surviving on the flanks as gently sloping plateaus bounded by steep infacing escarpments. The streams follow circular courses around the dome, conforming to the weaker outcrops; this is known as *annular drainage*. The highest point is the granitic Harney Peak (2207 m, 7242 ft).

Faulted structures Faulting, whether it is a single fault or a broader rift-valley (pp. 44–6), may commonly determine the line of a valley and the direction and pattern of drainage. Streams take advantage of the weaker rocks in the shatter-zone of a fault, and may follow zig-zag courses due to the intersection of faults (p. 39). Waterfalls are commonly associated with steep faulted edges (pp. 150–1).

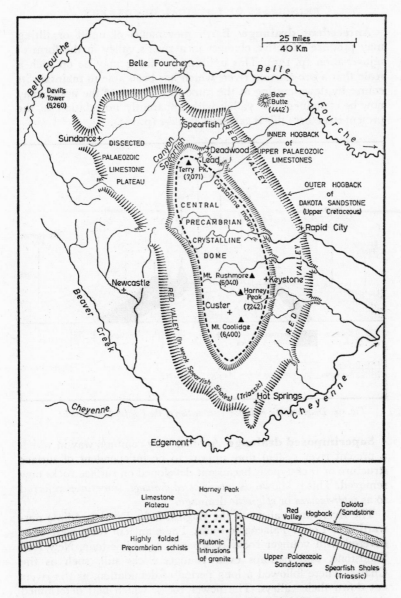

Fig. 93 *The dissected dome of the Black Hills of South Dakota*

Based on a map in a pamphlet issued by the Black Hills State Park Service.
Heights are in feet.

Antecedent drainage Earth movements of uplift or tilting may produce extensive changes in a river's valley in the form of rejuvenation (p. 135). This uplifting or tilting may be on such a scale that a pre-existing river, which has been able to maintain its course by down-cutting at the same rate as that of the uplift, may now be contained in a valley with apparently no relation to the present structure; this is *antecedent drainage* (p. 155).

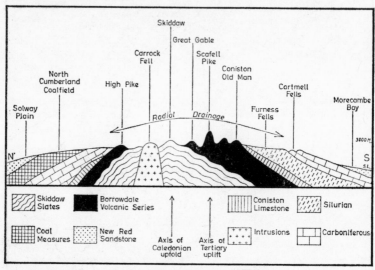

Fig. 94 Diagrammatic geological cross-section of the English Lake District

Superimposed drainage Another very common way in which a present river system may appear to be independent of surface structure or rock-type is because it developed on surface rocks now removed. This is known as *superimposed drainage*, sometimes referred to as *super-induced* or *epigenetic* drainage.

The rivers and valleys of the English Lake District radiate outward over Lower Palaeozoic rocks. These ancient rocks were once covered with younger rocks (Carboniferous Limestone, New Red Sandstone, and perhaps even younger rocks still, such as the Chalk). There followed a long period of denudation, as the rivers cut down their valleys. The newer rocks, which had determined the direction of the drainage, were stripped away from the higher parts of the dome, though some remain as a discontinuous frame around the margins (fig. 94). The rivers maintained their directions over the complex older rocks, crossing the different outcrops at all angles; thus the predominant pattern of the drainage is a legacy of a vanished cover. The Derwent flows to the west; the Ehen, Esk and Duddon to the south-west; the Crake, Leven and Kent to the south; and the Eamont and Lowther to the north-east via the Eden in its well-defined trough between the Lakeland massif and the steep edge of the Pennine escarpment. The dome was elongated along a west–east axis, and as a result the drainage is not truly radial except in the west. A watershed can be traced which curves eastward from Pillar over Great Gable to Esk Hause, Dunmail Raise and Kirkstone Pass to Wasdale Pike in the Shap Fells. North of this line lie Borrowdale, Thirlmere, Ullswater and Haweswater, draining more or less northward; south of it are the Duddon valley, Coniston, Windermere and the Kent valley, draining southward.

The rivers of eastern Glamorgan and Monmouth—the Rhondda, Rhondda Fach, Cynon, Taff, Rhymney, Sirhowey and Ebbw, flow in deep steep-sided valleys in a broadly south-south-easterly direction to the Bristol Channel. They cross almost at right-angles the varied succession of Devonian and Carboniferous rocks of the coalfield syncline and its margins; in places they are flowing against the dip. The drainage was superimposed on gently dipping newer beds which covered these structures; the newer strata, which included the red marls of the New Red Sandstone and the clays and limestones of the Lias, have now been removed, although they still cover the neighbouring Vale of Glamorgan. The river Wye, too, follows a course which would appear to be quite unrelated to the present surface rock-types, as its valley crosses and re-crosses outcrops of Old Red Sandstone, Carboniferous Limestone and Coal Measures.

In the Hampshire Basin rivers such as the Meon, Test and Avon on the mainland (fig. 95) and the Medina on the Isle of Wight have cut across east–west periclinal folds of Alpine age in a strikingly discordant manner. It has been suggested that the original drainage of the area was orientated along west–east, not north–south trends, but that in Pliocene and early Pleistocene times a transgression of the sea to a height of 200 m (650 ft) led to the planing-off by wave action of the periclinal crests and the initiation of the existing river system. The re-orientation of the present consequent streams, which follow closely the slope of the old uplifted sea-floor and largely ignore the important fold-lines, occurred after the regression of the sea in the Pleistocene period.

The Meuse crosses the Ardennes between Charleville and Namur in a valley transverse to the trend of the Hercynian structures (p. 59). The river developed on a pre-existing cover of Cretaceous and Tertiary rocks now almost wholly removed, so that the valley

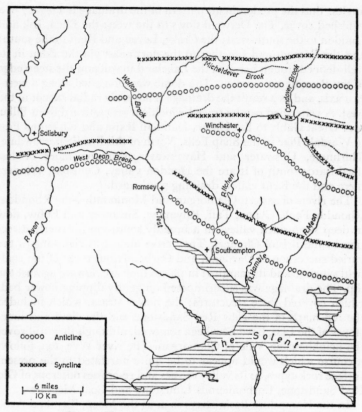

Fig. 95 *The superimposed rivers of the Hampshire Basin*

is cut through Lower Palaeozoic limestones, sandstones and quartz-ites, forming in places a tortuous gorge 90–150 m (300–500 ft) below the surface of the plateau of the High Ardennes. The tributaries of the Meuse—Lesse, Ourthe and Amblève—are similarly super-imposed, but their courses reveal subsequent adaptation to structure; longitudinal sections have developed in less resistant Upper Devonian shales, leaving ridges of more resistant sandstone.

It will be noted, then, that discordance with structure is a very important symptom of a history of drainage superimposition. How-ever, it should not be thought that all discordant rivers are super-imposed, for discordance can arise through antecedence, through glacial diversion (pp. 207–9), and occasionally even through river-capture (p. 176). Nor should it be assumed that superimposed drainage is always discordant. For example, the rivers of the North Downs, such as the Mole and the Medway, appear to follow in a simple and direct fashion the dip of the Chalk and of the Tertiary rocks over large sections of their courses. Yet detailed study of this area has revealed that these rivers, like their counterparts in the Hampshire Basin, did not come into existence until after the retreat of the sea which covered much of the area in Plio-Pleistocene times mentioned above. They too were superimposed from a thin cover, now largely removed by denudation, of marine deposits resting on a surface of marine planation.

Anteposition It may happen that the present course and valley of a river are the results of a combination of superimposition and antecedence, for which the term anteposition has been suggested. This applies to several of the Appalachian rivers, and specifically to the Colorado. This, flowing from its snow-fed sources in the state of Colorado into the Gulf of California, has maintained its course during a long period of uplift, entrenching itself deeply, notably in the Grand Canyon (p. 155, fig. 71, plate 58). But the rocks at the surface of the Canyon are Permian in age; all the Mesozoic and younger rocks have been stripped away as the river superimposed itself upon the landscape. Perhaps as much as 3000 m (10,000 ft) of strata have been removed.

CHAPTER 7

Lakes

A LAKE may be defined as a hollow, more or less extensive, in the earth's surface which contains water. Some of the largest are virtually inland seas (such as Lakes Superior, Victoria and Baikal), and some (notably the Caspian, the Aral and the Dead Seas) are given the name of 'sea'. The Caspian covers an area of 450,000 sq. km (169,000 sq. miles) and Lake Superior is 560 km (350 miles) in length, 260 km (160 miles) in width; its maximum depth is 400 m (1300 ft), its floor 210 m (700 ft) below sea-level, and it could comfortably cover the Benelux countries. At the other end of the scale are tiny sheets of water, such as *tarns* (English Lake District), *llyns* (Wales), *lochans* (Scotland), *étangs* (France) and *stagni* (Italy).

The permanence of a sheet of water depends first on the depth of the hollow, i.e. its storage capacity, and second on the relationship between the amount of water received (either directly from rain on to the water surface or more substantially from inflowing streams which drain the surrounding slopes) and the amount lost (by evaporation, by way of an outflowing stream, and by seepage through the bed). A large shallow sheet of water will rapidly show the effects of a prolonged drought, and when a natural lake is to be used as a reservoir for drinking-water or to supply a head for hydro-electricity generation, it is nearly always necessary to deepen it and increase its capacity by a dam, as at Haweswater and Thirlmere in the English Lake District (fig. 96), and Lochs Laggan and Sloy in the Scottish Highlands. The largest man-made lake is Bratsk, dammed up in the valley of the Angara in the U.S.S.R., followed by Kariba, ponded up behind the Kariba Dam on the Zambezi river.

The lake may be of a seasonal nature; a large, shallow sheet of water may accumulate during a period of rains or following the spring snow-melt on surrounding mountains, but during a succeeding hot dry season the water area will shrink, degenerate into a swamp and perhaps disappear completely. This is especially the case in areas bordering the hot deserts; Lake Chad on the border between Nigeria and the Republic of Chad fluctuates between 10,000 and 50,000 sq. km (6000–30,000 sq. miles) in area. Lake Eyre in South

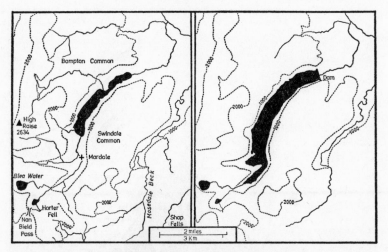

Fig. 96 *Haweswater in the English Lake District*

The left-hand map shows the extent of Haweswater before the building of the dam at the lower end of the valley by the Manchester Corporation. The right-hand map shows its present extent, a substantial man-made modification of the landscape.

Australia (fig. 97) is normally almost wholly dry, its bed consisting of a salt-crust, but in 1950–1, following exceptional rain, a shallow sheet of water 5000 sq. km (3000 sq. miles) in area accumulated, though by 1953 it had dwindled once more into an area of salt-pans. Inland drainage lakes, those which have no river outlet to the sea (i.e. forming *aretic* drainage systems), tend to become increasingly saline, for the salts brought in by streams remain while water is lost by evaporation; thus the Dead Sea has a salinity of 238‰ (p. 348).

Classification of lakes Lakes may be classified according to the mode of origin of the hollows which contain their waters. The most important categories are produced by erosion, by deposition, and by earth movements and volcanic activity. It will be appreciated, however, that some lakes are due to more than one cause working in conjunction; a sheet of water can accumulate in an erosion hollow and be deepened by a natural barrier caused by deposition.

LAKE-HOLLOWS PRODUCED BY EROSION

(i) **Glaciation** The erosive processes of valley-glaciers and ice-sheets (pp. 230–7), creating cirques, U-shaped valleys and the irregular surface of a glaciated lowland, will provide depressions in which water can accumulate. Lakes are one of the most charac-

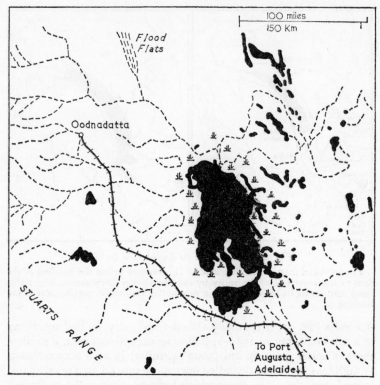

Fig. 97 Lake Eyre

teristic features of a formerly glaciated landscape; much of the charm of the English Lake District (fig. 98), North Wales, the Scottish Highlands, and the Italian Alps (with Lakes Maggiore, Como and the upper part of Garda) is the result of the presence of sheets of water among the mountains. One of the most striking groups of 'finger lakes' are in New York State (fig. 99). The low hill country in which they lie once drained south to the Susquehanna river, but ice-lobes scoured deep valleys in which the lakes now lie and breached the pre-glacial divide, so that they now drain northward, ultimately to Lake Ontario. The map reveals some extraordinary drainage patterns. By contrast to these long narrow lakes, the Lake Plateau of Finland (fig. 90 and plate 65) and parts of the Canadian Shield (plate 76) are lake-studded; many lie in irregular hollows carved by ice in the hard crystalline rocks.

Few glacial lakes are due entirely to erosion, because the effects

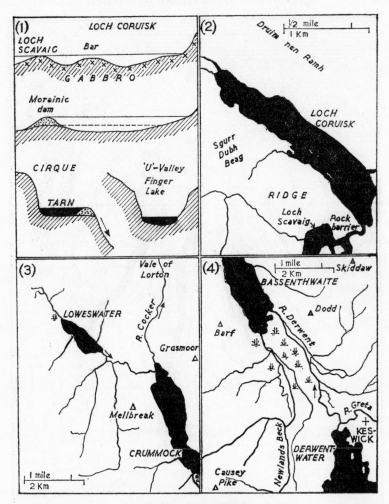

Fig. 98 *Glacial lakes*

The permanent water areas are indicated in black, the area liable to temporary inundation by a marsh symbol.

1. Profiles of a rock-basin lake (Loch Coruisk), a moraine-dammed lake, a cirque lake, and a lake lying in a glaciated valley.

2. Loch Coruisk, a true rock-basin lake, is separated by a rock barrier from the open sea at Loch Scavaig.

3. Loweswater (English Lake District) drains inland, rather than directly to the sea in the west.

4. Bassenthwaite and Derwentwater were originally one lake, which was cut in two by material brought down by streams (plate 64).

193

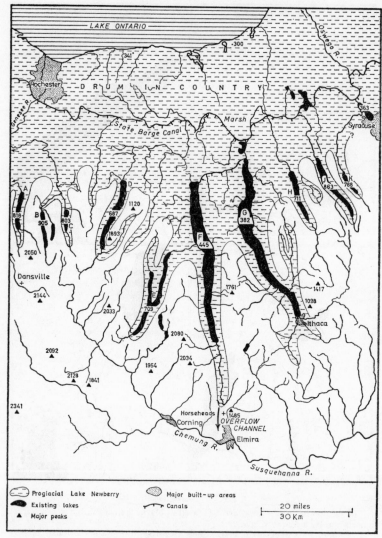

Fig. 99 *The Finger-lakes of New York State*

Based on O. D. von Engeln, *The Finger Lakes Region* (New York, 1961).
The height of the surface of each lake above sea-level is given (in feet).

of erosion and deposition are nearly always superimposed in the
same district. A glacial lake may lie in a U-shaped valley, the lower
part of which is gouged into solid rock, while a morainic dam

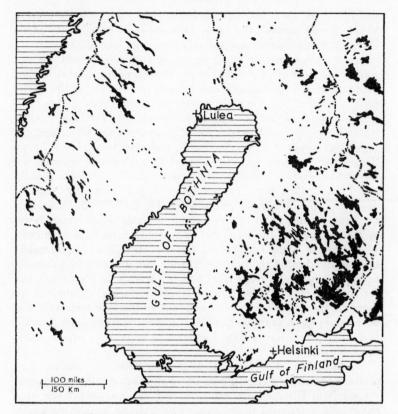

Fig. 100 *The Glint-line lakes of Scandinavia and the Lake Plateau of Finland.*

across the mouth of the valley increases both the depth and area of
the water (plate 79). The surface of Wastwater in the English Lake
District lies at 60 m (200 ft) above sea-level, while its depth is 79 m
(258 ft).

A large number of lakes is to be found in Scandinavia and
Canada across the boundary of the ancient shields (p. 28), known
as the 'Glint-line'. They are partly due to erosion, both by rivers
and glaciers, partly due to the blocking of these eroded valleys by
moraines. In Scandinavia (fig. 100) a series of these long, narrow
lakes is strung out along the valley of each stream flowing in more
or less parallel courses to the Baltic Sea. It seems that the steep edge
of the Baltic ice-sheet lay to the east, so that water was ponded up
between this ice-edge and the highlands to the west. Some of the

melt-water escaped westward, wearing out deep overflow channels. When the ice melted, the rivers resumed their pre-glacial courses toward the east, and the deepest valleys became lake-filled.

Some lakes undoubtedly lie wholly in rock basins, and one can stand on a shore of solid rock, heavily scarred with glacial striations, dropping steeply into clear water (plate 25). Llyn Dulyn lies on the eastern side of the Carnedd massif in North Wales; the rock-precipice of Craig-y-Dulyn drops so sheerly into the lake that the water is 15 m (55 ft) deep only a metre from the edge, and the maximum depth is 58 m (189 ft). On the other hand, Llyn Ogwen lies on the floor of a glacially eroded through-valley between the Carneddau and the Glyders, and is only 3 m (10 ft) deep. Glaslyn and Llyn Llydaw (fig. 114) are 39 m (127 ft) and 58 m (190 ft) deep respectively; it has been estimated that morainic dams account for some 12 and 15 m (40 and 50 ft) of water depth respectively, so that the bulk of the water lies in a true rock basin in each case. Loch Coruisk in the Isle of Skye (fig. 98), about 2·4 km (1·5 miles) in length and some 550 m (1800 ft) wide, lies in a rock basin almost divided in two by a ridge from which a few rocky islets reach the surface. The lake level lies 8 m (26 ft) above the sea; the floor of the more northerly basin is 27 m (90 ft) below the lake surface, of the other basin 38 m (125 ft). Coruisk is separated from Loch Scavaig and the sea by a bar of solid rock, 300 m wide, over which rushes its outlet stream.

(ii) **Solution** The removal of certain rocks in solution may produce hollows which can contain small lakes. Some of the meres ('flashes') of Cheshire are probably due to local subsidence as a result of the removal in solution of underlying beds of rock-salt. Lough Derg in western Ireland (fig. 101) is a large, shallow-water area, the effect of the widening of the bed of the Shannon by solution of Carboniferous Limestone. Solution often forms great underground caverns (p. 121), which near the base of the limestone beds may contain subterranean lakes. The collapse of the limestone roof may produce a long, narrow surface lake; the Lac de Chaillexon, which lies in the Jura Mountains near the boundary between France and Switzerland, is an example, forming a winding sheet of water about 2 km (1·25 miles) in length.

In the limestone region of the Jugoslavian Karst, many lakes occur on the floors of the polja, which are due, in part at least, to solution. Their floors may carry seasonal lakes, most of which degenerate into salt-marsh or disappear altogether in summer. A few, however, like Lake Skadar on the Jugoslav–Albanian frontier, are perennial.

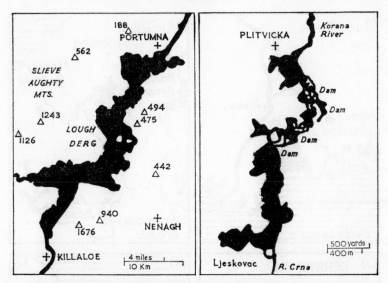

Fig. 101 *Lough Derg in Ireland* (left) *and Lake Plitvice in Jugoslavia* (right)
Lough Derg is a solution lake; Lake Plitvice consists of a series of 'stepped lakes', formed behind the dams built up by the deposition of calcareous tufa.

(iii) **Wind** The deflating action of the wind in desert regions (p. 264) can produce great hollows reaching ground-water. Shallow salt-lakes or swamps may result; those in the Qattara depression in Egypt, and the Shott el Jerid and Shott Melrhir on either side of the Algerian–Tunisian frontier, may be due in part to deflation.

LAKE-HOLLOWS PRODUCED BY DEPOSITION

Lakes ponded up by deposition are known as 'barrier lakes', since their waters are contained, in part at least, by some natural dam, though often of a very short-lived nature. A landslide or avalanche may block a valley, so that a river's course is checked and the water ponded up; such a dam is usually unstable, and when the pressure of water ultimately sweeps it aside, disastrous floods may surge down the valley. This happened in north-western Wyoming in 1925, when a huge rock-slide dammed the Gros Ventre river. The following year this natural dam collapsed, and the resultant flood destroyed the town of Kelly. In 1959 another vast slide (p. 34 and fig. 11) blocked the Madison valley in Montana, ponding up Earthquake Lake, 5 km (3 miles) in length. The small, attractive Mirror Lake in the Upper Yosemite Valley in California is held up behind a rock-avalanche.

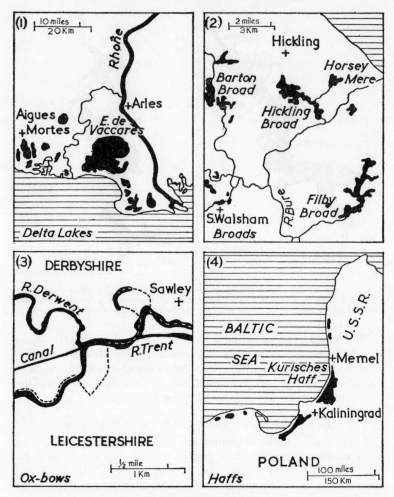

Fig. 102 Deposition lakes

Note. Recent work on the Broads has shown that their origin may be ascribed to a variety of contributory causes—estuarine deposition, the growth of peat, post-glacial submergence along the coast, and, especially, artificial causes such as peat-cutting and the construction of embankments. In (3) it can be seen that the county boundary between Derbyshire and Leicestershire was drawn before the ox-bow had been formed.

Ox-bows (cut-offs) in the flood-plain of a slowly meandering river (p. 157 and plate 60), wide shallow sheets of water produced by the blocking of an estuary or by deposition in a delta (the Etang

64 Derwentwater, in the English Lake District
In the foreground is the U-shaped Borrowdale valley, in the right background
the Skiddaw range, and in the left background Bassenthwaite.
(*G. P. Abraham*)

65 The Lake Plateau of Finland, near Punkaharju.
(*Paul Popper*)

66 The Gemindener Maar in the Eifel, West Germany
(*Plan und Karte G.m.b.H., Münster*)

67 The Märjelen See, ponded by the Aletsch Glacier, Switzerland
(*Paul Popper*

de Vaccarès at the mouth of the Rhône and the Grand and Salvador lakes in the Mississippi delta) provide examples on various scales (fig. 102). Brackish coastal lakes are enclosed by bars (p. 293) built up by coastal deposition, such as the Haffe of the Baltic Sea and the Florida Lagoons, or by wind-formed sand-dunes (p. 301), as along the Landes coast to the south of the Gironde estuary.

Morainic deposits As already stated, few glacial lakes are due wholly to erosion. More commonly, a glacial valley or a cirque is blocked in part by a crescentic terminal moraine (p. 238).

In undulating lowland country, covered with uneven deposits of glacial debris, thousands of oddly shaped lakes, ranging from pools to extensive sheets, may be found. The lake-plateaus of Mecklenburg in East Germany, and of Pomerania and East Prussia (now in Poland) are striking examples. A special variety of lake due to glacial deposition is known as a *kettle-hole*, where a large ice-block was buried in drift (p. 245), and so left upon melting a depression which now contains water or may be filled with peat; there are some good examples among the glacial deposits around Brampton to the north-east of Carlisle, in the neighbourhood of Lancaster, and in the Vale of Pickering. Some of the meres of Cheshire are undoubtedly due to this cause. Numerous examples occur in the Kettle Moraine country of Wisconsin, from which the name was derived (p. 244).

Ice dams Another unusual type of barrier lake is where ice forms the barrier, a phenomenon probably very common during the maximum periods of the Quaternary glaciation. The Märjelen See lies in an angle between the Aletsch Glacier and its rock walls (fig. 109 and plate 67), while a series of glacial lakes is ponded up in tributary valleys around the edge of the Vatnajökull ice-sheet in Iceland (fig. 103). At the margin of the Greenland ice-cap numerous fjords are glacier-dammed to form ribbon lakes 15–30 km (10–20 miles) in length, and over fifty named ice-dammed lakes are concentrated along the coastal strip of southern Alaska, most within 80 km (50 miles) of the ocean.

Vegetation dams One category of small depositional lake is due to vegetation. The blocking of slowly flowing streams near their mouths to form meres may be accelerated by the growth of sedges, rushes and other aquatic plants. Such are the *gooren* and *vennen* in the southern Netherlands and north-eastern Belgium. The 'mosses' and moors of upland areas such as the Pennines and the heathlands of northern Germany, notably the Lüneburg Heath, have numerous shallow lakes and pools lying on the irregular surface of the thick layer of peat which covers these moors, separated by hummocks or 'islands' of peat.

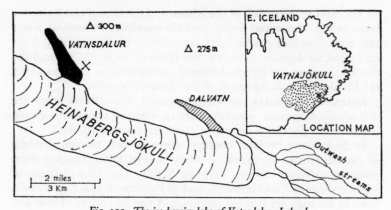

Fig. 103 *The ice-barrier lake of Vatnsdalur, Iceland*
After S. Thorarinsson and B. B. Roberts.
Vatnsdalur is a small lake (in black) ponded against the edge of an ice-tongue;
Dalvatn (with diagonal shading) was a similar lake, now filled with sediment.

The hills of the English Lake District and North Wales show numerous examples of these 'peat lakes', especially on cols, such as Three Tarns between Bowfell and Crinkle Crags above Langdale. The hollows in the peat are partly caused by uneven growth, partly by wind erosion of the peat during periods of drought.

Calcareous dams Some unusual deposition lakes in the Jugoslavian Karst have been formed by the growth of a barrier of calcareous material across a river; Lake Plitvice in central Jugoslavia is an example (fig. 101).

LAKE-HOLLOWS DUE TO EARTH MOVEMENTS
AND VULCANICITY

Hollows in the earth's crust are produced on a large scale by warping or fracturing, and the depressions so formed may contain either salt- or fresh-water lakes. Lough Neagh lies within a 'sagging' on the Antrim basalt plateau (fig. 34), while Victoria in East Africa, Titicaca on the high intermont Andean plateau in South America, and the Caspian Sea are other major examples of tectonic lakes. Where down-faulted troughs occur among the mountains, hollows capable of holding lakes are formed; the Great Salt Lake and other smaller lakes in Utah (known as *playas* or where highly saline *salinas*) are fragments of the once huge Lake Bonneville, which must have covered 50,000 sq. km (20,000 sq. miles) in the down-faulted Great Basin (fig. 106). The basin containing Lake Baikal in central Asia is defined by a whole series of faults, and the

Tarim Basin, enclosed by the inward-facing fault-line scarps of the Altyn Tagh on the south and the Tien Shan on the north, once occupied by a vast lake, is now covered with sands, gravels and the Lob Nor marshes.

Rift-valleys Rift-valleys (p. 44) afford obvious sites for lakes in their steep-sided linear depressions. The Jordan–East African rift-valley is a major example, containing the Dead Sea, Lake Tanganyika, Lake Nyasa (Malawi) and a series of smaller lakes shown in fig. 15. The Dead Sea occupies the deepest part of the northern section of this trough; its surface is 394 m (1294 ft) below sea-level and its greatest depth 400 m (1300 ft). It exemplifies the elongated character of a rift-valley lake, for it is 88 km (55 miles) in length and only about 16 km (10 miles) in width.

Crater-lakes The craters of dormant or extinct volcanoes may form sites for circular lakes (fig. 31, p. 75). These usually occur where an explosion crater has been hollowed out of solid rock, rather than those built up by ash, which cannot be water-holding. Examples include the Maaren in the Eifel (plate 66), Crater Lake in Oregon (fig. 104), Öskjuvatn in Iceland (fig. 31), Lakes Avernus and Bolsena in central Italy, and Keloed and Kawah Idjen in Java. Keloed has erupted six times since 1811, and on each occasion the lake water was ejected, causing great damage by the floods which poured down its flanks; in 1905 and 1907 dams were built across breaches in the crater rim in an attempt to check future flooding, but these were swept away in the major eruption of 1919. Finally, tunnels were driven through the rim to drain off the greater part of the water in an effort to prevent a recurrence of these disasters.

One of the largest crater-lakes in the world is Lake Toba, which lies in a caldera 1900 sq. km (750 sq. miles) in area among the Batak Highlands of northern Sumatra. Steep walls 6000 m (2000 ft) high encircle the lake, except for its river outlet in the south-east, where the Soengai Aasahan plunges through a series of gorges and over a waterfall 135 m (443 ft) high.

A small crater-lake, 550 m (1800 ft) in diameter, lies under the rim of Ruapehu, a volcanic peak rising to 2798 m (9175 ft), the highest point in the North Island of New Zealand. This lake is warmed from below by volcanic activity and rarely freezes. When Ruapehu erupted in 1945, a mass of lava was ejected, forming first an island, then filling in the crater. This was blown out by a later eruption, forming a new lake 300 m (1000 ft) deep, but this was soon partially filled with volcanic material, leaving a depth of water of about 80 m (260 ft). On Christmas Eve 1953 a natural dam of ash

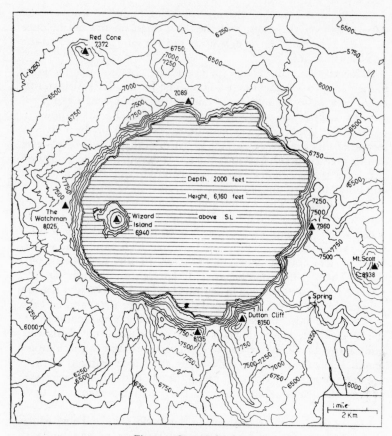

Fig. 104 *Crater Lake, Oregon*

Based on the United States Department of the Interior, Geological Survey,
1:62,500, Crater Lake National Park Quadrangle. Heights are in feet.

Crater Lake, about 600 m (2000 ft) deep, lies in the caldera of an ancient volcano, the rim of which rises in places to over 2400 m (8000 ft). The peak probably
once exceeded 4200 m (14,000 ft), on the basis of extrapolating (projecting upward) the slopes below the existing rim, together with the evidence of deeply
carved, glacially eroded valleys, which are now cut off at the rim by distinct
notches. These glaciers must have been on a considerable scale (similar to those on
Mount Rainier, as shown on fig. 110), and would imply a considerable and lofty
area of accumulation. This peak is known to geologists as Mount Mazama. The
crater was probably formed by cauldron subsidence (p. 77) about 6000 years
ago, dated by radiocarbon examination (p. 14) of wood from trees killed by the
eruption. It has been calculated that of an estimated 70 cubic km (17 cubic
miles) of material which disappeared from Mount Mazama, only 8 cubic km (2
cubic miles) are spread around; the rest collapsed within the underlying 'chamber'.

and snow resting on the solid lava rim of the crater collapsed, releasing much water, which rushed down to the Whangaehu river with a load of mud and boulders. Unhappily, the flood weakened a bridge over the river 40 km (25 miles) away, and an express train crashed through to the river-bed, with a death roll of 151.

Occasionally a lava flow may block a valley and so form a lake basin. One such has made a barrier across the Jordan valley, damming up the Sea of Galilee, and another in the East African rift-valley is responsible for Lake Kivu. A smaller example is Lac d'Aydat, in the Auvergne area of the Central Massif of France. Lake Van in eastern Turkey, at a height of about 1700 m (5600 ft) above the sea, was formed by outpourings of lava from the large volcano of Nemrut, which has a caldera 10 km (6 miles) in diameter. This lava blocked the valley of a headstream of the Euphrates, so forming a lake.

FORMER LAKES

It will be realized that lakes are a very temporary and ephemeral feature of the landscape. They may be filled up by river-borne alluvium, especially where in mountainous districts heavily laden streams rapidly build lacustrine deltas (p. 166), and so eventually fill the lake (plate 63). Bassenthwaite and Derwentwater are separated by an alluvial flat 6 km (4 miles) across (fig. 98), deposited by the river Greta and the Newlands Beck, and now above normal lake level because of the lowering of the lip at the northern end of Bassenthwaite. In times of heavy rain the flat may be flooded and the valley returns to its former appearance. The Derwent is pushing out its delta and has filled up the southern part of Derwentwater in Borrowdale. Most of the other English Lakes show similar indications; the head of each is now a marshy 'bottom'.

In several valleys this filling-up process has resulted in the complete disappearance of the lakes. The Kentmere valley, in the English Lake District to the north-east of Windermere, formerly contained two lakes; the surface of the more northerly lay at about 226 m (740 ft) O.D., that of the southern at about 158 m (520 ft) O.D. The northern lake was contained on the south by a spur of solid rock, with a morainic dam of boulders (fig. 105). This lake had an over-flow channel to the west of Rook Howe, but the river Kent, originally forming a second minor overflow in the south-eastern corner, cut down through the morainic dam in a gorge into the bed-rock to form the attractive falls of the Jumb. This lowering of the outlet, together with extensive sedimentation, converted the lake into a broad level valley-floor, crossed by drainage channels and by the

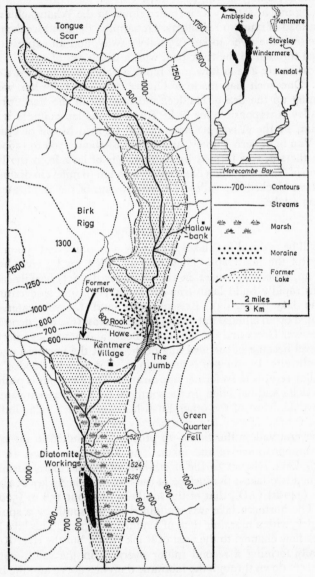

Fig. 105 *Former lakes in the Kentmere valley*
The gorge through the morainic barrier separating the former upper and lower lakes is indicated by hachures. (Heights are given in feet.)

Kent within its embankments. The lower lake received relatively little sediment, as a result of the filtering effect of the upper lake. There was extensive growth and accumulation of diatoms (p. 8), so that the lake floor was covered with diatomaceous earths. The lower lake had become a shallow mere by the beginning of the nineteenth century, and in 1840 the Kent was deepened artificially and the lake drained, though there are still extensive tracts of marsh. A company is working the diatomaceous earths and has excavated a considerable hollow which is now water filled; they are, in fact, resurrecting the lake.

On a larger scale the eastern wing of Lake Geneva is being slowly filled in by the Rhône (fig. 80). Various estimates by Swiss and French scientists put the life of the lake at between forty and fifty thousand years if the present rates of sedimentation are maintained.

Lakes which were once of considerable size, but have now either disappeared completely or are represented only by fragments, have nevertheless frequently left a striking impress on the landscape (fig. 106). Some may have disappeared due to climatic changes, such as increasing aridity, as happened in the Great Basin lakes of North America. Many were ponded up along the edge of an ice-mass towards the end of the Quaternary glaciation; these are referred to as *proglacial lakes*. Thus vast lakes, of which the present Great Lakes are the remnants, lay between the continental water-parting and the ice-sheet edge. 'Lake Agassiz', of which Lakes Winnipeg, Winnipegosis and Manitoba are the remnants, lay farther west, probably covering 260,000 sq. km (100,000 sq. miles).

Possibly the great weight of the continental ice-sheet at its maximum caused an isostatic 'down-warping' of the crust (p. 23), and as the ice-sheet melted lakes accumulated in these marginal depressed areas. Glacial scouring and morainic damming also contributed to their formation. These lakes drained at first southward to the proto-Mississippi by way of what is now the valley of the Illinois river, then as the ice receded northward they drained through the Hudson–Mohawk valley, and ultimately the St Lawrence outlet was uncovered, when the lakes attained more or less their present outlines. The progressive melting of the Arctic ice-cap may have caused a gradual isostatic up-tilting in northern Canada, so that it is possible that should this trend continue in two millennia or less a southern outlet may again be used.

The ice-sheets over parts of Britain ponded back rivers and formed a number of proglacial lakes. In the West Midlands a lake was created in the Shrewsbury area between the ice-front and the

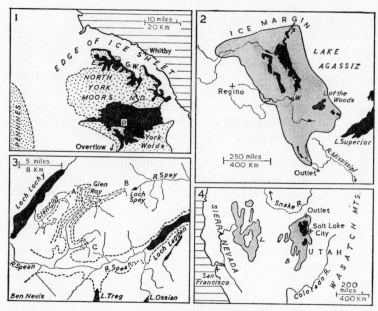

Fig. 106 *Former lakes*

1. During one stage of the Quaternary glaciation, various ice-sheets surrounded the hill-country of the Cleveland Hills and York Moors (stippled), which protruded through the ice. Between these hills and the ice-edge, several lakes (in black) were ponded—'Eskdale' (**E**), 'Glaisdale' (**G**) and 'Wheeldale' (**W**) to the north of the Moors, and 'Pickering' (**P**), between the Moors and the Wolds. The northern lakes discharged southward, the water cutting the striking overflow channel now known as Newton Dale (**N.D.**) (fig 107). Lake 'Pickering' drained southward to the Vale of York by means of an overflow channel near Kirkham Abbey. The Derwent, although rising only a few km from the North Sea, still drains southward, ultimately to the Humber, through this channel.

2. During the retreat of the continental ice-sheet, an immense lake was dammed up in central Canada between the edge of the ice-sheet to the north and the continental divide to the south. To this lake the name Agassiz was given, after the famous glaciologist. Its overflow channel lay to the south, by way of what is now the Minnesota valley. As the ice retreated and other overflow channels to the north-east were developed, much fine sediment was deposited on the former lake floor, now forming the wheat-lands of the 'Till Plains'.

3. The Parallel Roads of Glen Roy are clearly defined shore-terraces, 12–15 m (40–50 ft) wide (shown by pecked lines). Ice spreading northward from the slopes of Ben Nevis blocked the south-facing exits of Glens Gloy and Roy, ponding up lakes in the two glens. The former lake drained through the col **A** into Lake Roy, and that in turn overflowed at **B** into Glen Spey. Later the ice gradually withdrew and so uncovered overflow **C**, and the waters then flowed eastward along Glen Spean to find an outlet over another col east of Loch Laggan through Strath Mashie. Ultimately the ice retreated sufficiently to uncover the mouths of all the glens and the waters drained away directly through the Spean and Lochy rivers to the open sea at Loch Linnhe, as happens today.

4. The Great Basin lies in western U.S.A. between the Wasatch Mountains and the Sierra Nevada; part of the basin is an area of inland drainage. It is an arid

206

Shropshire Hills, to which the name 'Lake Lapworth' has been given; its overflow to the south cut a gorge at Ironbridge. Other lakes were formed to the east of the Pennines, ponded between these uplands and the ice-sheet over the east coast and the North Sea. In the Vale of York lay the extensive 'Lake Humber' and farther to the south 'Lake Fenland', probably linked by an overflow channel through the Ancaster Gap near Sleaford, and their combined overflow escaped eastward to the ice-free southern North Sea along the line of the broad Waveney valley. Of particular interest are the proglacial lakes on the margins of the North York Moors (fig. 106).

Glacial lakes were ponded up in many of the Yorkshire dales by morainic dams, and survived long after the disappearance of the ice. In these lakes much material was deposited—coarse in summer when the streams were active and vigorous, fine in winter when there was less melt-water, so producing laminated clays. The lakes were slowly infilled, and at the same time their outlets were lowered as the rivers cut down into the morainic dams. Today Airedale contains an extensive lake-flat of at least 30 m (100 ft) thickness of these soft lacustrine deposits, whose margins are skirted by the railways and roads to avoid flooding and engineering difficulties. Similarly, a large lake-flat can be seen in the Ribble valley below Settle, across which the railway runs on an embankment to Giggleswick station.

Former lake beaches are sometimes visible, representing different levels of the lakes at stages in their history; the most striking example is the Parallel Roads of Glen Roy in Inverness-shire, and a series of former lake beaches exists in Utah around the Great Salt Lake, marking the various levels of former lake Bonneville. Several outlets were used by the ponded water at different stages as the ice-sheets withdrew. The most striking results are steep-sided valleys eroded as overflow channels or 'spillways', often cutting through pre-glacial watersheds. One of the most notable examples is Newton Dale (figs. 106, 107). Water was held up in Lakes 'Eskdale', 'Glais-

---

region, with numerous salt-lakes and salt-flats; those of Bonneville (west of Salt Lake City) are so extensive and level that they have been frequently used for attempting world speed records by cars. Two lakes, Bonneville (**B**) and Lahontan (**L**), came into existence during the Quaternary glaciation. Their oscillations in extent are shown by wave-cut terraces, deltas and shingle spits at various levels; the highest deposits occur at 300 m (1000 ft) above the present Great Salt Lake, for at this level the water found an outlet to the north through Red Rock Pass into the Snake River valley. Since this maximum extent of the lakes, the climate has become increasingly arid, and the Great Salt Lake and smaller salt-swamps now represent the remnants of the former vast lakes.

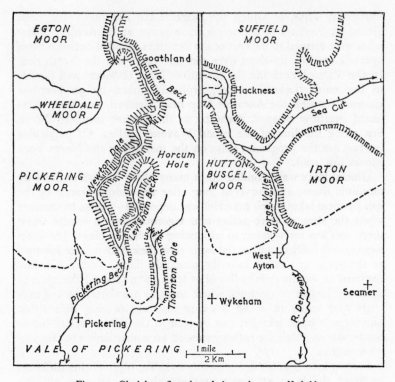

Fig. 107 *Glacial overflow channels in north-eastern Yorkshire*

The pecked line defines the edge of the uplands. Hachures are used to emphasize particularly steep-sided parts of the valleys.

The left-hand example shows Newton Dale. It is a through-valley, but there is a low watershed two miles south of Goathland; the Eller beck flows northward to join the Esk, the Pickering beck southward to join the Derwent.

The right-hand example shows Forge Valley, through which the Derwent flows southward. The river formerly flowed east to reach the sea just north of Scarborough, but when the ice-sheet created a barrier, it was forced to turn southward. It maintained this course in post-glacial times. The Sea Cut, an artificial channel made to take off some two-thirds of the water of the upper Derwent to relieve flooding in the Vale of Pickering to the south and south-west, more or less follows the pre-glacial valley.

dale' and 'Wheeldale' at a surface height of 230 m (750 ft) O.D. As the overflow to the south remained the principal exit for a considerable time, it attained unusual size.

Some of these overflow channels are now dry, but others still used by streams have produced, when one considers the general relief, some curious permanent effects in post-glacial drainage (fig. 107). Thus the river Derwent, rising only a few miles from the North Sea, to which it flowed directly in pre-glacial times, now follows a great curve westward through the Vale of Pickering and then southward for 160 km (100 miles) to join the Ouse and so the Humber. Again, the upper Severn in pre-glacial times probably flowed northward to the Dee estuary, but when this was blocked by ice the overflow from the ponded 'Lake Lapworth' (mentioned above) cut the Ironbridge gorge southward through the former watershed; this outlet is still utilized, and so the Severn makes a sweeping curve through the Midlands, almost reversing the direction of the upper river. These glacial overflow channels are to be found in many upland areas, and the reconstruction of past events in the light of these present features is a fascinating problem to the student of land-forms.

CHAPTER 8

Glaciation

SNOW

The snow-line Snow may fall in any latitude—even on the Equator in the mountains of East Africa and in the Andes—but rarely at sea-level within about 30° North and South. The snow-line indicates the lowest edge of a more or less continuous snow-cover. Its position depends both on climatic factors (p. 449) and on the nature of the terrain, that is, the presence of gentle slopes upon which snow can lie or of basin-shaped hollows sheltered for most of the year from the sun and wind. Many of the Alpine peaks stand out boldly from the swathing snow-fields round their bases because their faces and ridges are too steep for snow to accumulate.

The *permanent snow-line* represents the level at which wastage of snow due to melting in summer fails to remove the winter accumulation. Occasionally this may be higher than usual, as in the Alps in 1949, when an exceptionally small winter snowfall was followed by a hot, dry summer. The height of the permanent snow-line varies with latitude, altitude and aspect; at the Poles it lies at sea-level, in southern Greenland at 600 m (2000 ft), in Norway at 1200–1500 m (4–5000 ft), in the Alps at about 2700 m (9000 ft), and in East Africa at 5000 m (16,000 ft). These heights are very approximate, and differ considerably according to local physical conditions. The permanent snow-line on the south side of the Himalayas is at 5000 m (16,000 ft), while on the north it is 1200 m (4000 ft) higher, because the monsoon air-streams bring a much heavier precipitation to the southern slopes. Continental interiors, such as northern Siberia and northern Canada, which experience extremely low winter temperatures, have usually only small falls of powder blizzard-snow, soon disappearing in spring.

The *winter* (or *temporary*) *snow-line* in middle latitudes varies markedly from place to place and from year to year, particularly on oceanic margins, as in the British Isles. In the Highlands of Scotland above 900 m (3000 ft) snow lies on more than 80 days in the year on an average. The summit of Ben Nevis is only about 100 m (300 ft) below the estimated permanent snow-line, and the mountain usually carries heavy snow in its north-facing corries; one of these patches is almost permanent, lying for years on end and disappearing only during an exceptionally warm August. On the summit the amount varies enormously, but it always vanishes during May; in 1885 there was a record maximum winter depth of 361 cm (142 in) on the flat summit plateau. In Snowdonia and the English Lake District exceptionally snowy winters (1961, 1969) occur among others with only sporadic falls or none at all; such unreliability is the despair of British skiers.

Avalanches At and above the permanent snow-line, winter snowfall equals or exceeds the summer loss, but the snow depth does not increase indefinitely. On steep slopes the snow is removed in large masses by gravity, producing avalanches which may occur either in winter, when fresh non-coherent snow slides off the older snow, or in spring, when wet partially thawed masses of enormous size fall down the valley slopes. Sometimes old snow may become consolidated and compacted into slabs, by partial thawing and re-freezing or by the effects of the wind (hence *'wind-slab'*), resulting in an *en masse* movement of the whole slab; this is particularly dangerous to skiers. Avalanches can be very destructive; during the spring of 1951 a sudden thaw after heavy winter snowfall caused widespread avalanches in Switzerland and Austria, with loss of life and destruction of property. An avalanche is always a potential danger in the high mountains, both to skiers and climbers; even the expert may be involved, as in July 1964, when fourteen experienced guides were swept to their deaths by an avalanche down the face of the Aiguille Verte in the French Alps. One of the largest falls ever reported occurred in January 1962 in Peru, when a mass of snow and ice estimated to total 2·1 million cubic m (2·8 million cubic yds) fell from Mount Huascaron in the Peruvian Andes on to the village of Ranrahirca, with a heavy loss of life.

In such ranges as the Alps the probable avalanche tracks are mostly known, and potentially dangerous areas are avoided. Villages, roads and railways are carefully sited, natural breaks such as rock spurs and thick pine-woods are utilized, while steel sheds and galleries protect roads and railways at critical points.

FIRN AND GLACIER ICE

Firn Where snow is able to accumulate in a hollow or basin (a process known as *alimentation*), both from direct fall and from avalanches down the surrounding slopes, it is compressed by the addition of successive layers and is gradually changed into a more compact form. Air is retained between the individual snow particles, forming a mass of whitish granular ice. During surface melting on summer days, water percolates into the mass and refreezes at night. *Sublimation* also assists, whereby molecules of water vapour escape from the snow-flakes and reattach themselves, so that the crystalline granules become progressively more tightly packed. A section through a snow-field will usually show some degree of stratification, in which it is possible to distinguish each year's contribution of snow. This mass is known as *Firn* (in German) or as *névé* (in French), though the former is now adopted by glaciologists, partly because other terms can be derived from it; these include *firnification* (the creation of the firn), the *firn equilibrium line* where supply of new snow exactly balances wastage, the *firn-field*, the actual mass of accumulated snow, and the *firn-basin* in which it accumulates.

The density of firn varies; new snow has a density of less than 0·1; old more closely packed snow near the surface of 0·3; firn, when the pore spaces have been sealed off, 0·55 (the arbitrary lower limit) to 0·82.

Glacier ice If a basin of accumulation is sufficiently large, and if the seasonal addition of firn is adequate, a tongue-like mass of ice may move out and down from the edge of the snow-field. This varies considerably in its physical properties and appearance. Firn has intercommunicating air vesicles and is thus porous, while ice has the remaining air vesicles non-intercommunicating and is thus impervious. Layers of clear blue or green 'glassy' ice, which have scarcely any contained air, often alternate with the white granular ice derived from firn as a result of the pressure of the overlying layers. In glacier ice the density attains about 0·9, approaching the figure of 0·917 for pure ice.

Glaciers move slowly downhill from their basins of accumulation, following the line of least resistance, usually a pre-glacial river valley. The glacier is said to 'flow', a term which covers the complex physical processes involved; in other words, it acts as a *rheid* (p. 35). In Greenland a rate of flow exceeding 30 m (100 ft) per day in summer has been measured and 18 m (60 ft) per day is common, while in the Alps the rate seems to be about 0·3 m (1 ft) or less per day. The record movement measured during one year was 1800 m (5610 ft) by the Storström Glacier in Greenland; this high figure is the result of the huge ice-reservoir supplying the glacier, its enormous thickness, weight and pressure (estimated to be 320,000 kg per sq m for each 300 m of thickness), and the confined channel of its exit between the rock peaks near the coast. Occasional sudden short-lived glacier advances produce some remarkable velocities; for example, the Black Rapids Glacier in Alaska attained 76 m (250 ft) a day for a short period in 1937. The cause of these short-term variations is not understood; it may possibly be related to general meteorological conditions.

It must be emphasized that these rates refer to surface ice, and there is a great deal of differential movement within the mass. The sides of a glacier move less rapidly than the middle, sometimes at only half the rate. On a slope there appears to be an oblique movement, toward the margins; in the upper part the ice descends towards the bed of the glacier, while nearer the tongue the movement seems to be upward. It varies too in velocity with the gradient of the bed and the width of the valley in which it is moving. The ice slows down over a gentle stretch, which causes an increase in thickness, caused by the accumulation from ice higher up which is moving more rapidly. Conversely, a steep section will cause an acceleration, with a thinning of the bulk of the ice, but at the foot of this section the thickness will build up again. These facts have important results on glacial erosion. One other motion within the ice is a downstream passage of a bulge of ice at a considerably greater speed than the normal glacier flow, graphically called a *surge*. Such a surge has been watched for 15 years moving down one of the glaciers on the flanks of Mount Rainier in the Cascade Mountains, and during 1966 surges were closely followed on a dozen Alaskan glaciers.

It is difficult to understand the physical mechanism of glacier flow, and much active research is in progress in the Alps, Norway and North America, particularly in Alaska and Greenland, while an increasing amount of work is being carried out in Antarctica. The basic research requirement is the sinking of a borehole, which is lined with plastic or aluminium; the distortion of this vertical hole at various depths gives indications of the rate and nature of the ice-movement. Other research is carried out in low temperature laboratories. The point is that the glacier-mass seems to behave like a viscous liquid, yet it is a crystalline solid. Its study involves a varied combination of physical change, resulting from differences in temperature and pressure, crystalline structure, gravity movement, molecular change among the particles of ice, cracking and shearing. In recent years there has been investigation of the temperature changes within glaciers, which can be broadly divided into 'warm' and 'cold' categories. It may seem a contradiction in terms to talk of a 'warm glacier', but there is a very great difference in the physical character of such a type, where the ice-mass is at or near 0° C (obviously it cannot be warmer than that), and a 'cold glacier' where the temperature may be —20° or —30° C. A 'warm glacier' is mainly warmed by the percolation of melt-water produced by conduction heat at the surface (p. 388); as the water passes down through the mass it is cooled by contact, giving up latent heat as it refreezes and so raising the internal temperature of the entire mass. Thus a 'warm glacier' may approach 0° C throughout its mass in summer, though in winter superficial cooling will produce a very much colder crust, while at depth the temperature may remain near 0° C. A 'cold glacier' remains at a very low temperature throughout the year, with no surface melting, as in parts of Greenland and Antarctica. H. W. Ahlman has proposed the terms *temperate glacier* and *polar glacier* for the two types. One further complexity is that many glaciers are of the cold variety at their heads and upper portions, warm near their tongues.

With these complexities in mind, the movement may be regarded as either *gravity flow*, involving the processes of regelation, intergranular translation, plastic deformation and laminar flow, or *extrusion flow*.

Regelation is the result of pressure within the ice-mass, which produces a minute local fall in the melting-point of ice, and so liberates molecules of water. Not only do these molecules move to a point where pressure is less before recrystallization takes place, but they also form a lubricating film which helps the ice-grains to move relative to one another. Thus there is a gradual movement of and within the ice-mass in a downhill direction. Associated with this is the concept of *intergranular translation*, which believes that ice-grains behave as mechanical units (such as in a mass of lead shot), and slide over each other, lubricated by an intergranular film containing chlorides and other salts. A rapidly flowing glacier has more mobile molecules within its crystalline mass, either because the supply of firn from above is greater or because the slope is steep, thus increasing the strains within the mass. The narrowing of the rock-bed enclosing the glacier increases the rate of flow for the same reason.

Little is known yet about the *plastic deformation* of ice, but it has been ascertained by exploration beneath glaciers that under certain conditions the ice becomes plastic, as a result of intermolecular and intergranular movement. This slow deformation due to internal stress is sometimes known as *creep*.

Laminar or *lamellar flow* involves a movement produced by a definite thrust along the line of slope, caused by the solid character of the ice and by its weight higher up. There is much slipping, shearing and gliding along fracture planes, both *shear-planes* and *glide-planes*, on a major scale (so producing complex ice-falls) and on a minor scale

68 Mount Erebus (3790 m, 12,450 ft), the world's most southerly volcano, and
the edge of the Antarctic ice-sheet
(*Paul Popper*)

69 The Recherche Glacier, Spitsbergen
(*Paul Popper*

70 The Aletsch Glacier, Switzerland
This photograph should be carefully compared with Fig. 98.
(Swissair-Photo A.G.)

71 The Svartisen Glacier, northern Norway
This glacier moves down from an extensive ice-cap to the Svartisen Lake.
(Widerøs Flyveselskap og Polarfly)

among the individual ice-grains. Movement over the rock-floor is known as *basal slip*. This thrust may even push the 'snout' (the termination of a glacier) for a short distance uphill. Within a cirque-glacier (p. 226), and possibly also at a steep ice-fall, the slipping may be rotational in character.

The surface of a glacier usually forms a crust, which has no flow-movement itself, but which is carried along above the deeper moving ice.

It is clear that each of these processes of glacier-flow has, in the words of J. K. Charlesworth, '. . . a grain of truth; it is valid in part and in certain circumstances'. Each involves to some extent the force of gravity, and so they are grouped as gravity flow, the joint result of internal deformation, slide and creep.

Extrusion flow The large continental ice-masses must have had very slight gradients over the extensive lowlands which they formerly covered. The gradient between Fennoscandia and the southern Netherlands, little enough as it is in a distance of more than 1600 km (1000 miles), may well have been lessened still further by isostatic depression due to the weight of ice in the north. The suggestion has been made that the ice-sheets accumulated to such a thickness that the ice moved as a result of its own weight by extrusion. The thickness and pressure resulted in a downward movement within the ice, which became towards the margins a more horizontal and outward compressive movement. Plastic deformation in the basal layers would probably facilitate this. A boring in 1968 through the Antarctic ice-sheet at Byrd Station by American scientists revealed the presence of a layer of water between the base of the ice and the rock-floor, kept liquid by the immense pressure. This would obviously facilitate ice-movement through its lubricating quality.

Extrusion flow is also likely in glaciers issuing from the margins of ice-sheets, as in Greenland, and in valley-glaciers where the ice is compelled to move uphill. The fact that extrusion flow is more rapid than gravity flow, and has a more potent eroding ability, may well explain the formation of deep rock basins.

Diffluence One important category of glacier-flow involves the lateral branching of a glacier, so that part of the ice flows away from the main stream. This usually results from some down-valley blocking, either by a narrowing of the valley profile or the junction of a tributary. The ice in the main valley builds up, and when thick enough it flows over a col into a neighbouring valley. This too has important erosional results, including the breaching of preglacial watersheds. A large-scale and striking example of this is shown in the

Karakoram Mountains. The North Rimu Glacier sends off a lobe of ice to the north-east, its melt-water flowing to the Yarkand river and on into the Lop Nor basin. The other lobe joins the main Rimu Glacier, its melt-water flowing to the Shayok river, hence to the Indus.

Ablation The wasting or consuming of snow and ice is commonly referred to as ablation, in contrast to the process of alimentation. It involves melting and evaporation, though abrasion and the calving of bergs also cause loss. Melting may be caused by solar radiation, acting more by conduction via solid debris on the neighbouring rock walls than by direct radiation, since so much of the light rays is reflected from an ice-surface (p. 388). Melting is more active in summer and in the day-time upon ice-masses in middle latitudes, and warm rainfall and surface melt-water will also help. The result is the creation of streams on the surface (*superglacial*), within the ice (*englacial*) and on the rock floor below (*subglacial*). These streams issue from the ice-margins; many glacier-tongues are the sources of rivers, as in the case of the Rhône.

Evaporation, the direct transference of water from the solid ice to the gaseous state, depends on the strength of the winds and the temperature and humidity of the atmosphere, and is particularly active at high altitudes and in winter. Powerful winds in Polar areas, blowing hard crystals of ice, also cause loss from the surface by direct abrasion, rather similar to wind-blown sand in deserts (p. 264). Wastage also occurs when the ice margins reach tide-water, causing masses to break off as *bergs* (p. 350).

If there is a balance between supply and wastage, the margin of an ice-mass will remain constant, and a glacier or ice-sheet is said to be stationary. If the temperature should rise, wastage will increase, and if precipitation in the area of accumulation decreases, so also does the supply of ice; in either case the ice-margin shrinks, and the glacier is said to be retreating, though this is not a good term as it seems to imply actual movement in a reverse direction.

Glacier fluctuations Glaciers are extremely sensitive to climatic variations, which produce changes in the net result of accumulation and ablation in their collecting areas, and so reflect fluctuations in their varying lengths. A close watch is maintained on the Alpine glaciers by the Swiss Commission for the Study of Glaciers, and on those in America by the Committee on Glaciers of the Geophysical Union, both by direct measurement and by aerial survey.

During the postglacial climatic optimum of the Atlantic Stage

(p. 260), the glaciers shrank and even disappeared from many ranges. They persisted in the higher parts of the Alps and in mountains farther north in both America and Europe, but disappeared from more southerly ranges such as the Sierra Nevada of the U.S.A. The Sub-Boreal brought a renewal of glacial conditions, which enabled the present cirque-glaciers and glacierets of the Sierra Nevada to re-establish themselves; it is important to realize that these are not shrunken remnants of the main Quaternary Glaciation, but wholly new ones, the product of what used to be called 'The Little Ice Age', now referred to as the *Neoglacial*.

These fluctuations continued during the historic period. The early Middle Ages were distinctly mild, when the Norsemen colonized Iceland and Greenland, and forests grew on the shores of Greenland; tree-roots penetrate coffins now enclosed in permanently frozen ground. These milder conditions hardly lasted beyond the twelfth century, and in the Alps the glaciers seem to have advanced to another maximum by the end of the sixteenth century. Records show that some villages were overwhelmed by ice, summer pastures were no longer usable, and former passes became blocked. A once flourishing silver-mine near Argentière in the Chamonix valley is still buried under ice. Other glacial advances took place in 1719, 1743, 1770–80 and 1818–21; the last major one was in 1850, perhaps the most marked in historic times. During the last fifty years a general shrinkage has occurred, the result of a slight rise in air temperature. But in very recent years it has been reported from Switzerland that one-third of the glaciers under observation are again increasing in length. This is shown even more strikingly in Alaska, where a number of glaciers have advanced the positions of their snouts; the Bering Glacier, the largest in North America, advanced 1200 m (3900 ft) between 1963 and 1966.

TYPES OF ICE-MASS

There have been numerous classifications of ice-masses, some of them involving dozens of types. Many are, however, minor variants of one another, and it is adequate to consider three main groups. The first consists of *ice-sheets* and *ice-caps*, the second *valley-glaciers* (sometimes called mountain-glaciers or Alpine glaciers), and the third *piedmont* or *expanded-foot glaciers*. These three groups correspond respectively to the zones of predominant supply of ice, of movement of ice and of wastage of ice.

(i) **Ice-sheets and ice-caps** Large ice-sheets, covering considerable proportions of continental areas, are the maximum result

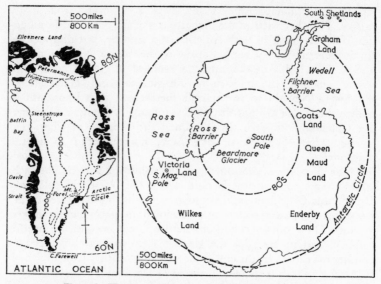

Fig. 108 *The Greenland* (left) *and Antarctic* (right) *ice-caps*

On the Greenland map, the areas free from ice are shaded solid, leaving the ice-surface white. The 7000 and 9000 feet (equivalent to 2100 and 2700 m) contours of the ice-surface are shown.

On the Antarctic map it is not possible to show the areas of rock. An indication of the extent of the major ice-barriers is given by the curved pecks.

of the onset of a widespread glacial period (p. 246). Antarctica and Greenland represent the only surviving examples of these on a continental scale (fig. 108).

Antarctica The Antarctic ice-sheet covers an area of about 12·7 million sq. km (5 million sq. miles) (plate 68). Near the coast in some parts, ranges of mountains with individual peaks known as *nunataks* project above the ice-sheet, between which separate glaciers make their way to the sea to form shelf-ice. In places the ice-sheet itself extends over the sea, particularly where the floating mass of the Ross Barrier, 520,000 sq. km (200,000 sq. miles) in area and 400 m (1300 ft) thick, extends seawards, to end in ice-cliffs from which tabular icebergs break off at intervals (p. 350). One of the main objects of research has been to determine the thickness of the ice-sheet, using echo-sounding methods, and to obtain some impression of the rock surface beneath. In the coastal regions the thickness was found to vary between 250 and 760 m (800 and 2500 ft) but as physicists moved inland so did they find the thickness increase, and a figure of 4300 m (14,000 ft) has been recorded, beneath which the rock-surface lies at 2500 m (8100 ft) below sea-level, probably the result of isostatic depression (p. 23). One of the most interesting discoveries was that the land surface under the ice seems to be extremely rugged, with deep fjord-like valleys alternating with steep-sided ridges, the crests of which near the edge of the ice-sheet sometimes protrude as nunataks. In 1968 the first actual boring down through the Antarctic ice-sheet was made by U.S. engineers, who at a depth of 2100 m (7000 ft) reached bedrock covered with a layer of volcanic ash.

Greenland The Greenland ice-sheet is a low flat dome which covers about 1,800,000 sq. km (700,000 sq. miles), and except for individual nunataks near the margins, no rock is visible away from the coast. Part is surrounded by a rock rim of mountains, inter-rupted by fjords (p. 306), but along sections of the coast the ice-sheet reaches the sea, either as vertical and even overhanging ice-cliffs (known as a 'Chinese wall') or as a more gently sloping face. The centre is a great dome of ice covered with powder-snow, attaining about 3000 m (10,000 ft) above sea-level. The highest actual point of Greenland is probably Mount Forel in the south-east (3360 m, 11,024 ft). It is believed that the ice is about 3000 m (10,000 ft) thick, and recent researches, using echo-sounding methods, indicate that the solid rock beneath is in parts actually below sea-level. Glaciers flow between the rock ridges to reach the sea, known as 'tide-water glaciers'. The largest, in fact the largest glacier in the northern hemisphere, is the Storström in the north-east, 130 km (81 miles) in length, although the longest is the thin and narrow Petermann Glacier in the east, 201 km (125 miles) long, of which the last 40 km (25 miles) float on the sea. The Humboldt Glacier in the north-west ends in a line of cliffs 64 km (40 miles) wide and 90 m (300 ft) high, from which icebergs detach themselves, to float southward into the Atlantic.

Ice-caps Smaller masses of ice may be subdivided into *island ice-caps* and *plateau ice-caps*. Examples of the former include Franz Josef Land, Novaya Zemlya and Spitsbergen (plate 69). About an eighth of Iceland is covered with plateau ice-caps, each known as a *jökull*, of which thirty-seven individuals are distinguished. The largest is Vatnajökull, which covers about 8800 sq. km (3400 sq. miles); its gently rolling surface probably conforms to the nature of the land beneath, while glaciers flow from the margins in the form both of narrow valley-tongues and of broader lobes.

The ice-masses of the Norwegian fjelds afford examples of the plateau ice-cap; the largest, the Jostedalsbre, covers an area of about 1500 sq. km (600 sq. miles). Actually they display some features which place them intermediately between small ice-caps and the type of valley-glacier flowing from a snow-field.

(ii) **Valley-** (or **Alpine-**) **glaciers** Valley-glaciers are a characteristic feature of the greater mountain ranges of the world. They comprise tongues of ice moving from basins, in which firn accumulates, downhill along pre-existing valleys. Their fluctuating size and length depend on the extent of the snow catchment area, on the amount of precipitation, and on the temperatures experienced in their valley courses. The 'snout', the convex end of the glacier, extends to a point where the amount of ice ablation is finally equalled by the supply of ice which is being brought down by the glacier movement. Most Alpine glaciers have shrunk very appreciably during the last century (p. 217).

The Aletsch Glacier in the Bernese Oberland, some 16 km (10 miles) in length, is the longest in Europe (fig. 109 and plate 70). It originates in a group of snow-fields, surrounded by such peaks as the Jungfrau (plate 73) and the Mönch; these snow-fields merge into the extensive firn-field of the Konkordia Platz at a height of about 2800 m (9200 ft), from which the glacier-tongue pushes its way southward between long rock ridges. Note also the remarkable development of small radiating valley-tongues around the peak of Mount Rainier (4391 m, 14,408 ft), the highest peak in the Cascade Mountains of western America (fig. 110).

The Bergschrund A firn-field is separated from the steep ice-walls of surrounding peaks by a gaping crack, which can be seen round the head of the basin. This is a bergschrund (fig. 114), and its crossing (usually by means of a slender snow-bridge) is often a major problem in climbing an Alpine peak. It represents the point where the moving ice-mass is drawing away from the enclosing walls of the basin.

The Randkluft This feature is the gap between the rock face at the back of the cirque and the firn or cirque-glacier. It develops because of melting caused by radiation of heat from the rock wall.

Crevasses As the slope increases, the ice surface is split by cracks or crevasses, for differential movement within the ice causes tension and shearing (fig. 111). They may be *transverse* (i.e. they lie across the glacier), as a result of increasing slope, or *longitudinal* or *marginal* (i.e. parallel to the direction of flow). Refinements of classification include *chevron* crevasses (where examples near the ice-margins have been rotated or twisted), and *splaying* crevasses, which start as longitudinal and splay towards the sides. Where the slope increases markedly, crevasses intersect in all directions, forming an *ice-fall*, a confused labyrinth of deep clefts and of isolated ice pin-

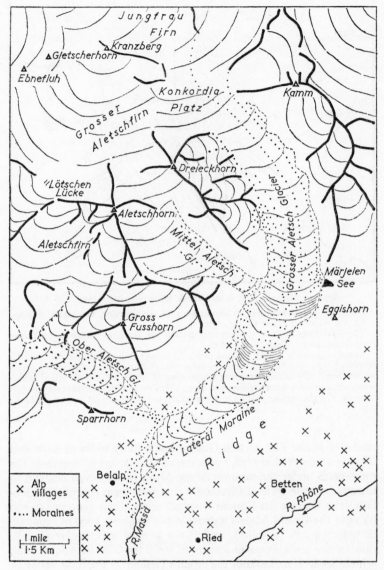

Fig. 109 *The Aletsch Glacier*

The main ridges are shown by heavy black lines, the peaks by triangles, and the areas of ice and snow by generalized form-lines. See also plates 67, 70.

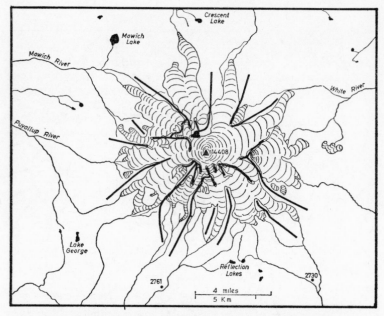

Fig. 110 Mount Rainier, in the Cascade Mountains, Washington, U.S.A.

Based on the United States Department of the Interior, Geological Survey, 1:62,500, Mount Rainier Quadrangle.

The form-lines indicate glaciers and snow-fields, the heavy lines the radiating rock-ridges. Heights are given in feet.

Mount Rainier is a lofty isolated peak, an almost extinct volcanic cone, which receives heavy snowfall and nurtures numerous short radiating glaciers, of which 26 are specifically named, covering a total area of 100 sq. km (40 sq. miles); melt-water streams issue from their snouts. Steep-sided rock-ridges separate the glacier-troughs. The glaciers are shrinking appreciably, the Nisqually Glacier on the south side at 21 m (70 ft) per annum.

nacles or *séracs*. The Aletsch Glacier has a fairly uniform slope and there is no marked ice-fall, but some of the glaciers issuing from the snow-fields under the summits of the Mont Blanc massif descend so rapidly into the Vale of Chamonix that they form complicated ice-falls. The Glacier des Bossons descends from about 3000 m (10,000 ft) near the Grand Mulets mountain-inn to about 1000 m (3000 ft) in only 3·2 km (2 miles), and forms a most spectacular ice-fall. Avalanches are a frequent feature of steep ice-falls; in August 1965 a huge ice-avalanche from the Allalin Glacier in the Saas valley in Switzerland entombed almost a hundred workers at the Mattmark dam-site.

Foliation and ogives The ice-mass of a glacier is characterized by remarkable structural details, not wholly understood. A banding in the ice is common, involving layers representing annual increments of snow in the accumulation area, in which the crystals are arranged in different ways; the layer may be clear and glass-like, or coarse and bubbly in structure. The bands may be near-horizontal, or along planes at any angle to the surface, or arranged in a wavy foliated pattern. Sometimes the layers are distinctly folded on both large and small scales, the result of local compression, possibly due to a powerful tributary stream of ice, or to an adaptation of the ice to irregularities in gradient and velocity.

A feature which has aroused much research interest is the presence of bands of light and dark ice, arcuate downstream, on the surface of a glacier or within its mass, and similar patterns of 'ice-waves', depressions and dirt-bands. The light bands contain many air-bubbles within the ice, the dark bands are virtually bubble-free and contain much dirt. Probably the white bubbly ice was formed from winter snow, when melting and refreezing were negligible, while the dark ice represents partial melting, the accumulation of many dirt particles and then refreezing. The waves on the surface of the glacier are due to the fact that the white ice reflects much insolation (p. 388) and there is little melting, while the dark bands experience greater melting because of the increased thermal conductivity, become etched in the ice-surface as troughs, and therefore accumulate more melt-water and dirt in summer. The arcuate downstream pattern obviously results from the more rapid flow of the centre of the glacier by comparison with that at the margins, where it is retarded by friction. These bands are known as *ogives*.

The surface of the glacier The surface of a glacier during winter and spring is swathed in snow, which masks the crevasses, so providing considerable risk to a mountaineering party unless its members are roped and proceed with caution. In summer the surface is extremely irregular; the crevasses are mainly open and visible, and can be jumped or crossed by substantial snow-bridges.

Long wave-like ridges of ice, forced up by compression, occur at intervals. Small pools, even lakes, lie on the surface of the ice during the day, and streams flow in deeply cut runnels. They take the first opportunity of cascading down crevasses, and so wear out a sort of sink-hole in the ice, known as a *moulin* or 'glacier mill'. At night all this surface water freezes, and the glacier is still except for an odd curious 'groan' from the ice as it moves in its bed, and for the occasional crash of a collapsing sérac.

Moraines 'Moraine' is an ancient word used by peasants in the French Alps in the eighteenth century for banks of earth and stones, and it gradually crept into Alpine literature to become an accepted term. A glacier is an extremely important agent of erosion, transport and deposition. Its transporting function is shown by the various forms of moraine (fig. 111). Frost action is rampant on the ridges and buttresses above the snow-fields and glaciers; from them angular blocks of all sizes fall on to the ice below. Many of the smaller stones are warmed in summer by the sun's heat and slowly sink into the ice, forming a pitted surface. Larger blocks may protect a stalk of ice from the sun's rays, so producing 'glacier-tables' (plate 74). This debris is slowly carried away, lying on the surface in a line more or less parallel and near to the edge of the glacier, forming a *lateral moraine*. Often the lines of moraine protect the ice below them from melting, and so form a prominent ridge on the glacier surface.

Where a tributary glacier joins the main ice-stream, as in the case of the Mittel-Aletsch and Ober-Aletsch Glaciers, two of the lateral moraines may join to form a *medial moraine*. A medial moraine may also stretch away from a rock peak projecting from the upper snow-field. The Aletsch shows half a dozen of these parallel medial moraines, faithfully following the curves of the valley. Some of them are 12–15 m (40–50 ft) high, linear mounds of rocks of all sizes, others are just a discontinuous line of isolated blocks.

Towards the end of the glacier the whole surface is covered with debris; in fact, it is often difficult to say where the ice actually ends, although many glaciers have a terminal 'ice grotto' from which rushes a torrent of water, heavily laden with fine glacial debris. Round the snout of the Aletsch Glacier is a crescentic mound known as a *terminal moraine*, consisting of material varying in size from enormous blocks to finely powdered 'rock flour'. A series of such moraines can be traced down the valley, showing the past halting-places in the gradual retreat of the ice-tongue.

In addition to this material carried on the surface, much debris finds its way down crevasses, to be frozen into the ice as *englacial moraine*. The glacier also removes fragments from its bed and sides, and this, together with surface material which has reached the rock floor by way of crevasses, forms the *subglacial moraine*.

All this debris is deposited at or beyond the glacier snout, the heavier in the terminal moraine, the lighter carried farther by the melt-water stream which issues from the snout, to be laid down as fluvioglacial deposits (p. 245). If the glacier retreats rapidly, the terminal moraine is less clearly defined, and a more or less horizontal sheet of *ground moraine* is deposited.

The Aletsch is a typical valley-glacier, but although it is the longest in Europe it is small by Antarctic standards; the world's

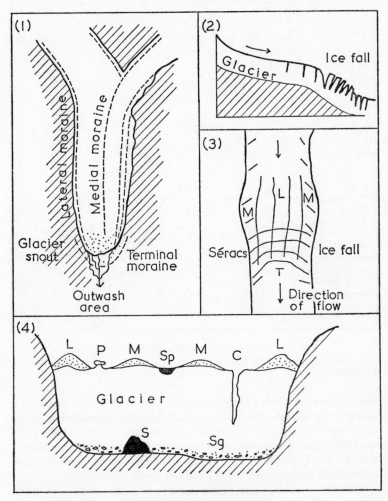

Fig. 111 *Features of valley glaciers*

1. Types of moraine. 2. The occurrence of transverse crevasses and an ice-fall.
3. **L** = longitudinal crevasses formed where a glacier widens; **T** = transverse
crevasses where the bed steepens; **M** = marginal crevasses formed because of the
more rapid rate of flow in the centre. The ice-fall consists of a complicated mass of
séracs, or ice-pinnacles, separated by crevasses intersecting in all directions.
4. Cross-section of a glacier: **C** = crevasse; **L** = lateral moraine; **M** = medial
moraine; **P** = perched block or 'glacier-table' (see plate 74); **S** = subglacial
stream; **Sg** = subglacial or 'ground' moraine; **Sp** = superglacial stream.

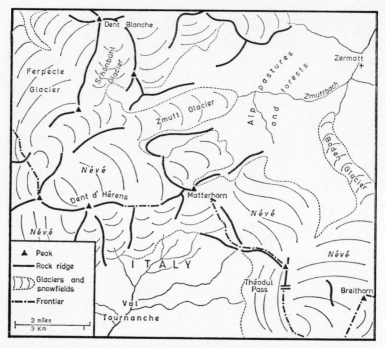

Fig. 112 *The Pennine Alps in the neighbourhood of Zermatt*

Drawn diagrammatically from sheets 283, 284 of the *Landeskarte der Schweiz,*
1:50,000.

longest, the Lambert Glacier (400 km, 250 miles), was discovered
in 1957. In Alaska and in New Zealand, where snowfall is heavy
as a result of the high mountain ranges near and parallel to the
sea from which moist onshore winds flow, there are some very large,
though short, glaciers. The Franz Josef Glacier in New Zealand
descends to within 120 m (400 ft) of sea-level.

Varieties of valley-glacier Some consist of short tongues or lobes
known as *cirque-glaciers* or *glacierets*, which barely protrude from the
basins in which the firn accumulates. An even smaller ice-mass,
lying on a steeply sloping hollow, gully or bench high in the mountains, is called a *niche-glacier*; it has obviously developed from a
compacted snow-patch. Others emerge from basins high on a moun-
tainside, forming *hanging glaciers*, from which great masses periodically break off as ice avalanches. Where a glacier opens on to a
lowland area, tongues of ice flow downward, unconfined by any
marked valley, known as *wall-sided glaciers*. These, however, are

uncommon, since they tend either to spread out or to excavate their beds by erosion so as to contain themselves and become valley-glaciers. Where a glacier emerges from the edge of an ice-cap, not from a cirque, it is known as an *outlet glacier*; these are commonly found in northern Norway and Iceland. If glaciers reach the sea, as in Alaska and Greenland, so discharging floes or bergs, they are known as *tide-water* or *tidal glaciers*.

The valley-glacier is the most common form of ice-mass today, and except at the maxima of the past Ice Age, when the ice-sheets must have swathed virtually everything within their margins, they have exercised, or are still exercising, a potent influence on the landscape of many upland areas (fig. 112).

(iii) **Piedmont glaciers** Piedmont glaciers are formed when individual tongues of ice flow down from between mountain ranges and spread out on the plains or 'foreland' beyond. In its simplest form, a lobe of ice expands beyond the mouth of a valley to form an *expanded-foot glacier*. Such ice-masses are common on the edge of Vatnajökull in Iceland; Skeidarajökull, on its southern flanks, is an example.

The term 'piedmont glacier' is confined to the coalescence upon a foreland of several individual valley-glaciers. The Butterpoint Piedmont and Wilson Piedmont Glaciers in South Victoria Land in Antarctica, the Bering Glacier in Alaska and the Frederikshaab Glacier on the western coast of Greenland are examples. The Malaspina Glacier in Alaska (fig. 113) covers an area of 4200 sq km (1600 sq. miles). Snowfall in the St Elias Range of southern Alaska is heavy, the result of moist air-streams from over the Pacific Ocean meeting a high mountain barrier lying parallel to the coast, and vast firn-fields have accumulated among the peaks, the highest of which, Mount Logan, reaches 6050 m (19,850 ft). Towards Yakutat Bay flow four glaciers, supplying lobes of ice which spread out over the coastal plain. One reaches the sea, forming ice-cliffs, the others disappear under an irregular morainic mass. The lobes have an almost horizontal surface 460 m (1500 ft) above the sea, with an ice thickness of probably about 300 m (1000 ft). The ice has spread out so much that the rate of movement is very slow; in fact, much of the frontal ice is motionless, and trees even grow on the morainic surface above. In the past the glaciers must have brought down much more ice than they do at present, to account for the great accumulation, but today supply is just exceeded by wastage.

Piedmont glaciers are now uncommon, since valley-glaciers in mid-latitude mountain ranges have shrunk well inside their con-

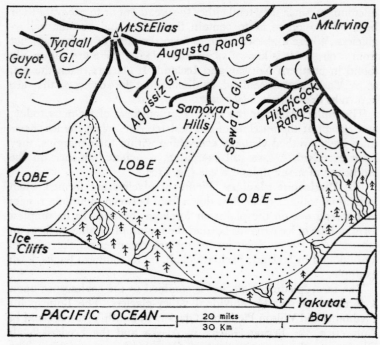

Fig. 113 The Malaspina Glacier

Based on a map in I. C. Russell, 'Second Expedition to Mount St Elias', in *Annual Report of the U.S. Geological Survey* (Washington, 1893). The morainic surface is stippled, ridges are shown by heavy black lines, and forest by tree symbols.

Three lobes of sluggish, virtually stationary ice extend over the coastal plain from the glaciers issuing from the snow-fields among the St Elias range, near the south coast of Alaska. Parts are thickly covered with moraine, on which trees grow, and it is difficult to say where the edge of the ice lies.

taining valleys, but during the maximum stages of the Quaternary Ice Age such ice-masses must have been widespread. The Bavarian Foreland (fig. 120) between the Bavarian Alps and the Danube, the Swiss Plateau, the Lannemezan on the northern flanks of the Pyrenees, and the northern part of the Plain of Lombardy in Italy, were once occupied by piedmont glaciers.

GLACIAL EROSION

Glaciation has made a tremendous impress on a large part of the land surface in middle and high latitudes. With the potent ability of glaciers to erode, to transport and to deposit material—functions which can still be seen in activity at high latitudes and altitudes—

glaciation has been a major agent in moulding the landscape. Generally speaking, the uplands have been chiefly affected by glacial erosion, the lowlands by glacial deposition, although glacial deposits are to be found among the hills, and many evidences of ice-planed rock surfaces occur in the lowlands.

There was much dispute in the nineteenth century concerning the effectiveness of the eroding power of ice. The crux of the controversy was whether many of the characteristic land-forms in formerly glaciated uplands were due to erosion by ice or water; in the latter case it was thought that the snow-fields and glaciers formed a protective covering to the underlying rock, while other forms of denudation attacked those parts not so protected. Indeed, it is probable that a completely static glacier is protective, though freeze–thaw effects will be potent around its margins.

It is now agreed that ice has considerable erosive power. A thick glacier will result in more down-cutting, a thin one will allow more widening by freeze-thaw action (p. 255) along the sides. Weathering also affects the rocky peaks and ridges standing above the snow-fields. The glaciated mountains, even though post-glacial rivers and weathering may have introduced some modifications, are characterized by much bare rock, sharp 'fretted uplands', cirques, arêtes, prominent peaks, and deeply cut and straightened valleys where the subsidiary streams are far from being graded to the main valleys. In the unglaciated areas the relief is more rounded, the river systems are well developed, and the gentle slopes are covered with weathered rock waste.

The mechanics of glacial erosion A glacier erodes in two main ways. The first is by *plucking* or *exaration* (when the ice at the base of the glacier freezes on to rock protuberances, particularly well jointed rock), and tears out blocks which are then removed by the movement of the ice. The mechanism of plucking is, however, more complex than this would imply. It involves both vertical pressure by the overlying ice, and downhill drag resulting from flow pressure in the direction of glacier movement. It may well create stresses in the rock, thus enlarging existing joints and even creating new ones in little jointed rock, thus stimulating freeze–thaw activity. It seems too that the overlying ice can induce friction cracks in certain types of rock, particularly in those of medium and coarse grain. Rock under the ice can be affected in other minor (though in the sum important) ways. The ice can produce a kind of bruising or scarring (as distinct from striations), which may be curved, conchoidal or crescentic in pattern, on the rock-floor. These are possibly caused by a vibratory 'knocking' or percussion of loosely embedded boulders near the base of the ground-moraine (p. 224), and are known as '*chatter-marks*'. Pressure-release, resulting from the melting of the overlying ice-load, is particularly important in making rocks more subject to plucking.

The second method is by *abrasion*, when debris frozen into the base of the glacier is dragged over the rock floor, which is scraped, polished and scored with deep scratches or striations. These can be seen everywhere in a glacially eroded districts, from those on the volcanic rocks of Snowdon to those on the ancient mica-schists which make up Manhattan Island and are visible on the rocks in Central Park, New York. The load itself is ground down. Thus erosion produces on the one hand finely ground rock, known as *rock flour*, and on the other hand fragments of all shapes and sizes, for unlike river pebbles the masses of rock are not worn uniformly.

These processes are still in progress in the higher mountain ranges, but their results can be clearly seen in the British uplands.

Land-forms due to glacial erosion (i) *Cirques* The upper end of a glaciated valley commonly consists of a steep-sided rock basin, known variously as a *cirque* (French), a *corrie* or *coire* (Gaelic), a *cwm* (Welsh) and a *combe* (in Cumberland) (plate 77). Cirques vary in size from tiny rock basins in the British uplands to the huge Walcott Cirque in the Antarctic continent, said to have a back-wall 3000 m (10,000 ft) high. The Western Cwm on Mount Everest, though still ice-filled, probably has a back-wall almost as high.

The most satisfactory theory of its origin is that a shallow pre-glacial hollow has been progressively enlarged. A patch of snow produces alternate thawing and freezing of the rocks around its margins, which causes them to 'rot' or 'disintegrate', a process to which has been given the term *nivation*. Melt-water helps to move the resulting debris, thus forming a *nivation-hollow* (p. 255). As this grows it nourishes a small firn-field or even a cirque-glacier, which plucks rocks from its bed. Melt-water, especially that which makes its way down both the bergschrund and the randkluft, helps by alternate freezing and thawing to eat both into the back-wall of the cirque (a process known as *basal sapping*), thus maintaining its steepness, and also into its floor, thus maintaining the basin shape, and moreover it provides debris which freezes into the base of the ice and so acts as an abrasive. Investigations indicate that the ice movement seems to pivot about a point situated centrally in the cirque, a process known as *rotational slip*, which also tends to emphasize the basin shape.

When the ice finally disappears the rock basin remains as a

72 Mount Assiniboine in the Canadian Rockies, near Banff, Alberta
This fine example of a pyramid peak, rising to 11,870 feet, is carved out of almost horizontal sedimentary strata, an indication of vertical uplift rather than of acute folding.

(Canadian Pacific)

73 The Jungfrau, Bernese Oberland, Switzerland
The ascent of the Jungfrau (13,668 feet) can be made easily from the Jungfraujoch (the col between the Jungfrau summit and the Mönch in the extreme left, where the terminus of the mountain railway is situated). On the south-west the faces and ridges fall sheerly to the glacier at the head of the deeply cut Rotthal. Note the steep wall of the huge trough-end, cutting off the upper snow-fields from the valley.

(Swissair-Photo A.G.)

74 A 'glacier-table' on the Ober-Aletsch Glacier, Switzerland
(*Paul Popper*)

75 An erratic boulder of Shap Granite resting on Carboniferous Limestone,
on Orton Scar, northern Pennines
(*R. Kay Gresswell*)

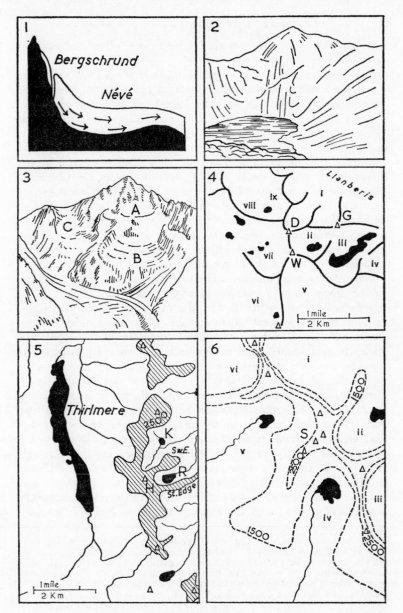

Fig. 114 *Cirques*

1. Cross-section of a typical cirque. **2.** A pyramidal peak, with a small lake in the cirque below (based on Snowdon). **3.** An alpine peak (based on the Aletschhorn); **A, B, C** indicate positions of firn-fields, where cirques are developing.

striking feature of the post-glacial landscape, often containing a small lake (p. 196, fig. 114, plate 77).

(ii) *Arêtes and peaks* If several cirques develop in a mountain mass, they will jointly produce striking relief forms. Steep-sided ridges (*arêtes*), with cols at their lowest points, form when two cirques cut back to back, a phenomenon known as *head-wall recession*. If three or more develop, the surviving central mass becomes a pyramidal peak, later sharpened by frost action (fig. 114). The Matterhorn in the Pennine Alps (figs. 22, 112, plate 13) is a famous example of such a peak, Mount Assiniboine in the Canadian Rockies stands out magnificently from the near-horizontal strata out of which it is carved (plate 72), and many of the Black Cuillins of Skye show similar characteristics on a smaller scale.

(iii) *Glacial valleys* (fig. 115) When a glacier occupies a pre-glacial valley, the erosive effects considerably modify its form, depending on the height above the valley floor to which the upper surface of the glacier attains. Before the onset of direct glacial erosion, active freeze–thaw processes (p. 255) ahead of the advancing glacier will break up the rock on the valley floor and lower walls, making it more susceptible to erosion. The cross-profile tends to become U-shaped, with a flat floor and steep sides (plates 64, 78). The valley is straightened, and any projecting spurs are planed off or *truncated*, as shown on the southern slopes of Saddleback near Keswick in the English Lake District. Where the glacier did not entirely fill the preglacial valley there is a prominent change of slope, leaving *benches* or *alps* above the steep walls. Similarly, high tributary valleys, formerly graded (p. 145) to the preglacial river valley, are left *hanging*, so that their streams fall abruptly into the main valley in a series of cascades (plate 57). In the Yosemite valley, shown in this photograph, there are three sets of hanging valleys at different levels, each graded to an uplift of the Sierra Nevada (p. 43).

Some valleys end abruptly at their heads in a steep wall, known as a *trough-end*, above which lies a group of cirques. Probably a whole

---

4. The Snowdon peaks, ridges and cwms: **W** = Y Wyddfa (main summit); **D** = Crib-y-Ddisgl; **G** = Crib Goch. The cirques are: **i.** Cwm Glas; **ii.** Cwm Glaslyn; **iii.** Cwm Llydaw; **iv.** Cwm Dyli; **v.** Cwm Tregalan; **vi.** Cwm-y-Llan; **vii.** Cwm Clogwyn; **viii.** Cwm Brynog; **ix.** Cwm D'ur-arddu. This cirque-erosion into several sides of a mountain-mass produces what is sometimes called '*biscuit-board relief*'. **5.** The Helvellyn cirques. Red Tarn (**R**) lies between the steep arêtes of Striding Edge (**St Edge**) and Swirral Edge (**Sw. E.**), and the now dry Keppelcove Tarn (**K**) between Swirral Edge and the Raise ridge. **6.** The south-western cirques of the Cuillins. Sgurr Alasdair (**S**) 991 m (3251 ft) above sea-level, is the highest peak in Skye. The six cirques are: **i.** Coireachan Ruadha; **ii.** Coire an Lochain; **iii.** An Garbh-choire; **iv.** Coir' a Ghrundda; **v.** Coire Lagan (plate 25); **vi.** Coire na Banachdich.

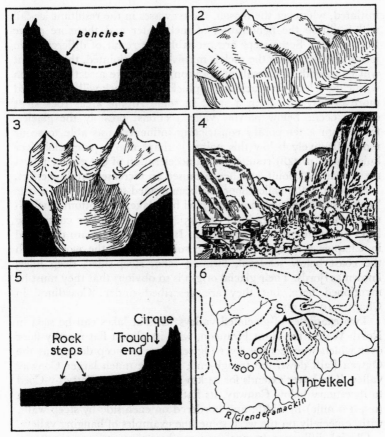

Fig. 115 *Features of a glaciated valley*

1. Cross-profile of a glaciated valley. **2.** Sketch-diagram of a glaciated valley. **3.** The 'trough-end'. **4.** The Lauterbrunnen valley, Switzerland. **5.** Long-profile of a glaciated valley. **6.** The truncated spurs of Saddleback (**S**), English Lake District.

series of cirque-glaciers developed, and their ice coalesced at one point to form the main valley-glacier; the thickness of the ice, its weight and its eroding power were therefore suddenly enormously increased. An examination of a longitudinal valley profile some-times reveals a series of *rock steps*, due partly to unequal eroding power on the part of the ice (thus a step commonly occurs where a tributary glacier formerly joined the main one and so the extra mass of ice was able to erode more vigorously), and partly to the degree of resistance to erosion of the valley floor. Once a step is initiated, whatever the reason, the crevasses in the resultant ice-fall will facilitate the penetration of melt-water and therefore active frost erosion. Some steps are probably the result of differences in jointing; thus where the rock is massive and unjointed, the gradient will be smooth, the result of abrasion, but where a section is riven with close joints, thus facilitating plucking, the rock will be removed more rapidly and form a step. In some Alpine areas rock steps clearly occur below narrow valley-sections; possi'ly the glacier, freed from a temporary constricting influence, was able to scour more effectively below this point. Perhaps also a preglacial river knickpoint (p. 148) resulted in an acceleration of the glacier movement near that point, and hence caused an increased eroding effect.

Sometimes the valley floor has been eroded so deeply that a true 'rock basin' is formed, now occupied by a lake, the bottom of which may well be below sea-level (p. 195). Glaciated valleys containing lakes are widespread in the British Isles; these lakes are described in Chapter 7. Glacial troughs in some parts of the world reach the coast below sea-level, forming long, steep-sided, deep inlets known as fjords. Their glacial origin is so obvious that they must be mentioned here, but they are described under 'Coastlines' in Chapter 10 (p. 305).

Two examples of glaciated valleys without lakes can be seen in North Wales: the Nant Ffrancon, with its almost flat marshy floor over which wanders the river Ogwen after its steep descent as the Ogwen Falls down the trough-end, and the much larger Conway valley. The latter extends for 24 km (15 miles) from Bettws-y-Coed to its estuary below Conway, its flat floor between 0·8 and 1·6 km (0·5–1·0 mile) in width, and bordered on each side by steep walls, with, especially on the west, some fine examples of hanging valleys.

Glacial diffluence (p. 215) can produce very striking results on the postglacial valley landscape, notably the breaching of former water-divides. In North Wales, for example, the diverging Nant Ffrancon, Llanberis, Gwynant and Nantlle valleys are due in part at least to diffluence; from Langdale in the English Lake District a glacier divided several times, the main stream eroding the Windermere trough, other portions being responsible for the valley in which Esthwaite Water lies and several other valleys; and most complex results of diffluence can be seen in the neighbourhood of Loch Lomond in Scotland.

While undoubtedly the characteristic profile of a glaciated valley is U-shaped, recent work indicates that some examples reveal at intervals distinct V-profiles. The reason for this is not clear, though the V-section usually seem to be associated with steeper portions of the valley floor, separating U-sections above and below. This un-usual profile may be the result of fracturing or jointing. The Tenaya Canyon, in upper Yosemite, California, is narrow and keel-shaped in section, because vertical glacial erosion was favoured by a narrow belt of longitudinal fractures, while lateral erosion was limited by the flanking masses of massive unjointed rock. Possibly the V-shape may be emphasized by rapid vertical erosion either by fast flowing surface streams during interglacial periods, or by subglacial streams under enormous hydrostatic pressure.

(iv) *Other erosion features* As triking result of glacial erosion, which in fact demonstrates its nature, is the moulding of masses of rock which project above the general level. The upstream (or 'onset') side of such a mass is smoothed and polished, although often deeply striated as well, and its profile is rounded; here abrasion is the main eroding force. The downstream (or 'lee') side, on the other hand, especially if well jointed, is made rougher and more irregular by the plucking action of the ice. The resulting form is known as a *roche moutonnée*, so called originally because of the similarity of these residual hummocks to wigs made of sheepskin which were once worn in France, and, as so often happens, the name has stuck (fig. 116). An alternative term is *'stoss-and-lee'*.

Another feature, known as *crag-and-tail*, is due to some hard obstructive mass of rock, the crag, which lies in the path of oncoming ice (fig. 117). This mass protects softer rocks in its lee from glacial erosion, for the ice seems to have moved over and around the 'crag', leaving a gently sloping 'tail' in its lee. A striking example is the hard igneous plug of the Edinburgh Castle Rock, with its 'tail' of Carboniferous Limestone lying on Old Red Sandstone, sloping eastward, along which now lies the 'Royal Mile'. Sometimes the 'tail' may consist of drift, preserved in the 'dead-space' or area of stagnation in the lee of the obstructing mass, as in the case of Arthur's Seat (p. 84), also near Edinburgh. In areas of low relief, glacial activity may result in a landscape of low knobs, separated by shallow basins, now water-filled; in Scotland this is known as 'knob and lochan relief'. This is especially marked where the rock floor is diversified by faults, joints, intrusions and other lines of weakness, thus stimulating differential erosion.

Erosion by ice-sheets While most of these erosional features have been produced by valley-glaciers in upland areas, the erosional effect of the continental ice-sheets has been widespread. Such regions as the Canadian Shield and Finland have been planed smooth, the soil removed, and striated or fluted 'pavements' of bare rock remain. Many small hollows of irregular size and shape, probably initiated by plucking in the crystalline rocks of these areas, have been worn; most are now lake-filled (fig. 100, plate 65).

Fluvioglacial erosion Streams issuing from the edge of an ice-sheet or a glacier-tongue can exercise a considerable erosive effect. Moreover, subglacial streams may flow on the rock-floor

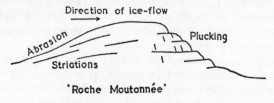

Direction of ice-flow

Abrasion Plucking

Striations

'Roche Moutonnée'

Fig. 116 *A 'roche moutonnée'*

In a roche moutonnée the upstream side is polished by ice-abrasion (although deep striations may be cut), while the downstream side, particularly if the rock is well jointed, will be roughened by plucking.

beneath the ice, sometimes under considerable hydrostatic pressure, which can create melt-water channels of considerable dimensions and apparently anomalous positions. Frequently the results are quite independent of the normal drainage development, although they often materially contribute to the postglacial drainage systems.

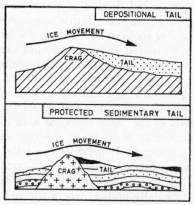

Fig. 117 *'Crag-and-tail'*

A prominent mass of resistant rock may form a shelter behind which softer sedimentary strata are preserved, or in the lee of which material may be deposited.

While northern and central Europe was under the continental ice-sheets, their southern edges lay more or less west–east. As the uplands of central Europe formed a barrier farther south, a great volume of melt-water was forced to flow westward into the North Sea (when it was not frozen) along the front of the ice-sheets, thus carving west–east depressions known as *Urstromtäler* ('ancient river valleys') in Germany and as *Pradoliny* in Poland. As the ice-sheets retreated, a series of these broad, shallow troughs was successively formed. Five main lines can be traced: four of them lie south and west of the Baltic End Moraine (p. 241), the fifth is farther north-east across Pomerania. The present rivers which cross the North European Plain flow in a general direction from south-east to north-west, but take advantage of sections of the Urstromtäler, which helps to explain the frequent 'elbows' in their courses. Parts of the Urstromtäler which are now streamless have facilitated the construction of Germany's west–east 'Mittelland' canal system.

Glacial overflow channels were cut by melt-water escaping over a pre-glacial watershed, usually from a marginal glacial lake. These, too, have contributed to the postglacial drainage system, and are described in connection with glacial lakes in Chapter 7 (pp. 207–9 and figs. 106, 107).

GLACIAL DEPOSITION

Glaciers and ice-sheets are able to transport vast amounts of material ranging from the finest rock flour to enormous boulders—on the surface, frozen into the body of the ice, and at its base. Streams issuing from the margins, formed of the melt-water, and those within or below the ice itself, are also powerful transporting agents. The deposition of this load at and beyond the point of melting can modify the landscape very markedly. Moreover, after the maximum of a glacial advance the edge of the ice decays, either steadily or in stages interrupted by long standstill periods and so a great area covered by glacial material is gradually exposed. The name of *drift* is given to these deposits; it includes not only unstratified glacial material (or *till*) but also stratified fluvioglacial debris, the outwash sands and gravels deposited by the melt-water streams. It is estimated that about 36 per cent of the area of Europe, 23 per cent of North America and 8 per cent of the world's land-surface are drift-covered. The glacial material may lie in fairly continuous lines of hills, or in uneven but roughly horizontal sheets. Fig. 118 shows diagrammatically the typical arrangement of the drift deposits near a former ice-sheet margin.

The glaciation of Quaternary times affected the land surface for a long period of time; consequently in Europe a distinction is made between the Older Drift, laid down during the earlier glacial advances, and the Younger Drift, the product of subsequent advances which did not extend as far. The Older Drift has been much altered: it has been eroded, re-sorted and re-deposited by postglacial rivers, the distinctive moraines have been largely destroyed, lake-hollows have been naturally drained or filled with sediment, and more mature drainage systems have developed. The Younger Drift consists of much less weathered material, while the drainage systems are im-

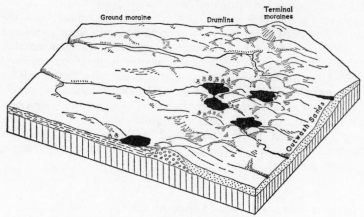

Fig. 118 *Block-diagram of glacial and fluvioglacial deposits*
(After Seydlitz)

mature and indeterminate. The same difference exists between the Older and Younger Drifts in Great Britain and in North America.

Deposition in glaciated uplands Glaciated lowlands reveal the effects of deposition far more strikingly than do glaciated uplands, and the small deposits in the latter are not very significant compared with the more pronounced erosive features. But most of the Welsh, English Lake District and Scottish valleys, and many of the Alpine valleys (now clear of ice in their lower parts), contain patches and hummocks of ground moraine, crescentic terminal moraines (plate 79), and sheets of outwash sand and gravel. These deposits are irregular and patchy, and are on a small scale compared with the lowland sheets, but they may have an important effect on drainage, particularly on the formation of past and present lakes (p. 199).

The U-shape of some glacial valleys is only in part erosional. Their conspicuously flat floors (plate 78) are sometimes the result of infilling with till, and with rock-waste distributed in level sheets by melt-water streams after the glacier had retreated some distance. The 'filling' may lie deeply over an irregular rock floor, as shown by test borings put down during searches for stable sites for barrages for hydro-electricity reservoirs in Switzerland, Austria, France and the U.S.A.

Land-forms due to glacial deposition (i) *Erratics* The widespread occurrence of rocks of all sizes, far from the outcrop from which they were derived, furnishes invaluable evidence in tracing the directions of ice movement. Sometimes subsequent erosion has left them precariously balanced, when they are known as *perched blocks*. These erratics include boulders of Shap granite found in the Scarborough area, in the Ribble valley and near Wolverhampton. The bluish igneous microgranite (known as rie-beckite-eurite) from the little island of Ailsa Craig off the Ayrshire coast of Southern Scotland has been found in Merseyside, Anglesey, the Isle of Man, in the Fishguard area of southern Wales, and even around the coast of south-eastern Ireland as far as Cork. Most striking, because of their colour contrast, are the granite boulders lying on Pennine limestone (plate 75). Some erratics are of gigantic size; one in Alberta is said to weigh over 18×10^6 kg (18,000 tons) and the Madison Boulder in New Hampshire, 25 m (83 ft) long, is claimed to weigh approximately 4700 tons, though it has been moved only 3 km (2 miles) from its located origin; in Huntingdonshire an erratic of chalk bears the village of Catsworth; and near Monthey, in the Rhône valley to the south of Lake Geneva, each of several granite erratics contains its own quarry. In the corner of Boone County, Kentucky, boulders of a beautiful red jasper, a hard flint-like rock, lie 1000 km (600 miles) from the nearest bed-rock of this type, north of Lake Huron.

(ii) *Till* This consists of an unstratified mass of material with a matrix of clay and sand, containing stones of all shapes and sizes. The term till is preferable to boulder-clay, since it does not imply any definite constitution, though the latter term is still widely used. Till may be divided into two main types: *lodgement till*, deposited beneath the active ice, and *ablation till*, left *in situ* as the ice decays and melts. Its composition varies enormously according to the source of its constituents. In England, for example, various tills are distinctive enough to have local names; in East Anglia the Chalky Boulder-clay is characteristic—a tough, greyish clay containing flints and angular chalk fragments; the Contorted Drift of the Cromer area contains both chalk rocks and igneous boulders from northern England; and the Cromer Till also contains boulders from Scandinavia. In Lincolnshire and the East Riding of Yorkshire the Hessle Boulder-clay and the Purple Boulder-clay can be distinguished. Most of this has been laid down in sheets, into which the postglacial rivers are cutting. While the sheets are more or less evenly spread, in detail they are gently undulating, even hummocky. In parts of the North European Plain the shallow irregular depressions are lake-filled (p. 199).

A *boulder-train* consists of a series of erratics, usually worn from the same bed-rock source, carried forward by the moving ice and

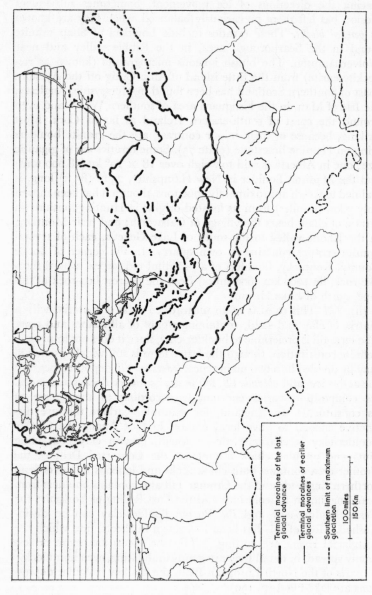

Fig. 119 *The terminal moraines of the North European Plain* (after P. Wolstedt)

Terminal moraines of the last
glacial advance

Terminal moraines of earlier
glacial advances

Southern limit of maximum
glaciation

100 miles
150 Km

deposited either in a fan pattern, with the apex pointing to the origin of the material, or in a more or less straight line (they can be mapped to indicate the exact movement of the ice)—e.g. the train of dark Silurian boulders across the limestone Craven District in Yorkshire.

(iii) *Moraines* The various moraines associated with valley-glaciers are on a relatively small scale. But where the ice-sheets lingered for a considerable length of time, morainic hills can be traced across the lowlands for many miles (fig. 119). The main lines produced by long standstills of the ice are true *terminal moraines*, the others, the result of brief pauses or even of slight advances during a retreat, are *recessional moraines*.

The older moraines on the North European Plain have been largely destroyed by post-glacial erosion, but those formed by the last major standstill and decay of the ice can still be traced southward through the Jutland peninsula, then eastward to the Oder valley, and on into Poland and the former East Prussia. There are many broadly parallel lines, but the most clearly defined is the Baltic Heights or Baltic End Moraine, which sometimes consists of a sharply continuous ridge, elsewhere of a belt of irregular mounds separated by lake-filled depressions. Many of the hills are over 180 m (600 ft) above sea-level; the Turmberg, near Gdansk reaches 331 m (1085 ft) (the highest point in the North European Plain), and two hills in the former East Prussia also exceed 300 m (1000 ft). Beyond the area shown in fig 119, more morainic lines indicate still later stages in the final retreat; across southern Finland, parallel to the coast and 64 km (40 miles) inland, is the steep double rampart of the Salpausselka end moraine, and across southern Norway lies the Ra moraine of the same age.

Morainic hills are not common in the glaciated lowlands of the British Isles. In the Vale of York a moraine 15 m (50 ft) in height extends from the Wolds to the Pennine foothills between the Ure and Wharfe valleys. It is cut through by the Derwent at Stamford Bridge and by the Ouse at York, which is largely built on the moraine. The largest terminal moraine in Britain is the Cromer Ridge in Norfolk, a belt of hummocky hills of sand and gravel, 8 km (5 miles) in width, 24 km (15 miles) in length, and over 90 m (300 ft) high.

A type of moraine which can be seen at the outer margins of ice-sheet movements consists of masses of sand and gravel, with little clay. In sand-quarries in the eastern Netherlands the layers seem to have been folded and even overthrust in a remarkable way, if on a miniature scale. These may have been caused by ice-pressure from the north 'rucking-up' or 'bull-dozing' the drifts, hence the terms *stuwwallen* or *moraines de poussée* (*push moraines*). They form low hills

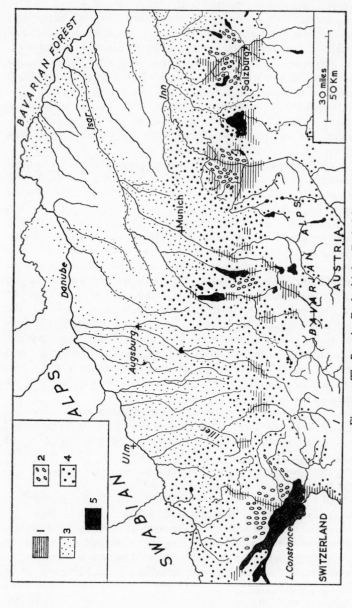

Fig. 120 *The Bavarian Foreland* (after E. de Martonne)

1. Former lake basins; 2. drumlins; 3. fluvioglacial deposits; 4. moraines; 5. lakes.

extending interruptedly from Utrecht into West Germany and similar features may be seen in New York State and in southern Iowa.

(iv) *Drumlins* In some areas, particularly where a valley-glacier opens out or where a piedmont glacier debouches on to a plain (so causing a thinning of the ice-mass), the till has been deposited as swarms of rounded hummocks, from small mounds a few metres long and high to considerable hillocks 1·6 km (1 mile) or more in length and as much as 90 m (300 ft) high. They are found in Northern Ireland (note the drumlin-islands of Strangford Lough on plate 93), the Midland Valley of Scotland and many parts of northern England (in the Solway Plain, the coastal plain near Lancaster and the Aire Gap). Drumlins are, however, wholly absent from East Anglia. In the North European Plain and in the Bavarian Foreland of southern Germany they lie within the successive lines of terminal moraines (fig. 120). They commonly occur *en échelon* in a sort of rhythmical pattern, and as a result the term 'basket of eggs relief' is sometimes applied to a drumlin landscape.

These drumlins usually consist of a sandy rather than a clay-till, and reveal a long axis which lies more or less in the line of the ice movement. The ice deposited each mass (possibly from part of the ice base which was locally more heavily loaded with material) because friction between the till and the underlying floor was greater than that between the till and the overlying ice. Its shape was then streamlined by ice movement. Thus level ground moraine was deposited as the result of the decay of more or less stagnant 'dead ice' and drumlins were fashioned by active ice. It is also possible that some drumlins were produced by the pressure of active ice, which modified sheets of sandy till laid down by an earlier glacial advance.

Some drumlins contain a rock core, around which the till has been plastered. The cover may be so thin that the result is called a *rock-drumlin* or *false-drumlin*.

(v) *Eskers and kames* A rather confusing series of terms has been used to denote ridges and mounds of glacial sands and gravels, since, unfortunately, the usage of the several terms has differed with various writers. One convenient distinction is that an esker consists of an elongated ridge of sand and gravel, longitudinal or parallel to the direction of ice-flow and at right-angles to the ice-front, while a kame lies parallel to the ice-front. Frequently, however, the mounds are so irregular and amorphous in shape that such a classification cannot readily be applied. It would seem desirable to follow the terminology defined by J. K. Charlesworth, who gives the general name of *esker* to all these deposits, and divides them into three groups: *osar* (singular *os*), *kame-terraces* and *kame-moraines*.

The *os* is in fact an esker in a strict sense, a winding ridge of coarse sand and gravel. Osar are common in Finland, the former East Prussia and Sweden, where they wind across country among the lakes and marshes; they are also found in parts of northern England and Scotland. Their origin is not certain; it was formerly thought that they were the 'casts' of stream courses within the ice, which survived when it finally melted. More probably they represent a continuously receding 'delta' formed at the edge of an ice-sheet or glacier by an englacial or subglacial stream, while the ice-sheet decayed rapidly. Owing to the enclosed nature of the subglacial stream, hydrostatic pressure was considerable, causing the flow of water to be rapid and enabling the stream to carry a heavy load. On issuing from beneath the ice the pressure was relieved, the stream abruptly slowed down and much of the load was deposited. When larger humps of material occur at intervals along it, the term *beaded esker* is sometimes used. The 'bead' is formed during the brief but intensive activity of summer, when increased melt-water means the deposition of more material, whereas in winter deposition is less but more regular.

Kame-terraces are formed along the edge of the glacier ice, 'the faithful cast of a former ragged ice-margin, affected by slumping and dissecting by later streams' (Charlesworth). They were laid down by streams occupying a trough between the ice-tongue and the valley-side, forming narrow flat-topped terrace-like ridges (fig. 121). They sometimes look like the shore-lines of former glacial-lakes, but are much less regular. Examples can be seen along the edges of valleys in the Lammermuir Hills in eastern Scotland, in places with a series of four, one above another.

Kame-moraines (called *kame-deltas* by some geomorphologists) consist of undulating mounds of bedded sands and gravels arranged in a chaotic and complicated pattern. As in the case of kame-terraces, their inner faces represent moulds of the ice-contact slope. They are in effect groups of alluvial cones or deltas deposited unevenly along the front of a long-stagnant and slowly decaying ice-sheet, while the more linear os was formed when the ice-sheet was retreating rapidly. One of the most characteristic features of kame-moraine landscape is a small shallow hollow or 'kettle', hence the American term of 'kame-and-kettle-moraine'. The word 'kettle', incidentally, is an interesting example of derivation. It has nothing to do with the kitchen utensil, but comes from the Kettle Range in southern Wisconsin near Lake Michigan, where these hollows are particularly numerous. Most were due to deposition around individual blocks of ice broken off from the main sheet, which finally melted and left the hole each had occupied, usually containing a small lake (p. 199). It is possible to see blocks of ice *in situ* in this manner in Iceland and Spitsbergen.

Kame-moraines are widespread on the lowlands of North America and north-western Europe, covering many square km of country. In North America a belt of kame-moraines has been traced in loops from Long Island away west into Wisconsin, and in Europe they lie more or less parallel to the Baltic Heights. There are numerous examples on a smaller scale in Britain, notably near Carstairs on the northern flanks of the Southern Uplands of Scotland.

Gravel kame-moraines occur widespread among the bogs of the Central Plain of Ireland, in the words of J. K. Charlesworth, '. . . either as tangles of billowy mounds, with rolling and knobby surface, or as ramparts, steep-sided, narrow-crested and up to 20 m (60 ft) high, which wind serpent-like across the plains'. In Ireland they are called *eiscir*, from which the anglicized word esker was derived. The Eiscir Riada extends across Ireland from near Dublin into Galway, though interrupted occasionally and breached completely by the Shannon near Athlone.

Land-forms due to fluvioglacial deposition Melt-water flowing from the edge of an ice-mass washes out and deposits stratified sheets of clays, sands and gravels, hence the term *outwash materials*. If material is deposited over an extensive lowland from a broad ice-margin, the result is termed an *outwash plain*, or its Icelandic name of *sandur*. The coarsest material is deposited near the ice-margin, the finer material farther away. Over the surface of the sandur, the melt-water flows in braided streams which constantly migrate from one side to another. By contrast, in narrow valleys, e.g. the Rhône, the outwash may cover the floor to considerable depths; these deposits are known as *valley-trains*.

The finest materials may be laid down in lakes along the margins of the ice-sheets, as in 'Lake Agassiz' in North America (p. 205 and fig. 106), forming fertile lacustrine plains. In the North European Plain, sheets of sand with patches of gravel have been laid down outside the terminal moraines. In places these sheets are as much as 75 m (250 ft) thick, overlying the ground moraine of earlier glaciations. Both the rivers of the Urstromtäler (p. 236) and the postglacial rivers, such as the Weser and the Elbe, have cut the sheets into blocks, separated by alluvium-floored valleys. The sandy areas are known as *Geest*, and form the characteristic German heathlands, such as the Lüneburg Heath. The present river terraces are commonly covered

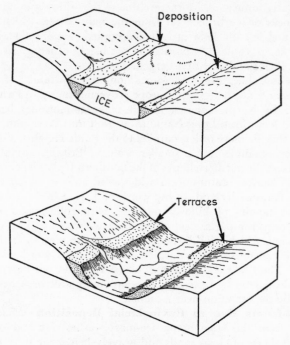

Fig. 121 *Block-diagrams of a kame-terrace*

with layers of outwash material, re-sorted and re-deposited; this is strikingly shown on the Bavarian Foreland between the Bavarian Alps and the Danube (fig. 120).

One form of fluvioglacial deposit, the *varve*, has been mentioned on pp. 13–14 in connection with the dating of geological time.

PAST ICE AGES

Several references have been made to periods in the past when glaciation was much more widespread in the world than at present. Evidence of even Pre-Cambrian ice ages has been discovered in several parts of the world in the form of 'fossil till', known as *tillite*. Near Lake Superior this deposit is as much as 180 m (600 ft) thick, and covers many thousands of square km, indicating a long period of glaciation starting about a thousand million years ago. Pre-Cambrian tillites and boulder-beds have also been found in places as distant as Scotland and northern Michigan.

During late Carboniferous and Permian times, ice-sheets affected

76 The Canadian Shield in the Keewatin District of the Northwest Territories, Canada

The continental ice-sheet scoured the ancient rocks of the Shield, removing most of the soil. The depressions are lake-filled.

(Royal Canadian Air Force)

77 Blea Water, above Mardale, English Lake District

Blea Water lies in a small cirque a mile west of the head of Haweswater. The ridge behind the tarn is High Street.

(Aerofilms Ltd

78 The upper Langdale valley, English Lake District
The glaciated U-shaped valley is largely infilled with boulder-clay.
(*R. Kay Gresswell*)

79 A terminal moraine at the southern end of Lake Windermere, English
Lake District
(*R. Kay Gresswell*)

parts of South America, South Africa, India and south-eastern Australia; in the last, five distinct beds of tillite, with an overall thickness of 600 m (2000 ft), are separated by intervening deposits containing coal-seams, indicating warm interglacial periods. This occurrence of contemporaneous tillites in the four southern continents is taken by some authorities to afford further proof of the former existence of Gondwanaland (p. 25). *Roches moutonnées*, striated pavements and erratics have also been discovered of this age. After the Carbo-Permian glaciation, temperatures were appreciably higher over all the world, and probably the entire globe was devoid of glaciers and ice-sheets until the Pleistocene.

It seems that at the beginning of Pleistocene times began what is known as the Quaternary Ice Age or the Pleistocene glaciation. Until recently its onset was dated as about 600,000 years ago (a 'short time-scale'), based on the astronomical theory of periodical perturbations in the orbit of the earth to explain climatic fluctuations. But the interpretation of new evidence by radiocarbon dating of interglacial organic deposits, the study of 'cores' of material obtained from the ocean floor and other evidence, have led some workers to push back the onset to 1·8 to 2 million years ago (a 'long time-scale'). Many causes have been suggested, including changes in the eccentricity of the earth's orbit or in the inclination of its axis, changes in solar radiation as indicated by sun-spot activity, changes in the positions of the poles, changes in the amount of water-vapour in the atmosphere, changes in the distribution of land and sea, perhaps a slight increase in altitude as a result of the broad up-arching of the continents, or some change in the nature and direction of ocean currents. A theory that has aroused much interest is the possible reduction of the amount of carbon dioxide in the atmosphere, which at present contains about 0·03 per cent by volume; this gas absorbs long-wave solar radiation (p. 382) and therefore tends to warm the atmosphere. If the proportion were reduced by half, the surface temperature would drop by about 7° C. The difficulty in all these theories is why the particular change to which the onset of glacial conditions is attributed should have taken place. The change in temperature need not be very large; a drop of 11° C over Britain throughout the year would be sufficient.

Whatever the cause, the Quaternary ice-sheets at their maxima covered about 30 million sq. km (12 million sq. miles) in North America, north-western Europe (figs. 122–4), the Alps and the Himalayas. As a result of the smaller land-masses in the southern hemisphere, the ice-sheets there were less extensive, but the New Zealand Alps, Tasmania, southern Chile and Patagonia were

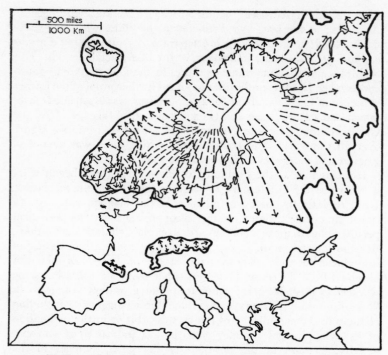

Fig. 122 *The movement and extent of the Quaternary ice-sheets in Europe*
Based on W. B. Wright, *The Quaternary Ice Age* (London, 1914).
The southern limits and the main directions of the advance of the ice-sheets are shown diagrammatically.

probably ice-covered, while the Antarctic ice-sheet was of considerably greater extent. In fact, all mountain ranges with present snow-fields and glaciers must then have developed much more extensive ice-masses.

The Quaternary glaciation followed a definite sequence of events. Over the pre-glacial land areas the temperature decreased and precipitation may have increased. In the higher parts the snow steadily accumulated to form firn-fields from which valley-glaciers issued. At first the ridges and peaks projected, then the uplands between the valley-glaciers were perhaps completely covered, forming ice-caps. As these increased in area they gradually coalesced, and as thawing in the lowlands became less able to cope with the ice debouching into the plains, in time there formed continental ice-sheets covering huge areas. At the climax period, all land but the highest peaks must have been obliterated by large, smooth ice-domes. Over continental Europe the chief centre from which the ice-sheets moved was Fennoscandia (a convenient term which denotes Finland, Sweden and Norway).[1] In Britain there were at first several centres—over the Scottish Highlands, the Southern Uplands, northern and western Ireland, the English Lake District, the Pennines and northern and central Wales. The main British ice-sheet developed in the Scottish Highlands, from which ice moved both south over the Irish Sea and along the east coast, where it met and partially merged with the Scandinavian ice.

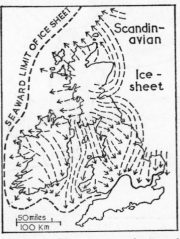

Fig. 123 *The movement and extent of the Quaternary ice-sheets in the British Isles.* Based on W. B. Wright, *The Quaternary Ice Age* (London, 1914).

The continental ice-sheets of North America moved out from several centres over northern Canada. Some of the large Cordilleran valley-glaciers flowed to the east, coalescing as a piedmont glacier on the High Plains. Others moved to the west over the plateau of British Columbia, probably overriding all but the highest summits of the Coast Ranges; distinct glacial traces have been discovered at heights of 1500 m (5000 ft). The ice moved on to the ocean as a shelf, but did not advance far because of mass wastage in the ocean. In the east it was long believed that the ice dispersed from three main centres, the Keewatin, Patrician and Labradorian (fig. 124) which ultimately merged. Now the concept of a single Laurentide sheet, originating in the mountains of Labrador, is regarded as more likely. It spread out in each direction, at its maximum reaching as far south as where St Louis now stands. In the north it merged with the Greenland ice-sheet, then enlarged compared with the present, over the northern Arctic islands. Part of Alaska in the lee of the Pacific mountain ranges probably remained ice-free. The

---

[1] Recent research has led some authorities to believe that the ice-sheet did not remain over the source-region in Fennoscandia for the whole glacial period, but that the glaciation comprised two main phases, an earlier, longer and intense one, and a later, short-lived one. Evidence for the double glaciation rests on the discovery in Norway and Sweden of fossiliferous remains, including various plant pollens, lying between two thick layers of drift.

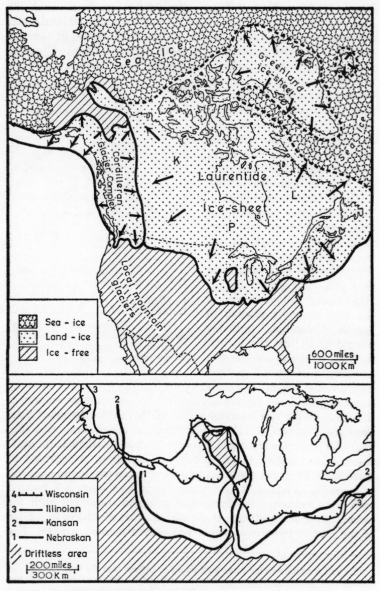

Fig. 124 *The Quaternary glaciation in North America*
K = Keewatin; **L** = Labradorian; **P** = Patrician.
Based on R. F. Flint, *Glacial and Pleistocene Geology* (New York, 1957).

fluctuations of the several advances in the south were very great, and had one curious result, the leaving of a 'Driftless Zone' in western Wisconsin and adjoining parts of Minnesota, Iowa and Illinois. Here the slow extrusive flow of the ice-sheets was channelled aside by low hills, and while the area was never encircled by ice at any one time, it is surrounded by land that at one stage or another was covered with ice (fig. 124).

THE CHRONOLOGY OF THE QUATERNARY GLACIATION

It is clear from the evidence of the deposits left by the Quaternary ice-sheets that there was not just a single advance to a glacial maximum, followed by a steady uninterrupted shrinkage. The chronology of the Quaternary ice-sheets was first patiently unravelled in the Alpine Foreland to the north-east of Lake Constance. In 1909, A. Penck and E. Brückner published what has become a classic piece of work. They found four distinct series of fluvioglacial out-wash gravels, occurring at different heights above the present river-floors, each one associated with a series of morainic deposits. Thus they inferred that four distinct glacial advances occurred, to which they gave the names *Günz*, *Mindel*, *Riss* and *Würm*, from the names of four Bavarian tributaries of the Danube. The Riss stage was that of maximum glaciation. Between these advances occurred inter-glacial phases during which the climate was milder, and the ice-sheets accordingly shrank some distance polewards. From the evidence of plant remains contained within the interglacial deposits, the climates of some interglacial periods may have been appreciably warmer than at present. The longest interglacial period was the Mindel-Riss, sometimes known as the 'Great Interglacial', which lasted about 190,000 years, if the 'short time-scale' of the Quaternary glaciation be used. The concept of this fourfold Alpine glacia-tion is still fundamentally accepted, although more recent workers suggest that the main glacial advances may have had several stages separated by minor phases of recession. In Switzerland, for example, two additional stages have been postulated occurring between the Mindel and the Riss and interrupting the Great Interglacial, to which the names *Kander* and *Glütsch* have been given. In addition, it is believed that a still earlier glacial advance occurred before the Günz, to which has been given the name *Donau*.

The glacial cycle has also been investigated in detail in northern Germany and Poland, over which spread the Scandinavian ice-sheets, leaving ample evidence in the form of the terminal moraines and fluvioglacial deposits already described. At first three stages were distinguished, named *Elster*, *Saale* and *Weichsel*; careful correlation with the Alpine stages indicate that these correspond to the Mindel, Riss and Würm respectively.

Subsequent evidence has been brought forward suggesting that a further phase intervened between the Saale and Weichsel advances, known as the *Warthe*, whose main moraines can be traced in the Fläming, an area of heathland to the south-west of Berlin, and on into southern Poland. Some authorities contend that the Warthe advance was an early stage of the Weichsel glaciation, but the consensus of opinion maintains that it represents a temporary halt and slight re-advance during the retreat of the ice-sheets towards the end of the Saale glaciation. Other evidence indicates that a still earlier advance of the Scandinavian ice-sheets may have preceded the Elster stage. This has been called the *Elbe* glaciation, and is tentatively correlated with the Günz in the Alps.

Further detailed work has investigated the stages of recession of the fourth glaciation across the North European Plain, each stage reproducing the features of the zones of the ice-margins. Three main stages (the *Brandenburg, Frankfurt* and *Pomeranian*) have been defined; the most prominent is the Pomeranian, the moraines of which are the Baltic Heights or Baltic End Moraine (p. 241).

The multiple nature of the Pleistocene glaciation has also long been recognized in Great Britain. While in northern Germany the glacial stages were mainly defined in terms of terminal moraines, in Britain these features are by no means so evident, and other forms of evidence have had to be utilized. Thus at an early date four distinct tills were distinguished in East Anglia, namely the Norwich Brickearth (now known as the Cromer Till), the Great Chalky Boulder-clay (the Lowestoft Till), the Little Chalky Boulder-clay (the Gipping Till), and the Hunstanton Boulder-clay. At first it was assumed that these corresponded to the Günz, Mindel, Riss and Würm periods, but recent detailed examination of the interglacial deposits has partly disproved this correlation. Of particular importance in this respect is the Cromer Forest Bed, a peaty interglacial deposit outcropping along the Norfolk coast and underlying the Cromer Till and the Lowestoft Till. As the latter is itself overlain at Hoxne, in Suffolk, by deposits of the second interglacial period (Mindel-Riss), the two tills must be regarded as the product of one glacial period (the Mindel), and the Cromer Forest Bed itself as of the first interglacial period (Günz-Mindel). The first glacial period (the Günz) thus appears to be something of a mystery in East Anglia; its one clear representative would appear to be the Weybourne Crag, a shelly deposit containing the remains of creatures accustomed to live under cold conditions. A still earlier glaciation, which may be contemporaneous with the Donau, is possibly indicated by the Red Crag, a marine bed with fossils of creatures which obviously lived under very cold conditions.

As elsewhere, the glacial deposits may be divided broadly into the 'Younger Drift' and the 'Older Drift'. The southern limit of the Newer Drift may be traced from the coast of northern Norfolk through the Vale of York (where it is indicated by terminal moraines), the West Midlands and South Wales to the mouth of the Shannon.

The later stages of this final glacial advance strikingly affected the uplands of Britain before the progressive amelioration of climate led to the disappearance of the ice-sheets from these islands. Most of northern Britain was covered by the continuous North British Ice-sheet, and later still Snowdonia, the English Lake District, the Southern Uplands and the Scottish Highlands nurtured individual ice-caps long after the overall swathing ice-sheets had retreated. From the ice-caps, outlet-glaciers (p. 227) pushed downward and outward, producing many striking erosive forms.

The chronology of the Quaternary glaciation has also been patiently unravelled in the North American continent. Despite some serious and unsolved problems, a four-fold glaciation has been established, to which is given the names *Nebraska, Kansan, Illinoian*, and *Iowan-Wisconsin*, tentatively correlated with the four Alpine glaciations. Like the Riss in Europe, with which it was contemporaneous, the Illinoian extended farthest south. The four glacial periods were separated by interglacials known as the *Aftonian* (the earliest), the *Yarmouth*, and the *Sangamon*. Possibly a still earlier *Pre-Nebraskan* glaciation may have occurred.

THE PERIGLACIAL FRINGE

In addition to the areas of land covered by the advances of the Quaternary ice-sheet, there must have been a zone or belt, both latitudinal and altitudinal, known as the *periglacial fringe*, lying marginal to the ice-front. This belt must have been of considerable width, although fluctuating greatly during Pleistocene times; during interglacial periods it would have migrated poleward and towards the summits of higher mountains, while during glacial advances it must have affected even low-lying areas in middle latitudes, such as southern England and northern France. The study of periglacial phenomena has developed extensively during the last 25 years, and a formidable jargon has been introduced, not all of which is universally accepted. The thickness of the surface layer liable to alternate freezing and thawing is known as the *active layer* or *mollisol*,

THE GLACIAL CHRONOLOGY

The following table, based on the reseaches of P. Woldstedt, F. E. Zeuner, R.G. West and others, hazards an outline correlation of the Quaternary Glaciation, using a 'short time-scale' in which the onset of the glaciation occurred about 600,000 years ago. If a 'long time-scale' (p. 247) is adopted, the various stages must be increased proportionally; the Great Interglacial may on this reckoning have exceeded 400,000 years.

Estimated duration (*thousand years*)	The Alpine Area	North Germany and Poland	British Isles	North America	
100	Würm 3 Würm 2 Würm 1	Weichsel { Pomeranian Frankfurt Brandenburg	North British York Line Hunstanton Boulder-clay	Wisconsin Iowan	YOUNGER DRIFT
65	THIRD INTERGLACIAL PERIOD				
60	Riss	Warthe Saale	Gipping Till	Illinoian	OLDER DRIFT
190	SECOND ('GREAT') INTERGLACIAL PERIOD				
50	Mindel	Elster	Lowestoft Till Cromer Till	Kansan	
65	FIRST INTERGLACIAL PERIOD				
50	Günz	Elbe ?	Weybourne Crag	Nebraskan	
?	Donau?	?	Red Crag?	Pre- Nebraskan?	

while the permanently frozen thickness beneath is the *permafrost layer* or *pergelisol*. The first alternative in each case has remained the more generally used. The study of all the features of frozen ground is known as *cryopedology*. A detailed classification system of processes has been published, of which three may be mentioned: *congelifraction* (frost splitting), *congeliturbation* (frost heaving and churning), and *congelifluction* (earth flow under frost conditions).

This periglacial zone was characterized in most areas by deep-freezing of the soils, sub-soils and even bedrock (*permafrost*), producing tundra characters; as a result, important changes in hydrological conditions occurred. Firstly, many rocks had their joints and interstices sealed by ice, and were thus rendered impermeable. On such rocks runoff and erosion may have been much greater than under normal conditions, and it is possible that many dry valleys of chalk and limestone uplands may have been formed in this way (pp. 126–8). Secondly, the relative importance of surface runoff was increased owing to a reduction in direct evaporation (the result of the low prevailing temperatures) and to the almost complete absence of plant transpiration.

A very important form of weathering under periglacial conditions is *freeze–thaw*, particularly during months when the day-to-night temperature changes fluctuated on either side of 0° C. Jointed rocks would be especially prone to this form of breakdown, which would quickly produce a mantle of debris for powerful melt-water streams to remove. Some of the most striking effects of freeze–thaw were associated with *snow-patch erosion*, or *nivation*. On north-facing slopes in particular, small hollows were occupied by semi-permanent snow-accumulation. Beneath the snow, freeze–thaw led to the weathering or comminution of the bedrock, and the particles thus loosened were washed out during warmer periods by rivulets of melt-water. The snow-patches thus slowly 'ate' into the hillsides, producing large and distinctive nivation-hollows.

Another very important process of the periglacial fringe was rapid soil-creep, *solifluction* or *sludging*, one of the special categories of mass-movement (p. 103); under periglacial conditions this is known as *congelifluction*. During short summer periods, some thawing of the upper layers of the soil occurred, so that above the surface of the permanently frozen layer (the *tjaele*) a layer of detritus became heavily charged with melt-water which was unable to percolate away. The superficial layer was able, therefore, to creep downhill, frequently in the form of lobes, a half-metre or so in thickness, particularly on steep slopes where its progress was speeded up by the slipping plane of the tjaele itself. Evidence of past solifluction in Britain may be found in such forms as the Coombe Deposit and Coombe Rock of much of the chalk country, the Head which covers raised beaches in the western coastal areas, and the Tjaele Gravels of East Anglia. The Coombe Deposits consist of chalk-rubble and flints, with some sand, and in places are as much as 24 m (80 ft) thick; where they have been compacted and cemented they are known as Coombe Rock. An excellent example is revealed in the cliff-section at Black Rock near Brighton, where the Coombe Deposits mask the face of a degraded sea-cliff of Pleistocene age cut in the Chalk. At the base of this cliff is a raised beach (p. 312), at about 8 m (25 ft) above present sea-level, its stratified shingle deposits largely hidden by the Coombe Deposits. Further, the Clay-with-flints (p. 131) is in part a solifluction deposit developed from former cappings of Eocene sands and clays on the Chalk. It has been suggested that solifluction may under certain conditions produce terrace-like forms and flattened summits. While some of these

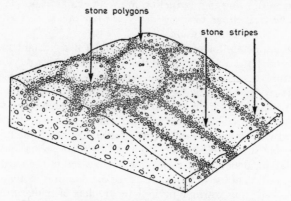

Fig. 125 *Patterned ground formed under periglacial conditions*

may be accumulated features of weathered rock-material, it is believed that erosion of the solid rock may occur, a process known as *altiplanation*. The recognition of features of this kind on the chalklands of southern England and on Dartmoor has been claimed, though they are evident on a vastly larger scale in Alaska.

A most interesting result of freeze–thaw is the development of 'patterned ground', a term coined by A. L. Washburn for more or less symmetrical features such as circles, polygons, nets and stripes (fig. 125), characteristic of areas now or in the past subject to intensive frost action. The ground material is sorted into polygonal forms up to a metre or more in diameter, with stones around the perimeter and finer material in the centre. There is no certainty as to the mode of these forms; Washburn reviewed nineteen possible hypotheses relying on freeze-thaw processes, such as contraction following temperature change, moisture controlled movements and solifluction. In fact, patterned ground probably has a number of causes and what these are is the subject of much research.

If the ground freezes under periglacial conditions to low temperatures (below −5° C), especially where it is covered with a thick mantle of loose material, cracks will develop. In summer these fill with melt-water, which freezes to form an ice-wedge, tapering downwards. Each winter the cracks and wedges may enlarge, and ultimately they may penetrate up to 10 m (35 ft). Repeated melting and freezing not only causes an increase in size, but also helps to shatter the surrounding material. The sites of former ice-wedges can often be recognized by curving masses of gravel outlining their shape (fig. 126).

Terminal curvature of
surrounding beds

ice

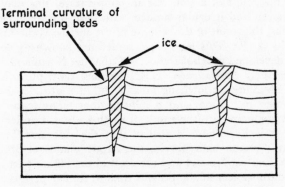

Wedges preserved by infilling

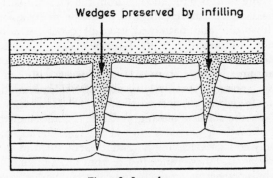

Fig. 126 Ice-wedges

Finally, the factors must be noted which favour the formation of climatic river terraces under periglacial conditions. In some areas, particularly of high relief and steep slopes, the processes of freeze–thaw weathering and solifluction were so active that the rivers were supplied with more detritus than they could effectively transport. Thus, instead of increased erosion, the infilling of valley bottoms took place. With the establishment of interglacial and postglacial conditions, however, weathering processes were slowed down and soil movement was hindered by the anchoring effect of vegetation, so that the now underloaded streams were able to cut into the infillings to leave them upstanding as terraces. Such terraces are, of course, to be distinguished from those due to the numerous sea-level changes of the Pleistocene period.

When masses of ground-ice lying among a drift-cover finally decay, the result of a slow rise in temperature, this may leave a very uneven and irregular surface, with basin-shaped hollows, pits and sinks, the result of the collapse of the surface as the underlying ice melts *in situ*. This landscape, sometimes known as *thermokarst*, is well developed in Alaska and the Canadian Northland.

Contemporary periglaciation This preceding description is of conditions obtaining in lowlying areas in middle latitudes during the Quaternary glaciation, and it may be that many features of the present landscape are the result of earth-sculpture under cold climates. It is important to realize that the extensive Tundras of northern Canada and Siberia, the Arctic islands of Spitsbergen, Novaya Zemlya, Iceland and others, and the higher parts of Scandinavia are subject at the present time to similar periglacial influences. In recent years much work has been done on the features resulting from frozen ground. Borings indicate that permafrost may extend more than 600 m (2000 ft) from the surface. All the features mentioned above (the creation of a mantle of rock-waste, blockfields, frost-soils and patterned ground) occur on a huge scale. The study of ground-ice, in the form of wedges, lenses and needles (*pipkrakes*), is of particular interest. One fascinating feature results when autumn freezing traps a layer of water between the newly frozen surface and the underlying permafrost. Hydrostatic pressure, resulting from the expansion of water on freezing, may raise a 'blister' on the surface, leaving an isolated dome-shaped or conical mound of earth or gravel, known as a *pingo*. The top of the pingo commonly collapses, leaving a 'crater' partly filled in summer with a shallow pond. Pingos range in size from small mounds 6 m (20 ft) high to hills of over 90 m (300 ft) in height and with a basal diameter of 0·8 km. Many of these features have been recently or currently formed, though some large ones, from the evidence of radiocarbon dating, developed 5–6000 years ago. American workers distinguish between 'closed system pingos', where the layer of water thus trapped is isolated, and 'open system pingos', where the hydraulic head is related to an adjacent slope; the latter are especially common in eastern Greenland, where plentiful water from summer snowmelt on the mountains flows downwards towards the coast over the layer of permafrost. Some workers have sought to call these features *hydrolaccoliths or cryolaccoliths*, restricting the term pingo to a small isolated mound with an actually existing ice-core, but there appears to be no unanimity about this usage.

We can then, hardly avoid the conclusion that under peri-glaciation the physical landscape is in a state of rapid evolution. Indeed, many geographers are today of the opinion that the present surface features of much of Britain, including such fundamental forms as slopes, were carved under these abnormal conditions, and that the present period is by contrast one of relative stagnation in terms of land-form development.

POSTGLACIAL TIME

Much attention has been devoted to working out the chronology of postglacial time, although of course, this refers to the lowlands in middle latitudes, not to high latitudes and altitudes where per-manent ice is still present. It will be also appreciated that as the Quaternary glaciation began and ended at different times in various areas, according to their position relative to the main centres of ice-dispersal, no single date can be ascribed to the beginning of post-glacial time. In geological terms, this is the division between the Pleistocene and the Holocene or Recent; an arbitrary date for East Anglia would be about 15,000 B.C., for southern Scandinavia about 8000 B.C., for southern Finland about 6500 B.C.

The withdrawal of the ice-sheets has had profound effects on the surface features of the land. As the ice withdrew, so did the peri-glacial fringe, with its diverse phenomena. Moreover, as the ice melted, quantities of water were returned to the sea, resulting in rises of relative sea-level. This change was complicated by isostatic recovery (p. 23) as the weight of the overlying ice-sheets was reduced. The net result has been a fluctuating relative level of land and sea with resultant modifications on the shapes of the continental seas, on the nature of the coast-line (p. 303), and on the base-levels of rivers, with all that is involved therein (pp. 145–8).

Several methods have been used for working out the postglacial sequence. The most important are the study of varved clays, mainly in Scandinavia (p. 13), and pollen analysis, the determina-tion from the nature of the pollen preserved in peat of the plants growing at the time; much work has been done by H. Godwin on these lines. If a particular plant association can be recognized in the peat, an indication is given of the contemporary climatic conditions which allowed it to flourish. Other evidence includes the shells of molluscs, the archaeological remains of culture periods (often found in river gravels and peat-bogs), the counting of tree-rings (*dendro-chronology*), and radiocarbon dating (p. 42).

Much work has been done in various parts of the world, but two summary examples must suffice: the time-charts for East Anglia and for Scandinavia. It must be remembered that much research is still in progress, that the dates ascribed are constantly being modified, and also that they vary with latitude, i.e., they are *time-transgressive*.

I. EAST ANGLIA

(mainly after H. Godwin)

A. *Pre-Boreal* or *Sub-Arctic*. Before 7500 B.C. Dry and cold climate. Birch-pine flora. Relative sea-level 600 m (200 ft) lower than at present. Britain joined to Europe.

B. *Boreal*. 7500–5500 B.C. Dry climate, cold winters, warm summers. Pine-hazel flora. Steady rise in relative sea-level, and development of North Sea.

C. *Atlantic*. 5500–3000 B.C. Climate much milder and damper, attaining what is called 'the climatic optimum', with temperatures about 2° C above present. Mixed oak forest. Widespread growth of peat. Continued rise of sea-level. Breaching of Straits of Dover about 5000 B.C.

D. *Sub-Boreal*. 3000–500 B.C. Renewal of cooler climatic conditions, probably somewhat drier. Oak giving place to pine.

E. *Sub-Atlantic*. Since 500 B.C. Moister climatic conditions. Alder-oak-elm-birch-beech flora, with the beech generally dominant. Renewed growth of peat.

II. SCANDINAVIA

(mainly after E. H. de Geer)

Four main stages have been postulated, with special reference to the evolution of the Baltic Sea.

1. *Daniglacial*. 18,000–15,000 B.C. The stage of ice-retreat from Denmark, though Scandinavia itself was still ice-covered.

2. *Gotiglacial*. 15,000–8000 B.C. Ice-retreat from southern Sweden. The 'Baltic Ice-Lake' was enlarged into the Yoldia Sea by an extension of the North Sea across the southern Baltic region.

3. *Finiglacial*. 8000–6500 B.C. Retreat of the ice-sheet from Finland, after a pause allowing the formation of the Salpausselka (p. 241). Much removal of ice, hence uplift of land by isostatic recovery. The Yoldia Sea was reduced to the land-locked Lake Ancylus, with overflows across central Sweden and through the Sound.

4. *Postglacial.* Since 6500 B.C. Widespread oceanic transgression, forming the Littorina (sometimes spelt Litorina) Sea (slightly larger than the present Baltic), and spreading over lowlands of Western Europe (the 'Flandrian Transgression'). Maximum extension of sea at about 4000 B.C. Since then, an oscillation of about 3–4·5 m (10–15 ft) on either side of present sea-level: 0·6 m (2 ft) above in 1600 B.C., 4·5 m (15 ft) below 700 B.C., 1·5 m (5 ft) above B.C./A.D., 2·4 m (8 ft) below A.D. 800.

CHAPTER 9

The Desert Lands

ABOUT a third of the land surface of the globe experiences desert or semi-desert conditions, not including the polar and sub-polar lands which are sometimes called 'cold deserts'. The basic fact responsible for the existence of a desert is aridity (p. 449, 481), expressed in the scantiness of the vegetation cover (pp. 530–3). The climatic definition of deserts and semi-deserts is discussed in Chapter 18; it will suffice to say that there are two main groups of arid lands.

The tropical, sometimes called 'Trade-wind', deserts lie between about 20° and 30° N and S of the Equator, and are especially marked on the western sides of continents. The greatest continuous extent is the 'Old World Desert Belt', which stretches from Morocco in north-western Africa, through the Sahara, Arabia and Baluchistan into north-western India. These deserts gradually merge through transitional climatic and vegetation zones into areas with more humid conditions. The other group of arid lands comprises continental interiors in middle latitudes, particularly high basins lying in the rain-shadow of the surrounding mountain ranges, such as the Gobi and Turkestan Deserts in Asia, and the Colorado Desert and Great Basin in North America.

One important feature of these desert lands is the partial or almost complete absence of vegetation cover, and therefore the exposure of an unprotected surface to denudation. A second feature is that water, the absence of which would seem to be an inherent feature of a desert by definition, is in the opinion of some geomorphologists the dominating agent of earth sculpture there as in humid lands; the difference lies in its intermittence, intensity and spasmodic quality. A large part of the desert lands are, in fact, not arid but semi-arid. A third feature is that wind is an agent of importance, though perhaps not as relatively important as earlier observers, impressed with the dramatic quality of sandstorms, thought; as E. Blackwater said as long ago as 1934, the effectiveness of wind action ". . . is one of the most important unsettled problems of desert geomorphology". A fourth feature is that weathering, both of a mechanical and chemical nature, is extremely potent (p. 97) in preparing a mantle of unconsolidated material on which other agents can work. Of course, these various factors operate in lands other than deserts. The erosional effect of wind can be seen on any unconsolidated surface, such as a newly ploughed field in East Anglia after a period of drought; its transporting and depositional effects are shown by the sand-dune belts of the European coastline, and by the inland sand-dunes of the heathlands of Belgium, the Netherlands and Germany.

Arid conditions are temporary phases over various parts of the earth's surface. The red Triassic rocks of the English Midlands were, in part at least, laid down under arid or near-arid conditions. The individual sand grains comprising the characteristic sandstones are almost perfectly rounded, similar to the 'millet-seed' desert sands of today. Moreover, the Pre-Cambrian rocks of Charnwood Forest— which rises like an island from the Triassic rocks covering most of the ancient surface—show distinct evidence of wind erosion, in the shape of scoring, polishing and fluting.

On the other hand, many present-day deserts are the result of increasing desiccation, even during historic times. Explorers and archaeologists have discovered abundant evidence in central Asia of thriving civilizations and cities now sand-buried, and there are many other indications of what must have been fertile inhabited lands; Sir Aurel Stein, for example, discovered in 1906 the ruins of the city of Tunhuang in the Gobi Desert. The borders of the Sahara have advanced, reaching the Mediterranean coastlands which were once granaries of the Roman Empire. It is possible that when the Quaternary ice-sheets extended southward over the northern continents, the present 'rain belt' of middle latitudes extended farther south also, covering more or less the present deserts. As the ice-sheets withdrew northward, so did the rain belt, and the desert border advanced. It may be that the 'glacial period' in higher latitudes during Quaternary times produced a 'pluvial period' in the hot deserts. To the west of the Wad Rir' depression in southern Algeria is a rocky limestone desert from 580–700 m (1900–2300 ft) above sea-level, cut up by a maze of dry gorges and known as the Chebka or Shebka, which means a net. The pattern is much like that of the 'badlands' of the western U.S.A., which has a rainfall of about 40 cm (15 in), so the suggestion has been made that the erosion in the Chebka was carried out in a past pluvial period.

WIND EROSION AND TRANSPORTATION

The result of desert weathering (pp. 97–8) is the production of a mantle of rock waste in a form small enough to be moved by the wind. In addition, the infrequent and intermittent, yet powerful torrents resulting from localized downpours sweep down loose unconsolidated masses of mud, sand and gravel, which, after they have dried out, can be readily acted upon by the wind.

The erosive power of the wind involves a triple process: (i) deflation; (ii) abrasion leading to corrasion; and (iii) attrition.

(i) **Deflation** This involves the blowing away of any unconsolidated material, actually lowering the land surface. The finest material is borne away as a dust-storm, in which a cubic mile of air is estimated to contain 16,000 tons of dust. It may travel for many miles, as indicated by deposits of red Saharan dust which sometimes fall in Italy and southern France, even on rare occasions in southern England. Heavier material in the form of sand grains is swept along in sand-storms, while still coarser material moves in a series of hops near the surface. The wind is therefore a selective transporting agent. One of the most graphic results of deflation is soil erosion (pp. 506–8), as shown in the 'Dust Bowl' area of western U.S.A. Deflation is also a very important cause of the formation of both small and large depressions, clearly seen in the limestone plateau west of the Nile Valley. The largest is the Qattara, the floor of which reaches 134 m (440 ft) below sea-level; the smaller Faiyum depression is at 40 m (130 ft) below, though its rim rises 330 m (1100 ft) higher. Similar though shallower depressions are to be found in the Kalahari and West Australian deserts, and in Mongolia. Many contain oases or salt-lakes on their floors, indicating that the water-table, the limiting factor in the wind's deflating effect, has been reached.

The chief difficulty is to account for the beginning of localized deflation. Faulting may initiate a basin, since if a resistant surface cap is broken, the wind can get at the underlying softer rock, and when a small local 'blow-out' has been formed, it is gradually enlarged and deepened by eddy action (fig. 127). Some smaller basins in the North American deserts seem to be due to differential weathering of areas of less resistant level-bedded rocks, and are often surrounded by pronounced rims of more resistant rock. Another theory is that solution during a past wetter period helped to account for the beginnings at least of some depressions.

(ii) **Abrasion** The sand-blast effect of wind carrying a load of hard quartz grains is of great potency. It can smooth and polish a surface of homogeneous rock where horizontal beds of uniform hardness occur. Individual pieces of rock, broken off by mechanical weathering, but too heavy to be moved by the wind, are worn on the windward side, and are known as *ventifacts*. A particular type of ventifact, with three wind-faceted surfaces, is called a *Dreikanter*. The

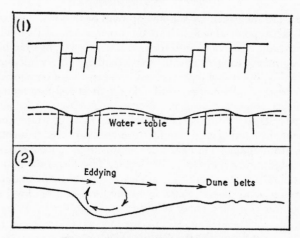

Fig. 127 Deflation hollows in the hot deserts

1. Minor fault displacements may initiate a depression if a resistant surface stratum is fractured. This allows the action of the wind to attack the less resistant strata until the water-table is reached. **2.** Erosion by wind-eddies.

surfaces of pebbles may be so wind-polished as to form a 'desert pavement' or 'desert mosaic'. If the rocks are of varying degrees of hardness, differential abrasion produces etching, grooving, fluting and honeycombing. Lines of weakness, such as joints, are attacked, producing a grotesquely pinnacled relief. The heavier particles are carried near the ground but their force is retarded at ground-level itself by friction, so that undercutting is most marked 0·5 m or so above the ground, leaving a rock 'pedestal'. Undercutting tends to steepen the faces of rocky hills, forming cliffs with shallow wind-worn recesses, almost deep enough to be called caves, at their bases.

Some fantastic rock structures are produced by wind abrasion. The characteristic mushroom blocks, undercut at the base, are known as *gour* (singular *gara*) in the Sahara. Where a hard horizontal stratum lies above a soft one, if weathering along the joints breaks into the hard cap, wind abrasion will carry on until separate tabular masses (*zeugen*) are left standing upon the softer underlying rock, until finally they are completely undercut. These zeugen often stand up prominently 30 m (100 ft) or more above the surrounding level.

When rocks of differing resistance occur in bands roughly parallel to the direction of the prevailing wind, a 'ridge and furrow' relief may be produced, as in the Atacama Desert. Well-worn ridges, with blunt rounded fronts facing the wind and with a sharp keel-like crest, are separated by shallow grooves or furrows. An extreme form of this type of abrasion is in the central Asiatic deserts; here the ridges have been worn into rock-ribs, 6 m (20 ft) or so in height, known as *yardang*, roughly parallel to each other. Thus wind abrasion and deflation, operating together, cut into an area of bare rock. The depressions gradually merge, until only the most resistant masses are left standing above the surface.

(iii) **Attrition** Wind-borne material is in a constant state of movement, impacting both against the rock surfaces it meets and also one grain against another. The result is to produce a characteristic rounded sand grain (or 'millet seed'). This forms the dominant 'end product' of desert erosion, and is the constituent material of the extensive sand deserts, whether in the form of great level sheets or in the ever-changing chaos of dunes.

WIND DEPOSITION

The load carried by the wind depends on the size of the material and on the strength of the wind. The finest dust-like grains may be moved right out of the desert areas, the heavier ones may simply be re-sorted within the deserts themselves. There are the spectacular dust-storms through which the sun shines as through ground-glass; there are the stinging sand-storms, a menace to desert travellers; and there is the endless movement of the surface, a hissing, whispering trickle down the face of a dune, which, if the wind rises, may become a 'smoking dune'.

Many factors account for the various depositional forms. The nature of the surface over which the sand is moved is important— whether it consists of deep sand, or of bare rock (over which sand will move with little to arrest its progress), or whether rocks and pebbles are strewn about. The presence of an obstacle may start a mound—a rock, a thorny shrub or an area of pebbles which may exercise a frictional drag. Once a patch of sand becomes higher than its surroundings, it will continue to grow. The main difficulty is to explain how a dune may start, especially on a relatively featureless surface with no obstructions; there is no satisfactory answer.

The direction and strength of the dominant winds is another factor; the correlation of dune alignment with the winds is obvious. The presence of vegetation (often deliberately planted to protect an oasis, and in fact the whole Nile valley is lined with eucalyptus trees, acacia and other drought-resisting bushes), and of ground-water reaching the surface, may exert considerable control on the movement and deposition of sand.

Large areas of the hot deserts are covered with level, gently rippled or slightly undulating sheets of sand; the surface may be in continual motion, but the general features remain the same. These level sand-sheets are nearly always pebble-strewn, for the wind seems to spread sand evenly between the pebbles.

Dunes The accumulation of sand into low hills of varying shape and extent is one of the most interesting features of desert relief (plate 83). According to R. A. Bagnold, the word dune in its true sense should be restricted to barkhans and seif dunes. Other features of accumulated sand are known variously as 'shadows', 'drifts', 'sheets', 'undulations' and 'whale-backs'. The term is here used in its more general sense. Detailed classifications include such fanciful names as 'star dunes', 'hairpin dunes', 'pyramid dunes', 'sword dunes' and 'smoking dunes'.

The simplest form is an *attached-dune* or *head-dune*, a sand-drift caused by an obstacle such as a rock in the path of wind; the sand accumulates in 'dead-air' spaces round the rock to form as nearly as possible a 'wind-line' shape. On the lee side of the rock a *tail-dune* may taper gradually away for a distance varying from 3 m (10 ft) to ¾ km (½ mile). On the windward side of the attached-dune, separated from it by a space kept clear by eddy-motion, is an *advanced-dune*, filling the dead-air spaces created as the air current rises to surmount the obstacle. Flanking the rock are *lateral-dunes*, with *wake-dunes* trailing away in the same direction. These lateral-dunes are separated from the main mass by a trough, probably kept clear by the slip-stream on either side of the obstacle. These dunes are shown in fig. 128, but this is an idealized pattern and all sorts of variations, the result of other factors, can be seen.

Barkhans To crescentic dunes has been given the name *barkhan* (also spelt *barchan* or *barchane*), derived from the deserts of Turkestan. These dunes occur transversely to the wind, their 'horns' trailed out in the direction towards which it is blowing, since there is less mass to be moved on the edges of the dunes. In Turkestan the wind changes seasonally, blowing alternately southward and northward and the dunes change likewise, their horns swinging right round.

The chief problem concerning barkhans is their initiation, how they start to form. This is not connected with large obstacles, for they seem to occur readily on fairly level open surfaces; possibly patches of pebbles, or a sudden wind fluctuation, may cause the accumulation of a low heap of sand, from which the features of the characteristic barkhan can develop.

Barkhans vary in height up to more than 30 m (100 ft). The slope is gentle on the windward side, steeper on the 'slip-face' or sheltered leeward side, where eddy motion assists in maintaining a slight

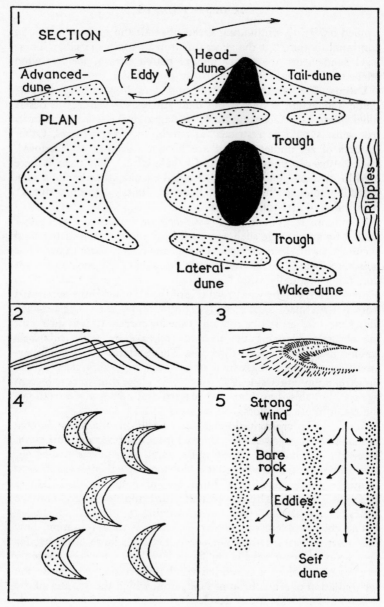

Fig. 128 Sand-dunes

concave slope. If the supply of sand is continuous, the dune may advance as a result of its endless movement up the windward slope and over the crest; such dunes may constitute a threat to oases unless their movement is checked. A similar threat occurs from the advance of coastal dunes (p. 301). Many experiments have been made in connection with the stabilization of dunes on the borders of deserts, lest they advance and overwhelm the oases and irrigated lands, as in Libya, Egypt and Israel. They involve the planting of shelter belts of acacias and eucalypts. Recently the ground has been sprayed between newly planted seedling trees with a carefully blended oil, which helps the surface sand to cohere, or is covered with a plastic film to prevent its blowing and thus baring the roots.

Barkhans are occasionally found as isolated hills, but more often they occur in groups, sometimes as a regular series, more often as a chaotic ever-changing pattern of partially coalesced ridges at right angles to the prevailing wind. Such a desert is among the worst country to cross; occasionally, however, it is possible to find long passages, known as *gassi* in the Sahara, through the dune country, permanent enough to be regular caravan routes. They probably originated as a fortuitous series of gaps between the sand-dunes more or less in a straight line, which the wind treated as a 'wind tunnel', concentrating its unimpeded force straight along it, and so keeping it free of sand.

Longitudinal dunes Sometimes sand occurs as long ridges, often many km in length, crossing the desert in the direction of the prevailing wind. They are known as *seif-* or *sif-dunes*. These may be formed by the coalescence of a line of barkhans, probably where in addition to the prevailing winds other cross winds occasionally blow at right angles, sweeping off the tails of former barkhans. The dominant wind blows straight along the depressions between the dune-lines, keeping them clear of sand, while eddies help to build up the sides of the dunes. Extensive lines of seif-dunes are to be found in the Sahara south of the Qattara depression, in southern Persia, in the Thar Desert and in the West Australian Desert; in the last-named they extend unbroken for hundreds of km as ridges 15 m (50 ft) high. In the words of A. D. Tweedie, '. . . the visible landscape is a flat pink disc ribbed from horizon to horizon by the red sand ridges and

---

1. Section and plan of a dune assemblage resulting from an obstacle in the path of the wind. **2.** The advance of a moving dune; the sand is blown up the gentle windward slope and falls over the crest into the 'wind-shadow' of the slip-face, where eddying gives it a concave form. **3.** Sketch of a *barkhan* or crescentic dune; the wind blows the sand over and round, hence the elongated horns. **4.** A series of barkhans. **5.** Linear *seif* dunes, parallel to the prevailing wind, but subject to cross-winds which bring plentiful sand.

streaked by the darker lines of acacia and spinifex which grow in the slightly moister areas between them.'

Dunes in western Europe Sand-dunes are not only a common coastal feature (p. 300), but they occur in middle latitudes wherever the surface consists of unconsolidated sands and particularly where the binding effects of a scanty vegetation are limited, as in some of the heathlands. In the Kempenland of north-eastern Belgium extensive dune patterns are orientated more or less from south-west to north-east. To the west and south of Lommel, near the Dutch frontier, the moving dunes vary in height from 5 to 15 m (15–50 ft), some of them showing a characteristic crescentic form. Their movement is not very marked, since the unplanted and therefore 'unfixed' areas are small in extent. The movement of sand is within the dune area itself; the wind destroys, piles up, levels and hollows out an ever-changing relief.

Loess The finest wind-borne material is finally laid down far beyond the limits of the deserts. A German geologist, von Richthofen, first studied these deposits in north-western China (fig. 129), where a sheet of this yellow, friable, porous material, covering nearly 650,000 sq. km (250,000 sq. miles), swathes the 'solid' landscape to a depth of from 90–300 m (300–1000 ft). It occurs at all elevations, from near sea-level to about 2400 m (8000 ft). To it the name *loess* has been given, from a village in Alsace where somewhat similar deposits occur. In section it reveals innumerable vertical tubes, often lined with calcium carbonate, which are thought to be the 'casts' of grass stems. The loess grains are very coherent; as a result of this fact and of its porosity, which allows water to sink rapidly through, not only rivers but also roads are contained between steep loess walls, while habitations are actually cut in the cliff sides (plate 101). In addition to north-western China, loess is found in many of the basins and depressions of central Asia.

Loess represents the accumulation, over a long period of time, of fine material carried by outblowing winds from the Asiatic winter high-pressure area of the Gobi Desert. It was deposited where the steppe vegetation exercised a binding effect, and possibly where increased rainfall helped to wash it down from the air to the earth, where the coherence of its grains helped to stabilize it.

The loess in Europe The name loess is applied in Europe to deposits of dust which were removed by the wind, probably during dry inter- and postglacial periods, from the unconsolidated sands and clays laid down by continental ice-sheets. Though extremely fine, when viewed microscopically it can be seen to consist of tiny angular pieces of felspar and calcite, obviously the result of glacial

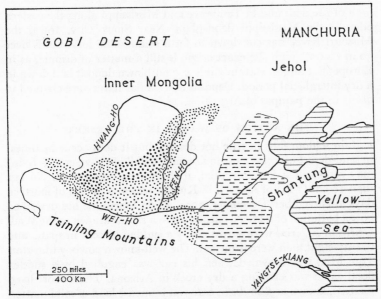

Fig. 129 *The loess area of northern China*

Areas with loess exceeding about 75 m (250 ft) in thickness are shown by large dots, areas with thinner deposits by fine dots, and areas where loess has been mixed with alluvium and re-deposited by pecked lines.

abrasive activity, although weathering soon dissolves the calcite and more slowly breaks down the felspar. It is found in those parts of Germany known as the Börde, which lie to the north of the uplands of central Europe, on the low plateaus of central Belgium, and in north-eastern and eastern France; in the last two countries it is known as *limon*. There is some dispute as to the exact mode of its origin, for it has been redeposited by rivers on their terraces, and is found at all altitudes up to 1500 m. (5000 ft). It is always found west and south of the Baltic End Moraine, that is, outside the area covered by the last great advance of the ice-sheets (p. 241). Water may be involved in its redistribution, but wind (perhaps blowing from a high-pressure area over the ice-sheets) must have been the primary agent.

The loess in America Thick deposits of a broadly similar wind-borne deposit occur in the western American states, in the broad valleys of the Mississippi and Missouri. This is known generally as *adobe*. It forms great sheets over Illinois, Iowa, Missouri, Kansas and Nebraska, and extends interruptedly southward almost as far as the

Gulf of Mexico; and in Tennessee and Mississippi along the eastern edge of the Mississippi flood-plain. Near Sioux City, Iowa, the Missouri River has cut down to form remarkable loess bluffs over 30 m (100 ft) high. Its exact origin is still a matter of dispute, as in Europe, but almost certainly it is a wind-blown deposit laid down in a dry interglacial period. Deposits of a similar nature are claimed to occur in the pampas of Argentina.

THE ACTION OF WATER IN THE DESERT

Rain seldom falls in true hot deserts, but it does occur at times; in semi-desert margins a mean annual rainfall of 12–25 cm (5–10 in) may represent a few torrential, if short-lived downpours. No less than 18 cm (7 in) of rain fell in Kuwait within a period of fourteen days in November–December, 1954. E. de Martonne has described how a battalion of soldiers, encamped in a ravine known as the Wadi Urirlu in Algeria, were surprised by a sudden torrent in spate, and twenty-eight men were drowned. The well-known cowboy film-star, Tom Mix, was drowned when his car was engulfed by a sudden flood as he was crossing a dry creek in Arizona. These short lived torrents, known as *flash-floods*, carry an immense load of solid matter, the product of desert weathering, and they sweep along so much material that they turn into mud flows and soon come to rest. At other times the stream divides into channels, so forming an alluvial cone or 'dry delta', usually at the mouth of the valley or at the foot of the hill-slope where the gradient eases (fig. 78). This mass of loose unconsolidated material can then be attacked by the wind.

Where a number of dry stream courses open into a depression which is more or less enclosed, its floor may be covered with water-borne sheet-flood deposits (fig. 130). A 'piedmont fringe' of angular scree, gravel and coarse material surrounds the edge of the basin. Where a number of alluvial cones coalesce along the edge of a depression, a gently undulating sloping surface, known as a *bajada*, is formed. This is succeeded by an expanse of sand, and then perhaps by a mud-sheet in the centre, which at times may become a swamp or even a temporary salt-lake, surrounded by sheets of rock-salt and gypsum. These basins of inland drainage occur in every desert, but especially in the large intermont depressions of North America (*bolsons*) and central Asia, where the mountain rims are high enough to receive appreciable precipitation in the form of rain or snow; this whole landscape in America is referred to as 'mountain-and-bolson desert'. Both the temporary lakes and the hollows they occupy are known as *playas*, *salinas* or *salars* in the U.S.A. and Mexico, and as *shotts* (in French *chotts*) in North Africa.

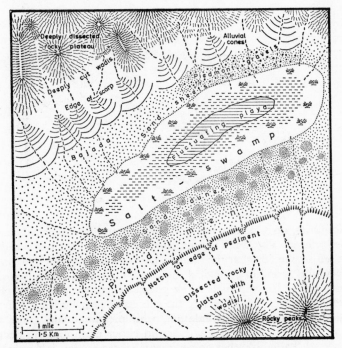

Fig. 130 *Diagrammatic representation of a desert basin*
Based on features shown in intermont depressions in the south-west of Nevada, in the United States.

Wadis Ravines, wadis, nullahs, and in America 'washes' and 'arroyos', are striking features of a rock desert. Their walls, steep and craggy, rise abruptly from the floors, mainly because of the concentrated vertical erosion by powerful, though short-lived, torrents, the result of episodic rainstorms. Wadis however, occur in extremely arid areas, where present stream erosion is so infrequent that it can hardly account for the deeply cut ravines, which, moreover, often show interlocking spurs and other features of river action. It may be, as discussed on p. 263, that the 'glacial period' in higher latitudes during Quaternary times may have been accompanied by a 'pluvial period' in the hot deserts. These deeply cut desert valleys are most strikingly developed where they are carved by *exotic rivers,* that is, those which flow across a desert but have derived the bulk of their water from outside it, commonly from snow-melt on distant mountains. Such a river is the Colorado (p. 155), the master-stream of south-western U.S.A., which with its

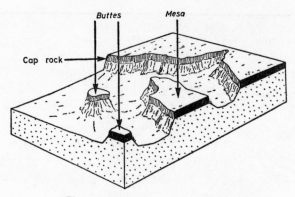

Fig. 131 *Block-diagram of a mesa and buttes*

tributaries, both permanent and intermittent, has trenched across the Colorado Plateau. The immense depth of the canyons is accentuated by the fact that the plateau has been undergoing slow uplift for a long period of geological time, and the river's vertical erosion has more or less kept pace. The plateau edges are steep and cliff-like, sometimes stepped where the successive near-horizontal strata vary in resistance to denudation, with piedmont fans along the flanks where deposition is active, and either more uniform deeply dissected 'badland' relief or pediments where erosion is dominant. Isolated residuals of the original plateau form mesas and buttes (fig. 131), prominent, flat-topped, tabular hills, usually capped with a resistant rock-stratum or a duricrust (p. 102), their profiles accentuated by desert weathering, causing the formation of slopes of angular debris. A mesa is more extensive than a butte.

Sometimes denudation may remove the overlying rocks until a horizontal resistant stratum is reached, leaving a stripped surface. If the strata of which it is composed are gently tilted, a whole series of steps or edges will develop where the resistant beds outcrop, as is so well shown in Utah and Arizona (fig. 132).

Pediments A feature of many arid and semi-arid areas is a gently sloping surface of rock, bare or with but a thin veneer of fine debris weathered from the rock face or mountain-front above, that stretches away from the foot of an upland mass. Its upper edge runs abruptly, with a marked change of slope, into the mountain-front or into embayments formed where the upland is dissected by deep steep-sided canyons; the lower edge dips very gradually (with an angle varying from 2° to 6°) beneath the accumulation of debris found in the lower part of a desert basin. Occasionally an accumulation of

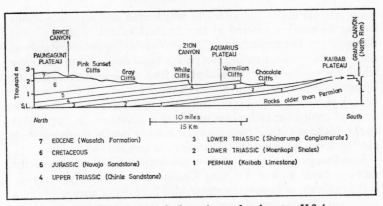

Fig. 132 Escarpments in the desert-plateau of south-western U.S.A.

Based on H. E. Gregory, *Geologic and Geographic Sketches of Zion and Bryce Canyon National Parks* (1956).

scree and alluvial material conceals the sharp nick at the base of the mountain-front, and in some instances the rock surface is completely masked by the bajada.

To this rock surface is given the name *pediment*. Superficially it is in many ways similar to the bajada, especially when it is strewn with alluvium. But while the latter is an undulating plain of aggradation formed by the accumulation of debris, the pediment is wholly the product of degradation, cut independently across rocks and structures of very varying characteristics.

There is much controversy about its origin, but it is undoubtedly the product of denudation under semi-arid conditions. One school of thought regards it as the result of lateral planation by streams, sheet-flow, rills and downwash caused by episodic rainstorms. Several types of evidence support this view. The pediments are always concave in profile, a fact which suggests the erosive action of running water; they are often scarred by numerous small water-channels showing signs of lateral migration; and the sharp change of angle at the upper margin of the pediment is explained in terms of the contrast between the concentrated runoff on the mountain-face in deep gullies and ravines, and the unconcentrated wash of sheet-flood across the pediment. It is significant that the change of angle gradually disappears near the margins of the semi-desert where rainfall increases in both duration and total, causing more consistent down-cutting.

A second school regards pediments as resulting from slope retreat (p. 107); the steep mountain front with an angle of slope of 30° or more retreats under the attack of weathering and erosion without having its gradient altered to any marked extent. The pediment therefore is a low angle (6° to 7°) slope developing independently at the foot of a steep retreating face; it is a basal slope of transport over which the weathered material derived from the face above is carried by occasional rainstorms. Some refashioning of the pediment by lateral planation may occur, but only as a subsidiary process. A series of coalescing pediments will form a pediplain (p. 109). Akin to the pediment is the *peripediment*, a level or gently sloping surface cutting across an area of infill rather than into the solid rock.

Inselbergs A feature of most deserts, and indeed also of the savanna-lands with a Tropical Continental climate (pp. 463–4), is the presence of prominent steep-sided hills rising abruptly from a pediment which extends on all sides into a monotonous plain; sometimes there is a distinct marginal depression around the flank of the *inselberg*, the general name given to these upstanding masses. The term *bornhardt* is sometimes used (after the person who first studied them), though some researchers have sought to restrict this term to a specific type of inselberg, those with a distinctly domed summit.

Most inselbergs are made of granite, gneiss or other crystalline rocks. They are believed to develop from the extension of pediments on each flank of a rocky ridge; thus the mountain front on each side recedes through slope-retreat (p. 107), ultimately reducing the intervening upland to an inselberg. This accounts for the presence of such a prominent feature in a landscape which otherwise shows many of the characteristics of a late stage of an erosional sequence. The spacing of joints and other lines of weakness plays a major part in determining the form of these residual masses; some claim that their margins are delimited by major joints. The dome-like surface of many inselbergs may be attributed, in part at least, to pressure release (unloading) and the development of curvilinear jointing (pp. 98–9), further affected by exfoliation (p. 97). A few writers suggest that some inselbergs are simply residual masses resulting from the exposure and perhaps some re-shaping by weathering of the cores of resistant rock revealed by the removal of a deep layer of weathered regolith, rather akin to one theory of the formation of tors (p. 100). This rotting is particularly active under conditions of constant high temperature and periodic humidity; the rotted layer may extend for hundreds of metres in depth.

Where the rocks are strongly jointed in a rectilinear pattern, the inselberg may develop into a large castellated pile of rectangular boulders; these are sometimes known as *castle-kopjes* or simply as *kopjes*.

An inselberg may, of course, be the product of several cycles of erosion, and several domed surfaces ('dome-on-dome residuals') may sometimes be recognized, superimposed upon each other. This is the result of what is known as *multi-cyclic bevelling*.

DESERT LANDSCAPES AND THE
CYCLE OF ARID EROSION

Desert landscapes, compounded of these various features which have been analysed, are thus of great variety. It is possible, however, to summarize five distinctive types.

The first is the true sand desert, known as the *Erg* in the Sahara and the *Koum* in Turkestan. It consists of vast, almost horizontal sandsheets, or of regular dune-lines, or of an undulating 'sand-sea'.

The second is the stony desert, where horizontal sheets of smoothly angular gravel cover the surface. This is known as the *Reg* in Algeria and as the *Serir* in Libya and Egypt.

The third is the rock desert or the *Hamada*, which consists of a bare rock surface, wind-swept almost clear of sand. It may form a wind-smoothed pavement, or it may be diversified by zeugen and yardangs.

The fourth is the area of desert plateau, crossed by canyon-like valleys of exotic rivers deriving their water from beyond the desert lands, with steep plateau-edges, isolated mesas and buttes, and piedmont fans at the foot of steep slopes.

The fifth aspect takes the form of rock peaks, as in the Tibesti and Ahaggar ranges of the central Sahara, the chaotic peaks of Sinai, the mountains of western Arabia and those of Baluchistan. With their harsh, serrated outlines, their steep, craggy faces cut into by wadis, rising from a swathing mantle of angular rock waste, they form most prominent features.

Is it possible to perceive any cycle of erosion in such varied landscapes? At the beginning of this century W. M. Davis postulated a theoretical cycle on the following lines. Earth-movements are assumed to produce a series of intermontane basins, in each of which a centripetal drainage system develops. The intervening mountain areas are gradually worn down by weathering and sporadic water action, with the basins providing a number of individual and local baselevels of erosion. During the youthful stage, the basins represent zones of accumulation (that is, the individual base-levels slowly rise) and the uplands are degraded, so that the overall relief of the area is at first diminished. However, with the passage of time some of the mountain barriers separating adjacent basins are broken through by headward erosion, and 'capture' of one basin by another occurs fairly frequently; such captures are associated with the rejuvenation of the higher basins. A high degree of such drainage integration is assumed to make the passage from youth to maturity. During the latter stage a progressively wider area becomes subject to the influence of a single base-level of erosion, and at the same time the original mountain masses are reduced to isolated residual hills. With the continued decrease of relief and thus of slope, rainfall and the action of running water becomes progressively less important, and in the old age stage wind action gradually becomes dominant. Deflation removes much of the previously accumulated alluvial infill, and abrasion attacks an increasing area of bare rock on the desert peneplain, forming a low-level erosion surface.

More recently L. C. King has formulated the concept of the *cycle of pediplanation*, which has received much wider acceptance than the scheme proposed by Davis. The dominant factors here are assumed to be slope retreat (p. 107) and pedimentation, that is, the process of formation of a series of concave rock surfaces; much less stress is placed on aggradation, which is probably important only under certain local or structural conditions. In youth, river action is assumed to carve the land surface into a series of steep-sided blocks. The valleyside slopes thus initiated retreat to leave pediments bound ing the streams; these are at first very limited in area but gradually extend as the uplands are cut into. By the stage of maturity the pediments are major landscape features, and the former uplands are reduced to inselbergs. In old age, the dominant feature is the *pediplain*, a surface resulting from the amalgamation of numerous concave pediments, above which rise widely separated inselbergs. Overall reduction of relief is not important until the ultimate stage of the pediplanation cycle. It is important to note that pediplains may be multicyclic in origin, that is, the effects of more than one cycle of erosion may be superimposed on the desert landscape.

Pediplains of a remarkable degree of perfection and of great age are now recognized as among the most important elements of the African landscape. The 'Gondwana' pediplain, which occurs at a height of 1200 m (4000 ft) or more above sea-level, is believed to date from Cretaceous times; it is so-called because it is thought to have been formed before the break-up of Gondwanaland through continental drift (p. 25). The 'African' pediplain, at 600 m (2000 ft) and above, is assigned to the early Tertiary. It has further been suggested by King that many of the so-called peneplains of humid-temperate and even polar lands are in reality pediplains, but his views on this by no means command acceptance.

Much of western Australia likewise reveals a late stage in the

80 Sub-aerial weathering and erosion in horizontal strata in Utah (left)

(Paul Popper)

81 A 'tor-stone' in current-bedded Millstone Grit at Brimham Rocks, Yorkshire (right)

(R. Kay Gresswell)

Note the thin remaining pillar in the right-hand photograph, supporting the mass of the rock. The tor was probably formed in the manner described on p. 124.

82 The rock desert of the Algerian Sahara
(*Paul Popper*)

83 Sand-dunes in Death Valley, California
(*Paul Popper*)

[279

cycle of arid erosion, for it consists mainly of a featureless desert-plain, merging to the east into the Great Artesian Basin. A few rounded and scoured rock-masses rise above the general surface, while in the west parts of the Hammersley Plateau reveal flat-topped fragments of an earlier higher erosion surface. In places differences in rock resistance afford other minor irregularities, as where parallel ridges of hard quartzite are separated by bands of more easily eroded limestone. But, for the most part, it forms a subdued and monotonous area of desert-plain; in the description of A. D. Tweedie, 'Such landforms as stand above the flatness of Arid Australia are stark and fretted forms, but the infrequency of their occurrence is a characteristic of a flat and almost featureless region'.

CHAPTER 10

Coastlines

THE term coast is used to indicate the zone of contact between land and sea. The *shore* comprises the area between low-water spring tides and the base of the cliffs (or, in the case of a low-lying coast, between low water and the highest point reached by storm waves), the *beach* consists of accumulations of sand and stones upon the shore, and the *coastline* itself is demarcated either by a cliff-line or by the line reached by the highest storm waves. The shore is divided into two zones: the *foreshore* extends from the lowest low-water line to the average high-water line, while the *backshore* extends from the high-water line to the coastline.

FACTORS OF COASTAL EVOLUTION

The character of any coastline depends on the interaction of a number of factors. In the first place there is the work of the waves, tides and currents, which act as agents of erosion, transportation and deposition. In the second place there is the nature of the land margin which is being subjected to these marine processes—whether the rocks are resistant or not, varied or homogeneous in character, and whether the land margin is high and steep or low-lying, straight or indented. In the third place there are the changes in the relative level of land and sea, sometimes called *positive* and *negative movements*, according to whether the result is to elevate or depress sea-level relative to the land margin. In the fourth place there are special factors involved in some coastal areas, such as the growth of coral, the effects of glaciers and ice-sheets which reach the sea, and volcanic activity (many volcanoes occur near the coast). In the fifth place there are many artificial features, for the work of man is of profound significance, as developments in marine civil engineering have shown; these include the dredging of estuaries, the creation of ports, the reclamation of coastal marshes, the construction of coastal defences against erosion, such as groynes, dykes and breakwaters, and the building of piers, moles and promenades. Since the Coast Protection Act of 1949, many local authorities in Britain have carried out schemes of coastal defence, with considerable help from government grants. Even so, this entails a great local burden, and many believe that the entire cost should be a national responsibility.

MARINE EROSION

Waves A wave in the open ocean is formed by the oscillation of water particles, which describe a circle as the wave passes; each particle moves slightly forward on the crest, and then returns almost to its original position in the trough. Unless the wind or a current actually drifts it, a piece of floating wood or cork hardly changes its position. The oscillation of the particles is caused by the friction of the wind upon the surface of the water. The effect of strong winds is to produce large waves, which may travel for thousands of miles over the open ocean as rounded *swells*, as much as 12–15 m (40–50 ft) vertically between crest and troughs. The longest swell ever measured was 1130 m (3700 ft) horizontally between two successive crests (i.e. its *wave-length*), the record height 21 m (70 ft), measured by an automatic wave recorder during hurricane 'Betsy' in 1961 in the western Atlantic.

When a wave passes into shallow water, however, its crest steepens, curls over and then breaks (when it is known as a *breaker*), so that a mass of broken water rushes bodily up the beach as the *swash* or the *send*. The water then returns down the slope of the beach as the *backwash*. This piling-up of water against the coast is counterbalanced by an undercurrent known as the *undertow*, which flows near the bottom away from the shore, or where localized as a *rip-current*.

A wave is driven on to the shore by the wind, and its height and hence its energy are determined by the strength of that wind and by the distance of open water over which it has blown; this is known as the *fetch*. Most important, therefore, in the evolution of a coastline is its position and aspect relative to wind direction and to the open sea, particularly to the direction of maximum fetch and of the largest or *dominant* waves (those capable of the most powerful work).

Of special significance are the storm waves, the product of exceptional gales blowing over a considerable fetch; they may affect a coastline more markedly in a single day than the ordinary prevalent waves during weeks of relatively calm weather. These storm waves are for the most part *destructive* in their results. Because of their frequency, one following another in rapid succession at a rate of about 12 to 14 a minute, and because of the almost vertical plunge of water (hence 'plunging breakers') as the wave breaks, the backwash is much more powerful than the swash. Thus these destructive waves tend to 'comb' down the beach and move material seaward. More gentle waves, rolling in at about 6 to 8 a minute, have a more powerful forward push of the swash, and because of frictional re-tardation a less powerful backwash; they therefore tend to move shingle up the beach. These are *constructive* waves, 'spilling' or 'surging' breakers.

The average pressure exerted by Atlantic waves in winter on the western coast of Ireland is nearly 11,000 kg per square m, and during a heavy storm this may be three times as great. The effect of storm waves on the coastline is most marked at high tide, for their force is felt higher up the beach or on the face of cliffs.

As waves approach the shore and the water shallows, their speed is reduced. If the coast consists of alternate promontories and bays, the shallowing is more rapid in front of a promontory than in the deeper water of a bay. The wave is therefore bent or *refracted* from the bay on to the side of the headland, accentuating the erosive processes there. Refraction may also take place on a straight coast should the waves advance obliquely, so that ultimately they will break almost parallel to it.

Waves as erosive agents The waves operate as agents of erosion in several ways. The *hydraulic action* of a mass of water has a direct shattering effect as it pounds the rocks. Air contained in cracks and fissures in the cliff-face is compressed until its pressure is equivalent to that exerted by the wave, so that the resultant ex-pansion when the wave retreats may have virtually an explosive effect. When this process is repeated constantly, these fissures are enlarged, particularly where the rock is much jointed or faulted.

More potent is the *corrasive action* of the mass of fragments, ranging from sand to boulders, which is pounded by the waves against the foot of the cliffs. This undercutting effect produces an overhanging cliff upon which the agents of weathering, such as frost action and rain-wash, can act, depending on the nature of the constituent rocks.

These fragments are pounded against the cliff and against each other, as the swash forces material up to the beach and the backwash drags it back again, so that the fragments are themselves worn down by *attrition*; the shingle on a steep beach during strong winds is in constant churning, grinding motion.

Finally, where such rocks as limestone are present (as in the Car-boniferous Limestone of the Tenby coast of Pembrokeshire), the chemical *solvent action* of sea-water may have a considerable effect. (It must be mentioned, however, that many of the features along this coast are now believed to be the result not of this solvent action but of the sea attacking a pre-existing underground river-system which had developed in the limestone.)

Thus the erosive work of the waves is fourfold—hydraulic action, corrasion, attrition and solution—and is analogous to the erosive work of a river.

Tidal scour A tidal current may have a powerful erosive effect, the importance of which in forming underwater channels and even submarine erosion surfaces is becoming increasingly apparent. A tidal scour may be experienced in the mouth of a 'bottlenecked' estuary such as the Mersey. It may also be felt offshore, as in the Solway Firth, with its extensive and ever-changing mud banks and sand flats, through which, about 1·6 km (1 mile) offshore, a constant deep channel runs parallel to the coast from Silloth to White-haven for about 30 km (18 miles). This is scoured by a 5-knot ebb-tide, supplemented by the discharge of river water into the head of the Firth. As another example, it is known that a tidal current of 4·5 knots off Hurst Castle Spit on the Hampshire coast can scour shingle at a depth of 60 m (32 fathoms). Sand ripples have been observed 160 m (90 fathoms) down on the Continental Shelf off south-western England.

The erosion of a cliff and a wave-cut platform Once again it is useful to envisage a sequence of events in analysing the results of marine erosion. If one imagines a smoothly sloping land surface (fig. 133) as an initial stage in the sequence, wave action begins to cut a notch in this slope, which is enlarged into a cliff rising from a wave-cut bench.

Cliffs The details of cliff formation depend on the nature of the rocks, their stratification and jointing, their resistance to erosion, their homogeneity or heterogeneity, and the presence of bands of weakness such as shatter-belts along faults. Solid massive rocks, notably Old Red Sandstone (plate 84), Portland Limestone (plate 85) and granite, are cut back slowly and form steep cliffs, standing out boldly as headlands. Less resistant rocks are more rapidly eroded to form bays. The effects of differential marine erosion are realized if fig. 134, showing the south-western coast of Corsica, is examined.

It is not only hard resistant rocks which form cliffs. Chalk, for example, along the coasts of Dorset (plate 85), the Isle of Wight (plate 86), Sussex (where Beachy Head exceeds 150m (500 ft) in height, eastern Kent, the East Riding of Yorkshire and the French side of the Channel, forms steep clean-cut cliffs. Sometimes the chalk strata are horizontal, as along the coasts of Sussex and Yorkshire. Elsewhere the Chalk has been folded, as in the Isle of Wight and Dorset; at Ballard Down near Swanage the strata have a high dip, at the Needles they are near-vertical, and near Lulworth the Chalk has been inverted. This folding has hardened and compacted the

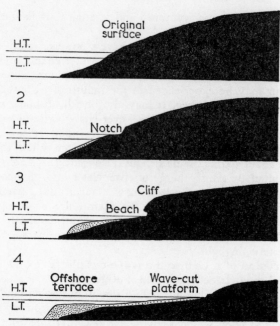

Fig. 133 *Stages in the development of a shore-profile*

1. The theoretical initial stage (i.e. an upland sloping uniformly and smoothly below sea-level).

2. A small notch has been cut by the waves, with a little offshore deposition.

3. The cliff has developed, a rock platform (with a veneer of beach deposits) has been formed, and an offshore terrace is extending.

4. A broad platform and terrace have developed, and the cliffs are decreasing in steepness under the attack of weathering.

Chalk in the case of the Needles, making it unusually resistant to marine erosion (concentrated along clean-cut vertical joints), so that the prominent stacks have been formed. Rapid undercutting at the base, coupled with the permeable nature of the rock (hence making it less subject to weathering) cause the Chalk to stand up boldly, although undercutting also produces frequent rock-falls, and the cliff usually recedes rapidly. Great cliff-falls are especially common in the Isle of Wight, and along the Kentish coast between Folkestone and Dover, where chalk overlies Gault Clay. The Chalk in Sussex has receded so rapidly that the mouths of the dry valleys (p. 126) are 15–30 m (50–100 ft) above the base of the cliff, now separating the high cliffs of the Seven Sisters. For about 5 km (3 miles) along the coast at Folkestone Warren, cliff-falls have occurred on such a scale that a distinct *undercliff* has developed. The exact cause of these

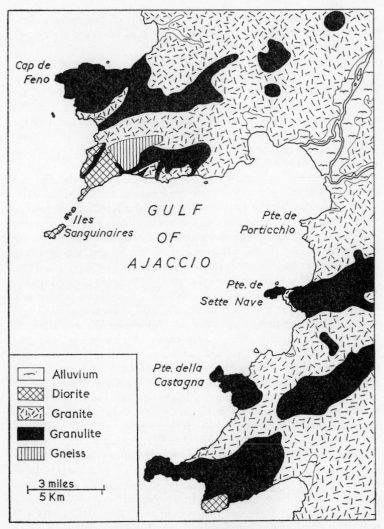

Fig. 134 *Differential marine erosion in south-western Corsica*

The effects of the hard resistant masses of 'granulite' (the name given by French geologists to a type of biotite-granite) and of diorite is shown by the promontories projecting boldly westward.

falls is not clearly understood, since the dip of the rocks is here inland, but they usually occur after heavy rain when the underlying Gault has become plastic.

Still larger examples of cliff-slipping can be seen along the Devon coast between Exmouth and Lyme Regis, where undercliffs of disturbed Cretaceous materials have accumulated at the foot of the now inland chalk cliffs; much of this undercliff is wooded. Another example, called the Undercliff, can be seen along the coast of the Isle of Wight south-west of Ventnor.

Again, the Tertiary clays and sandstones of the steep cliffs near Bournemouth are easily eroded at the base, and have been protected by concrete esplanades and by drain-pipes sunk behind the cliff edge, but in wet weather there are frequent slides and cliff-falls.

At Alum Bay, in the Isle of Wight near the Needles, the Eocene strata, from the Reading Beds to the Barton Clay, rise almost vertically in the cliff-face. The more compact coloured sandstones are separated by clay strata, from each of which, following a period of rain, a tongue of mud moves glacier-like on to the beach (cf. plate 36).

Even glacial clays may develop steep cliffs because of rapid undercutting, as along the coasts of Norfolk and Yorkshire, particularly between Scarborough and Filey. The till sometimes slips and slumps down the face of the cliff. In Holderness the low till cliffs have receded rapidly inland as a result of erosion; it has been estimated that if the present rate (some 2 m (6–7 ft) a year) has been maintained since Roman times, a strip of land with an average width of 4 km (2·5 miles) must have been removed since the fifth century. The sites of numerous villages are now well out to sea—Wilsthorpe, Hartburn, Cleton, Great Colden, Waxholme and many more are now just names in historical records and on old maps.

Finally, it is important to remember that not all sea-cliffs are the product of marine erosion. In areas where changes of relative sea-level have occurred (p. 303), a cliff may be simply the steep slopes of the land, though of course modified by subsequent marine erosion. Again, faulting may result in a steep face of rock dropping sheerly to the sea; there are many examples along the islands and mainland coast of western Scotland, notably the basalt cliffs along the sheltered eastern coast of Skye.

Caves and stacks. When cliffs in hard strata are subject to marine erosion, caves may be formed if there is some local weakness, such as a much jointed or faulted zone (fig. 135). The collapse of a cave roof can produce a long narrow inlet, known in the Orkneys as a *geo*, often 30 m long and as much in depth. A similar phenomenon is the famous Huntsman's Leap in southern Pembrokeshire. Many caves

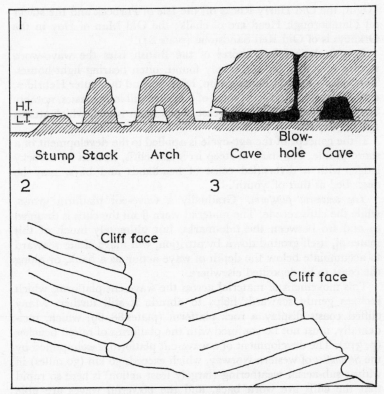

Fig. 135 *Cliffs, caves and stacks*

1 shows the gradual destruction of a cliff in jointed rocks. Stacks and arches, formed by resistant masses bounded by joint-planes, stand up as residual masses above the wave-cut platform, but are gradually attacked. Sea-caves are cut along joints or dykes, and may connect with the surface along a vertical joint, forming a blow-hole.

2 and **3** show how the steepness of a cliff may be affected by the direction of dip of the strata.

reach the surface some distance inland as a vertical pit, termed a *gloup* or a *blow-hole*; the latter name is derived from the spray thrown into the air up the hole by the compressional force of a wave surging into the cave below. If a cave is driven into the side of a headland, or occasionally one from each side, an *arch* may be formed, such as the well-known Durdle Door in Dorset (plate 85), the Needle Eye near Wick in the north of Scotland, and the Green Bridge in southern Pembrokeshire. The next stage is the collapse of the arch, leaving the seaward section standing as a *stack*; the Needles off the Isle of

Wight, the Old Harry Rocks off the Isle of Purbeck and the stacks off Flamborough Head are of chalk; the Old Man of Hoy in the Orkneys is of Old Red Sandstone (plate 84).

Round the western shores of the British Isles the wave-worn stumps of stacks are commonly found, often bearing light-houses. The 'stacs' of the St Kilda group, lying beyond the Outer Hebrides, are of hard igneous rocks; some of these are still lofty masses, notably Stac an Arnim, to the north-east of the main island, which attains 191 m (627 ft).

If the concept of the age-cycle is applied to the development of a shore profile, this stage of steep irregular cliffs, fronted by a rocky ledge, with a diversified coast of headlands and bays, may be described as that of 'youth'.

The wave-cut platform. Gradually a wave-cut platform grows, while the cliffs recede. The material worn from the cliffs is dragged to and fro between the tide-marks, but ultimately much of this material, itself ground down by attrition, will move either seaward to accumulate below the depth of wave action as a bank, or along the coast to be deposited elsewhere.

This movement of material across the wave-cut platform, which inclines gently seaward, helps to abrade it still further. Many cliffed coasts display a rock platform (plates 88, 89) which, incidentally, must not be confused with the platforms of raised beaches (p. 312). The development of a wave-cut platform is exemplified by the *Strandflat* of western Norway, which exceeds 50 km (30 miles) in width; sub-aerial weathering (largely frost action) is here so rapid that the cliffs are worn back, and the powerful waves are able rapidly to remove the products of erosion; possibly planation by shelf-ice is a contributory or even a major factor. A splendid example of such a platform is seen along the coast of western Malta, cut in a Tertiary shelly limestone (plate 88).

Usually, however, when the platform attains a certain width, it is covered with shallow water which reduces wave action, and so the rate of coastal erosion slows down, ultimately virtually ceasing. Pursuing the idea of a cycle, a stage of maturity is thus reached; a shore profile has been produced of a gently sloping cliff upon which sub-aerial erosion will continue to act, a rock platform with probably a thin layer of material forming a beach, and a wave-built terrace of accumulation beyond the depth of wave action. Processes of transport and deposition become increasingly important.

It is possible to pursue the cycle into the stage of old age. In theory, the cliff will be worn back by sub-aerial erosion until its gradient is so gentle as to be unrecognizable, while the offshore accumulations of material will steadily increase. It is possible that some widespread peneplains now above sea-level (such as the 180 m (600 ft) platform in Wales) may be due to extensive marine planation, probably carried on during a long period of widespread gradual subsidence in past geological times.

The shore profile of equilibrium. The net tendency of erosion and deposition is towards the production of a shore profile of equilibrium, so forming a slope where the amount of sediment accumulated is more or less balanced by the amount removed. Obviously, this balance is very temporary and can easily be disturbed, for example, by an exceptionally high tide with a strong onshore wind, but after the storm is spent the normal processes labour once more to restore a profile which is the net result of the average set of conditions obtaining along that stretch of coast.

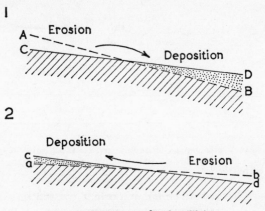

Fig. 136 *The shore-profile of equilibrium*

1. The steeply sloping surface **AB** is subject to marine erosion on the upper part of the shore, and to deposition on the lower, so producing a new more gentle profile **CD**, which is in a state of equilibrium (i.e. the slope is such that wave action is able to remove just as much sediment as is received).

2. The more gentle initial slope **ab** is built up on the landward side of the beach and eroded on the seaward side until the new profile **cd**, in a state of equilibrium, is attained.

The theoretical idea of the profile of equilibrium is shown in fig. 136. If the original coastline is steeper than the ideal profile, erosion will cut a rock platform and cliff, while deposition will build out a terrace (**CD**). But if the original coastline is gentler, the beach will be built up, and erosion will be effective farther seaward (**cd**).

MARINE TRANSPORTATION

The rate and direction of movement of sand and pebbles along the coast by waves and currents is a fascinating problem; its study is one of great practical importance for both hydraulic engineers and for geomorphologists interested in the shaping and formation of coastal features. Modern methods include aerial photography, skin-diving, underwater television, and the use of various 'tracers'. The last involves the marking of material by dyes, notably of a fluorescent quality, and by radioactive agents, which enable its movement to be traced with some precision.

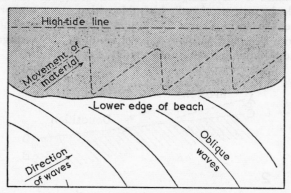

Fig. 137 *Beach-drifting*

When a wave approaches a beach obliquely, its swash runs obliquely up the beach, carrying material. The backwash runs back straight down the slope, but the next wave carries the material obliquely up again. Thus it moves along the shore.

Waves The first and most important transporting agent is a wave itself. A breaker sweeps material up the beach, the backwash drags some of it down again. If waves break obliquely on to the beach, there may result a drift along the beach, as shown in fig. 137, for material advances obliquely across the beach, but is dragged back straight down the steepest slope. This *longshore drift* (or *beach-drifting*) is shown in the west–east movement of material along the south coast of England, since the dominant winds and waves are from the south-west. This movement may be halted by a projecting headland, or by the deeper water of an estuary, or by man-made groynes of wood or concrete.

Currents In addition to the waves, currents are important transporting agents. The undertow moves shingle, sand and mud seaward, and during strong onshore gales, when water is piled high up the beach and so has to flow back, the beach may be heavily scoured. Tidal streams, particularly in a constricted estuary and especially when reinforced by a river current, may also 'flush out' material. Longshore currents often move fine material parallel to the shore well below low-tide level, as along the Belgian coast.

Wind Wind must be added as a transporting agent, since large areas of unconsolidated material are exposed to its effect on an open shore. Hence sand can be moved inland. At low tide, during a strong wind, the surface of an extensive sand-flat is in perpetual hissing motion, and where the supply is adequate the sand may be transported inland (pp. 300–1).

MARINE DEPOSITION

Material worn from the margins of the land ultimately finds its resting place in the sea. The coarser material moves to and fro, and may be deposited temporarily elsewhere along the coast. Attrition, will, however, ultimately reduce it to a fine consistency, and it will be deposited below the level of influence of wave action. There must always be an ultimate net loss, since gravity helps seaward and hampers landward movement. On the other hand, it is evident that some beach-material is derived from the sea-floor. In stormy weather powerful waves may disturb the offshore sea-floor, providing material which is flung up on to the beach.

Between the tide-marks occur littoral deposits of coarse sand and shingle (plate 94), sometimes in the form of a terrace or *berm*, and above high tide there is sand, often accumulated in dune-form by the wind (p. 300.) Some beaches are composed entirely of shelly material, e.g. the coast of Skye (notably on the shores of Lochs Bracadale and Dunvegan) and the nearby mainland at Morar. These dazzling white shell-sands are derived from the remains of nullipores which lived on the nearby sea-floor, comminuted and washed up by powerful waves. A remarkable shell-beach, over half a mile long, can be seen on the coast of Herm, one of the Channel Islands; many rare shells can be found, including some which normally occur in sub-tropical latitudes. Loads of shell-sand are carted inland from the Breton beaches to help improve the thin acid soils farther inland. Between low-water mark and the 180 m (100 fathom) line are fine sands and finely comminuted shelly material, and beyond the 180 m line various muds of terrigenous origin (p. 344). Other deposits include coral fragments and volcanic material, and immense quantities of fine alluvium brought down by rivers, forming mudbanks in estuaries and deltas. Important contributions are made by the growth of vegetation, especially in salt-marshes (p. 302–3).

The main features of coastal deposition are: (i) beaches and beach ridges; (ii) bars and spits; (iii) cuspate forelands; (iv) dune-belts; and (v) mud-flats and salt-marshes.

(i) **Beaches and beach-ridges** The term beach is applied to the accumulation of material between the low-water spring-tide line and the highest point attained by storm waves. On an upland coast, where erosion is dominant, the beach may be non-existent, or merely an unstable mass of boulders and shingle under the cliffs. A cove between headlands commonly has a small crescent of sand at its head, called a *bay-head* or *pocket beach*. On the other hand, on lowland coasts areas of sand may slope gently shoreward, perhaps with a narrow shingle beach; at low tide an immense tract of these sands may be uncovered, as in Morecambe Bay, along the coast of south-western Lancashire, and in the angle between Anglesey and the coast of North Wales (fig. 138). The most typical beach is one

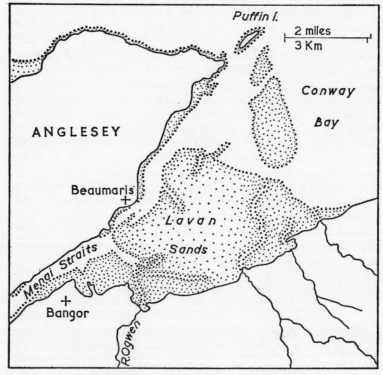

Fig. 138 The Lavan Sands off the coast of North Wales
The area of sand uncovered at Low Water Mean Ordinary Tides is stippled.

with a gently concave profile, the landward side backed by dunes, succeeded by a stretch of shingle, then sand, and perhaps rocks covered with seaweed just above the low-tide mark, indicating the underlying wave-cut platform.

Ridges sometimes develop along the beach, roughly parallel to the coastline; they are called *fulls*, and are separated by long, shallow depressions known as *lows* or *swales*. They are to be seen, for example, on the Lancashire coast north of Formby Point, and on Dungeness. They are formed by constructive waves approaching a coast, and tend to lie at right angles to the direction of their approach.

Beach-cusps A minor though puzzling feature of beaches formed wholly or largely of shingle is a series of projections, separated by more or less rounded and evenly spaced depressions, occurring near high-water mark, giving a kind of scalloped pattern to the edge of the beach. Their method of formation is not really understood, though they are the result of a powerful swash and backwash, and seem to form most readily when waves approach the coast at or near right-angles. Once a cusp is initiated, the fact that it breaks up the swash and compels it to rush with a scouring swirling action into the depression on either side will help to maintain and emphasize it. Coarser material is moved to either side of the depression, so building up the cusp, leaving finer material on the floor of the depression. This does not, however, explain their initiation. Cusps can usually be seen along the shingle beaches of the English Channel, notably at Chesil Beach, Hurst Castle Spit and Alum Bay in the Isle of Wight, but a period of storm waves may suddenly destroy the pattern by combing down the cusps to a uniform slope once again.

(ii) **Bars and spits** A diversity of coastal features is produced by the growth of bars and spits of sand or shingle, either offshore and parallel to it, or across the mouth of a bay or estuary, or at the point at which the direction of the coastline abruptly changes, or between the mainland and an island. The main conditions necessary for their growth is an ample longshore drift of material, together with an irregular coastline which the sea is smoothing off by building up these lines of sediment.

An *offshore bar* lies roughly parallel to the shoreline. The origin of such bars is disputed, and they are probably formed by the inter-action of a number of factors. They may be due to wave action, so that material drifted along the shore is deposited just inside the line where the incoming waves first break (hence the term *break-point bar*), or they may be formed from material 'combed' directly down the beach by the backwash. Possibly, too, the scour of longshore currents helps. The result is the production of a bar, behind which marshes,

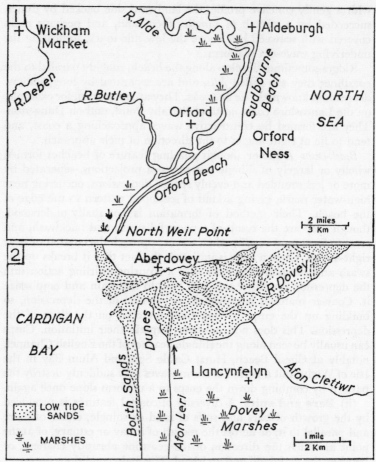

Fig. 139 *Orford Ness and Borth Point*

1. The upper map shows Orford Ness. A shingle spit has developed south-westward from a cuspate foreland (Orford Ness itself). The long tapering spit has diverted southward both the river Alde and the river Butley. In medieval times Orford was a port facing the open sea.

2. The lower map is of Borth Point, which projects northward halfway across the Dovey estuary. It consists of shingle covered with dunes, and behind lie the saltings of the Dovey marshes.

mud-flats and lagoons can accumulate. Sometimes the bar moves inland, driven by the attack of the waves; this is illustrated well in the south-east and east of the U.S.A., where a flat coastal plain is bordered by shallow water. Offshore sand-bars were originally built

84 The Old Man of Hoy, Orkney
This is a 450-foot pinnacle of horizontally bedded Old Red Sandstone.
(Paul Popper)

85 A stretch of the Dorset coast, showing the arch and promontory of Durdle
Door in Portland and Purbeck Limestone, linked to the main coast (of Chalk) by
a low ridge of Wealden Sands and Clays, Greensand and Gault Clay.
(R. Kay Gresswell

86 The Needles, Isle of Wight

These stacks of chalk are remnants of a once continuous ridge, breached by a post-glacial marine transgression and by prolonged marine erosion. The stacks consist of wedge-shaped masses of unusually hard and compact chalk with well-defined vertical joints.

(Paul Popper)

87 The Bow Fiddle Rock, on the Banffshire coast, Scotland

(Mustograph)

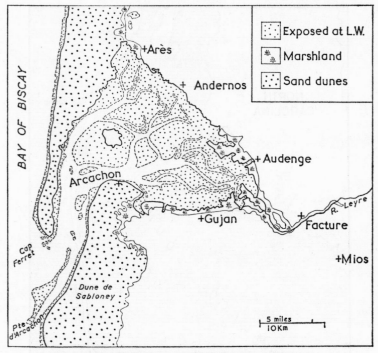

Fig. 140 *Cap Ferret, on the Landes coast of France*

The line of sand-dunes which borders the Landes coast of south-western France, is breached by the Baie d'Arcachon, into which flows the river Leyre. Across the mouth of the bay has grown the sand-spit of Cap Ferret. To the south is the Dune de Sabloney, the largest moving dune in Europe.

up far from the mainland, but they have advanced inland, enclosing marshes and lagoons. For long distances these bars have no connection with the mainland, hence their common name of *barrier-islands*, and enclose areas of shallow water known as *sounds*; this is well exemplified at Cape Hatteras off the coast of North Carolina, where Pamlico Sound is enclosed by two long barrier-islands which converge to form the blunted triangle of the Cape (fig. 141). Farther south the bars have advanced still more, and have filled up the lagoons in many places, forming extensive sandy beaches at Daytona, Palm Beach and Miami in Florida.

When material is piled up in a linear form but with one end attached to the land, the other projecting into the open sea or across the mouth of a river, a *spit* of sand or shingle or both is formed (plate 92). Sometimes the seaward end is curved or hooked,

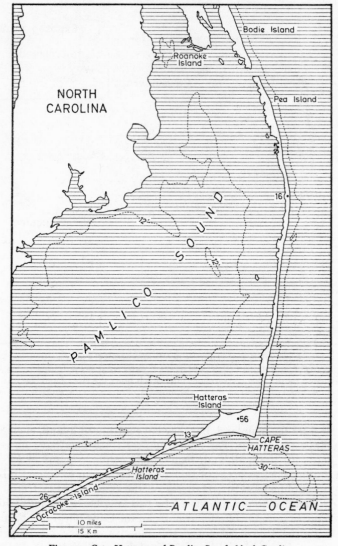

Fig. 141 *Cape Hatteras and Pamlico Sound, North Carolina*

due to waves advancing obliquely up the shore which tend to swing round the end of the spit (fig. 142).

Hurst Castle Spit, on the Hampshire coast opposite the Isle of Wight, has developed in a south-easterly direction across the mouth

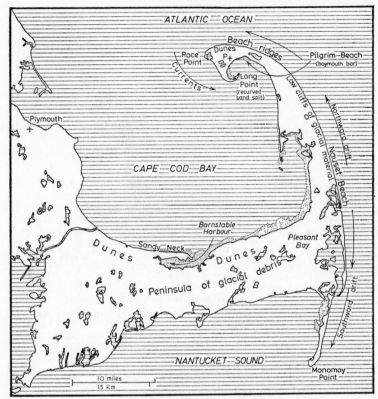

Fig. 142 *Cape Cod, Massachusetts*
P = Provincetown.

of the Solent, on a foundation of a clay bank. It has been built up of shingle drifted eastward, and its orientation is determined by the direction of the dominant waves from the south-west. Near its eastern tip a series of lateral shingle ridges join it almost at right-angles, giving it a recurved tip trending north-north-west. It is probable that waves coming down the Solent from the north-east are responsible for this change of direction.

Cape Cod on the coast of Massachusetts provides a remarkable example of both erosion and deposition in close juxtaposition on a lowland coast (fig. 142). The peninsula is a mass of easily eroded glacial drift, which at Nauset Beach experiences the full attack of the waves; one wonders how it has survived when powerful storm-waves pound at the base of the earthy cliffs. But the material is not

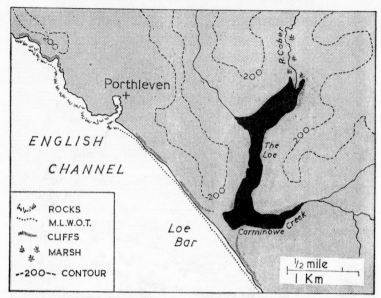

Fig. 143 Loe Bar in southern Cornwall

Loe Bar near Porthleven (not to be confused with Looe farther east) consists of flint-shingle, and is 400 m long and about 180 m wide. The Loe (also known as Loe Pool) is a fresh-water lagoon. Note the small lacustrine delta which the river Cober is building out into the head of the Pool. The origin of the bar, the surface of which is considerably above the level reached by the highest tides, is due to wave-action. **MLWOT** = Mean Low Water Ordinary Tides.

lost to the land; it drifts both north and south, for about here the direction of longshore drift changes, forming beach-ridges. The northern spit ends in Race Point, but currents sweep the material round in a recurve to form the hook-shaped Long Point, behind which lies an area of dunes. The southward drift has built a spit across the irregular indentation of Pleasant Bay and farther south has formed Monomoy Point.

In long funnel-shaped indentations of the sea, such as rias (p. 305), bars are sometimes built out at right-angles to the shore some distance up, where the force of the incoming waves is markedly weakened. In Dingle Bay in south-western Ireland a spit has developed from either shore, leaving on the landward side a stretch of tidal flats which is being naturally reclaimed by the deposition of alluvium from inflowing rivers and by the growth of vegetation.

If a spit continues to grow lengthwise, it may ultimately link two headlands to form a *bay-bar*. These bars are either of shingle, as in the case of Loe Bar in southern Cornwall (fig. 143), or of sand, such

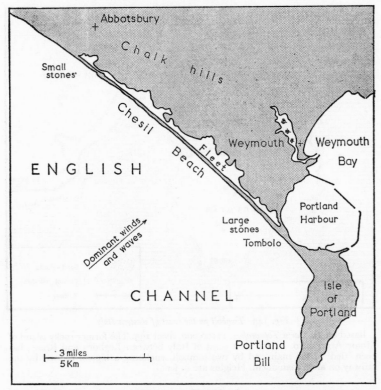

Fig. 144 *Chesil Beach*

Chesil Beach consists of a shingle ridge extending from Bridport (just north-west of the map) to the Isle of Portland, for about 25 km (16 miles). From a point south of Abbotsbury it forms a bar, enclosing the lagoon of the Fleet; this opens into Portland Harbour, the narrow connection being crossed by a swing-bridge. The size of the shingle is graded from that of a small pea near Bridport to that of a large potato near Portland. It consists mostly of flints. The beach reaches a height of 12 m (40 ft) above high water at Portland, but is lower to the west. The beach was breached in a number of places during the storms at the end of November 1954, and its permanence seemed threatened, but it is now once more solidly continuous.

as the *Nehrungen* on the Baltic coast (fig. 152). Lagoons and marshes are usually impounded between the bar and the mainland.

Chesil Beach, on the coast of Dorset (fig. 144), has 'tied' the Isle of Portland to the mainland; such a connecting bar is some-times known as a *tombolo*. An example from the coast of western Italy will suffice to show the remarkable coastal modifications pro-duced by tomboli (fig. 145).

PRINCIPLES OF PHYSICAL GEOGRAPHY

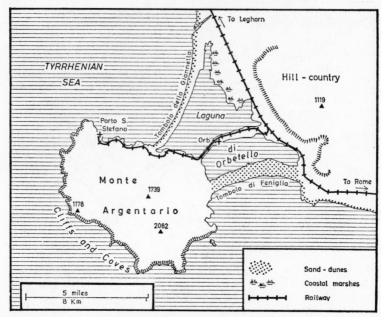

Fig. 145 *Tomboli on the coast of western Italy*

Based on the *Carta Topografica*, 1 : 100,000, sheet 135. The former rocky island of Monte Argentario, on the west coast of Italy between Leghorn and Rome, has been 'tied' to the mainland by two tomboli, enclosing a lagoon crossed by the railway on an embankment. Heights are in feet.

(iii) **Cuspate forelands** Striking examples of these features are Dungeness, Cape Kennedy in Florida, and the Darss on the coast of Mecklenburg, East Germany (fig. 146). They consist for the most part of shingle ridges, and seem to be the result either of the convergence of two separate spits, as in the case of the Darss, or of the combined effect of two sets of powerful waves—in the case of Dungeness from the south-west and the east.

(iv) **Dune-belts** Extensive sandy beaches are nearly always backed by sand-dunes because strong coastal winds can readily move some of the sand which dries out at low tide. Examples are the Culbin sand-dunes of the Moray Firth which have advanced inland between Burghead and Nairn, engulfing a number of farms; the Ainsdale (plate 109) and Formby sand-dunes in southern Lancashire; and the extensive Hayle dunes bordering St Ives Bay in Cornwall. Examples from the continent of Europe include the coast of the Landes (fig. 140), the North Sea coast from the French

300

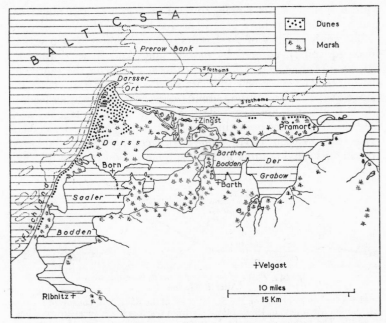

Fig. 146 *The cuspate foreland of the Darss, Mecklenburg*

The sandy peninsula of Fischland culminates northward in the cuspate foreland of the Darss in Mecklenburg, East Germany. This foreland consists of shingle ridges on which blown-sand has accumulated to form dunes. As each ridge was built up, the foreland grew out into the Baltic, and its apex, the Darsser Ort, has advanced 400 m (1300 ft) seaward in the last two centuries. A shallow bank extends beyond the point, indicating its future extension. The Pramort sand-spit has grown eastward, and so shallow brackish lagoons, known as *Bodden* (p. 312), have been impounded.

frontier round to the northern point of Denmark, and the southern shores of the Baltic. The movement inland of many of these dune-belts may threaten or actually engulf villages and farm-land. The fixation of dunes by planting grasses and pines (such as Corsican pine and Scots pine), or by facing them with brushwood fences, is often necessary. The dune grasses (p. 539) have intertwined roots which help to bind the loose sand, while their tufted growth assists in checking its surface drift. The outer dune-belt is often artificially stabilized to form a rampart which will act as a defensive barrier for the land behind.

A fascinating example of blown coastal sands is the *machair* in the Western Isles of Scotland, particularly in South Uist and Tiree. Behind the coastal dunes flat or gently undulating stretches of whitish shell-sand are largely covered with grasses and flowering plants, and in parts have been improved by the crofters.

The sandy country of the Landes occupies almost 16,000 sq km (6000 sq. miles), covered by marine sands laid down during an extension of the sea in Quaternary times; this forms the source of the dunes. The dune-belt stretches southward from the Pointe de Grave at the mouth of the Gironde estuary to Biarritz; its overall width is about 7 km (4·5 miles), and behind it lie lagoons and marshes. The dunes occur in lines separated by depressions, their crests rising to 45 m (150 ft), while the largest, the Dune de Sabloney (the biggest moving dune in Europe), exceeds 90 m (300 ft.) (fig. 140).

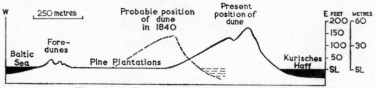

Fig. 147 *Profile of the moving dunes of the Baltic coast*
(After P. Gerhardt)

The Kurische Nehrung, on the Soviet coast of the Baltic Sea, consists of a sand-spit, 80 km (50 miles) in length, but only from 0·4 to 1·6 km (0·25 to 1·0 miles) in width. The eastern side of the Nehrung is bordered by a line of 'wandering dunes' 45–60 m (150–200 ft) high. These dunes have moved rapidly eastward, at an average rate of 6 m (20 ft) a year for the last two hundred years. Their movement is now much less, partly because pine forests have been planted on the seaward side to reduce the supply of sand, partly because the water of the lagoon (*Haff*) helps to arrest the sand.

Along the southern Baltic coast the moving dunes, formed from new-blown sand, are known as 'white dunes', while those which have been fixed by vegetation are called 'grey dunes'. On the Kurische Nehrung (figs. 147, 152) these dunes rise to 55 m (180 ft), and in one case to 66 m (217 ft). The dunes on the sand-spits do not now move much, since their movement is limited by the lagoons, but on the inland coast of the lagoons others menace agricultural land, so that belts of pines have been planted.

(v) **Mud-flats and salt-marshes** Coastal marshes are found either behind shingle-bars and sand-spits, or in the sheltered parts of estuaries and embayments (plates 110, 111). In Norfolk marshes extend between the coast and offshore bars such as Scolt Head Island; in Wales the Dovey marshes lie behind the Borth sand-spit (fig. 139), and the reclaimed Romney Marsh is behind Dungeness. Vast marshes lie between the mainland coasts of the Netherlands, West Germany and Denmark and the protective line of the West,

88 Wave-cut platform on the west coast of Malta

Malta consists almost entirely of Tertiary limestone. Along the west coast, this almost level platform of shelly limestone has been cut by the waves, backed on its landward side by an ancient cliff of massive limestone (note the huge horizontal joint-plane). In this unstable part of the Mediterranean basin, changes of sea-level have been frequent. The seaward edge of the platform is bounded by the present cliffs, 50 feet high.

(F. J. Monkhouse)

89 Wave-eroded platform in Carboniferous Limestone, Redskin Cove, Northumberland

Berwick-upon-Tweed is in the background.

(Eric Kay)

90 The North Cape of Norway
The ancient peneplaned surface ends abruptly with these magnificent cliffs,
the northern tip of Europe.

(*Widerøs Flyveselskap og Polarfly*)

91 The coast of Cornwall near Fowey
The lower part of the valley of the Fowey, a river flowing from Bodmin Moor
across the gently undulating peneplain of southern Cornwall, has been inundated
by a positive change of sea-level, forming a ria-estuary.

(*Aerofilms Ltd*)

92 Spurn Head, at the mouth of the Humber estuary
View looking south towards the Lincolnshire coast in the distance. Material
moving southwards along the Yorkshire coast accumulates at Spurn Head. This
spit has changed its position frequently, but numerous groynes help to stabilize it.
Deposition of river alluvium takes place within the curve of the spit.
(Aerofilms Ltd)

93 Reagh Island in Strangford Lough, Co. Down, Northern Ireland
The slight positive change of sea-level has converted this drumlin-country into a
number of rounded, gently humped islands.
(Aerofilms Ltd

94 Shingle ridges off Spurn Head, Yorkshire, uncovered at an exceptionally
low tide
(*F. J. Monkhouse*)

95 Coral growing on the outer Barrier Reef, Queensland
(*Paul Popper*)

East, and North Frisian Islands. These marshes, known as *Watten* in Germany (fig. 152) and *Wadden* in the Netherlands, when seen from the air at low tide look like the surface of a leaf, with a maze of creeks and channels separating the sheets of mud. Similar extensive marshes lie behind the dune-belt along the Biscay coast of France.

The second type of coastal marsh, within sheltered estuaries and bays, includes extensive areas in Morecambe Bay, in the Cheshire Dee, in Southampton Water and the Solent, in the Essex estuaries and in the southern part of the Wash. Others occur at the mouths of the Jade and Weser in northern West Germany.

The basis of the formation of coastal mud-flats and marshes is the deposition of fine silt by the tides in a sheltered area, helped in some cases by alluvium brought down by rivers. Then vegetation begins to spread and help the process. Sea-grass (*Zostera*) and marsh samphire (*Salicornia*) and in some places perennial rice grass (*Spartina townsendii*) (plate 110) form increasingly dense communities which help to trap silt (p. 539); first discontinuous hummocks, then more continuous areas, are built up, while the tidal waters flow between them in more and more restricted channels or creeks. Gradually other plants establish themselves, and the surface may be further raised by the addition of wind-blown sand.

The natural reclamation of marshes may be accelerated by the work of man, such as the building of dykes or of railway embankments, as across part of Morecambe Bay. Along the North Sea coasts of West Germany and Denmark, wickerwork fences are nailed to piles. Between the island of Sylt and the Jutland mainland, rectangles of these fences, with one side open to the sea, have been constructed. Then more substantial dykes are erected, and the marsh gradually turns into 'saltings', as turf-forming grasses spread. The 'sea-washed turf' which grows on the Solway marshes off the Cumberland coast is renowned for its fine quality. Along the coast of the Netherlands the marshes have been reclaimed by surrounding them with dykes, and then draining them to form *polders*.

CHANGES OF SEA-LEVEL[1]

Changes in the level of the sea relative to that of the land may have a great effect on the form of the coast, since a vertical rise or

[1] The term 'sea level' is used extremely loosely, and in a variety of contexts. In its simplest form the term implies a general level of the sea as if it were uninfluenced by tides or waves. More specifically a 'mean sea-level', or 'datum', is inferred, a geodetic equipotential from which heights of the land-surface are computed.

The Old British Datum was based on a number of short-term tidal observations, carried out between 7 and 16 March 1844 at the Victoria Dock, Liverpool. Tidal observations for the ten days were taken at five-minute intervals for an hour fall of even a few metres on a low-lying coast can produce enormous changes in the configuration. Sometimes the change in level is world-wide and uniform, which indicates an actual movement of the sea itself; this is known as a *eustatic movement*. The most important eustatic movement is connected with postglacial changes, when a widespread rise was caused by a return of water to the oceans as the ice-sheets melted (p. 259); at their maximum the Quaternary ice-sheets must have lowered the water-level by about 90 m (300 ft) (excluding the effects of isostatic recovery (p. 23), which would probably reduce this by a third). It is believed that the remaining ice-sheets in the world still contain enough water to raise sea-level a further 30 m (100 ft). The average mean eustatic rise throughout the world is estimated at present to be from 1·12 to 1·18 mm annually. On the other hand, many movements are more local, due to warping or tilting of the crust, down-faulting, and isostatic depression and recovery.

Such changes in level have operated not only in geological but also in historical times, and may even be seen and measured in progress at the present. In southern Sweden such scientists as Celsius and Linnaeus in the eighteenth century made careful observations, and modern researchers have continued their work. It is often possible to find archaeological evidence such as Neolithic settlements and Roman pavements below present low-tide mark, botanical evidence such as submerged forests (p. 310), and physical evidence, particularly of raised beaches and ancient cliffs now standing well above present sea-level.

CLASSIFICATION OF COASTS

The types of coasts resulting from changes of sea-level may be summarized under both the nature of the movement (submergence or emergence) and also the nature of the former coast (upland or lowland) affected. Coasts produced by submergence (i.e. an *oceanic transgression*) are today much the more widespread, because of the general postglacial rise of sea-level.

Not all coasts fit neatly into these categories involving clear changes in the relative level of land and sea. Many sections may have been raised and depressed several times relative to sea-level, about high and low water. Thus the Ordnance Survey obtained a Datum, a Mean Sea Level, which held good until 1921. When the O.S. decided in 1911 to re-execute the primary level network for Great Britain, it was decided also to obtain a new datum. Newlyn Tidal Observatory, on a pier projecting into the English Channel, had virtually an open-ocean site. From 1 May 1915 to 30 April 1921 the mean of hourly records was computed, and after various corrections had been applied, a New Datum was determined as the basis for all heights in Great Britain.

and different results may be superimposed; these are known as *compound* coasts. When the Quaternary ice-sheets melted, this involved both a eustatic change, the result of the liberation of meltwater, and the directly opposed uplift, the result of isostatic recovery; the joint results are complicated.

Moreover, there are the *neutral* coasts, where little or no relative change has taken place, but where accumulation of material has created new land. Within this group are coasts of sedimentation, including mud-flats, marshes and deltas, volcanic coasts where the edges of lava-flows form the coastline, and coral coasts.

(i) **Submerged upland coasts** When the margin of an irregular upland area is submerged, a more or less indented coastline is produced, with islands and peninsulas representing the former uplands, and with inlets indicating the former valleys. The most important varieties are (*a*) rias, and (*b*) fjords, in which the mountains and valleys are transverse to the coast; and (*c*) Dalmatian or longitudinal, when the mountains are parallel to the coast.

(*a*) *Ria coastlines* A ria coastline is produced when submergence affects an upland area where the hills and river valleys meet the coastline more or less at right-angles. The rias are funnel shaped, decreasing in width and depth as they run inland, while into the head of each flows a stream originally responsible for the formation of the valley but now far too small for the size of the inlet. This type of coast is exemplified by the area south of Cape Finisterre in north-western Spain from whence the name *ria* is derived, in south-western Ireland (fig. 148) and along the west coast of Brittany. In each case the present submarine contours of the ria give a clear indication of the former valley that once continued farther seaward.

Along the south coast of Cornwall and Devon, and again in southern Pembrokeshire, a species of ria-estuary has been formed as the result of the submergence of the margins of low dissected plateaus (plate 91). The two largest indentations in the South-west Peninsula are the Tamar estuary, at the mouth of which is Plymouth Sound, and the Fal estuary. The mouth of the latter consists of Carrick Roads, a sheltered sheet of water 26 sq. km (10 sq. miles) in area, much of which exceeds 18 m (60 ft) in depth. At one time Falmouth was one of the most important ports in England, and is still a ship-repairing centre.

(*b*) *Fjord coastlines* Fjords occur in western Scotland, Norway (fig. 149), Greenland, Labrador, British Columbia, Alaska, southern Chile and New Zealand. The main reason for their existence is the submergence of deep glacial troughs, and so fjords have

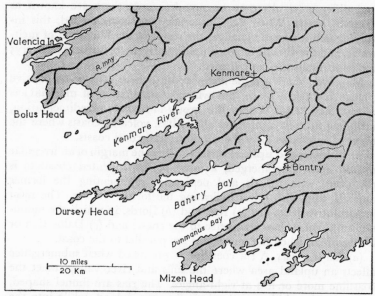

Fig. 148 *The ria coast of south-western Ireland*

The heavy lines indicate the approximate trend of the main ridges, most of which end in promontories.

many characteristics of glaciated valleys, such as U-shaped profiles, hanging valleys and truncated spurs. The Norwegian fjords, deeply cut in the plateau of Scandinavia, have much steeper walls than the Scottish lochs; the north wall of the Sogne Fjord slopes downward at an angle of 28° to 34° from the 1500 m (5000 ft) plateau of the Jostedalsbre to 900 m (3000 ft) below sea-level, while one of its branches, the Naero Fjord, has continuous slopes exceeding 50°. In plan these inlets are long, narrow and rectilinear, with branches joining at or near right-angles. The Sogne Fjord is 183 m (114 miles) long but rarely exceeds 5 km (3 miles) in width; the Trondheim Fjord is 120 km (75 miles) long; the Hardanger Fjord is 112 km (70 miles) long and has a 37 km (23 miles) branch, the Sör Fjord. The world's longest is Nordvestfjord 314 km (195 miles) in eastern Greenland.

Near the mouth of each fjord is usually a bar or threshold of solid rock, sometimes with a cover of glacial debris which may represent a terminal moraine. Many of the Scottish lochs (which are gentle forms of fjords) reveal this feature; Loch Leven, for example, opens into Loch Linnhe, hence the sea, over a rock bar between Balla-

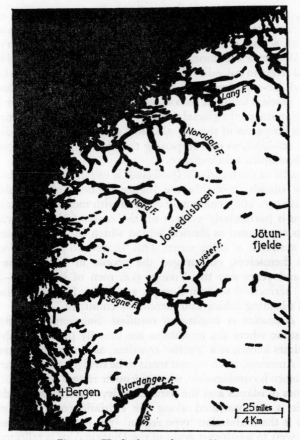

Fig. 149 *The fjord coast of western Norway*

chulish and North Ballachulish, which is sometimes uncovered at exceptionally low tides. Similarly, the Falls of Lorna are caused by a rock bar at Connel Ferry near Oban, at the mouth of Loch Etive. The Norwegian fjords have true thresholds, submerged to a depth of 45–60 m (150–200 ft). These thresholds are much shallower than the inner fjords; 900 m (3000 ft) soundings in the Sogne Fjord are by no means exceptional and 1245 m (4085 ft) has been recorded, deep water going right to the head. Possibly the threshold is due to the fact that the ground was less deeply frozen and therefore less shattered nearer the sea than further inland.

The origin of fjords has long been debated. Some correspond to faults, possibly even to rift-valleys; others follow a line of weak sediments, such as the Hardanger Fjord, which lies along a syncline of weak schists enclosed between two masses of hard, crystalline rock. Whatever the cause, these lines of least resistance enabled pre-glacial rivers to cut valleys in the ancient uplands to the sea. On these uplands ice-caps developed during the Quaternary glaciation, from which glaciers followed lines of least resistance (i.e. the river valleys) and so produced the glacial troughs. Finally, submergence turned the valleys into arms of the sea, although as glaciers can and did erode below sea-level, so overdeepening the valleys, it is not necessary to postulate much submergence.

Parallel to the trend of a fjord coast is usually a series of low hummocky islands, probably the higher portions of the submerged strandflat (p. 288), commonly covered with morainic material. This feature is particularly evident off the coast of Norway, where the islands are known as *skerries*, behind which the 'skerry-guard' provides a stretch of calm water. This 'inner-lead', together with the fjords themselves, encouraged the people to turn to the sea; a thousand years ago the Vikings (i.e. men of the 'viks'—bays or fjords) terrorized western Europe, and today Norway is one of the world's leading fishing and mercantile countries.

(*c*) *Dalmatian or longitudinal coastlines* Submergence may affect a coastline where the mountains are broadly parallel to the coast, sometimes known as a 'Pacific' coastline, as along the western coasts of the Americas. Such a coast tends to be straight and regular, unless subsidence is considerable, when the outer ranges become longitudinal lines of islands and the parallel valleys form long sounds.

This is demonstrated along the Adriatic coast of Jugoslavia, which is the Dalmatian 'type region' (fig. 150). The characteristic north-west to south-east grain of the Dinaric Alps is reflected in the shape and orientation of the islands, peninsulas and gulfs, known as *canali* and *valloni*. The submergence of this area is still in progress, as traces of human settlement, even of Roman antiquities, have been found 1·5–2 m (5–6 ft) below present sea-level.

Another example of these submerged longitudinal coastlines is shown by Cork harbour (fig. 151), on the coast of southern Ireland.

(ii) **Submerged lowland coasts** When submergence of a low-lying area takes place the results are usually extensive, since as slopes are gentle a very slight depression allows the sea to cover a considerable area. River valleys are converted into broad, shallow

---

Fig. 150 *The Dalmatian coast of Jugoslavia*
The land is in solid black, the submarine contours in metres.

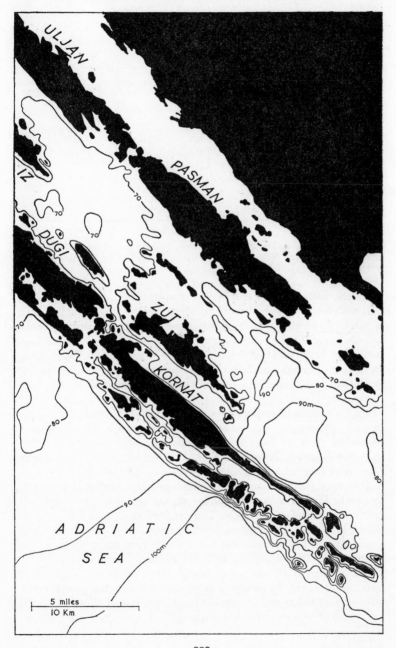

5 miles
10 Km

309

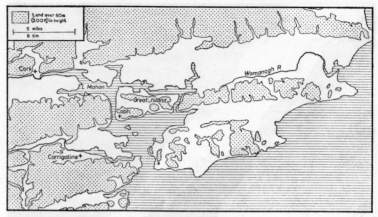

Fig. 151 Cork Harbour

The pattern of coastal indentations produced by the partial submergence of an area of low ridges and valleys parallel to the main trend of the coastline is clearly revealed.

estuaries, with tracts of marsh and mud-flats uncovered at low tide, and with a maze of creeks and winding shallow inlets. Such a coast is that of northern Essex. Just as a submerged upland coast affords scope for the work of marine erosion, such a drowned lowland coast is the scene of deposition—the formation of offshore bars, spits, coastal lagoons and marshes. Deposition slowly straightens out the coast and obliterates the indentations produced by the submergence.

An interesting example of a submerged lowland coast is afforded by plate 93, showing part of Strangford Lough in Northern Ireland. Drumlins (p. 243) form low, round hummocky islands. Another example of this is Boston Harbor, U.S.A.; on the seaward side of Nantasket Beach, the drumlins have been cut in half by the waves.

Submerged forests Along some coasts are layers of peat in which are embedded tree-stumps and roots, occurring between the tide levels or even below low tide. The excavation of dock basins, as at Barry in Glamorgan and at Southampton, have revealed several such features. Other examples are at Formby in Lancashire, on the coast of the Wirral Peninsula, near Harlech in North Wales, at Borth on the Dovey estuary, on the shore of the Bristol Channel at Pentuan in Cornwall (where a layer contains oak stumps and roots, 20 m (65 ft) below present sea-level), and at many points along the east coast. Peat is sometimes brought to the surface by North Sea trawlers.

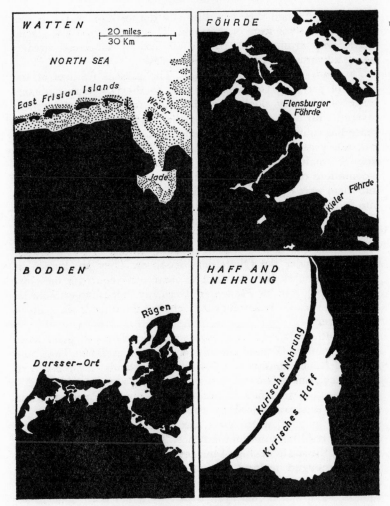

Fig. 152 *The coasts of pre-*1945 *Germany*

Fjärds Around the low-lying coasts of southern Sweden are numerous indentations to which are given the name of *fjärd*. Somewhat similar features are found in the Shetland Isles and along the coast of Nova Scotia. They are found on the margins of glaciated lowlands or peneplains, and consist of openings with parallel, gently sloping sides and numerous fringing islands. Apart from their lower shores and surroundings, they are broader and less regular then fjords, though they are distinctly deeper than rias, and frequently have some form of threshold. They are obviously of glacial origin, with erosion both by the ice and by subglacial streams taking advantage of preglacial river valleys.

The coasts of Germany (fig. 152) The present features of the coasts of pre-1945 Germany are due to the postglacial submergence of a gently sloping but irregular surface of young rocks covered with drift deposits. Along the North Sea section submergence has converted the former outer belt of dunes into a string of low, sandy islands, the East and North Frisian Islands, separated from the mainland by mud-flats, or *Watten* (p. 303). In addition, indentations out of all proportion to the size of the rivers flowing into them were formed; the Ems flows into the Dollart and the Jade into Jade-Busen.

The western Baltic coast is characterized by long, straight-sided inlets which penetrate far into Schleswig-Holstein. They represent the submergence of long valleys, probably eroded by rivers flowing in tunnels under the Quaternary ice-sheet. They are known as *fjords* in Denmark (which is apt to cause confusion with the Norwegian type), but as *Föhrden* in Germany. Flensburger Föhrde, Kieler Föhrde and Eckernföhrde Bucht are the three main inlets of this nature.

The southern Baltic coast shows a series of curiously shaped islands separating inlets of equally curious shape, known as *Bodden*. The whole island of Rügen consists of a few irregular islands (the remains of subsidence), linked by sand-spits and enclosing numerous bodden.

The south-eastern Baltic coast, which is now in Poland and the U.S.S.R., is characterized by the development of *Nehrungen*, or sand-spits (p. 295), which grow across the mouths of shallow embayments resulting from submergence. Behind the sand-spits, which represent the straightening of the coast, lie shallow lagoons (*Haffe*).

(iii) **Emerged upland coasts** The chief feature of such a coast is a *raised beach* or cliff-line now found well above the present zone of wave action (fig. 153). Much work has been done in an effort to correlate the various beaches which occur at different levels, both with other beaches in different localities and with other phenomena such as river terraces (p. 161). The old coastlines are revealed as distinct notches in the slope, backed by a cliff, often with distinct caves, and fronted by a wave-cut rock platform covered with beach material such as shell-banks and shingle. Round some parts of the coasts (at Holderness, along the English Channel, in the Isle of Man and in North Wales) is the pre-glacial 3 m (10 ft) beach. In western Scotland the 30 m (100 ft) beach is particularly marked in

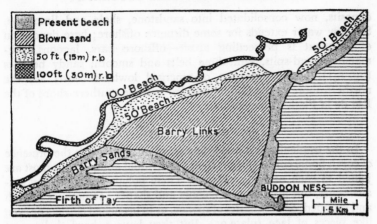

Fig. 153 *The raised beaches of the Firth of Tay*
The sea is indicated by horizontal shading.

the mainland coast opposite the Isle of Skye and in the Western Isles. Then, at a lower level and later in age, traces of beaches are found variously between 14–20 m (45–65 ft) above O.D., known rather misleadingly as the 15 m (50 ft) beach. Finally, the series known as the 7·6 m (25 ft) beach is again well developed in Scotland. The order in which these beaches have been listed is roughly that of age, but to fit them into a detailed chronological sequence, including also the movements of subsidence which produced the submerged forests, is highly complex.

Many parts of the world show evidence of this movement. Mention has already been made (p. 288) of the superb wave-cut platform along the western coast of Malta (plate 88). A negative change of sea-level has occurred in this unstable area of the Mediterranean Sea. The wave-cut platform is now 15 m (50 ft) above the present sea, and ends abruptly in cliffs, the base of which is being attacked vigorously by the sea; the former cliffs, flanking the landward edge of the platform, are visible on plate 88.

(iv) **Emerged lowland coasts** An emerged lowland coast has been produced by the uplift of part of the neighbouring continental shelf. The landward edge of such a coastal plain in the southeastern U.S.A. is formed by the 'Fall-line', where rivers descend from the Appalachians in a series of waterfalls (fig. 68). The coastal plain merges gently without any change of slope into the continental shelf, of which it was once part. Its rocks are formed of sand, gravel, clay and calcareous materials, both continental and shallow-water

313

deposits, now consolidated into sandstone, shale and limestone. Shallow water extends for some distance offshore; here deposition of sediment is proceeding apace—offshore bars, lagoons, salt-marshes, sand-spits, sand-dune belts and smooth, sandy beaches (fig. 141). Other examples of emerged lowland coasts are the northern shore of the Gulf of Mexico and the southern shore of the Rio de la Plata in Argentina.

CORAL COASTS

In the tropical seas a major feature of many coasts is the presence of coral. Some corals (*polyps*) live as individual animals, but most are joined to each other in colonies. These polyps resemble tiny sea-anemones, but are of many colours and shapes—sometimes smooth and rounded, or like masses of sponge, or in branched clusters or even like pieces of lace (plate 95). But they have a hard skeleton of calcium carbonate, and as each coral dies this accumulates to form coral limestone. In addition to corals, other organisms help to form masses of rock or *reefs*. Calcareous algae (*nullipores*), which also precipitate calcium carbonate within themselves, help to cement the spaces between dead coral, and the remains of molluscs and echinoderms (such as starfish, sea-urchins and sea-cucumbers) assist in building up the reef.

Conditions of coral growth Corals cannot live if the temperature of the water falls below 20° C. This means that they are confined to tropical and near-tropical seas within about 30° N and S of the Equator, although locally, as in the case of the Bermudas, they may extend a little farther from the Equator, because in this case they are situated in the path of the Gulf Stream, which has temperatures exceptionally high for the latitudes (p. 355). On the other hand, coral is often absent from coasts within the tropics on the western sides of continents, because of cool currents and the upwelling of deep water which is cool for the latitude (p. 353).

Corals cannot live for long out of water, and are therefore rarely found above low-tide level, while conversely they cannot grow at depths much exceeding 45–55 m (25–30 fathoms). They need clear oxygenated water, with plentiful supplies of microscopic life as food, and they cannot live in fresh or silt-laden water. Food supplies are most plentiful on the seaward side of a growing reef, so that the coral tends to grow more rapidly outward. Recent work by Dr Hans Hass in the Maldives indicates that as the coral grows outward as well as upward, its delicate open scaffolding tends to compress under its own weight in the centre, which helps to explain the development of the characteristic lagoon surrounded by reefs. As the reef extends,

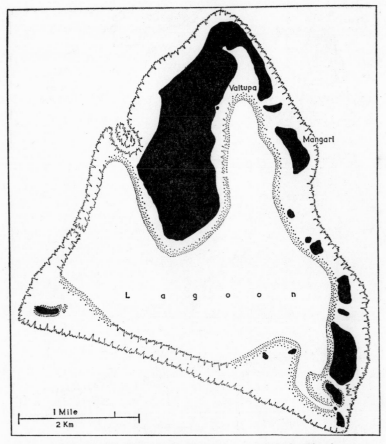

Fig. 154 *The island of Aitutaki* (based on an Admiralty Chart)

Aitutaki lies at 18° 15′ S and 159° 45′ W in the group of the Cook Islands, east of Fiji. The interesting point is that it is of volcanic foundation, but is bordered both by a fringing reef on the north and by a barrier reef on the south. The land area is about 1600 hectares (3900 acres), and it attains an altitude of 140 m (450 ft) near the extreme north.

The areas above the highest high tides are shown in solid black. Banks of coral sand are stippled, and the edge of the reef is indicated by a serrated line.

waves wash much broken coral in the form of boulders and sand over the crest, thus binding up a more compact reef-flat, often with sand-dunes, upon which vegetation, even palms, may ultimately grow, forming a typical 'coral island'.

Types of coral reefs Coral reefs form round the edges of

315

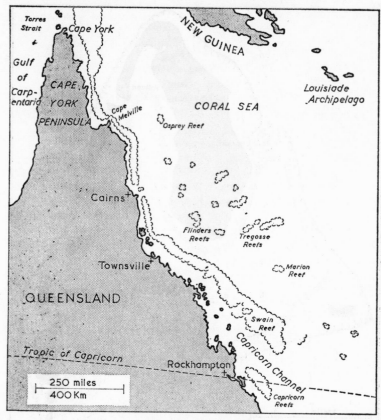

Fig. 155 *The Great Barrier Reef of Australia*

continents (Australia), and round the shores of islands (New Guinea and New Caledonia) and of volcanic peaks which, rising steeply from the ocean floor, account for many of the scattered Pacific islands, as Fiji and Samoa. They also form low coral islands rising from the ocean depths, such as the Gilbert and Ellice Islands and the Marshall Islands. Coral is most widespread in the western and central Pacific, but is also found in the Indian Ocean in the Laccadives and Maldives west of Ceylon, the Andamans, Seychelles and Mauritius. In the Atlantic, however, it is almost entirely confined to the West Indian archipelago.

Darwin divided reefs into three main forms—the fringing reef, the barrier reef and the atoll. The *fringing reef* consists simply of

316

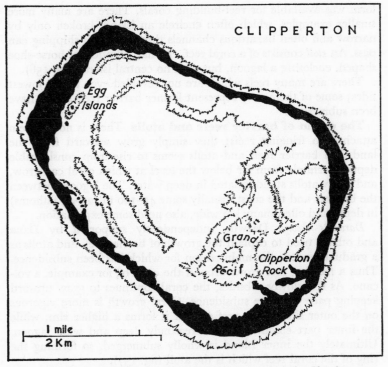

Fig. 156 *The island of Clipperton* (based on an Admiralty Chart)

The French island of Clipperton is isolated in the eastern Pacific at 10° 18′ N and 109° 13′ W. It consists of an elliptical atoll, the rim of which is made of hard coral conglomerate, in parts covered with coral shingle and wind-blown sand. The interesting feature is that in the extreme south-east is Clipperton Rock, a mass of volcanic rock (trachyte). Here is a case where part of the volcanic foundation of the atoll is actually revealed. The lagoon has a maximum depth of 100 m (55 fathoms), but contains many reefs. There is no channel through the atoll rim into the lagoon. The land above sea-level is in solid black, and the edge of the reef is indicated by a serrated line.

an uneven platform of coral fringing the coast, with a shallow, quite narrow lagoon between it and the mainland, and with its seaward edge sloping down into deep water (fig. 154). The *barrier reef* is separated from the mainland by a much deeper, wider channel. The largest in the world is the Great Barrier Reef off the Queensland coast (fig. 155), 2028 km (1260 miles) in length; it is thought by some authorities to be so extensive because the offshore zone was down-faulted, and the adjacent peneplaned surface of the land was let down to a fortuitously suitable depth, so that an extensive platform was available for reef-building corals. There are many more smaller examples, which often encircle an island, broken only by narrow and often hazardous channels through which shipping can pass. An *atoll* consists of a coral reef, circular, elliptical or horse-shoe shaped, enclosing a lagoon, but with no central island (fig. 156).

There are many reefs which are not revealed even at the lowest tides; some of them may represent former barrier reefs which have been submerged.

The origin of barrier reefs and atolls There is no problem attached to fringing reefs; they simply grow seaward from the land. But barrier reefs and atolls seems to rise from considerable depths, certainly from far below the level at which coral can grow, and many atolls stand isolated in deep water. The lagoons between the barriers and the coast, usually some 45–100 m (25–45 fathoms) in depth and often many km wide, also need some explanation.

Darwin's theory Darwin, independently supported by Dana and others, tried to explain the growth of barrier reefs and atolls as a gradual process, the main reason for which has been subsidence. Thus a fringing reef grows around the coast of, for example, a volcano. As this slowly subsides, the coral continues to grow upward keeping pace with the subsidence. Coral growth is more vigorous on the outer side of the reef so that it forms a higher rim, while the inner part comprises an increasingly deep and wide lagoon. Ultimately the inner island is wholly submerged, so forming the ring of the coral reef which is the atoll (fig. 157).

Some of Darwin's supporters have shown that submergence has indeed taken place, for neighbouring coastlines reveal distinct evidence of drowned valleys. This is shown along parts of the coast of Indonesia, and the edge of the coastal plain of Queensland, for example, has certainly been down-faulted. Some reefs, however, occur in areas where there is no evidence at all of submergence, even where it would have shown on neighbouring coasts had it so happened. In Timor and elsewhere, moreover, reefs have obviously been uplifted beyond present sea-level. More difficult still is the case where both atolls and raised coral reefs appear in the same group of islands.

Murray's theory An alternative hypothesis, put forward by Sir John Murray after his voyage in the *Challenger* in 1872, and supported by Agassiz, Semper and others, claimed that in most cases subsidence was not involved. His idea of atoll formation is that the base of a reef consists of a submarine hill or plateau rising from the ocean floor. These eminences, all reaching within about 55 m (30 fathoms) of the surface, consist either of sub-surface volcanic

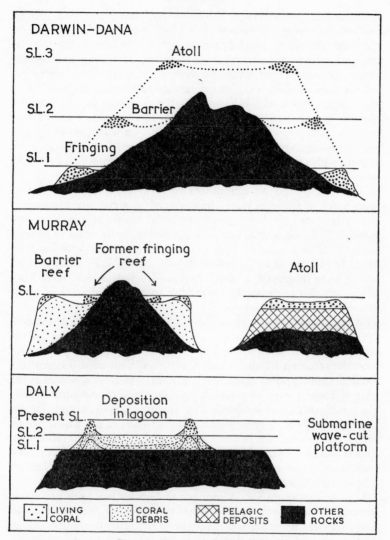

DARWIN–DANA

S.L.3 _____ Atoll _____

S.L.2 _____ Barrier _____

S.L.1 ___ Fringing ___

MURRAY

Barrier   Former fringing
reef   reef   Atoll

S.L.

DALY   Deposition
                      in lagoon

Present S.L.   Submarine
S.L.2   wave-cut
S.L.1   platform

LIVING CORAL   CORAL DEBRIS   PELAGIC DEPOSITS   OTHER ROCKS

Fig. 157 *Theories of atoll formation*

peaks and wave-worn stumps, or of an accumulation of various pelagic deposits which have been built up on deeper plateaus until they reach within 55 m of the surface. As pelagic deposits (p. 344) accumulate extremely slowly, and could not build up at the necessary steep angle, this latter possibility seems hardly credible.

According to Murray, a barrier reef has also been formed without involving subsidence. As a fringing reef grows, pounded by the surf, masses of coral fragments gradually accumulate on the seaward side of the reef, washed there by the waves, and become cemented into a solid bank. As coral tends to grow more strongly on the seaward side, it builds out farther and farther on banks of its own debris. While this is happening the corals on the inner side of the reef are deprived of food and so die. Murray also supposed that much of this dead coral was dissolved in the water, so forming a much deeper lagoon. But some scientists claim that filling-in of the lagoon with sediment and coral fragments will more than outweigh any possible loss by solution.

Daly's theory As an alternative to subsidence it was suggested that a rise of sea-level might be responsible, a rise which certainly did take place in late- and postglacial times as the ice-sheets melted. Daly discovered traces of glaciation on the sides of Mauna Kea in Hawaii, which means (a) that the water in those latitudes must have been so much cooler during these glacial times that all corals were destroyed, and (b) that sea-level must then have been about 90 m (50 fathoms) lower due to the withdrawal of water contained in the ice-caps (p. 304). It follows that all preglacial reefs and other islands were planed down by marine erosion to the sea-level of that time. The platforms thus formed provided bases for the upward growth of coral when the temperature of the sea increased and the receding ice-sheets slowly returned their melt-water to the oceans, causing a rise of sea-level. As coral can grow upwards at the rate of about 1 m in 30 years, there is no doubt of its ability to keep pace with this rising sea-level. This theory helps to account for the narrow, steep-sided reefs which compose most of the atolls. Sometimes their slopes are as steep as 75°, and it seems impossible that this could be a bank of fragments, or in fact could be anything but solid coral. The lagoons would be partially filled with sediments, particularly debris washed from the reef itself.

The evidence of borings A number of borings has been put down through reefs in order to determine the nature of the rock foundation. In 1904 a boring was made at Funafuti in the Ellice Islands to the north of Fiji to a depth of 340 m (1114 ft), as well as several other shallower borings in the lagoon floor. Several borings were put down through the Great Barrier Reef in 1937. In 1947 five borings, including one to the depth of 779 m (2556 ft), were sunk in the Bikini atoll in the Marshall Islands. On Funafuti at about 230 m (750 ft) the rock changed into a hard limestone, but it was still poorly consolidated at the base of the Bikini borings, showing that there is no definite single process for the conversion of coral into limestone rock; it may be physical, chemical or probably both. A boring on the Eniwetok atoll in 1954 went down through some 1300–1400 m (4200–4600 ft) of coral rock, where it reached basaltic rock, the base on which the coral had grown. More recent deep borings, together with seismic evidence, seem to confirm that most, if not all, atolls do rest on truncated volcanic cones.

A Classification of Land Forms

THE preceding chapters have been devoted to a study first of the initial structures produced by the internal or tectonic forces originating within the earth, and second of the effects of the external or gradational forces which ceaselessly modify these structures. It is necessary to summarize briefly their joint results in terms of what might be called the *sequential land-forms*. These represent merely a cross-section in time within the sequencies of gradual change and development.

The various land-forms, or, as some American geomorphologists call them, the *terrain elements* or *terrain types*, must be defined in terms of their outstanding characteristics and of the main factors, internal and external, responsible for their present forms. One resultant quality is their dimensions and scale, whether they are major or minor features. Involved is what may be called their *relative relief* or *local relief*, the amplitude between the altitude of the highest and lowest points in any particular district, such as the height of a ridge above a valley-floor. Another important characteristic is the nature of the gradients and slopes bounding a feature, whether these are steep, gentle or intermediate, concave or convex, continuous and uniform or interrupted by distinct 'breaks', bevels and changes of slope. Consideration is necessary of the material of which the features are formed: the nature and character of the bed-rocks, whether old or young, homogeneous or heterogeneous, how they have been arranged by the tectonic forces, and whether they are covered with superficial materials. Leading from this are the degrees of resistance to denudation, the stage reached in the cycle of denudation, and their resultant shapes and outlines, whether smooth and regular, or diversified and dissected. The way in which the distinctive land-form patterns are arranged or spaced relative to each other must also be considered.

Various methods of classification are possible. A. N. Strahler divided land-forms into two main groups: the *initial forms*, where the original features produced by tectonic forces have been only slightly modified, and the *sequential forms*, where the modifications are more pronounced, and where the initial forms may have been destroyed virtually beyond recognition, retaining only the merest vestiges of their foundations. The sequential forms include (i) *erosional types* (eroded valleys and depressions), (ii) *residual types* (surviving parts of a worn-down initial form), and (iii) *depositional types* (a new set of land-forms built up by deposition from the much altered fragments of the old).

Other classifications are based on the actual relief-forms. Preston E. James, requiring a physical background for his regional surveys, divided the land-surface of each continent into nine categories: plain, high mountain, mountain and bolson, hilly upland and plateau, hamada, intermont basin, low mountain, erg, and ice-covered areas. Using these land-form elements, he produced 'surface configuration maps'. Again, E. Raisz standardized a set of physiographic symbols to indicate the forty 'morphologic types' he required.

The classification that follows is essentially a compromise between initial tectonic cause, method and degree of external modification, and present relief form. The four major groups are: (i) mountains, (ii) plateaus, (iii) plains, and (iv) valleys and basins. The category of hill, used in some classifications, is not separately distinguished because this is a matter of relative scale. The South Downs, the Chilterns and the Cotswolds are hills, though they have some remarkably steep slopes on their escarpments. But the peaks of the English Lake District and Snowdonia are certainly mountains, yet they are very much lower than the Black Hills of South Dakota or the Nilgiri Hills of southern India. As A. A. Miller once said, 'We have elevated mere altitude to an unmerited eminence'. In this classification some hills may be included in the categories of mountains and plateaus, while others may form mere erosional remnants in distinctive lowland regions and therefore are included within the category of plains.

A. MOUNTAINS

These are features with usually steep slopes, rising to prominent ridges or individual summits (sharp or rounded), and marginally dissected by deep valleys.

I. Fold mountains

The result of compressive forces in the crust along well marked linear zones.

(*a*) *New (or young) fold mountains* (Alpine-Himalayan and Circum-Pacific systems). Characterized by complex rock-folding, associated faulting and vulcanicity; linear ranges, cordilleras and arcs; denudation into ridges and peaks by weathering, glaciation and running water. High ranges associated with snow-fields and glaciers (Alps, Himalayas).

(b) *Old fold mountains* (Caledonian and Hercynian systems of central Europe). Ancient folding, followed by prolonged phases of denudation. Later uplift with faulting and tilting, and vulcanicity. Denuded into residual forms: monadnocks (Pennine peaks), accordant summits (Highlands of Scotland), dissected fold-domes (English Lake District), irregular patterns of peaks and valleys (Snowdonia), and residual plateaus (see under B).

II. Mountains produced by vulcanicity

The major land-forms resulting from the processes by which solid, liquid or gaseous materials are forced into the earth's crust or escape on to the surface.

(a) *Volcanic cones:* ash (Vulcan), cinder (Paricutín), lava (Puy de Dôme, Mauna Loa), composite (Fujiyama, Etna). In many cases they form culminating points on fold-mountain ranges (Andean peaks), or on fault-blocks (Mount Rainier, Mount Shasta).

(b) *Residual forms of volcanic cones:* plugs or domes (Lassen Peak, California), calderas (Askja).

(c) *Residual forms of intrusive masses:* exposed batholiths (Dartmoor, Mont Blanc), dissected laccoliths (Henry Mountains of Utah).

(d) *Residual forms of extrusive lavas:* dissected volcanic rocks (mountains of central English Lake District, Snowdonia).

B. PLATEAUS

These are upland areas on various scales, with surfaces of some degree of altitudinal uniformity, bounded by one or more slopes falling steeply away, sometimes rising on one or more sides by prominent slopes to mountain ridges. Where the surrounding slopes are particularly steep and the upland surface is sharply defined, the term *tableland* may be used.

I. Tectonic plateaus

(a) *Major or continental blocks* (Africa, Arabia, Deccan), bounded by well-defined edges (Western Ghats, Drakensbergs).

(b) *Faults-blocks and horsts*, smaller and well defined by marginal faults (Harz Mountains, Morvan, Black Forest).

(c) *Tilt-blocks*, with one steeper out-facing edge (Meseta of Spain, with Sierra Morena).

(d) *Intermont plateaus*, enclosed within the ranges of fold-systems (plateaus of North American Cordillera, Anatolia and Armenia in Asia Minor, Iran). Basin-and-range structure (Great Basin of Nevada and Utah).

II. Residual plateaus

Derived from ancient fold-systems by extensive denudation (see A.I.(b)) (Ardennes, Middle Rhine Highlands, fjeld-plateau of Scandinavia, Alleghany-Cumberland Plateau of Appalachians). Where much dissected, accordant summits remain (Highlands of Scotland). In semi-arid areas, dissection into steep-sided tabular masses (mesas and buttes of south-western U.S.A.).

III. Volcanic plateaus

Derived from basaltic lava-flows (Antrim, Abyssinia, Columbia-Snake, north-western Deccan).

C. PLAINS

An extensive area of lowland, with a level or gently undulating surface.

I. Structural plains

(a) *Relatively undisturbed horizontal strata*, except for faint warping (Russian Platform, much of American Midwestern plains).

(b) *Emerged coastal plains* of marine deposition (east coast and Gulf coast plains of U.S.A.).

II. Erosional plains

(a) *Product of river erosion:* with valleys and gentle interfluves (peneplains and panplains). Development of river patterns, producing vales and escarpments, known in America as *cuestaform plains* (Midwest of U.S.A., e.g. Niagara Escarpment; south-eastern England).

(b) *Modification and 'surface trimming' by glacial erosion:* worn-down ice-scoured shield-lands of uneven surface (Canadian Shield, Finland).

(c) *Product of desert deflation* (the Reg, the Serir and the Hamada of the Sahara Desert).

(d) *Product of semi-arid denudation:* pediments (south-western U.S.A.) and pediplains (S.W. Africa, western Australia).

(e) *Product of marine planation* of continental margins (Norwegian Strand-flat), possibly followed by tectonic uplift (Pliocene platform of southern Weald).

III. Depositional plains

(a) *Product of river deposition:* alluvial plains (widespread), flood-plains (Mississippi), deltaic plains (lower Egypt), piedmont alluvial plains fringing mountain slopes (bajada-plains), infilled lacustrine plains (Vale of Pickering in Yorkshire), infilled gulfs of the sea (Plain of Lombardy, Tigris-Euphrates plain), infilled structural depressions (Central Valley of California, Hungarian Basin).

(b) *Product of glacial deposition:* drift- or till-plains, usually a veneer with minor diversities due to ground-moraine, terminal moraines, drumlins (northern Europe and south-central Canada).

(c) *Product of fluvioglacial deposition:* outwash plains of sands and gravels (southern Michigan, heathlands of western Europe).

(d) *Product of aeolian deposition:* sheets of sand, sand-ridges, dunes (Erg of Sahara, Koum of Turkestan). Loess-plains (NW China, Börde of western Europe, Argentine pampas, central U.S.A., especially Nebraska and Iowa).

(e) *Product of marine deposition* along shallow coastal margins: accretion of sand, mud, alluvium and vegetation, in the form of tidal flats, deltaic development, estuarine banks (coast of Netherlands, West Germany and Denmark, head of Adriatic Sea, Gulf of Mexico coast). Stimulated by uplift and by man's collaboration (dyking, draining, polders).

D. VALLEYS AND BASINS

Smaller lowlands lying within or among upland areas, sometimes alluvium filled, usually containing a river, frequently containing or formerly contained a lake. A valley is longer and narrower than a basin.

I. Tectonic valleys and basins

(a) *Synclinal valleys.* The low-lying valley still corresponds to a syncline (the *vaux* of the Jura, Central Valley of California).

(b) *Synclinal basins* of a gentle downfold (London Basin).

(c) *Fault-valleys* along the line of a fault (Eden valley in north-western England, Vale of Andalusia in Spain, Great Glen of Scotland).

(d) *Rift-valleys*, a trough let down between parallel faults (Vale of Clwyd, Central Valley of Scotland, Jordan Valley, East Africa, middle Rhine).

II. Erosional valleys

(a) *Normal river valleys*, features according to stage of cycle of erosion, widening ultimately to erosional plain.

(b) *Gorges and canyons*, deep in proportion to width, due to (i) vertical fluvial erosion more rapid than lateral weathering, frequently in desert climates (Grand Canyon), where river flows on very resistant rock (Aar Gorge, Switzerland); (ii) to rejuvenation (lower Wye); and (iii) to antecedence (Ganges and Brahmaputra).

(c) *Anticlinal valleys* cut along crests of anticlines (numerous in Appalachian Ridge and Valley region).

(d) *Glacially eroded valleys*, over-deepened U-shaped valleys, with trough heads, rock-steps, truncated spurs, hanging lateral valleys (Lauterbrunnen in Switzerland, Yosemite in California).

III. Lake-basins

A classification of the origin of lake-basins is given in Chapter 7 and need not be repeated.

IV. Ocean basins

See Chapter 12.

The Configuration of the Oceans and Seas

OCEANOGRAPHY involves the consideration of a wide range of oceanic phenomena, both physical and biological. From the point of view of the physical geographer, the most important feature is the extent and shape of the ocean basins, for the distribution of land and sea is for him a fundamental concept. The structure and relief of the ocean floor and of the marginal seas afford contributory evidence towards the structure of the earth, concerned as we are with the permanence or otherwise of the oceans and continents, the problem of continental drift, changes of sea-level, the distribution of volcanoes and earthquake zones, and the accumulation of sediments on the sea floor which may ultimately form sedimentary rocks. The movements of sea-water, in the form of waves, tides and currents, affect the coastline with which they come in contact (pp. 281–3), and warm and cool currents may powerfully modify the climates of coastal areas. Salinity and water temperature, both at the surface and at depth, must also be considered. The biological aspects are more out of the domain of the physical geographer; the economic geographer is concerned with the distribution of fish, whales and other life, and therefore with the minute plant and animal organisms known as *plankton*, which form much of their food.

Much research has been carried on during the last century into the various aspects of oceanography. Survey ships have charted the seas, involving the recording and plotting of millions of depths by means of sounding, so that the 'relief' of the ocean floor is remarkably well known. Numerous scientific expeditions have been sent out by various countries, by learned societies, and by the Permanent International Council for the Exploration of the Sea, which has its headquarters at Copenhagen. Some of the more notable expeditions were the voyage of the *Challenger* (1872–6), that of the *Tuscarora* (1874–6) in the Pacific, the work of the Scandinavian seamen (Nansen, Amundsen and Pettersen) especially in northern waters, the *Michael Sars* North Atlantic expedition in 1910, the German *Meteor* expedition of 1925–7, and the voyage of *Discovery II* in 1932 in the Southern Ocean. More recently there have been in 1947–8 the voyage of the Swedish ship *Albatross*, the 1950–1 voyage of

Discovery II, the 1950–2 *Challenger* expedition, the 1951–2 Danish expedition, and numerous voyages during the International Geophysical Year (1957–8). These are but a few of the many, and at the present time regular series of observations, both physical and biological, which are being recorded in a number of areas, such as the Falkland Islands sector of the Southern Ocean. Much research is being carried out by such organizations as the International Association of Physical Oceanography, the British National Institute of Oceanography, the Scripps Institute, the Woods Hole Oceanographic Institute, and the Lamont Geological Observatory of New York in the U.S.A., and the Special Committee on Oceanic Research set up in 1957 by the International Council of Scientific Unions.

In these two chapters the scope of the survey must clearly be limited to a number of topics which are the concern of the physical geographer: the configuration of the oceans and their marginal seas, the mantle of deposits on the sea floor, the movements of the ocean waters, and their physical characteristics.

The area of the oceans The general pattern of the distribution of the oceans and continents has been commented upon (p. 26). It has been calculated that the land surface of the globe totals 148 million sq. km (57 million sq. miles), while the water surface totals 363 million sq. km (140 million sq. miles); this gives a relative percentage of 71 to 29. The bulk of the water surface is accounted for by the four great oceans, as follows:

| | Million | |
	Km^2	Sq. miles
Pacific	165·5	63·9
Atlantic	82·1	31·7
Indian	73·6	28·4
Arctic	14·0	5·4
	335·2	129·4

This total excludes marginal seas such as the Caribbean, the Mediterranean and the Bering Seas. The first three oceans include the sectors of the Southern Ocean, which is sometimes individually defined as the water south of 40° S.

SUBMARINE RELIEF

The continental shelf Around the coasts between low-tide level and about the 100-fathom (180 m) mark is a platform known as the continental shelf, structurally part of the continent itself; its

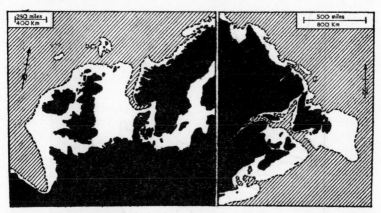

Fig. 158 *The continental shelf off north-western Europe* (left) *and north-eastern America* (right)

The pecked line indicates the 100-fathom (180 m) contour; depths below this are shaded with diagonal lines.

higher parts project as islands. The shelf slopes gently seaward with an angle of less than a degree. It is well developed off western Europe, where it extends westward for 320 km (200 miles) from Land's End, and off north-eastern North America (fig. 158), while off the Arctic coast of Siberia it is about 1200 km (750 miles) wide. Detailed hydrographic surveys in the Canadian Arctic since 1959 have outlined a continental shelf 120–160 km (75–100 miles) in width to the north of Canada, covered with water to 180 m (100 fathoms), and abruptly dropping by the continental slope to deep water. Around other continents it is much narrower, or almost completely absent, especially along coasts where fold mountains run parallel and close to the sea, as along the edge of the eastern Pacific.

Detailed soundings have revealed that the valleys of many rivers seem to continue across the continental shelf. This can be accounted for either by a rise in sea-level or by a sinking of the land; in other words, the continents really end at the edge of the shelf. This would explain the fact that the continental shelf is at its widest around the shores of lowland areas, where a slight change of level involves a considerable extent of land. Some continental shelves may be due in part to wave erosion (p. 288), to the building up of an off-shore terrace by deposition during long periods of geological time, and more potently to glacial erosion during a period of low sea-level; some authorities even claim that deposition by the Quaternary ice-sheets may have helped to build up the shelves in the Atlantic.

One difficult problem is that some submarine channels cross the

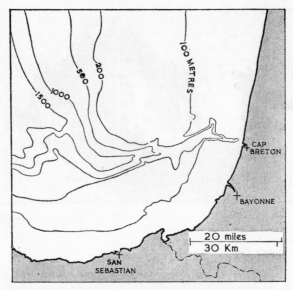

Fig. 159 *The Fosse de Cap Breton*

This submarine trench lies in the floor of the Bay of Biscay near the coast of south-western France.

shelf and then continue beyond its edge into deeper water. One of these lies off the mouth of the Congo, another off the Hudson in North America, a third in the Bay of Biscay (fig. 159). Some authorities ascribe them to faulting, others to former river erosion, involving therefore enormous changes in sea-level. Another explanation is that a river current, continuing into the sea, keeps its channel clear by depositing material at the sides, thus leaving a trough as sediment accumulates. All these explanations are open to criticism, particularly in the case of the deeper troughs. More difficult still to understand are troughs which do not cross the whole shelf, but are found near its oceanward margin; sometimes they form deep gorges cut into the edge of the shelf. A number has been surveyed in detail by echo-sounding off the New England coast. One theory attributes these gorges to the sapping action of submarine springs bursting out far down the continental slope. The most probable theory is that powerful, though sporadic, turbidity currents, carrying great loads of solid material, excavate the floors of these canyons. Commonly these currents are initiated by submarine earthquakes, as in 1929 when many trans-Atlantic cables to the south of Newfoundland were disrupted.

The continental slope At the edge of the shelf the seaward slope steepens considerably, forming the continental slope, which descends to about 3350 m (2000 fathoms). The slope varies, but is usually between 2° and 5°, although some surprisingly steep gradients (up to 15°) have been encountered by cable-laying ships.

The abyssal plain Almost two-thirds of the entire ocean floor lies at depths between 3350 and 5500 m (2000–3000 fathoms), forming an undulating deep-sea or 'abyssal' plain. It is by no means flat; the *Albatross* expedition found from continuous echo-sounding that the bottom profile is much more rugged in detail than was formerly thought. Its surface is covered with various pelagic oozes (pp. 344–6 and figs. 165, 166). Long curving ridges and more extensive submarine plateaus occur, while occasionally volcanic peaks rise steeply from the plain, known as *seamounts*, sometimes reaching the surface as isolated islands. Detailed soundings have recently revealed the presence in the Pacific of numerous flat-topped mountains, possibly wave-truncated volcanoes, rising to within 800 m of the surface. To them has been given the name *guyot*; there are estimated to be 10,000 guyots and seamounts in the Pacific.

The deeps Successive survey ships have sought to locate and sound the 'deeps' in the oceans. The soundings made by the *Challenger* expedition were taken by using a fine hemp line with a weight

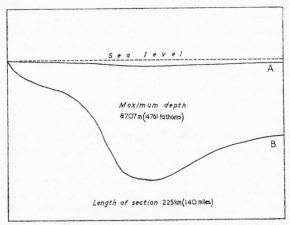

Fig. 160 *Sections across the Tonga Deep*

The upper section is true to scale, with no vertical exaggeration, the lower section has a vertical exaggeration of twenty times. The upper shows how misleading the term 'trench' or 'trough' can be.

on the end, and the depths were believed to be correct to about 45 m (25 fathoms); any inaccuracy was due to the difficult of obtaining a truly vertical sounding. Then piano wire was used, later attached to a 'sounding machine', which automatically checked the run-out of the wire as soon as the weight touched bottom. A 7000 m (4000 fathoms) sounding was a lengthy business. Today echo-sounding is employed, whereby either sonic or ultra-sonic vibrations are transmitted down through the water to the ocean floor, to return in the form of an echo which is electronically recorded as an *echogram*; it is a very speedy method. For very accurate survey work the ship is anchored, but for normal purposes a continuous profile is obtained while the ship is under way.

Most of these deeps form elongated 'troughs' or 'trenches', terms which however tend to give a false impression of the steepness of their sides, since they rarely exceed 7° (fig. 160). They usually lie near and parallel to coasts closely bordered by fold-mountain ranges, notably the Asiatic island arcs, hence are called *fore-deeps*. Most of the deeps occur in the Pacific Ocean. They are usually asymmetrical; the slope nearer the land is considerably steeper than that on the side of the open ocean; and they seem to be bounded by major tear-faults (p. 37).

THE PACIFIC OCEAN

Shape and size The Pacific Ocean, with its bordering seas, occupies about one-third of the area of the world, and exceeds the total land area by about one-eighth. It forms more or less a broad triangle in shape, with its apex in the north at the Bering Strait, enclosed on the west by the long broken line of Asia–Australasia, on the east by the Americas, and on the south by the edge of the Antarctic continent. Great distances are involved; thus from the Bering Strait to Cape Adare on the Antarctic continent is 15,000 km (9300 miles), while the width along the Equator exceeds 16,000 km (10,000 miles).

The Pacific depression The age and the mode of origin of this depression in the earth's surface have long been subjects of discussion. The volume of water occupies a space of no less than 174 million cubic miles. The basin seems to have begun to develop some 250–300 million years ago, when as a result of continental drift Gondwanaland in the south and Laurasia in the north broke up and the sialic 'continental rafts' moved apart (p. 25). These movements are still in progress, as shown by geophysical research into the East Pacific Rise (p. 334), into the linear deeps, and into the enormous tear-faults (p. 37) which define the ocean basin along

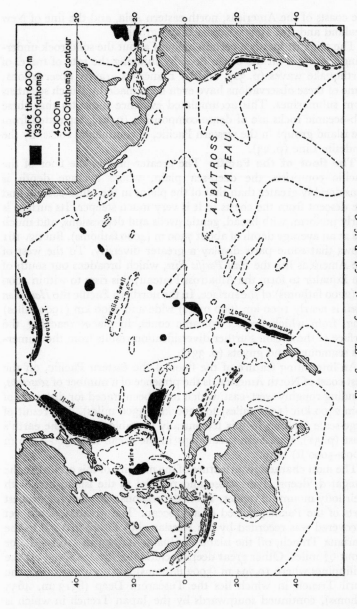

Fig. 161 The configuration of the Pacific Ocean

The following abbreviations used are : **Mar. T.** = Mariana Trench; **Ph. T.** = Philippine Trench. Small island groups are omitted.

Labels within map:
Atacama T.
ALBATROSS PLATEAU
Hawaiian Swell
Kermadec T. Tonga T.
Aleutian T.
Kuril T.
Japan T.
Mar. T.
Swire D.
Ph. T.
Sunda T.

Legend:
More than 6000 m (3300 fathoms)
4000 m (2200 fathoms) contour

333

the coasts of the Americas, north-eastern Asia, and the line of New Zealand and the Tonga-Samoa island groups.

The scientist knows a certain amount about the solid rock underlying the oozes on the ocean floor by studying the rate of travel of earthquake waves (p. 32), and by numerous gravity observations; some of these observations have even been made by Dutch scientists from submarines. This accumulated evidence indicates that these sub-oceanic rocks are of dense composition (p. 21), separated from the island groups in the western Pacific, of continental rocks, by the Andesite Line (p. 23).

The floor of the Pacific The greater part of the floor of the Pacific comprises the deep-sea plain, whose average depth is considerably greater than that of the plains in the other oceans, and the descent from the coasts to it is very much steeper. Its surface is fairly uniform, with broad, gentle swells and depressions, and much lies at an average depth of about 7300 m (4000 fathoms). But fig. 161 shows that some parts display a greater diversity. To the west of the Americas lies the *East Pacific Rise*, which broadens out south of the Equator to form the Albatross Plateau; this rises to within 4000 m (2200 fathoms) of the surface. In the northern Pacific the *Hawaiian Swell* is nearly 1000 km (600 miles) wide and 3000 km (1900 miles) long, from which project volcanic cones, in places reaching the surface of the ocean. Further diversification results from the numerous seamounts and guyots (p. 331).

An interesting feature of the floor of the eastern Pacific, off the west coast of North America, is the presence of a number of *seascarps*, trending roughly west–east, which have been traced for distances of from 2400 km (1500 miles) to 5300 km (3300 miles). The result of large-scale faulting in this tectonically unstable part of the earth's crust (p. 51), they form great steps in the ocean floor 300–1500 m (1000–5000 ft) high.

The most characteristic feature of the margins of this ocean are the elongated 'deeps', lying close and parallel to the island arcs with their lofty mountain ranges. These depressions represent the deepest parts of the Pacific, indeed of any ocean. The greatest depth yet discovered was recorded by the Russian ship *Vityaz* in 1959 in the Mariana Trench, off the island of Guam—11,033 m (36,198 ft) or about 6¾ miles. Other great deeps include the Emden Deep off the Philippines where 10,794 m (5902 fathoms) has been measured, the Kuril Trench in which lies the Tuscarora Deep (8513 m, 4655 fathoms), continued southwards by the Japan Trench in which is the Ramapo Deep (10,554 m, 5771 fathoms), the Mariana Trench (Mansyu Deep (9866 m, 5395 fathoms)), and the Tonga–Kermadec

Trench (Aldrich Deep, 9427 m, 5155 fathoms). In the north, parallel to the Aleutian Islands, is the Aleutian Trench with a maximum depth of 7682 m (4199 fathoms). None of these deeps occurs in the central part of the basin, but another line lies off the coast of South America, forming a trench parallel to the Andean upfold; the Atacama Trench attains a depth of 7635 m (4175 fathoms).

Pacific islands The Pacific Basin contains vast numbers of islands, one estimate putting the total at about 20,000, but their aggregate area is relatively small. The larger islands are 'continental', that is, they belong structurally to and represent a submerged part of the mainland, separated by foundered basins. In the east are the Aleutians, the islands off British Columbia, and the Chilean islands. In the west are the much more extensive island arcs of eastern Asia—the Kurils, the Japanese Archipelago, the Philippines, the Indonesian islands and New Zealand. Most are fold-mountain ranges, with numerous volcanic peaks, and their existence along lines of crustal weakness is evidenced by widespread earthquakes (p. 35, fig. 7).

Most of the smaller scattered groups of Pacific islands are situated in the south-western part of the basin. They are grouped under three names according to their racial groupings: Melanesia (including the Solomons, New Hebrides and Fiji), Micronesia (Carolines, Marshalls, Gilbert and Ellice), and Polynesia (Line Islands, Cook, Society and Tuamotu islands). In the northern Pacific lie the Hawaiian Islands. But most of the north-eastern and eastern Pacific consists of empty stretches of ocean, with but a few isolated island groups—Clipperton (fig. 156), 2400 km (1500 miles) off the Central American coast, the Galápagos Archipelago nearly 1000 km (600 miles) west of Ecuador, Easter Island formed by three extinct volcanoes rising from the Albatross Plateau, and Juan Fernandez, 580 km (360 miles) off the Chilean coast, reputed to be the setting for Alexander Selkirk's *Robinson Crusoe*.

Apart from the 'continental' islands of the fold ranges, these Pacific islands consist of two types: the 'high' volcanic islands and the 'low' coral islands. Hawaii consists of five volcanoes of different ages, rising to 4213 m (13,825 ft) in the cone of Mauna Kea and to 4168 m (13,675 ft) in Mauna Loa. Coral islands and their origin have been discussed; the true 'low' island is an atoll (p. 318).

Marginal seas Marginal seas in the Pacific Basin are almost entirely confined to the western side. The longitudinal nature (p. 308) of the American coasts accounts for their absence in the east; the only partially enclosed water areas are the Gulf of California

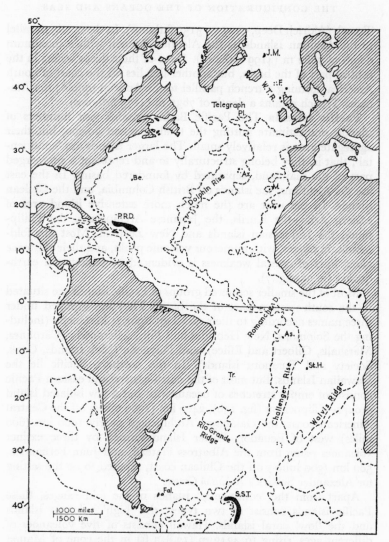

Fig. 162 *The Configuration of the Atlantic Ocean*

The 2000-fathom (about 3600 m) contour is shown by a pecked line, while the two deeps (Puerto Rico (**P.R.D.**) and South Sandwich (**S.S.T.**)) exceeding 7300 m (4000 fathoms) are in solid black. The Romanche Deep, which interrupts the Mid-Atlantic Rise, attains 7370 m at one point. **W.T.R.** = Wyville-Thomson ridge.

and the sounds between the foundered coast ranges and mainlands of British Columbia and Chile.

In the western Pacific semi-enclosed seas lie between the Asiatic mainland and the island festoons. These comprise the Bering Sea, enclosed by the Aleutian Islands; the Sea of Okhotsk within the Kamchatka Peninsula; the Sea of Japan between Korea and the Japanese Archipelago; the Yellow Sea between Korea and the Chinese mainland; the East China Sea between China and the line of the Ryuku Islands; and the South China Sea enclosed by the Philippines, Borneo, Malaya, Indo-China and southern China. Among the Indonesian islands lie the Celebes, Banda and other seas. None of these, except the Yellow Sea, which is mostly under 180 m (100 fathoms), is really shallow even by comparison with the rest of the Pacific, and most have basins exceeding 2700 m (1500 fathoms). In the Celebes Sea 5112 m (2795 fathoms) has been sounded, and the Sea of Japan has a maximum depth of 3576 m (1955 fathoms). Round Australia the Gulf of Carpentaria, the Arafura Sea and the Bass Strait lie on the continental shelf.

THE ATLANTIC OCEAN

Shape and size (fig. 162) The Atlantic Ocean, without its marginal seas, occupies rather less than one-sixth of the total area of the world, that is, approximately half the extent of the Pacific Ocean. Its general outline is that of a letter S, for as the coast of Saharan Africa bulges westward so does the north coast of South America recede into the Caribbean embayment; conversely, while Cape São Roque projects eastward, so does the Gulf of Guinea recede in the same direction. This complementary nature of the shape of the respective coasts helps, with other geological and biological evidence, to suggest that the continents on either side of the present ocean basin were once part of a single land-mass (pp. 24–5).

The Atlantic Basin narrows towards the Equator; the Liberian coast of Africa is only about 2600 km (1600 miles) away from Cape São Roque, forming the North Atlantic Basin, which is about 4800 km (3000 miles) wide in latitude 40° N, and the South Atlantic Basin, about 5900 km (3700 miles) wide in latitude 35° S. The South Atlantic opens broadly into the Southern Ocean, but the Northern Basin is much more enclosed by the presence of Greenland and Iceland.

The floor of the Atlantic The most striking feature about the floor of the Atlantic is the presence of a longitudinal 'rise', known as the Mid-Atlantic Ridge (fig. 163), or as the Dolphin Rise in the north and the Challenger Rise in the south. This was discovered by the *Challenger* in 1873, mapped by the *Meteor* in 1925-7, the first occasion on which echo-sounding was used and again in detail in 1953 by the U.S. research ship *Vema*. This submarine rise,

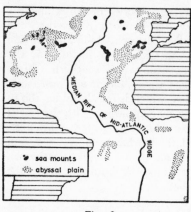

which slopes gently toward the deep-sea plain on either side, is S-shaped, following the general trend of the coastlines with remarkable fidelity. It is covered by an average depth of about 3000 m (1700 fathoms) of water. In the North Atlantic the ridge widens to form the Telegraph Plateau, which extends across the ocean from Ireland to Labrador. Its structural significance is discussed on pp. 25-6.

Fig. 163

There are several transverse ridges in the Atlantic. The Walvis Ridge runs in a north-easterly direction from the neighbourhood of Tristan da Cunha to the African coast, and the Rio Grande Ridge trends, less continuously, from the same area towards the South American coast. In the north a broad ridge rises from the Telegraph Plateau and runs north-westward from northern Scotland to south-eastern Greenland; the water over this ridge averages 1000 m (550 fathoms) in depth, and the Faeroes and Iceland represent its projecting higher parts. Between northern Scotland and Iceland lies the Wyville-Thomson Ridge.

'Linear deeps' or 'trenches' are uncommon in the Atlantic, which can be related to the fact that lines of recent folding near the Atlantic coasts are rare. Significantly, the chief deeps lie off the island arcs of the West Indies; immediately north of the island of Puerto Rico a depth of 8800 m (4812 fathoms) has been sounded, so far the greatest depth recorded in this ocean. Another deep, the Romanche, lies across the Mid-Atlantic Rise, and in fact cuts through it to a depth of 7370 m (4030 fathoms). The only other great deep was measured in the South Sandwich Trench (8312 m, 4545 fathoms), a curvilinear deep parallel and close to the South Sandwich Islands.

The continental shelf is of considerable extent in the North Atlantic, although not in the South, except off the east coast of

South America where the Falkland Islands rise from it. Both off the coast of western Europe and off north-eastern America it is a most striking and important feature (p. 329 and fig. 158). Off the Brazilian and African coasts of the South Atlantic, both of which are formed by the steep edges of plateaus, the continental shelf is virtually absent.

Atlantic islands If such 'continental' islands as the British Isles and Newfoundland, which are merely the slightly higher parts of the continental shelf, are excepted, the Atlantic has singularly few islands. The West Indies consist of a series of island arcs not far from the mainland, while Iceland and the Faeroes form the higher parts of the ridge between northern Scotland and Greenland. Similarly, the island groups in the extreme south (the Falklands, South Orkneys, South Shetlands, Georgia and the Sandwich Islands) are the higher parts of the complicated ridges and plateaus which extend between the tip of South America and the Grahamland Peninsula of Antarctica.

The true oceanic islands project from the Mid-Atlantic Ridge, notably the Azores in the north and lonely Ascension and Tristan da Cunha in the south. St Helena lies just east of the ridge, and appears to rise steeply from the deep-sea plain, as does the tiny Brazilian island of Trinidad to the west of the ridge in latitude 20° S. The coral islands of the Bermudas are built on submerged volcanic cones in the north-western Atlantic Basin. Madeira, off the Moroccan coast, consists almost entirely of volcanic material, built up by a long series of eruptions; the highest of numerous rugged peaks is Pico Ruivo, rising to 1846 m (6056 ft).

The other Atlantic islands for the most part rise from plateau-like extensions from the mainland, notably the Canaries, the Cape Verde Islands and several small islands in the angle of the Gulf of Guinea.

Marginal seas Just as the continental shelf is largely absent in the South Atlantic, so too are marginal seas. On the other hand, the submergence of the continental margins of Europe, producing a remarkably indented 'peninsular Europe', has resulted in a number of extensive marginal seas—the Baltic, the North and the Mediterranean Seas, with their several subsidiaries. The first two are shallow, less than 180 m (100 fathoms), and the passages between the Danish islands into the Baltic are only 20 m (11 fathoms) deep. The Mediterranean, with its several basins interrupted by peninsulas and islands, represents the foundering of a complicated structural area, part of the Alpine folded system. The depth in the Straits of Gibraltar is only about 360 m (200 fathoms), forming a submarine sill which slopes quite steeply on either side (fig. 171 and p. 352). The Mediterranean Basin has some areas exceeding 3600 m (2000 fathoms), and the deepest sounding (4632 m, 2533 fathoms), is found between Crete and Greece. The Black Sea, in which the deepest sounding is 2244 m (1227 fathoms), is separated from the Mediterranean by narrow straits (the Dardanelles, Sea of Marmara and the Bosphorus).

The Adriatic Sea affords a striking example of the formation of these deeper marginal seas. It consists of a narrow, somewhat elongated depression lying between the roughly parallel folds of the Italian Apennines and the Jugoslavian–Grecian Dinaric system (fig. 150). The foundering of the area was the result of earth disturbances in late Tertiary times, which affected the whole Balkan peninsula, as well as the neighbouring Aegean, Adriatic and Black Sea basins.

There are other marginal seas on the American side of the ocean. Baffin Bay and Hudson Bay are for the most part under 180 m (100 fathoms), and Davis Strait, between Greenland and Baffin Island, forms a shallow link between the Atlantic and the Arctic, since its maximum depth is only 205 m (112 fathoms). Within the Gulf of Mexico 3804 m (2080 fathoms) is the maximum sounding, while the Caribbean Sea consists of a complicated series of basins and ridges, including the Bartlett Deep, where 7200 m (3937 fathoms) has been measured.

THE INDIAN OCEAN

Shape and size (fig. 164) The Indian Ocean is rather smaller in area than the Atlantic. It differs from the two other oceans in shape in that it is enclosed by land in the north, and only just extends beyond the Tropic of Cancer. Its shores consist for the most part of ancient plateaus (Africa, Arabia, the Deccan and western Australia), the remnants of Gondwanaland (p. 25), except in the north-east, where it is bounded by the island festoons of Indonesia and the fold ranges along the coast of Burma. To the south lies part of the coast of Antarctica between about longitudes 20° E and 115° E.

The floor of the Indian Ocean In depth this ocean is far less diverse than the other two great oceans. About 60 per cent of the total area forms the deep-sea plain, with a depth of between 3600 m and 5500 m (2–3000 fathoms). Linear deeps are absent, except in the Sunda Trench where a maximum depth of 7454 m (4076 fathoms) has been sounded; this lies south of and parallel to the Java–Lesser Sunda Islands arc.

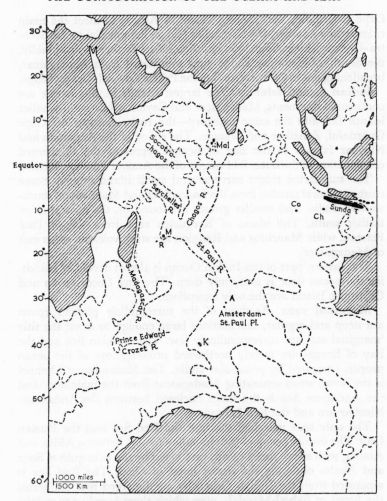

Fig. 164   The configuration of the Indian Ocean

The 2000 fathom (about 3600 m) contour is shown by a pecked line, and the Sunda Trench is in solid black.

A number of broad submarine ridges separates the several individual basins of the deep-sea plain. One runs more or less southward without a break from Cape Comorin, the southern tip of India, to the Antarctic continent. This ridge (fig. 164) widens out in the south to form the extensive Amsterdam–St Paul Plateau. The transverse Socotra–Chagos Ridge trends south-eastward from

Cape Guardafui, the 'Eastern Horn' of Africa, to meet the main ridge, while another, the Seychelles Ridge, lies parallel to the Socotra–Chagos Ridge about 1300 km (800 miles) farther south. Finally, the South Madagascar Ridge runs southward from Madagascar, broadening out in a transverse rise, the Prince Edward–Crozet Ridge.

Indian Ocean islands The largest islands in this ocean are continental fragments, Madagascar and Ceylon, while many smaller islands fall into the same category—the rocky Socotra off Cape Guardafui, Zanzibar and Comoro. The string of the Andaman and Nicobar Islands in the Bay of Bengal represent the submerged continuation of the outer fold ranges of the Arakan Yoma in Burma.

The submarine ridges carry several small island groups. Some clusters of coral patches form the Laccadives and Maldives off southwestern India, and smaller groups are found on the central ridge farther south. The island of Kerguelen rises from the St Paul Plateau, while Mauritius and Réunion are steep volcanic cones east of Madagascar.

The eastern part of the Indian Ocean is almost devoid of islands, for the ocean floor is uniformly deep. The tiny Cocos group and Christmas Island are the only exceptions.

Marginal seas As most of the surrounding plateau coasts are steep and regular, indentations large enough to merit the title 'marginal sea' are correspondingly few. The Arabian Sea and the Bay of Bengal are merely northward prolongations of the ocean proper, separated by peninsular India. The Mozambique Channel is the broad strait separating Madagascar from the mainland, and the Andaman Sea is the basin enclosed between the Andaman–Nicobar arc and the Kra Isthmus.

The only true marginal seas are the Red Sea and the Persian Gulf. The former occupies the rift-valley (p. 45) between Africa and Arabia, with steep, rocky coasts, prolonged by the twin gulfs of Suez and Akaba outlining the desert horst of Sinai. The Red Sea is separated from the Indian Ocean by a submerged rock sill across the Strait of Bab-al-Mandab, over which there is only 370 m (200 fathoms) of water (fig. 167). The Persian Gulf is a shallow trough which is being slowly filled with sediment by the Tigris–Euphrates. It is nearly shut off from the Gulf of Oman and the Indian Ocean by the northward projecting Oman Peninsula, which restricts the Strait of Hormuz to about 80 km (50 miles).

THE ARCTIC OCEAN

Shape and size The Arctic Ocean is roughly circular in shape, with the North Pole situated much nearer the Greenland margin (about 83° N) than the Alaskan–Siberian (70° N). Its area is some 14·2 million sq. km (5·5 million sq. miles), or roughly one-twelfth that of the Pacific. The basin is almost land-locked by the coasts of the great land-masses of the northern hemisphere, except for the narrow Bering Strait at 170° W, and for the passages already mentioned between Greenland, Iceland and the British Isles, with their submarine ridges. Much of the Arctic is, of course, permanently frozen. The *Ice Atlas of the Northern Hemisphere*, published by the U.S. Navy, distinguishes between the permanent unnavigable polar ice and the maximum (spring) and minimum (autumn) extent of pack-ice.

The floor of the Arctic Ocean. Knowledge of depths within this ocean is obviously limited, and except on the margins is almost entirely derived from echo-soundings. It seems, however, that there is a single extensive basin, the North Polar Basin, which probably averages 3600 m (2000 fathoms), with a maximum of 5625 m (3076 fathoms) recently found at point 78° N, 175° W by echo-sounding, while round this lie marginal seas for the most part less than 1800 m (1000 fathoms) deep.

The marginal seas Shallow marginal seas border the northern coasts of the land-masses—the Beaufort Sea off Alaska, the East Siberian and Laptev Seas off Siberia, the Kara Sea between the mouth of the River Ob and Novaya Zemlya, and the Barents Sea between Norway and Spitsbergen. Among the chaos of the Canadian islands are numerous sounds, straits and channels.

Arctic islands There is a large number of islands round the edge of the Arctic Basin. Most of them, like the Canadian archipelago, the New Siberian islands and the long curving Novaya Zemlya, are parts of the submerged edge of the land-masses, which have extensive continental shelves. Other islands, such as Spitsbergen, Bear Island and Jan Mayen, are the higher parts of submarine ridges.

MARINE DEPOSITS

The term 'marine deposits' includes all the material which is accumulating on the floor of the sea or ocean. The ultimate destination of the sediments worn from the land-masses is the ocean floor, where it accumulates in great thicknesses, together with other material such as the remains of plants and animals that either live on the sea floor or float in or on the surface of the water.

Most sedimentary rocks were laid down in former seas, and in course of time were uplifted to form the rocks of new land areas, particularly those laid down in shallow seas where only a small

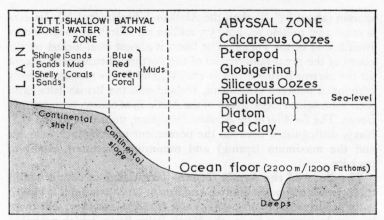

Fig. 165 Generalized profile of the ocean floor, showing the general distribution of marine deposits

For a description of each type, see fig. 166. **Litt.** = Littoral.

movement of uplift resulted in the emergence of the sea floor. Thus the Chalk consists of a great thickness of the remains of minute foraminifera, deposited in clear water which was free from other sediment such as mud or sand. The Millstone Grit, on the other hand, consists of consolidated beds of coarse sands and gravels, derived from a period of rapid erosion, and laid down in shallow waters near the margin of the land in deltas.

Classification of marine deposits Marine deposits may be divided into four groups, according to the part of the sea floor upon which they are accumulating. *Littoral deposits* are found between the high- and low-water spring-tide lines; *shallow-water deposits* occur from the low-water mark to about the 180 m (100 fathom) line, i.e. about the edge of the continental shelf; *bathyal deposits* occur on the continental slope; and *abyssal deposits* on the deep-sea plain and in the deeps (fig. 165). These groups merge gradually into one another.

A second distinction can be drawn between deposits which are *inorganic* and are derived from the wear and tear of the land (hence *terrigenous*), and those consisting of the remains of marine organisms (hence *organic*). A subdivision of the organic group is made into those deposits derived from the remains of shell-fish, sea-urchins and corals, found in the littoral and shallow-water zones, known as *neritic deposits*, and those which accumulate in deeper waters from the remains of plankton, both animal and vegetable, which are called *pelagic oozes*. Information about these pelagic deposits is limited, but each research voyage adds to knowledge. The scientists

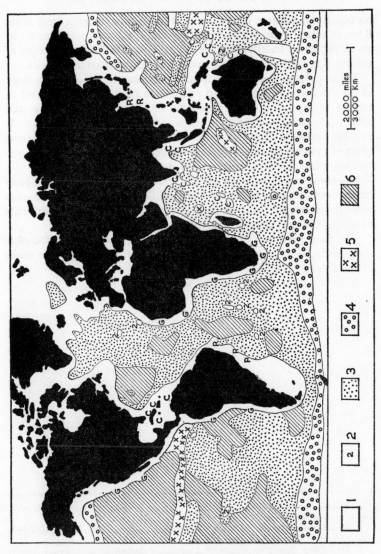

Fig. 166 *The world distribution of marine deposits*

Based on a map by Sir John Murray, *The Ocean* (London, 1928).

1. Terrigenous deposits (shingle, sand, shelly-gravels and shelly-sands).

2. Pteropod ooze (calcareous, thin fragile conical shells).

3. Globigerina ooze (calcareous, consisting of various foraminifera, globigerina being the most common).

on board the *Albatross* used a coring-tube suspended on a steel cable 8 km (5 miles) long, with which they could take out a section up to 20 m (65 ft) in length from the ocean bed. The skeletal remains of nekton (i.e. fish) contribute sporadically to the marine deposits.

These pelagic deposits accumulate extremely slowly. The floor of the Pacific is being built up at the rate of about 2·5 cm (1 in) in 20,000 years, the Atlantic about 10 times more rapidly. Yet the total thickness is considerable; in parts of the Atlantic, echo-sounding has revealed that the sediment upon the solid floor is about 3600 m (12,000 ft) thick; it must therefore have taken 250 to 300 million years to accumulate. Few sediments are older than the Cretaceous, indicating that the present basaltic floor of the ocean basins was formed some 250 to 300 million years ago. This agrees with the concept of the present pattern of oceans and continents as a result of continental drift (p. 24).

The dual classification of marine deposits, according to location and origin, is set out diagrammatically in fig. 165, and the main areas of occurrence are shown on fig. 166.

---

4. Diatom ooze (siliceous, skeletons of diatoms, i.e. microscopic plants flourishing in cool oceans).

5. Radiolarian ooze (siliceous, skeletons of foraminifera, with lattice-like structure, occurring only in deep water).

6. Red Clay (hydrated silicate of alumina, coloured by iron oxide, largely formed of volcanic dust, also meteoritic dust, iceberg debris, insoluble relics of organic life, e.g. sharks' teeth, etc.).

C. Coral sand and **mud**, derived from reefs.

R. Red mud, stained by ferric oxide.

G. Green mud, contains glauconite, i.e. dark green hydrated silicate of iron.

Note. **Blue mud** is by far the most widely spread of the terrigenous muds; it is earthy and plastic, and darkish blue in colour. It is found very widely in all the oceans along the continental slope, and is not indicated on this map.

CHAPTER 13

The Waters of the Oceans

SALINITY

SEAWATER contains a number of mineral substances in solution. The bulk of these must have been present since the oceans first accumulated on the geologically young earth, since the amount contributed by the rivers each year is negligible in proportion to the amount already there. Moreover, the actual mineral content of river water differs materially from that of the sea; although river water varies in different parts of the world, calcium salts (particularly calcium carbonate) are present in much greater quantity than are sodium salts, which are by far the most important minerals in seawater.

The degree of salinity is commonly expressed in terms of the number of parts of salt per thousand of seawater; if in 1000 grammes of water there are 35 grammes of dissolved salts, the salinity is said to be 35‰, the average salinity of the whole ocean. If samples of water from widely separated parts of the oceans and seas are analysed, the proportions of the salts remain virtually constant, even though the total salinity varies widely. The most important salts present in seawater are sodium chloride (78 per cent), magnesium chloride (11 per cent) and sodium sulphate (5 per cent), but many others are present in measurable quantities, and there are also 'trace elements' of great importance to marine plants and animals. When speaking of the ionic proportions of seawater, 55 per cent of the salt-content consists of chlorine, 31 per cent of sodium. As these proportions are virtually identical throughout the seas, the standard method of determining salinity is by precipitating the halides present (which are mainly chlorides) by adding silver nitrate, thus forming silver halides.

Distribution of salinity *Isohalines* (isopleths of salinity) can be drawn to show the salinity at the surface or at any intermediate depth. Surface salinity varies according to temperature (causing evaporation and concentration), the supplies of fresh water by rivers, rainfall or melting ice (therefore causing dilution), and the degree of mixing by surface and sub-surface currents. In the open ocean differences in salinity are relatively small. In the Atlantic, for

347

example, the areas of highest salinity lie near the Tropics (about 37‰), where the clear skies, constant high temperatures and brisk Trade Winds maintain active evaporation. Salinity decreases towards the Equator (35‰), where rainfall is heavier and evaporation less because of the higher relative humidity, greater cloudiness and calmer air-masses of the Doldrums. There is also a decrease towards the Poles (less than 34‰), the result of melting ice and of decreasing evaporation.

A detailed isohaline map shows much more variety than this simplification would imply, the result of surface currents and the presence of great rivers such as the Amazon and the Congo. Variations of salinity are most marked in partially or wholly enclosed seas. The Baltic Sea, for example, reveals decreasing salinity with increasing distance from the North Sea; off the south coast of Sweden it is about 11‰, off Bornholm about 8‰, and it decreases to about 2‰ at the head of the Gulf of Bothnia. This sea receives water from a number of rivers, such as the Oder and the Vistula, and much melt-water, while evaporation is low. The Black Sea also has many large rivers (including the Danube, Dniester, Dnieper and Don), and its salinity is between 18 and 20‰. The Red Sea, which receives no rivers and experiences great evaporation, has a summer salinity exceeding 40‰. The salinity of the Mediterranean increases eastward from about 36‰ off Gibraltar to about 39‰ in the eastern angle between Israel and Egypt.

Inland seas and lakes have a much higher salinity than the open sea, for while their rivers may bring down only a small quantity of salt, it accumulates there while water is removed by evaporation. The Great Salt Lake in Utah has a salinity of 220‰, the Dead Sea of 238‰ and Lake Van, in Asia Minor, of 330‰. Salinity may vary locally even within a lake or inland sea. The northern part of the Caspian Sea, into which flow the Volga and Ural Rivers, has a salinity of only 13‰, while the Gulf of Kara-Bogaz in the south-east, which is almost separated from the sea by sand-spits, has a salinity of 195‰.

THE TEMPERATURE OF THE OCEAN WATERS

The oceanographer is concerned with the temperature of the water, both at the surface and at depth, for its physical and biological results. The geographer is more interested in the effects which may be felt directly on coastal areas (for example, the 'cold-water coasts' discussed on p. 353), or indirectly by the movement of air-masses from over the surface of the ocean on to the land-masses.

For taking surface-water temperatures, a recording thermograph is used, the thermometer bulb of which is fixed in the condenser intake of a ship. For taking temperatures at depth, the standard instrument is a reversing thermometer, lowered on a fine steel wire. When the instrument reaches the required depth, it is inverted by a pull, the thread of mercury is broken, and the record of the temperature is preserved while it is being hauled to the surface.

Surface temperatures Differences between the temperatures of land and sea represent a fundamental concept in climatology, and the results upon the climates of the continents are developed below (pp. 389–90).

The distribution of surface-water temperatures in the oceans is a complex matter. Broadly, there is a gradual decrease in the annual average temperature of the waters of the open ocean from 26° C (79° F) at the Equator to 23° C (73° F) at latitude 20°, 14° C 57° F) at latitude 40°, and 1° C (34° F) at latitude 60°. The seasonal change of temperature is much less than in the case of the land-masses. The range is greater in the Atlantic than in the Pacific because of its smaller size, and greater in the oceans of the northern hemisphere than those in the south, because of the effect of cool air-masses moving over the oceans from the northern land-masses in winter. Generally speaking, the range of seasonal temperature is only about 6° C (10° F) between 20° N and S and again south of 50° S. The greatest ranges are about 22° C (40° F) in the north-western Atlantic off Newfoundland, and about 25° C (45° F) in the north-western Pacific off the Asiatic coast near Vladivostock. The highest water temperatures are recorded in enclosed or semi-enclosed tropical seas; in the Red Sea temperatures exceeding 38° C (100° F) are occasionally reported, although the average summer temperature is about 29° C (85° F). The 0° C (32° F) isotherm forms a rough circle round the Polar areas, moving equatorward during winter.

The air temperature over the oceans does not quite correspond to these figures; the annual figures show that the air is slightly warmer than the water on which it rests at the Tropics, but elsewhere it is slightly cooler. Seasonal figures differ slightly in this respect; in summer the sea tends to be cooler than the air (except in the Indian Ocean), but in winter the sea is slightly warmer.

These are general concepts, however, and ocean isotherms, like continental ones, rarely run from east to west. The chief anomalies are due to surface currents, where masses of water of a particular temperature move into other latitudes, retaining to some extent the temperature characteristics of their place of origin (pp. 352–3).

Vertical ascending movements of water from great depths may produce exceptionally cool surface water for a particular latitude, as off some parts of the western coasts of Africa and the Americas.

Icebergs Much of the surface water of the Arctic Ocean and around the margins of the Antarctic continent is permanently frozen (plates 68, 69). The ice varies in thickness from 0·6–4 m (2–14 ft) and in winter covers an extensive area, while pack-ice extends farther still from the Poles, reaching as far south as 60° N along the east coast of Greenland. The pack-ice may be in close contact, so that it forms a continuous icefield, but in summer large floes (horizontal sheets or slabs of ice, often many km in diameter) separate and drift away from the edge of the icefield.

The large masses of ice which break off from the tongue of a glacier on reaching the sea, or from the edge of an ice-barrier, are known as *bergs* (plate 96). While the flotation factor (i.e. the ratio of the emerged to the submerged portions) depends partly on the relative density of ice (0·9) and seawater (1·025), their shape is also important. The old 'rule of thumb' that the number of feet above the water equals the number of fathoms below (i.e. 1:6) seems by recent investigations to be an exaggeration. Proportions of 1:4 or 1:3 seem to be more likely. As some bergs tower up for 90 m (300 ft) or more above the sea, the underwater mass is still enormous.

In the northern hemisphere about 10,000 to 15,000 bergs are formed each year, the glaciers of Greenland producing all but about 500–600; these mainly come from Spitsbergen, Edge Island, Novaya Zemlya and Ellesmere Land. They are carried southward by the East Greenland and Labrador Currents, sometimes as far as 40° N, into the North Atlantic shipping-lanes. The Canadian-American Ice Patrols were set up after the *Titanic* disaster of 1912 to study the drift of bergs, report their presence by radio, and predict their probable courses. Bergs are rarely seen in the North Pacific, as a result of the narrow and shallow Bering Strait.

The bergs of the southern hemisphere originate from the edge of the Antarctic ice-barrier (p. 218). They are of enormous size, virtually forming great horizontal ice-islands; some have been recorded over 200 km (120 miles) in length, known as *tabular bergs*, while most in the northern hemisphere are *castellated bergs*. The Antarctic bergs move northward under the influence of the Falkland, Benguela and South Australian Currents, reaching as far as 40° S in the Atlantic and about 50° S in the Pacific. The largest berg ever recorded was sighted in 1956; it was over 30,000 sq. km (12,000 sq. miles), 334 km by 96 km (208 by 60 miles).

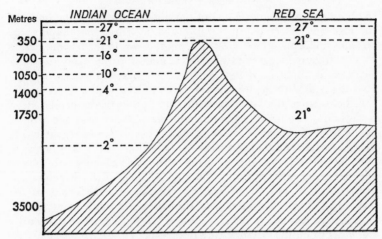

Fig. 167 *Temperature differences between the Red Sea and the Indian Ocean*
(after G. Schott)
The temperatures are given in degrees Centigrade.

The drift of the bergs is influenced both by the currents and the winds, together with a certain deflection due to the earth's rotation. Their typical speed is of the order of 7·5 km (4 nautical miles) a day with a wind of about 55 km per hour (30 knots). Cases have been known of bergs moving in the opposite direction to the wind and the surface current; this was due to a powerful sub-surface current, affecting the underwater mass of ice.

The temperature of ocean depths Generally speaking, water temperature decreases with depth, except in the Polar seas. From 370–730 m (200–400 fathoms) the fall in temperature is very marked, but below this it becomes extremely slight and gradual. Below a depth between 900 and 1450 m (500 and 800 fathoms), according to latitude, the decrease is scarcely appreciable; it is estimated that five-sixths of the total volume of ocean water has a temperature between 2 and 4° C (35–40° F).

In the Polar seas a small inversion of temperature is noticeable with depth. A thin layer of cold water, formed by the melting of Polar ice, floats on a layer of slightly warmer and more saline water, but below about 360 m (200 fathoms) the usual decrease takes place.

Partially enclosed seas, particularly those with a shallow submarine threshold, may have a very different temperature gradient from that in the open oceans. The Red Sea is separated from the Indian Ocean, at the Straits of Bab-al-Mandab, by a threshold which rises to about 350 m (200 fathoms) of the surface. Although the basin of the Red Sea descends to 2100 m (1200 fathoms), the temperature remains uniformly at 21° C (70° F) right down to the bottom, while in the Indian Ocean outside the temperature at 2100 m (1200 fathoms) is about 2° C (35° F) (fig. 167). The temperature of the Red Sea water throughout is the same as that of the water at the threshold, since the latter prevents the ingress of the cooler, deep oceanic water. Similarly, the temperature of water near the bottom of the Mediterranean is about 13° C (55° F), the coldest water that can pass over the sill at the Straits of Gibraltar, which is only 350 m (200 fathoms) deep (fig. 171). One of the most striking examples of such a barrier is the Wyville-Thomson Ridge (p. 338), which separates cold, stagnant Arctic water from the warmer water of the Atlantic Ocean. The water near the bottom of the Norwegian Sea on the northern side of the ridge is at 2° C (35° F), while that at the same depth on the Atlantic side is at 7° C (45° F).

SUBSURFACE MOVEMENTS OF OCEAN WATERS

It is clear that considerable movement of water takes place within the oceans, both vertically and horizontally. Vertical movements of water-masses are due either to differences in density at various depths or to the meeting of two converging currents, thus causing sinking which is counterbalanced elsewhere by ascending masses.

The *density of seawater* depends both on its salinity and on its temperature. For a particular salinity the density varies inversely with the temperature; that is, as the temperature falls the density rises. The mean density of the surface water for the whole ocean, calculated by Sir John Murray, is 1·0252, while from 3700 m (2000 fathoms) downward it remains virtually constant at 1·0280. On the other hand, for any particular temperature density increases with increasing salinity; at a temperature of 16° C (60° F), fresh water has a density of 1·0000, water with a salinity of 30‰ has a density of 1·0220, and water with a salinity of 40‰ of 1·0300. Thus increase of temperature, rainfall, river-inflow and melting ice reduces density, while decrease of temperature or increase of evaporation raises it.

In the Polar seas, therefore, cold surface waters of high density sink, while in the Tropics heating of the surface tends to make the water less dense, and it moves away poleward so that cooler water wells up to take its place. This theoretically involves the poleward movement of warmer surface water from the Tropics, while the cooler Polar water moves equatorward at depth, with sinking water-masses in the Polar regions and rising ones in equatorial latitudes. In fact, things are far from being as simple as this. One reason for variation from the theoretical concept is that the surface drift of water poleward from the Equator is considerably affected by the winds. Another reason is that while the cold, dense bottom water of the Southern Ocean can move northwards (even into the northern hemisphere), the Arctic bottom water forms a virtually stagnant pool, cut off from the rest of the oceans by the shallow sill in the Bering Strait and by the Wyville-Thomson Ridge. Most important of all is that water-masses can be distinguished within the oceans, each with a certain salinity and temperature, and of a particular region of origin. It frequently happens that individual water-masses are separated by a marked *discontinuity layer*, where temperature and salinity sharply change. In the North Atlantic the 'cold wall' is a discontinuity layer between the Labrador Current and the Gulf Stream, while in the Pacific another occurs between the Okhotsk Current and the Kuroshio.

In some parts of the oceans are to be found *convergences*, sharply defined lines separating converging masses of surface water, where, therefore, sinking occurs. Such a convergence is in the Southern Ocean, approximately around the 50° S parallel, where the cold, dense Antarctic surface waters meet the warmer but more saline waters spreading south. Another convergence is at about 40° S, and others, although less well defined because of the interrupting land-masses, are in the northern hemisphere. On the other hand, *divergences* counterbalance the water which has sunk elsewhere, as a result of which cold water wells up to the surface. The divergences are most clearly traceable off the west coasts of continents, where the offshore Trade Winds drift the warm surface water westward. The net result of these complications is that in parts of the ocean the movement of the water-masses (surface, intermediate, deep and bottom) is more or less horizontal, but at different levels and some-times in different directions. Where sufficient information can be obtained by survey ships, temperature-salinity (T-S) curves can be plotted on a graph with a horizontal scale of salinities against a vertical scale of temperatures.

SURFACE CURRENTS

Several references have been made to the movement of surface water. The general movement of a mass of surface water in a fairly defined direction is known as an ocean current.

The pattern of the oceanic circulation is produced by the inter-play of a number of factors. The importance of subsurface move-ment, the effects of density differences, and of convergences and divergences, must be taken into account. Many surface currents are, however, 'drifts', caused by friction between the winds and the surface water, and therefore moving more or less in the direction of the wind and varying in position and strength with the seasonal winds. Rotation of the earth, however, besides affecting the direction of the winds, tends to deflect the currents slightly obliquely, so that a wind in the northern hemisphere blowing toward the north-east will produce a drift with a more marked easterly component. In addition, the shape of the land-masses helps to determine the direction. If an ocean current is forced to flow through openings between islands, particularly if there is a marked difference in level on either side, the drift-current may increase in velocity and form a 'stream-current', such as the Florida Current between the Florida peninsula and Cuba. On issuing from this strait it becomes the Gulf Stream and flows northward along the east coast of North America. The Florida Current is probably one of the strongest major currents in the world; the moving mass of water is about 3·2 km (2 miles) deep and flows at almost 6 km per hour (3 knots).

The net result is the formation of a series of circulatory systems or 'cells', known as *gyres*, one in each of the major ocean basins. Situated between about 20° and 30° N and S, the movement of water therein is clockwise in the Northern Hemisphere and anti-clockwise in the Southern, though asymmetrical in plan, as a result of the Coriolis deflecting force (p. 404). The Indian Ocean forms an exception, where the triangular Indian peninsula and the monsoonal change of wind produce a double gyre moving in seasonally opposite directions.

The table below lists the average speeds of Atlantic currents; those

Velocity (per day)

North Atlantic	Nautical miles	km	South Atlantic	Nautical miles	km
West Greenland	15	28	South Equatorial	15	28
East Greenland	15	28	Brazil	15	28
Labrador	15	28	Falkland	10	18
Irminger	15	28	Guinea	25	46
North Atlantic Drift	5	9	Benguela	10	18
Norwegian	15	28	South Atlantic	10	18
Gulf Stream	20	37			
Florida	60	110			
Caribbean	15	28			
Antilles	15	28			
North Equatorial	15	28			
Guiana	25	46			
Canary	15	28			

in the Pacific have not been studied in so much detail, but are known to be markedly slower.

The terms 'warm' and 'cool' are applied to the various currents; these are, however, relative. In general, a poleward-moving current is warm as compared with general atmospheric temperatures, while an equatorward-moving one is cool. These currents may have considerable effect on the climates of the bordering land-masses (p. 482).

Atlantic currents (fig. 168) The land-masses divide this ocean into basins (p. 337), and there is a gyre in each, separated by a convergence. The Trades blow the tropical waters in a westerly direction, forming the North and South Equatorial Currents. Between the two is an easterly-moving Counter-equatorial Current in the Doldrums, which compensates to some extent for the piling-up of the water on the western side of the ocean. This is much more strongly developed in July, in fact in January it is hardly appreciable. Reinforced by that part of the South Equatorial Current which is diverted northward by the projecting Cape São Roque, the North Equatorial Current flows north-westward, part entering the Caribbean Sea, part passing east of the West Indies. The Gulf Stream issues from the Florida Strait and, reinforced by the rest of the North Equatorial Current, flows north along the American coast as far as Cape Hatteras. A detailed recent survey by the U.S. Navy Hydrographic Office shows that the Gulf Stream is from 30–60 km (20–40 miles) wide, and up to 600 m (2000 ft) in depth. It also revealed that the movement follows a sinuous course, "suggestive of a slowly swimming snake", and changes over periods of a few days. Beyond Cape Hatteras it is mainly diverted across the North Atlantic under the influence of the South-westerlies. One branch flows northward into Davis Strait, between Baffin Land and Greenland, and another approaches the south coast of Iceland. Part of the main drift continues along the Scandinavian coast as the Norwegian Current, on into the Barents Sea and toward Spitsbergen. The rest turns south between Spain and the Azores, and flows southward along the African coast as the cool Canary Current, so completing the clockwise circulation of the North Atlantic. The relative coolness of the southward-flowing Canary Current is emphasized by the up-welling of cold water off the west coast of North Africa. Within the relatively still waters of the area enclosed by this planetary circulation is the Sargasso Sea, where floating seaweed occurs in great amounts.

Similarly, in the South Atlantic part of the South Equatorial Current flows southward along the coast of South America as the

Fig. 168—*Surface currents in the Atlantic Ocean* (after G. Schott)

This represents average conditions in January. The abbreviations are: **Ant. C.** = Antilles Current; **E.G.C.** = East Greenland Current; **L.C.** = Labrador Current; **Fl.C.** = Florida Current; **N.C.** = Norwegian Current; **N.Eq.C.** = North Equatorial Current; **S.Eq.C.** = South Equatorial Current.

The heavy lines indicate the approximate position of the convergences or lines of separation (e.g. the 'Cold Wall').

356

Brazil Current and then the north-westerly winds drift the east-ward-moving South Atlantic Current. The anti-clockwise circulation of the South Atlantic is completed by the Benguela Current, flowing northward along the African coast, another current whose coolness is emphasized by up-welling cold water.

In addition to these two main circulations, cold surface currents creep into the Atlantic from the polar areas. The Labrador Current flows southward through Davis Strait and on past Newfoundland, the East Greenland Current flows through Denmark Strait along the coast east of Greenland, and the Irminger (or East Iceland) Current passes the east and south coasts of Iceland to lose itself in a series of 'whirls' where it meets the northern edge of the North Atlantic Drift. The cold water of the Labrador Current, being dense, sinks below the warmer waters south of Newfoundland, and has a considerable effect on the sub-surface circulation in the Atlantic. In the south the cold Falkland Current flows northward along the Patagonian coast until it also sinks.

A submarine current, similar in nature to the Cromwell Current in the Pacific (p. 358), was discovered in 1963; it is a highly saline undercurrent in equatorial latitudes, with its maximum flow in an easterly direction at 180 m (100 fathoms) depth, moving at a speed of about 4·5 km per hour (2·5 knots).

Pacific currents (figs. 169, 170) The same broad circulatory systems, clockwise in the northern hemisphere, anti-clockwise in the southern, can be discerned in the Pacific. The North and South Equatorial Currents flow westward, with a compensatory counter-current flowing in the reverse direction between them, along a line about 5° N. Inexplicably, the last is much stronger than in the Atlantic, although the Indonesian Archipelago forms a less complete barrier to the equatorial currents than does the west coast of South America; it may be that as the Pacific is so much wider in equatorial latitudes, a greater volume of water is involved. The circulation in the North Pacific comprises the Kuroshio or Japanese Current (analogous to the Gulf Stream), with its offshoot, the Tsushima Current, flowing northward into the Sea of Japan along the west coast of the islands, its continuation north-eastward as the broad North Pacific Drift, and the southward-flowing cool California Current. Part of the North Pacific Drift flows into the Gulf of Alaska, and then westward along the Alaskan coast, thus keeping it ice-free in winter as the current is relatively warm. The South Pacific circulation includes the East Australian Current, which skirts the north coast of New Guinea and continues southward along eastern Australia, then flows eastward in about latitude 40° S as the South Pacific

Current, and completes the circle northward along the west coast of South America as the Peruvian or Humboldt Current. Antarctic water and up-welling west-coast water, together with its northward direction, all make this latter current markedly cool.

The North Pacific is almost landlocked as compared with the Atlantic, and so there is a small additional circulatory system in the extreme north. The Alaskan Current, already mentioned as an offshoot of the North Pacific Drift, moves westward, continues as the Aleutian Current, and reinforced by cold water from melting ice, moves south as the Kamchatka Current. Part swings round eastward again, merging with the northern waters of the North Pacific Drift; part, reinforced by the cold melt-water of the Okhotsk Current, continues south past Sakhalin and Hokkaido as the Oyashio and then gradually sinks, like the Labrador Current, beneath the warmer waters of the North Pacific Drift.

It must be emphasized that, even more than in the case of the Atlantic, figs 169, 170 present extremely simplified pictures of the trends of the main planetary currents. The seasonal and monthly current maps produced by the U.S. Navy reveal many more complicated movements, particularly in the open ocean, involving great slow 'whirls'. There are also seasonal counter-currents, such as the warm El Niño, which in February and March creeps southward, sometimes even as far south as 12° S, between the coast and the Peruvian Current. For some unexplained reason, El Niño sometimes flows exceptionally far south beyond Cabo Blanca, even more inexplicably with a periodicity of about 7 to 8 years, though not invariably (1911, 1918, 1925, 1933, 1939, 1941, 1953, 1965); it then kills plankton and fish by the sudden change of temperature, so that beaches are littered with dead fish. It also involves remarkable climatic effects, for with it the tropical rain-belt also shifts south; for example, in March 1925 more than 38 cm (15 in) of rain fell at Trujillo, whereas the average monthly rainfall for the past eight years had been only 0·4 cm (0·17 in). Similarly, in the North Pacific the Davidson Current flows as a counter-current from November to January in a northward direction along the coast of California between the mainland and the California Current, reaching as far as 48° N.

In 1951–2 the Cromwell Current (originally called the Pacific Equatorial Underwater Current) was first discovered while fishing for tuna with long lines; it is a 'ribbon of water', 300 km (190 miles) wide, flowing eastwards along the Equator beneath the South Equatorial Current. It has been traced as far as the Galápagos Islands.

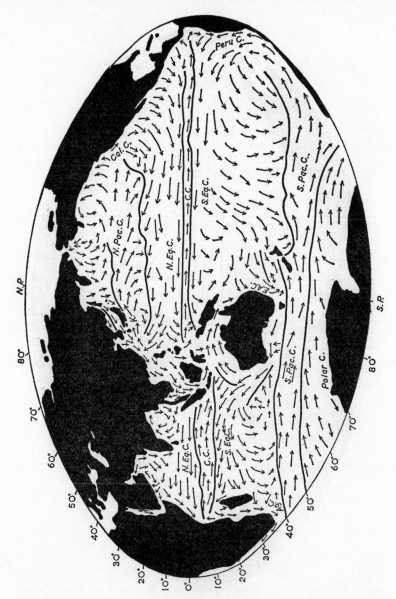

Fig. 169 *Surface currents in the Pacific and Indian Oceans during the northern winter*
(after G. Schott)

This represents average conditions from February to March. The abbreviations
are: **C.C.** = Counter-Current; **Cal. C.** = California Current; **E.A.C.** =East
Australian Current; **K.** = Kuroshio; **N.Eq.C.** = North Equatorial Current;
S.Eq.C. = South Equatorial Current; **S.Pac.C.** = South Pacific Current.

The heavy lines indicate the approximate position of the convergences.

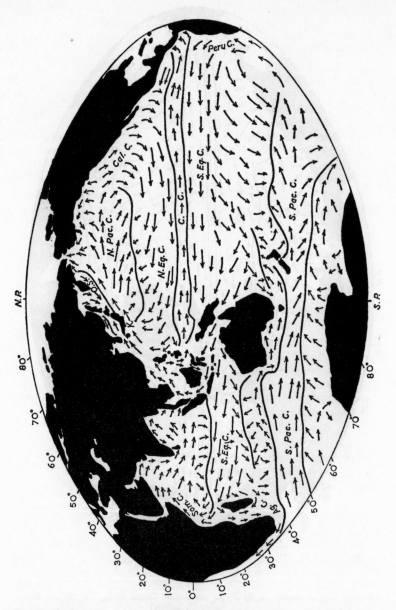

Fig. 170 *Surface currents in the Pacific and Indian Oceans during the northern summer* (after G. Schott)

This represents average conditions during August and September. The abbreviations used (in addition to those on fig. 169) are: **O.S.** = Oyashio; **Som. C.** = Somali Current.

Indian Ocean currents (figs. 169, 170) The seasonal changes in the position of the planetary wind systems tend to displace the major drifts a few degrees in either direction in the Atlantic and Pacific Oceans. In the northern Indian Ocean, however, there is a complete reversal as a result of the monsoonal change of air-streams (pp. 409-10).

In the southern Indian Ocean the circulation broadly resembles that of the other southern oceans, in that it is anti-clockwise. The South Equatorial Current, probably strengthened by water from the corresponding Pacific current, which has made its way through the Indonesian Archipelago, flows westward towards the African coast and turns southward along both coasts of Madagascar; that part passing between the island and the mainland is known as the Mozambique Current, and its southern section off Cape Province is sometimes called the Agulhas Current. It then turns eastward, merging with the South Pacific Drift. The northward-flowing West Australian Current, which completes the circulation, is much less well marked than the Peruvian and Benguela Currents. In the southern summer the current can be traced, but in winter, when the bulk of the West Wind Drift passes well south of Australia, it is reversed in direction, flowing southward from the neighbourhood of the Indonesian Archipelago.

In the northern Indian Ocean there is a distinct reversal between winter and summer. In winter the North Equatorial Current flows westward just south of Ceylon, and a distinct counter-current flows between it and the South Equatorial Current (fig. 169). The North-east Monsoon causes an eastward and northward drift along the eastern shores of India and along the Arabian coast; this, in fact, forms part of the usual east–west movement in tropical latitudes.

In summer (fig. 170), however, from July to late September, the South-west Monsoon is dominant. The North Equatorial Current is replaced by an easterly movement of water, which gives off branches into the Arabian Sea and the Bay of Bengal, there producing more or less clockwise circulations. In the former the current moves along the Horn of Africa, Arabia and western India; this causes an up-welling of cooler water, as the surface water moves away, which helps to explain the aridity of Somaliland and neighbouring countries.

Currents of the Southern Ocean The movement of water in the Southern Ocean is in one sense relatively simple: a west–east circumpolar drift, mainly under the influence of the north-west winds, with offshoots flowing northward into the other oceans. But from what has been said earlier concerning the Antarctic conver-

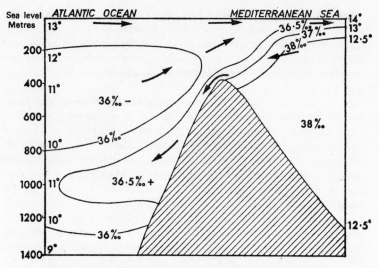

Fig. 171 *Cross-section of the Straits of Gibraltar* (after G. Schott)

This shows in diagrammatic form the effects on water temperature and salinity produced by the presence of a 'threshold' between the Mediterranean Sea and the Atlantic Ocean, with the resultant deep saline under-current and the compensatory surface-current.

The temperatures are given in degrees Centigrade. The isohalines are numbered.

gence, and the fact that water moves equatorward on the surface and at great depths but poleward at intermediate depths, it is clear that the surface and subsurface currents must have a very complicated relationship with each other.

Arctic currents The Arctic is a virtually stagnant, almost enclosed basin of water. However, there does seem to be a slow surface drift across the Pole from the Siberian coast to that of eastern Greenland; it was this drift which bore Nansen's ship the *Fram*, bound fast in the pack-ice, from near the New Siberian Islands in 1893 almost to Spitsbergen by 1896. The vessel drifted with the ice in an oscillatory manner through high latitudes, but did not, as he hoped, cross the Pole itself. Some cold surface water creeps through the channels between the Canadian islands into Baffin Bay, contributing to the Labrador Current, and other Arctic water flows southwards as the East Greenland Current.

Minor currents A number of currents, although of small importance compared with the planetary drifts, are caused by slight differences in level, salinity or temperature between adjoining seas. It has been calculated that the water received by the Mediterranean through river inflow and direct rainfall is only about a quarter of the amount lost by evaporation. A surface current therefore flows east-ward through the Straits of Gibraltar to make good both this loss and also that caused by the deeper movement westward of saline water out of the basin (fig. 171). Similarly, it is estimated that the Red Sea is lowered by evaporation as a result of prevailing high tempera-tures by amounts varying from 3–8 m (10–25 ft) per annum, and, moreover, a deep saline under-current flows out over the sill at the Straits of Bab-al-Mandab. A strong compensatory surface current therefore flows into the Red Sea from the Indian Ocean.

Conversely, partially enclosed seas which receive much river-water are raised in level. Many rivers flow into the Black Sea and, moreover, into it creeps a more saline current from the Mediter-ranean along the bottom of the Bosphorus. There is therefore a compensatory outward-flowing surface current.

THE TIDES

The periodic rise and fall in the level of the sea is known as the tides. In the open ocean the difference in height between high and low tides, the *range*, may be only a half-metre, but in shallow mar-

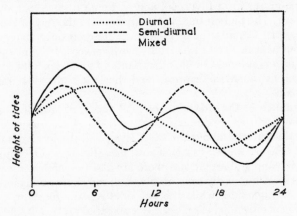

Fig. 172 *Types of tides*
Based on the *British Admiralty Tide Manual* (London, annually).

ginal seas it can increase to 9 m (30 ft), and in a constricted tidal estuary it may exceed 12 m (40 ft). An average high tide rises about 3·7 m (12 ft) above low tide at Southampton, about 5 m (17 ft) at Sheerness in the Thames estuary, about 7 m (23 ft) at London

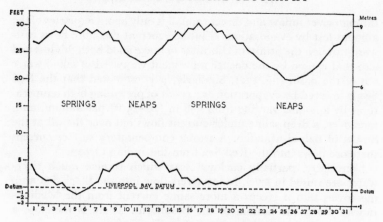

Fig. 173 *The heights of the tides at Liverpool, March,* 1942

Based on information supplied by the Liverpool Tidal Institute. The upper line joins the heights of successive high tides, the lower the heights of successive low tides. The highest high tide (spring) was 9·2 m (30·4 ft) above the Liverpool Bay Datum on 17 March, while the succeeding low tide was only 0·2 m (0·7 ft) above. Liverpool Bay Datum, used by the Mersey Docks and Harbour Board for their charts and for navigational purposes, is 4·4 m (14·54 ft) below Ordnance Datum.

Bridge, 9 m (30 ft) at Liverpool and 13 m (44 ft) at Avonmouth on the Severn. The highest tidal range known is in the Bay of Fundy in north-eastern Canada; at its mouth the range is only about 2·5 m (8 ft), but near its head 21 m (70 ft) has been recorded and 15–18 m (50–60 ft) is common. On the other hand, in some partially enclosed seas, such as the Mediterranean and the Baltic, the tidal range is very small.

Over much of the Atlantic two tidal rises and falls during each lunar day occur approximately every 12 hours 25 minutes; each pair of high tides attains about the same level, and the low tides do the same. These are known as *semi-diurnal tides* (fig. 172). Over much of the Pacific and Indian Oceans there are also two high and two low tides per day, but of different amplitudes; they may reveal inequalities in the high tides while the low tides remain constant, or vice versa. These are called *mixed tides*. A few special areas, notably in the Gulf of Mexico, in the waters around the Philippine Islands, off the Alaskan coast and off parts of the coast of China have only single *diurnal tides*—one high and one low during each twenty-four hours. It is difficult to explain these variations, for the forces responsible for the tides are uniform over the earth's surface, but the shapes of the oceans, the positions of the land-masses and the nature of the shallow marginal seas are mainly responsible.

The tide-producing forces A heavenly body exerts a gravitational effect on the earth's surface, the strength of which varies directly with its mass and inversely with the square of its distance. Thus the sun's mass is 26 million times that of the moon, but it is 380 times farther away, and its tide-producing effect on the earth is only 0·46 times that of the moon. When the moon, sun and earth are in the same line, either in conjunction or in opposition (positions known as *in syzygy*), their attraction is complementary and so the highest high tides and lowest low tides occur; these *spring tides* are experienced twice a month. When, however, the three bodies are at right angles, with the earth at the apex (a position known as *in quadrature*), the tide-producing forces are in opposition, and the tidal range is reduced; there occur low high tides and high low tides (*neap tides*) (fig. 173). Further variations are produced by the position of the moon relative to the earth. When it is at its nearest point (or *perigee*) its tide-producing effect is more pronounced, and these *perigean tides* are about 20 per cent higher at high tide. If they coincide with spring tides, very large tidal ranges occur. Conversely, *apogean tides*, when the moon is at its farthest distance, are lower than usual at high tide, and if these coincide with neap tides the range is small. The highest spring tides of all occur at the equinoxes, known as the Equinoctial Springs.

The moon is travelling in its orbit in the same direction as the earth is rotating. A lunar day implies the time that elapses between the moon passing twice over any one meridian on the earth, a period of 24 hours 50 minutes, during which most parts of the earth experience two high and two low tides; any high tide is thus about 50 minutes later than the corresponding tide of the previous day.

What is not clear is how the actual tides develop on a globe with land-masses and oceans. The old 'progressive-wave' theory held that the tide-producing forces resulted in the formation of two tidal waves in the Southern Ocean, one following but lagging slightly behind the moon as the earth rotates, the other on the opposite diameter of the earth. These corresponded to the progressive movement of high water, with low water between. From these major waves, branches with the same periods passed northward into the Atlantic, Indian and Pacific Oceans and successively into their marginal seas. But with an increasing number of tidal observations in different parts of the world, it became clear that this could not hold for the open ocean. For example, the 'age' of the tide between Cape Horn and Cape Farewell (the south-eastern corner of Greenland) hardly varies, instead of being progressively later. Moreover diurnal and semi-diurnal tides occur in different

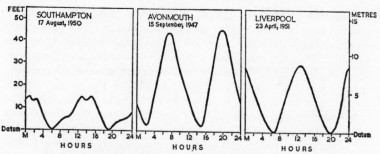

Fig. 174 *Daily tidal graphs for Southampton, Avonmouth and Liverpool*
Based on *British Admiralty Tide Manual* (London, various years).

parts of the oceans, instead of revealing a uniform and regular periodicity.

The *oscillatory* or *stationary wave theory* implies that the ocean surface can be divided into 'tidal units', each with a node or centre, in some cases with two nodes (a *binodal* system). The shape, size and depth of the waters within one of these tidal units has a profound effect on the pattern of the co-tidal lines. In each unit the surface of the water is set oscillating, the amplitude of oscillation varying with changes in the relative positions of the earth, moon and sun, together with a gyratory movement produced by the rotation of the earth. Maps have been drawn locating a series of nodal (or *amphidromic*) points in the oceans, from which co-tidal lines radiate. One point occurs in the North Sea almost equidistant from East Anglia and the Netherlands, and another in the North Channel of the Irish Sea. At these points the water remains at approximately the same level, and the height of the tidal rise increases outward along the co-tidal lines to their extremities. In some cases the theoretical amphidromic point is located over the land; that responsible for English Channel tides is actually in Wiltshire.

It is possible to predict the times and heights of tides with precision. The *Admiralty Tide-tables* provide these details for major ports around Great Britain for each year in advance (fig. 174). Tables of time- and height-differences are used to derive the figures for minor ports; thus the figures given for Liverpool are the basis for predicting tides at any point between the Solway Firth and Llandudno, together with the Isle of Man, while the Holyhead figures are used from the Menai Straits to Fishguard. These tidal predictions are made at Bidston Observatory and Tidal Institute in Wirral, which also supplies seaside resorts and newspapers with the predictions.

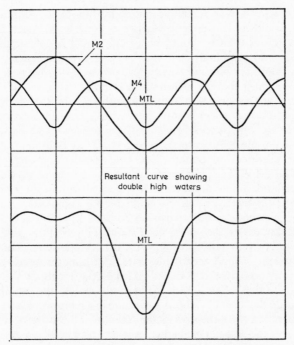

Fig. 175 *The basis of the prolonged high tide at Southampton*
Based on a diagram by D. H. Macmillan.

MTL=mean tidal level; **M2, M4**=semi-diurnal and quarter-diurnal tidal curves.

Anomalous tides Some tidal regimes do not fit into the
regular pattern, perhaps the most well-known being the 'double',
or rather 'prolonged', high tide at Southampton and in neighbouring
parts of the English Channel. This may be regarded as a more
or less rectangular opening into which is introduced the semi-
diurnal 'pulse' of the North Atlantic, which penetrates to the
Straits of Dover. This is, in fact, a progressive wave with the character
of a *Kelvin Wave*, in which the tidal flow is alternately east and west;
it is high tide at Dover when it is low tide at Land's End and vice
versa, and the tidal range increases on the right (i.e. south) of the
direction of the wave and decreases on the left, mainly because of the
Coriolis Force (p. 404). Thus when the tidal wave is flowing to the
east, high tide is higher on the south but when it is ebbing to the
west low tide is higher on the north. This largely explains why the
tidal range at Southampton is only about 3·7 m (12 ft), but on the
coast of Normandy and Brittany it is 12–13 m (40–44 ft). If the

Channel were a true rectangle the nodal line of the amphidromic system would be in the centre between the Isle of Wight and the eastern side of the Cherbourg peninsula, but it is not; actually the amphidromic point lies well inland in Wiltshire, and hence is known as a 'degenerate point'. Over this area of instability between Wight and Cherbourg there is a distinct deviation from normal, midway between high and low tide, that is, four times a day; in other words, a quarter-diurnal effect is superimposed on a normal semi-diurnal pattern (fig. 175), and when one of the lows is in phase with the semi-, a double high water is produced, as at Southampton. The reverse produces a double low tide as at Portland, Dorset, known as the *gulder*. Thus in the *Admiralty Tide-tables* Southampton and Portland, alone round the British coasts, require three columns, instead of the usual two for the single high and low tides.

A further anomaly experienced in Southampton Water is a period of slack (known as 'the young flood stand'), interrupting the normal tidal rise, though not the ebb. This occurs about 1·5 hours after low water at around mean tide level, and lasts for nearly 2 hours before the tidal rise is resumed. This is the result of the double entrance to the Solent, which is not in phase with the tidal rise and fall outside the Isle of Wight. A hydraulic gradient prevails at certain times between the entrances at the Needles and Spithead; at spring high water the respective heights are 1·07 and 2·0 m (3·5 and 6·5 ft), and this 'gradient' causes a flow from east to west. The reverse applies at spring low water, when the level at the Needles is a metre above that at Spithead. These interruptions to the normal tidal flow produce a pause in the tidal rise until the hydraulic head has been obliterated. Added to the prolonged high water, it gives 7 hours of slack water in each 24, to the great advantage of the port of Southampton.

Tides in estuaries The progressive wave concept, however, applies in a narrow estuary, where a tidal wave increases in height as the channel constricts and shallows. It is retarded at its base by friction, particularly when it is opposed by a river current, and may ultimately 'break' and move forward as a wall of foaming water, gradually diminishing in height and ultimately dying out. This *bore* is the result of a critical combination of slope and section of the bed, where a shallow-water estuary has an appreciable tidal range. The Severn bore, sometimes a metre high at spring tides, is an example (plate 97); the same phenomenon is known as the *eagre* on the Trent. It also occurs at the mouth of the Amazon (the *pororoca*), the Seine (the *mascaret*, though now largely obliterated by dredging), the Hooghly and several Chinese rivers. On the Tsien-tang-kiang, a river in northern China, the front of the wave attains a height of 3 m (10 ft), and advances upstream at a speed of 16 km (10 miles) per hour.

Tidal streams One result of tides in shallow water on the continental margins is the formation of tidal streams, which may take several forms. In river estuaries and harbours, tidal streams depend on the nature of high and low tide and on the configuration of the coast. Some flow strongly for about three hours before high tide and ebb for about three hours after it, with a period of slack water at high and low tide. But in some cases there are anomalies not altogether explained, as in the case of the 'young flood stand' in Southampton Water already described.

Tidal currents A distinction is sometimes made between a tidal stream, the normal movement in and out of an estuary, and a hydraulic current set up by differences of water-level at either end of a strait caused by differing tidal regimes. In the Menai Straits, between Anglesey and the mainland, high tide occurs at different times at either end, with the result that a 'hydraulic tidal current' is formed to compensate for the difference in height, flowing through the straits with considerable force. The same thing occurs in the Pentland Firth, characterized by its rough seas. In the Seymour Narrows between the northern end of Vancouver Island and the mainland of British Columbia, there can be as much as 4·0 m (13 ft) difference in water-level at either end, with resulting strong tidal currents. When a tidal current flows through irregular channels between islands, eddies or whirlpools may be formed, as in the case of the Maelstrom in the Lofoten Islands.

Storm surges It sometimes happens that the height of the sea along the coast may be raised above predicted tidal values through meteorological causes, ranging from offshore winds with a markedly local effect to a widespread 'storm surge'. Sea-level is raised over a considerable area, the results depending on the nature of the coast, the magnitude of the surge, and its occurrence relative to the time and type of the normal tidal oscillation. Such surges have been recorded on a number of occasions in the North Sea, notably in 1897, 1916, 1921, 1928, 1936, 1942, 1949 and 1953.

During the night of 31 January–1 February 1953 a very strong pressure gradient (p. 404) was set up in the rear of an intense depression (968 mb.) which had moved from near the Faeroes southeastward into the North Sea, resulting in the strongest northerly to north-westerly gale as yet experienced in the British Isles; gusts of over 160 km ph (100 mph) were recorded. This caused a surge or piling-up of water against the east coast of England from the Tees to the Straits of Dover, and also along the coast of the Netherlands. The vertical rise of water-level above the predicted values was between 2–2·5 m (6–8 ft) between the Wash and the Straits of Dover, and as much as 2·7–3 m (9–10 ft) along the coast of the Netherlands. The surge was associated with exceptionally powerful storm-waves, the sea defences were over-topped and waves, able to advance farther inshore than normal, attacked and in many places breached the coastal defences, both the natural sand-dune lines and the sea walls. This resulted in widespread flooding, damage and the loss of over 300 lives in England and nearly 1500 in the Netherlands.

One outcome of the disastrous flooding was the introduction of a flood-warning service in September 1953. Meteorologists and hydrographers at the Central Forecasting Office at Bracknell (p. 398) receive reports from harbour-masters at places along the east coast of Britain, and consider these in relation to the meteorological forecasts, then issuing if required a 'yellow' or a more urgent 'red' warning.

CHAPTER 14

Climate: General Features

CLIMATE directly affects the daily life of every person on the earth's surface, and forms an important feature of the environment. There is admittedly much controversy concerning the relationship between climate and racial characteristics, but no one can deny that some relationship does exist. Climate helps to determine food, clothing, housing and general mode of life; it plays a great part in promoting mental and physical vigour on the one hand, inertia on the other; its effects encourage or discourage the diseases and pests to which man is exposed in various parts of the world. The trade routes which opened up the world in sailing-ship days were controlled to a large extent by air currents, which in the modern age of aircraft are still of great importance.

Moreover, climate affects other features of the environment. Climatic elements are largely responsible for the detailed sculpture of the face of the earth—weathering, the work of running water and of glaciation, wind and water action in desert lands, even the storm-waves which pound the coasts, are all the results of climate. The mantles of soil and vegetation, both natural and cultivated, owe much of their character to the climatic environment in which they have developed.

Climatic changes The various climates of the world today differ considerably from those in the past. Many parts seem to have experienced both 'warm periods' and 'cold periods' (commonly called 'Ice Ages') (pp. 246–53), and also 'arid periods' and 'pluviose periods', as has been determined from such geological evidence as the nature of the rocks themselves and the character of the remains, plant and animal, found therein. The general distinction between Pole and Equator seems to have obtained for a very long time, but at times polar influences extended equatorward, at others the reverse took place. A mid-latitude location, such as the British Isles, has experienced a succession of changes—a tropical climate in the Carboniferous period, a desert climate in the Triassic, a cool temperate climate in the Cretaceous, and an arctic climate in the Quaternary, succeeded by minor fluctuations in post-glacial times.

These climatic changes are not limited to long-distant geological time. A fascinating line of research is the tracing of climatic changes within historical times by piecing together evidence from a variety of sources, many items scanty in themselves but often supplementing and corroborating some other information. There is much evidence of ruined settlements in what is now desert, of ancient irrigation works, of shrinking glaciers, of the width of the annual growth rings in trees (*dendrochronology*) (p. 14), and of both legends and eyewitness records of floods, droughts and great frosts. One example of such evidence is the discovery of coffins in Greenland, exhumed from what is now permanently frozen terrain, yet with remains of tree roots penetrating the coffins. When all the available evidence is analysed and put together with care, a climatic time-chart for the last few thousand years can be constructed (pp. 260-1).

CLIMATOLOGY AND METEOROLOGY

Weather and climate The term weather is used to describe the condition of the atmosphere at a certain place at some specific time; it is essentially a day-to-day, or even an hour-to-hour, phenomenon. Climate comprises a description of the condition of the atmosphere over a considerable area for a long time. Thirty-five years was usually considered to be the minimum period of time which must elapse before an adequate summary of a particular climate can be made.[1] In some parts of the world this 'climate' is harder to describe in average terms than others; thus it is sometimes said that the British Isles experience weather rather than climate.

Climatology The science of climatology falls largely within the province of the geographer, since it involves an analysis of the spatial distribution of atmospheric phenomena. A climate has been called a 'geographical entity', and much of the work of the climatologist lies in defining and describing different climates in their geographical settings. The terms 'climate' and 'climatic type' are commonly used synonymously; the description of the major climatic types forms the basis of Chapter 18. Climatology is in part descriptive, presenting an actual survey of pressure, temperature, rainfall and the other elements, but it must to some extent be explanatory as well. Here, then, the geographer comes into contact with meteorology, and he has realized in recent years that he must draw more and more from this science to assist his understanding of the climatic elements, hence the terms 'synoptic' and 'dynamic' climatology, whereby the subject is studied from the view-point of its relationship

[1] The International Meteorological Committee has laid down that the period 1901-30 is where possible to be used for the calculation of averages.

to the general circulation of the atmosphere, and to atmospheric dynamics and thermodynamics.

Meteorology Meteorology is the scientific study of the physical processes constantly at work in the atmosphere. Modern meteorology is to a large extent the field of the mathematical physicist, for a training in thermodynamics and hydrodynamics is essential if atmospheric processes are to be fully understood. One function of meteorology is its application to the forecasting of the future trend of the weather, a development of vital importance in this age of aviation. The basis of conventional forecasting is the construction of synoptic charts from simultaneous observations of atmospheric phenomena at a number of stations. Modern developments include mathematical forecasting, using a computer to solve sets of 'forecasting equations'; the recognition of *periodicities* and *singularities*; and in long-range forecasting the use of *analogues*, that is, the deduction of a series of future repeating patterns of weather situations by analysing and comparing sequences of past years. The trouble is that patterns or sequences are rarely exactly identical.

More important from the point of view of the climatologist is the contribution of scientific method and explanation which meteorology makes to his descriptive distributions. The analytical study of individual masses of air with distinctive features of temperature and humidity, and of the conditions of the upper air (pp. 375-7), have both yielded much information which the geographer can accept and use to assist his climatological descriptions.

The data The data of climatology are provided by stations which have recorded temperature, pressure, wind, rainfall and the other elements over a period of years. In western Europe and eastern North America there is a reasonably close network of such stations, elsewhere they are much more sporadic; there are about 2500 stations in the Northern Hemisphere. There are many parts of the land-masses (and much of the oceans) for which there are only short-term records or none at all; much of the work of the climatologist, in describing climatic types, must be in the form of cautious deductions from general principles and short-term observations. In recent years, knowledge of oceanic meteorology has been substantially increased by the systematic compilation of records by nine permanently stationed 'weather-ships', maintained by the U.S.A., Great Britain, Norway and other European countries in the eastern Atlantic, and by the constant flights of specially equipped aircraft, such as the Research Flight for the Meteorological Office. Numerous weather-satellites in orbit transmit information (pp. 441-2). In January 1967 a new high-speed weather communication link was opened between North America and western Europe, transmitting information from satellites in orbit via the U.S. national meteorological centre at Suitland (Maryland) to the West German meteorological centre at Offenbach (near Frankfurt), which is the disseminating centre for all western European weather bureaux. This link forms part of the World Weather Watch established in 1962 by the World Meteorological Organization.

The climatological data are given in terms of means, the arithmetic average, annual, seasonal and monthly. Occasionally, however, absolute figures, such as record maxima or minima, are required, together with the variability and frequency of various significant features and events.

'LOCAL CLIMATE' AND MICROCLIMATOLOGY

In recent years there has been considerable development in the detailed study of 'local climates'. This involves the careful examination of slight, yet significant, contrasts in climate that may result from small differences of slope and of aspect, from the colour and texture of the soils, from the proximity of water surfaces, from the nature of the vegetation cover, and from the effects of buildings.

Its application involves the appreciation of the significance of these small climatic variations for plants, animals, insects and human beings. The vital question of frost incidence, for example, has occupied the attention of climatologists and horticulturists alike. Various lines of research include the routes taken by the downward creep of cold air, the position of 'frost pockets', and the creation of 'frost-breaks' in the form of thick hedges. There has been much research into urban climates—the study of the interference of buildings with air movement, the questions of atmospheric pollution and of fog occurrence, and the degree of warming of the mass of air resting in and above a city (e.g. London's 'heat-island'). The siting of new towns, villages and airports calls for detailed preliminary work in this field. The work involves mainly the recording of the temperature of the air layer, a few metres thick, resting on the ground, and also includes records of soil temperature at certain depths.

A distinction is made between the study of a 'local climate', that is, the climate of a small area, and of a 'microclimate', that is, the detailed and very small-scale study of meteorological conditions within natural and cultivated vegetation, principally from an ecological point of view. A further term now employed, logically enough, is *micrometeorology*, which implies the detailed scientific study of the lowest layer of the atmosphere.

CLIMATIC FACTORS AND ELEMENTS

The term *climatic element* is a convenient label for each of the constituents which make up the sum total of climate—temperature, air pressure, wind, humidity and precipitation. While they are obviously interrelated, it is convenient to describe each in turn.

The elements result from the interplay of a number of *factors* or *controls*, that is, determining causes. In the description of each element which follows, the applicable determining factors are also considered. Several of these factors appear over and over again.

Latitude is an important factor, since it determines both the length of day throughout the year and the intensity and possible duration of sunlight received. *Altitude* has very definite results on temperature, pressure and precipitation, and mountain ranges sometimes form clearly defined climatic barriers and therefore boundaries of climatic types. The *distribution of land and sea* is a third factor; 'continental' and 'marine' influences can be very important. *Ocean currents* exercise cooling or warming effects on the margins of the land-masses near which they flow if onshore winds blow over them (pp. 353–62). Of more local importance are the presence of *large lakes*, the influence of *physical features* involving differences in aspect, shelter from or exposure to cold winds, the presence of deep valleys and basins, and the influence of *soils* and *vegetation*.

THE ATMOSPHERE

The atmosphere is a thin layer of gas held to the earth by gravitational attraction, three-quarters of which lies within 11 km (7 miles) of the earth's surface, 90 per cent within 16 km (10 miles) and as much as 97 per cent within 27 km (17 miles). The 'weathermaking' layers are limited to a height of a few km, particularly as a large proportion of the water-vapour is contained in the lowest 3000 m (10,000 ft); more than half, in fact, is below 2300 m (7500 ft).

In recent years much research has been carried on into the physical conditions of the upper air. The *balloons-sondes*, carrying self-recording instruments, have been gradually replaced by *radio-sondes*, by which radio transmitters, giving out signals indicating changes in pressure, temperature and humidity, are borne by balloons up to a height of 18–30,000 m (60–100,000 ft), when they burst. The *rawinsonde* (an abbreviation for 'radar wind-sounding') carries a radar target enabling the course of the balloon to be followed directly and so the upper air currents to be measured. More recently much information has been obtained by means of rockets and satellites. Observations derived from these ascents may be plotted in various ways, the most useful graph being the *tephigram*. The interpretation of this can reveal much information about the vertical structure of the atmosphere, particularly concerning its stability or instability and the possibility of rainfall.

It has been known for a long time that the temperature of the static atmosphere falls at an average rate of 0·6° C per 100 m (3·3° F per 1000 ft) of ascent; this is termed the *environmental* (or *static*) *lapse-rate*. It is now realized that this decrease of temperature is limited to a certain height, which seems to be about 18 km (11 miles) at the Equator, 9 km (5·5 miles) at latitude 50°, and probably 8 km (5 miles) at the Poles, although these figures vary with the season (the point of change seems to be higher in summer) and with general atmospheric conditions. The lower atmosphere can be divided into two parts—the *troposphere* up to this level of change, the *stratosphere* up to about 80 km (50 miles); the discontinuity plane between them is known as the *tropopause*. At the latter, temperatures over the Equator vary during the year only from about −79° C (−110° F) to −90° C (−130° F), though over the polar regions the seasonal difference is more marked, from about −40° C (−40° F) in summer to −79° C (−110° F) in winter. In the stratosphere this progressive reduction of temperature with height ceases, and until recently temperatures were regarded as being virtually constant. Now it is believed that there is a rise in temperature to about 15° to 20° C (60° to 70° F) at about 50 km (30 miles). Here occurs a belt of concentration of ozone (hence the *ozonosphere*), which causes the absorption of the shorter ultra-violet rays (p. 382) from the sun's radiant energy, with the resultant heating of the thin atmosphere and the formation of a 'hot layer'. The longer ultra-violet rays however, can pass through the ozonosphere to the earth's surface, producing the bronzing or 'sun-tan' which so many people desire. Beyond this zone the temperature falls again to about −80° C (−112° F) in the upper stratosphere. This is the region of the silvery white *noctilucent clouds* at a height of about 80 km (50 miles), clearly visible in high latitudes during summer nights. It is not known whether these clouds are of water-vapour, ice-particles, or volcanic or ice-coated meteoric dust; when a giant meteorite fell in Siberia in 1908, an unusually brilliant display of these clouds was seen the same evening, which seems to confirm that they are of dust.

Above the level of these noctilucent clouds at about 88 km (55 miles) up, lies the *ionosphere* (which is sometimes further divided into the *mesosphere* and *thermosphere*); this however, is the concern of those who deal with short-wave radio transmission, rather than that of the climatologist. In this zone the Aurorae Borealis and Australis can be observed in their respective hemispheres; they have been recorded as far as 1000 km (600 miles) up. In the ionosphere temperatures decrease towards absolute zero in outer space.

Constituents of the atmosphere Air is a mixture of gases, consisting mainly of about 78 per cent nitrogen and 21 per cent oxygen. A minute proportion of the oxygen exists in an allotropic form (O_3), known as *ozone*; its zone of maximum concentration seems to be at about 50 km (30 miles). Small quantities of carbon dioxide, hydrogen, and argon and other inert gases make up the remaining 1 per cent. These gases play a vital part in the radiation balance of the earth-atmosphere system.

In addition, air contains a variable quantity of water-vapour which plays a major role in the various weather phenomena. There is also much dust, specks of carbon in the form of soot, the spores of plants, salt particles evaporated from spray blown from the surface of the ocean, and minute pieces of disintegrating meteors, known as cosmic dust. One function of these particles is to serve as nuclei for the condensation of water-vapour.

CHAPTER 15

Temperature

MEASUREMENT AND RECORDING

Temperature scales Temperature can be measured in terms of the expansion or contraction in length of a column of liquid as the result of heating or cooling. These changes have been standardized in the form of certain scales; on the *Fahrenheit* scale the boiling-point of water is fixed at 212°, its freezing-point at 32°. The equivalent points on the *Centigrade* (since 1948 officially called *Celsius*) scale are 100° and 0°, on the *Réaumur* scale 80° and 0°. A simple formula expresses the relation between each:

$$\frac{C}{100} = \frac{F - 32}{180} = \frac{R}{80}, \text{ or, alternatively,}$$

$$F = \tfrac{9}{5} C + 32, \text{ and } C = \tfrac{5}{9}(F - 32).$$

Scientists sometimes use the *Absolute* or *Kelvin* scale, in which temperature is measured in degrees Centigrade from the point at which molecular movement ceases,—273·16° C = 0° K (*absolute zero*).

Instruments The column of liquid in a *thermometer*, the instrument used for measuring temperature, is of either mercury or alcohol. Thermometers must be carefully sited; very different readings will be obtained on the ground ('grass temperatures'), at various depths beneath the surface (usually 10 cm, 20 cm, 30 cm and 1 m), on a brick wall, or in the shade. The records usually required are of actual air temperatures—that is, 'shade temperatures' —measured under standardized conditions with precautions to exclude the direct rays of the sun; these are the figures generally quoted by climatologists. The thermometer is placed in a meteorological screen (a wooden box with a double top and bottom and louvred sides, in which the bulb of the thermometer is 1·2 m (4 ft) from the ground). The screen is carefully sited away from buildings.

Maximum and minimum thermometers are of great value, in which small dumb-bell-shaped markers indicate the highest and the lowest temperatures attained since the last setting, which is done each 24 hours with a small magnet. Most convenient is the *thermograph*, the principle of which is the expansion and contraction of a bi-metallic strip in the shape of a coil; one end is fixed, the other actuates a pen which traces a continuous record on a chart fixed to a revolving drum. Another type of thermograph is the *Bourdon tube*, used in aircraft, in which a curved metal tube is filled with spirit; its curvature changes with the temperature, and these changes are recorded on a chart or dial.

Presentation of temperature records Temperature figures may be presented in several ways. Extreme data are sometimes interesting (as when the Meteorological Office announces that it has been the warmest day in January for seventy years). Extreme maxima and minima, the incidence of frost, and other crucial figures are also of value.

The temperature figures of most value to climatologists seeking to present a long-term view of average conditions are *means*, with varying corrections and weightings. Diurnal, monthly and annual means are all used, but the most useful are the mean temperatures of the warmest and coolest months respectively over a minimum thirty years of observation (p. 372). From these can be computed the *mean range*, that is, the difference between the means of the warmest and coolest months. Annual means are of little value, as an examination of the following mean monthly figures for Peking will show:

J	F	M	A	M	J	J	A	S	O	N	D	
−5	−2	5	14	20	24	26	24	20	12	3	−3	°C
23	29	41	57	68	76	79	76	68	54	38	27	°F

The range is $31°$ C ($56°$ F), indicative of the great seasonal climatic contrasts in northern China. The annual mean is $12°$ C ($53°$ F), yet ten of the twelve months have means far from this figure. The annual mean for Verkhoyansk in Siberia is $−16°$ C ($2·7°$ F), but this is of little significance when we learn that the means of the coldest and warmest months are $−51°$ C ($−59°$ F) and $16°$ C ($60°$ F).

Temperature graphs and maps Monthly mean temperature for a station can be represented in the form of a simple graph (fig. 176); each climatic type discussed in Chapter 18 is illustrated in this way.

The distribution of mean air temperature over a considerable area is shown by drawing an *isotherm map*. All available stations are located on a base-map, temperatures plotted, and then lines drawn to represent a selected temperature. Rarely will they pass through a station; usually they must be interpolated proportionally. The values plotted may be either actual means, or reduced to sea-leve by adding a correction for the altitude of the station; thus a map either of 'actual isotherms' or 'sea-level isotherms' can be con-

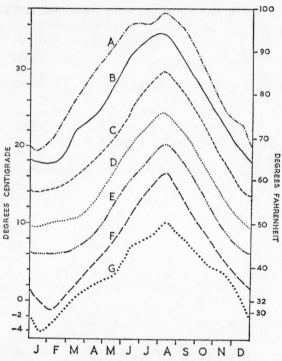

Fig. 176 *Monthly temperature figures for Ajaccio, Corsica*

This series of graphs illustrates the types of temperature figures available for each month at Ajaccio, in south-western Corsica, over the period 1851–1917.

A = extreme maxima; **B** = mean extreme monthly maxima; **C** = mean daily maxima; **D** = monthly means; **E** = mean daily minima; **F** = mean extreme monthly minima; and **G** = extreme minima. **D** represents the figures usually quoted in climatic tables.

structed. The disadvantage of the first is that it closely resembles a contour map, of the second that it portrays a hypothetical state of affairs.

Another very revealing temperature map is one which depicts the difference between the mean temperature (reduced to sea-level) of a station and the mean temperature for all stations in that latitude. This difference will give either a *positive anomaly* (i.e. it is warmer than the average for the latitude) or a *negative anomaly*. If these anomalies are plotted, and lines of equal anomaly (*isanomals*) are interpolated, the map will show clearly such features as the winter cold of continental interiors (p. 389), the summer heating of the land-masses, and the effects of the oceans.

TEMPERATURE FACTORS

These include: (i) insolation; (ii) latitude; (iii) the nature of the earth's surface; (iv) distance from the sea; (v) relief; (vi) winds; and (vii) ocean currents.

(i) **Insolation** The main source of heat affecting the atmosphere and the earth's surface is the radiant energy (*solar radiation*) which travels through space from the sun (a mass of intensely hot gases, with a temperature estimated to be at its surface about 6000° C), in the form of electro-magnetic energy. The earth receives only a very small part of the total radiation, perhaps one two-thousand-millionth, but this is vital for the earth. This solar or radiant energy is known as *insolation*. At the outer limit of the atmosphere the solar radiation consists of visible light rays (about 41 per cent of the total), very short gamma rays, alpha rays, X-rays and ultraviolet rays (9 per cent), and the longer infrared and heat rays (50 per cent).

The measure of the intensity of insolation per unit area received at the outer limit of the atmosphere is called the *Solar Constant*, which is calculable. This is equal to nearly 2 gramme-calories per square cm per minute; i.e., the radiation received each minute on a surface area of a square cm could raise the temperature of 1 gramme of water through 2° C (or alternatively equivalent to 1·5 kilowatts per sq. m). In point of fact, this term seems to be a misnomer, since radiation does seem to vary, as evidenced by sun-spot activity, though only by about 1–2 per cent. Moreover, the earth has an elliptical orbit around the sun, so that it is at its farthest point on July 4 (a position known as *aphelion*, at a distance of 152 million km (94·5 million miles), and at its nearest on 3 January (*perihelion*, at a distance of 147 million km (91·5 million miles)). The amount of solar energy received is therefore about 6 per cent more at the perihelion, and it would seem that as this occurs in the northern hemisphere winter, that season should be slightly milder than in the southern hemisphere. Actually other factors, notably the greater speed of the earth along its orbit in December–January due to the curvature of its ellipse, and the great area of land in the northern hemisphere, completely mask this effect.

What really matters is the effect on the insolation when it enters the earth's atmosphere and in part passes to the earth's surface. Some 19 per cent is directly absorbed by the atmosphere, particularly by the carbon dioxide and water-vapour present; this proportion is so low because the short light-waves pass, with little check, through the atmosphere. A small amount is lost as a result of 'scattering' caused by air molecules, dust particles and water-vapour; this scattered radiation is not converted into heat. The scattering effect is greatest for the rays of light at the short-wave end of the spectrum, the blue and violet end, which helps to explain the blue colour of the sky. A more serious loss (29 per cent) is that reflected directly back into space by clouds and dust; the drop in temperature when a cloud passes across the sun can be appreciable, and a very thick 'overcast' can locally reduce the incoming insolation by as much as 80 per cent.

The balance (rather less than 45 per cent) of the incoming insolation reaches the surface of the earth. Ten per cent is reflected directly back into space from that surface, still as short light-waves, leaving only about 35 per cent to heat the surface. As the surface warms it re-radiates some of the energy, but in the form of long-wave heat radiation; if this re-radiation did not take place, the surface would become constantly hotter. These longer waves are more readily absorbed by the atmosphere than is the incoming short-wave insolation, and so the atmosphere is indirectly heated. In addition, the heated surface of the earth warms the layer of air resting upon it by direct conduction, which then rises as a current or updraught of air. Where this is localized, the rapidly rising updraught is known as a *thermal*, which both birds and glider-pilots use to assist their ascent.

The surface of the earth therefore both absorbs and gives out radiant energy at the same time, although in the form of energy of different wavelengths; as a result the heating depends on the balance between the two (fig. 177). During a clear, sunny day in summer there is a net gain and the temperature rises, reaching a maximum in the early afternoon. During the night, incoming insolation ceases, while the outgoing radiation continues. There is, therefore, a loss of heat during the night, and the coldest period is just after dawn.

The atmosphere acts rather like a greenhouse roof, since long-wave radiation cannot pass out through it as readily as incoming short-wave insolation. Something like seven-tenths of the outgoing radiation is absorbed by the atmosphere, otherwise temperatures would drop more markedly during the night and in winter. This insulating effect is increased by a cloud-cover. Hard frosts in winter occur during a clear, starlit night, when the outgoing radiation into space is at its maximum. Similarly, in the tropical deserts with their cloudless skies, terrestrial radiation at night causes a sharp fall in temperature. In California and elsewhere 'smudging' is used to combat nocturnal cooling; fires giving off dense palls of smoke check radiation and so reduce frost incidence at blossom-time.

(ii) **Latitude** So far insolation has been discussed in general

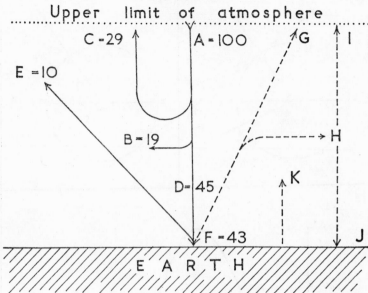

Fig. 177 *The balance of solar radiation*

A = incoming radiation (short-wave), i.e. insolation; **B** = loss by atmospheric absorption; **C** = loss by 'scattering' and direct reflection from clouds and dust back to space; **D** = effective radiation reaching the earth's surface; **E** = direct reflection from the water or ground; **F** = solar radiation converted into heat at the earth's surface; **G** = outgoing terrestrial radiation (long-wave) into space; **H** = outgoing terrestrial radiation (long-wave) absorbed by the atmosphere; **I** = heat radiated directly from the atmosphere to space; **J** = heat radiated from the atmosphere to the earth; **K** = direct heating of the atmosphere by convection and conduction from the earth.

Figures represent approximate mean percentages of the incoming insolation over the earth as a whole. Solid lines indicate short-wave radiation, pecked lines the long-wave outgoing terrestrial radiation.

terms, without considering how it may vary for astronomical reasons. This variation is a function of latitude.

The earth both *rotates* on its own axis, which is responsible for the alternation of day and night, and *revolves* in its elliptical orbit around the sun for a period of about 365¼ days, which is responsible for the seasons. The axis of the earth about which it rotates is tilted at an angle of 66½° to the plane of the path it traces during its revolution (known as the *plane of the ecliptic*). As a result, the position of the overhead midday sun varies throughout the year; the sun appears to 'decline south' in the northern winter, and to return

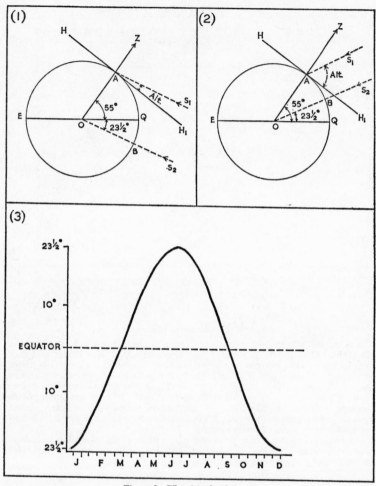

Fig. 178 *The altitude of the sun*

S_1, S_2 represent rays of the midday sun. **Z** = zenith (overhead point).

1. *22 Dec.* Place A is in latitude 55° N, the sun's declination is $23\frac{1}{2}$° S (i.e. overhead at B).

Then $\angle AOB = 55° + 23\frac{1}{2}° = 78\frac{1}{2}° = \angle ZAS_1$ (corresponding angles).
Altitude of sun at A = $\angle S_1AH_1 = 90° - \angle ZAS_1 = 11\frac{1}{2}°$.

2. *21 June.* The sun's declination is $23\frac{1}{2}$° N.

Then $\angle AOB = 55° - 23\frac{1}{2}° = 31\frac{1}{2}° = \angle ZAS_1$ (corresponding angles).
Altitude of sun at A = $\angle S_1AH_1 = 90° - \angle ZAS_1 = 58\frac{1}{2}°$.

3. *Declination.* The angular distance of the overhead sun north and south of the Equator is plotted throughout the year at weekly intervals (using the declination as given in the *Nautical Almanac*), or calculated from the formula $23\frac{1}{2}$° × *sine of the number of days from the nearest equinox.*

north in summer. The results of this change in *declination* are shown in fig. 178. The North Pole on 21 June is tilted towards the sun (the *June solstice*), on 22 December it is tilted away from it (the *December solstice*), and on 23 September and 21 March the two poles are just reached by the sun's rays (the *Equinoxes*).

These astronomical facts have two important results upon the degree of insolation. They cause very marked differences (*a*) in the *angle of incidence* of the sun's rays, and (*b*) in the *length of day and night* during the several seasons at various parts of the earth.

(*a*) The angle of incidence at which the sun's rays fall upon the surface of the earth influences their heating effect. Not only is the length of their path through the atmosphere increased in high latitudes (if the length of the path through the atmosphere is doubled, the amount of heating per unit of area is decreased to one quarter), but the amount of solar energy per unit of area is decreased, as shown in fig. 180. Difference in the angle of incidence therefore causes a proportional difference in the *intensity of insolation*.

(*b*) At the Equinoxes every parallel is half in daylight, half in darkness, and day and night are almost equal throughout the world. On 21 June, however, there is constant darkness within the Antarctic Circle; the length of day increases gradually to almost exactly 12 hours at the Equator, and at the Arctic Circle the sun does not sink below the horizon. These facts are summarized in the following table, derived from the *Nautical Almanac*:

Duration of Daylight
(in terms of monthly mean hours)

	Equator	50° N	60° N	70° N
Jan.	12·13	8·55	6·70	0·00
Feb.	12·13	10·00	9·02	4·98
Mar.	12·13	11·81	11·70	9·55
Apr.	12·13	13·69	14·47	14·65
May	12·13	15·33	17·03	18·96
June	12·13	16·22	18·59	No night
July	12·13	15·83	17·88	No night
Aug.	12·13	14·40	15·55	21·18
Sept.	12·13	12·60	12·86	15·50
Oct.	12·13	10·73	10·12	11·20
Nov.	12·13	9·03	7·51	6·4.
Dec.	12·13	8·09	5·93	0·00

The effect of the seasons is therefore to modify the duration of insolation, and the net result is that both *intensity* and *duration* of insolation must be taken into account. At the Equator the value of

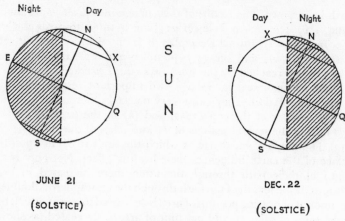

Fig. 179 *The Solstices*

Half of the globe is illuminated at any moment of time.
At the *June Solstice*, point X (50° N) has approximately 16½ hours of daylight.
At the *December Solstice*, point X has approximately 7½ hours of daylight.

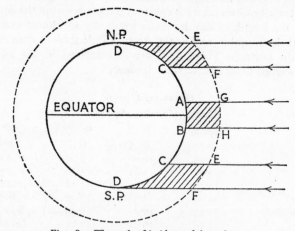

Fig. 180 *The angle of incidence of the sun's rays*

Area of surface **AB** is less than that of **CD**. Distance **AG** is less than **DE**. Therefore the concentration of heating is greater at **AB** than at **CD**.

the insolation varies little; the day is about 12 hours long and the sun at midday is never more than 23½° from overhead. There are two slight maxima of insolation at the Equinoxes, two slight minima at the Solstices. With increasing distance from the Equator, the longer are the summer days; during the Solstice at the Tropic the day is of about $13\frac{1}{2}$ hours and the midday sun is overhead, so that insolation is greater than it ever is at the Equator. Beyond the Tropic the length of the summer day continues to increase with latitude, but the angle of incidence of the sun's rays decreases. For a while the former more than compensates for the latter, and the amount of insolation increases to a maximum at $43\frac{1}{2}°$ latitude. But poleward of this the increasing length of the day fails to counter-balance the diminishing angle of the midday sun, and the value of the insolation therefore decreases.

When all the calculations of the theoretical value of insolation are completed and the effective incoming insolation is balanced against the outgoing terrestrial radiation, there is a net heat loss in latitudes poleward of 40°, a net heat gain between the 40° N and S parallels. As the earth's surface as a whole seems to be maintained at a more or less constant temperature, the total gain and loss must balance. In other words, there must be a *heat transference* from tropical to polar regions, by the movement of both air and water-masses. This is emphasized, since it is the basis of atmospheric movements.

Twilight. Particles of dust and water-vapour in the earth's atmosphere reflect the sun's light before it has appeared above the horizon in the morning and after it has disappeared below it in the evening. *Astronomical twilight* lasts from when the sun is 18° below the horizon until dawn, and again from sunset until the sun is 18° below the horizon. The more limited *civil twilight* extends only from 6° and to 6° respectively. In high latitudes the low slant of the sun's path as it dips or rises causes twilight to be prolonged as compared with low latitudes. At the North Pole twilight persists continuously from 23 September to 14 November, and then again from 29 January to 21 March, that is, during the time when the sun's declination is changing from 18° to 0° and back again; in between is a period of complete darkness.

(iii) **The nature of the surface** So far the value of insolation has been discussed as if it were affecting a homogeneous surface. The basic difference as far as the world is concerned is between land and water surfaces.

The *specific heat* of a substance is defined as the number of calories required to raise the temperature of 1 gramme of that substance through 1° C; a calorie represents the amount of heat required to raise the temperature of 1 gramme of water through 1° C. On this basis the specific heat of seawater is 0·94 and of granite about 0·2; in other words, water must absorb nearly five times as much heat in order to increase in temperature by the same amount as earth. Land

387

surfaces therefore heat up more rapidly and intensely than do water surfaces. On the other hand, land cools more rapidly when the source of heating is cut off.

Other factors emphasize this contrast. Solid earth is a bad conductor, hence a shallow layer is more intensely heated. The earth is opaque, while water allows the rays to penetrate to a greater depth and so affect more water to a less extent. The daily fluctuations in temperature may be detected 15–18 m (50–60 ft) below the surface of water, but not more than a metre below the surface of the ground. A water surface allows evaporation, which is a cooling process, and there is also a considerable heat transference through mixing by convection currents readily set up in water. It reflects a high proportion of the light rays (compare burnished silver, which remains cool under the hottest sun because it reflects 95 per cent of the rays), while darker earth substances absorb more. A piece of black paper left on the surface of a snowfield will in a few hours have sunk to a depth of several cm because it has absorbed heat and so melted the snow around. A snowfield reflects a large proportion of the rays, so that the snow remains while skiers become bronzed in the brilliant sunshine; the polar ice-caps are virtually perfect reflectors. The ratio between the total solar radiation falling upon a surface and the amount reflected, expressed as a decimal or a percentage, is known as the *reflection coefficient* or *albedo*. The earth's average albedo as seen from space (including also the albedo of the clouds) is about 0·4; i.e. four-tenths of the solar radiation is reflected back into space. Surface albedo varies from 0·03 for dark soil to 0·8 for a snowfield. Water has a low albedo (0·02) with near vertical rays, though a high one for low-angle slanting rays, and the figure for grass is about 0·25.

Dry sandy soils have a low specific heat and warm up rapidly at the surface, although the insulating effect of the air contained between the particles causes the heat to be retained at the surface, so that the sand only a few cm below may be as much as 15° C cooler. Swamps or waterlogged soils act like a water surface, and forests modify heating by casting shade. Microclimatology presents many examples of these variations; compare these records, taken on Salisbury Plain in June 1925, of air temperatures just above various surfaces:

Temperature

	°C	°F		°C	°F
Shade (air)	22	71	Bare earth	35·5	96
Macadam road	43	109	Grassland	29	85
Sand	35	95	Brick rubble	31	88

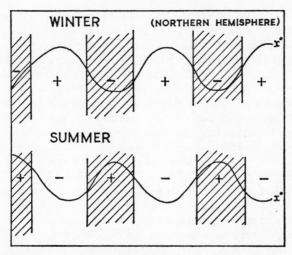

Fig. 181 *The effects of land masses and oceans upon temperature*

The shaded area represents land, the unshaded area water.

In winter the land is cooler than the sea; the isotherms therefore bend equatorward over the land.

In summer the land is warmer than the sea; the isotherms therefore bend poleward over the land.

The climatological results of differential heating of land and water result in both diurnal and seasonal contrasts. The decrease in temperature during a summer night as a result of outgoing radiation is only one-quarter as great in a given volume of water as in the same volume of earth. Consequently a water surface is subject to a much smaller diurnal range.

(iv) **Distance from the sea** One of the most fundamental concepts in climatology is the distribution of seasonal temperatures over land and sea. The continents tend to be warmer in summer than oceans in the same latitudes, but appreciably colder in winter; the larger the land-mass, the greater the contrast (fig. 181). Thus extreme temperatures, with a large seasonal range, are the result of continentality, which obviously affects the northern hemisphere much more than the southern. On the other hand, the oceans and those land margins affected by oceanic influences tend to have more equable temperatures, with low seasonal ranges. In the discussion of climatic types (Chapter 18), *continental* and *marine* factors are major criteria.

Large lakes have the same general effect as an ocean, but on a less marked scale, depending on the extent of the water area.

Microclimatologists have shown that the lakes of Switzerland affect temperatures a km or so from their shores. The Great Lakes of North America cause a marked northward deviation of the winter isotherms. A strip of country 50 km (30 miles) wide along the eastern shore of Lake Michigan is known as the 'Fruit Belt', and the southern shores of Lake Erie form the 'Grape Belt' of Pennsylvania and New York State; such fruit as grapes and peaches can be successfully ripened because of the warmth retained by the water-body in autumn. Paradoxically, the fact that the lake-waters warm up less rapidly in spring than the land causes a slight chilling of the coastal belts, and so premature blossoming of the fruit-trees is avoided.

(v) **Relief** One immediate relief influence on temperature has been mentioned—the effect of altitude. The environmental lapse-rate is about 0·6° C per 100 m (3·3° F per 1000 ft), but there are considerable variations from this figure. During the period 1884–1903, when an observatory was maintained on the summit of Ben Nevis at 1343 m (4406 ft), the mean annual temperature was − 0·25° C, compared with 8° C (47° F) at Fort William near sea-level. In the British Isles the lapse-rate averages 0·6° C per 133 m (1° F per 410 ft) in winter, but increases to 0·6° C per 90 m (1° F per 270 ft) in summer. It must of course be remembered that there may be a considerable difference between a 'free-air' lapse-rate and a 'relief' lapse-rate. Both tend to decrease by night and increase by day. Much more important variations, from the meteorologist's point of view, are found when vertical movements of air occur (pp. 429–32).

However, the environmental lapse-rate is of practical importance. Thus Delhi, in the Indo-Gangetic plain at a height of 219 m (718 ft) has a mean June temperature of 33° C (92° F), but at its hill-station, Simla (2204 m, 7232 ft), the corresponding temperature is only 19° C (67° F).

Other relief controls of temperature involve shelter and aspect. Shelter protects an area from cold winds; compare, for example, the bleak cold which may be experienced at the mouth of the Rhône when the mistral (p. 422) is blowing, with the mild Riviera coast to the east sheltered by the Maritime Alps. On a larger scale, southern China experiences abnormally low winter temperatures because it is exposed to the cold outblowing monsoon from which India is protected by the wall of the Himalayas. America has no real transverse barrier, so that New Orleans may experience 'cold snaps' brought by air streams from the Arctic, while very warm humid airstreams from the Gulf of Mexico may bring heat-waves to the northern states and even into Canada. Aspect, including both the direction and degree of slope, is important, since a southern aspect means greater insolation. A striking contrast in temperature, vegetation and human use is seen on the shady and sunny sides of an Alpine valley (p. 484-5).

Inversion. One striking exception to the decrease of temperature with altitude is where an inversion occurs. During clear, settled weather, as radiation of heat takes place during the night, the air on hill-slopes is rapidly cooled, and this dense air drains downward, filling valleys or basins with cold air, possibly at temperatures below freezing, when the upper slopes are markedly warmer. Miles City, Montana, at a height of 703 m (2371 ft) in a deep hollow in the Great Plains, has experienced the lowest temperature ever recorded in the U.S.A. (excluding Alaska), $-54°$ C ($-65°$ F). Yet the summit of Pike's Peak (4339 m, 14,255 ft) has never recorded a temperature below $-40°$ C ($-40°$ F).

Frost may cause serious losses to fruit-cultivators if it occurs during the critical blossom-time; the plum-growers of the Vale of Evesham, the viticulturalists of the Gironde and the Champagne districts, and the citrus-fruit growers of California alike face the problem. Smudging with dense clouds of smoke, carefully maintained oil-heaters among the trees, and even flooding with water to check the rate of cooling and to stimulate light fogs, are used. In southern France, where early vegetables (*primeurs*) are grown, the fields are covered before sundown with light straw mats, though not to exclude frost so much as to keep in ground warmth. The best method is to avoid 'frost pockets' by siting orchards on slopes away from lines of cold air drainage.

Inversions also occur as the result of the relative position of air-masses of different temperatures associated with local pressure systems (p. 413).

In addition to what might be called 'surface' or 'ground' inversions, similar phenomena sometimes occur at high altitudes. Their, importance is that they tend to prevent vertical movement of air, hence inducing a state of stability (p. 431), and so there is unlikely to be any rainfall.

The isothermal layer Under very stable high pressure conditions (p. 416), the temperature may remain the same within a layer of air extending to a considerable height.

(vi) **Winds** The effect of winds on temperature is to 'transport' temperatures prevailing in the areas over which they blow, either from land or sea. The result of exposure to, or shelter from, cold outblowing winds from continental interiors has been mentioned.

Onshore winds in the Tropics, blowing from over the cooler ocean, tend to modify temperatures on the coastal margins. On the other hand, onshore winds such as the Westerlies may in winter carry mild temperatures from over the oceans on to the continental margins. Local winds (the warm sirocco, the cold mistral and bora (fig. 192), and the foehn or chinook winds on the lee side of mountains) may produce very rapid local temperature changes (pp. 423–4).

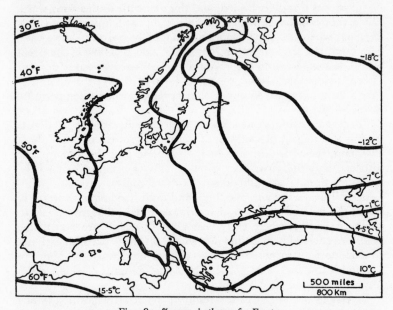

Fig. 182 *January isotherms for Europe*

These isotherms represent temperatures reduced to sea-level.
Note: (i) the 'Winter Gulf of Warmth' over Scandinavia and the North Sea; (ii) the north–south trend of the isotherms over the land, indicating the gradual increase eastward of winter cold.

(vii) **Ocean currents.** Ocean currents share with winds the ability to 'transport' temperatures (pp. 354–62). Where onshore winds blow over these currents, they convey similar temperature conditions to the land margins. Especially important are the 'cold-water coasts' off the western sides of the continents in the Tropics and Sub-tropics. The 'winter gulf of warmth' which affects the North Atlantic (fig. 182) is largely the result of the North Atlantic Drift; northern Norway is ice-free, though in the same latitude as central Greenland.

WORLD TEMPERATURE DISTRIBUTION

The seasonal distribution of temperature forms one criterion in the delimitation of climatic types (Chapter 18). Sea-level isothermal maps show for the northern hemisphere in winter a marked northward bend in the isotherms over the sea and a southward bend over the land; trace particularly the course of the 0° C (32° F) isotherm. In summer the isotherms bend northward (but not so markedly) over the land. In the southern hemisphere, where the land areas are small, the parallel east–west trend of the isotherms is more in evidence, particularly south of 35° S.

The highest shade temperature recorded in the world was 58° C (136·4° F) at Azizia in Tripoli (1922) and again in San Luis Potosí in Mexico (1933), the highest in Britain 38° C (100·5° F) at Tonbridge, Kent (1868). Conversely, the world's lowest recorded shade temperature was — 90° C (— 130° F) at Vostok in the Antarctic (1962), the lowest in Britain — 27·2° C (— 17° F) at Braemar (1895).

'Sensible temperatures' are the sensations of heat and cold felt by the human body. Temperature in this connection is intimately related to humidity; dry heat mitigates high temperatures by accelerating evaporation from the skin surface, while moist heat makes the body feel distinctly uncomfortable. An index of this is the *wet-bulb thermometer* (p. 428); if the wet-bulb reads much above 24° C (75° F), sustained manual labour is extremely difficult. Damp raw cold is much more trying than dry cold; damp air is a better conductor of heat than dry air, and allows the escape of heat from the body in cold weather. Very low temperatures can be endured (as in central Canada) when the air is dry and still and the sun shining. A *temperature–humidity index* can be constructed by various formulae which affords an indication of the effects of the weather on human comfort; on one such scale the index of 60 to 65 represents ideal conditions, while 80 can be regarded as so uncomfortable (as happens in New York and elsewhere) offices and factories may be closed.[1] Another physiological index in cold climates is *windchill*, obtained from a formula involving both temperature and wind force.

The facts are of prime importance to the geographer; they involve problems of tropical and low-latitude settlement, of white acclimatization in the tropics, of the exploitation of the Northland, and of the efficiency of labour.

Accumulated temperatures The calculation of accumulated temperatures is a statistical device for indicating the accumula-

---

[1] One such index used by the U.S. Weather Bureau is an empirical one: 0·4 × dry bulb temperature + wet bulb temperature) + 15.

tion of warmth over a basic critical value for a certain period of time. This has considerable significance in the study of relationships between plant activity and temperature conditions during the period of growth, in the classification of climatic types, and in the analysis of climatic cycles. The unit used is called either the *degree-day* or *day-degree*, which gives an exact measure of the departure of the mean daily figures from the selected base-temperature. Thus if 6° C is the datum, and 9° C is the mean for a specific month, it will count as $3 \times 31 = 93$ degree-days towards the final total. More accurately, though more tediously, the accumulated temperature for a specific day may be given in degree-hours.

SUNSHINE

The duration of sunshine is partly a function of latitude, for the hours of light (that is, of possible sunshine) vary with the seasons in different latitudes (p. 385). It is also a function of daytime cloudiness.

Measurement and recording The duration of bright sunshine is measured by means of a Campbell-Stokes recorder, a solid sphere of glass 10 cm in diameter. This focuses the rays of sunshine on to a sensitized card graduated in hours, and so burns a line during the time the sun is shining. Faint sunlight, near dawn or dusk, or when the sun is partially obscured, is not recorded. Tables of sunshine data are prepared from these records, in the form of either absolute duration in hours per day, or as a percentage of possible sunshine per day or month. When mean figures have been obtained over the requisite number of years, the values for each station can be plotted, and lines of equal mean duration of sunshine (*isohels*) can be interpolated.

Distribution of sunshine The sunniest parts of the earth are the hot deserts (p. 481) in the Trade Wind belt. At Helwan Observatory, in the Nile valley south of Cairo, the mean sunshine for the year is 3668 hours, which represents about 82 per cent of the possible amount for that latitude. From June to September the duration is about 90 per cent of the possible amount, and even the month with most cloud (January) still receives 70 per cent of the possible.

The lands around the Mediterranean Sea and other parts of the world with broadly similar climates (p. 467) enjoy an extremely sunny summer, with a mean of about 90 per cent of the possible duration. Even in winter, the rainy season, days without any sunshine are few, since the rain tends to be concentrated and short-lived.

The equatorial belts and much of the cool temperate latitudes have considerably less sunshine. The equatorial climate (p. 458) is characterized by rainfall and considerable cloud throughout the year; however, there is usually bright sunshine in the mornings, and the sky becomes overcast in the afternoons and early evenings. In cool temperate latitudes the skies are frequently cloudy and drizzle or rain is common; thus Valentia in western Ireland has a mean of only 1·3 hours of sunshine per day, or 17 per cent of the possible total. The periods of sunshine and cloud are as irregular and variable as the changeable weather of these latitudes would indicate, but May and June with about 40 per cent of the possible are generally the sunniest months. The interiors of the continents in these latitudes are drier and less cloudy, and so experience more sunshine than the western margins.

Note. The U.K. Meteorological Office at present gives temperature figures on the Centigrade (Celsius) scale, followed by the Fahrenheit conversion. In due course, consistent with the U.K.'s metrication policy, Centigrade alone will be used, which is already the practice among scientific workers. This book now follows the practice of using the Centigrade figures, but including the Fahrenheit equivalents where helpful. A conversion table is appended.

°F	°C	°F	°C	°F	°C	°F	°C
100	37·7	73·4	23	46·4	8	24·8	− 4
98·6	37	71·6	22	45	7·2	23	− 5
96·8	36	70	21·1	44·6	7	21·2	− 6
95	35	69·8	21	42·8	6	20	− 6·6
93·2	34	68	20	41	5	19·4	− 7
91·4	33	66·2	19	40	4·4	17·6	− 8
90	32·2	65	18·3	39·2	4	15·8	− 9
89·6	32	64·4	18	37·4	3	15	− 9·4
87·8	31	62·6	17	35·6	2	14	− 10
86	30	60·8	16	35	1·6	12·2	−11
85	29·4	60	15·5	33·8	1	10·4	− 12
84·2	29	59	15	33	0·5	10	− 12·2
82·4	28	57·2	14	32	0	8·6	− 13
80·6	27	55·4	13	31	− 0·5	6·8	− 14
80	26·6	55	12·7	30·2	− 1	5	− 15
78·8	26	53·6	12	30	− 1·1	3·2	− 16
77	25	51·8	11	28·4	− 2	1·4	− 17
75·2	24	50	10	26·6	− 3	1	− 17·2
75	23·8	48·2	9	25	− 3·8	0	− 17·7

Note: It is interesting that the temperature of − 40° is common to both scales; i.e. − 40° F = − 40° C, the only point at which they have a common value.

CHAPTER 16

Pressure and Winds

MEASUREMENT AND RECORDING

THE atmosphere has a definite weight and so exerts pressure upon the earth. The weight of air in a vertical column extending from the upper limit of the atmosphere to the earth's surface exerts a pressure which averages about 1 kg per sq. cm (14·7 lb per sq. in). This is, of course, more or less in balance with the force of gravity.

Instruments The weight of this vertical column of air is measured by a *barometer*, first done by Torricelli in 1643; in effect he balanced the weight of a column of mercury in a glass tube, sealed at one end, against the pressure of a column of the atmosphere of equal cross-section. Mercury, the heaviest known liquid, is used because a tube of water, for example, would need to be 10 m (34 ft) in length. Modern mercury barometers, such as the Kew or Fortin pattern, have various refinements and adjustments for applying corrections, and vernier scales for making exact readings. These barometers are read in terms of the length of the column of mercury, either in inches or millimetres; 29·92 in (the average sea-level pressure over the world) is equivalent to 760 mm.

During this century an alternative and more logical system of pressure measurement has been evolved, although the mercury column is still commonly used. Physicists use a unit called the *dyne*, the force which, when applied to a mass of 1 gramme for 1 second, will produce a change in velocity of 1 cm per second.[1] A *bar*, representing a pressure of a million dynes per sq. cm, is now the standard unit of pressure measurement, divided into a thousand *millibars* (*mb*). There is no universal formula for conversion, except at a constant temperature and latitude; at 45° N latitude, at sea-level, and at a temperature of 0° C, 1 bar equals 29·5306 in or 750·1 mm of mercury, or 30 in of mercury is equivalent to 1015·9 mb. The average sea-level pressure of 29·92 in is equivalent to 1013·25 mb.

In addition to the mercury barometer, other pressure-recording instruments have been devised. The *aneroid barometer* comprises

---

[1] This is according to the CGS (Centimetre-Gramme-Second) system, which will be rendered obsolete by the metricated *Système Internationale* (SI). The recognized unit of force is the *newton*, which equals 10^5 dynes, and 1 bar = 10 newtons.

essentially a metallic box, almost exhausted of air, the sides of which are flexible so that they can expand and contract with the external air pressure. These movements are connected by a spring to a needle which moves round a calibrated circular dial. A *barograph* operates on a similar principle, with the spring connected to an inked pen which makes a continuous record on a chart fixed to a revolving drum. These aneroid recorders are neither as accurate nor as reliable as the mercury patterns.

Corrections applied to barometer readings Corrections are applied to recorded pressures in order to attain comparable standards. These corrections are made for (i) latitude; (ii) temperature; and (iii) altitude.

(i) *Latitude* The earth is not a perfect sphere, and therefore the force of gravity varies according to the latitude, attaining a maximum at the Poles and a minimum at the Equator. Readings for precise comparisons are therefore standardized to 45° latitude.

(ii) *Temperature* Errors in pressure readings are introduced as a result of the different coefficients of expansion of both the mercury column and the metal scale. Tables are supplied with each instrument to enable the correction to be made; the bar is officially standardized at 0° C (32° F).

(iii) *Altitude* There is a considerable change of pressure with altitude, since the proportion of the overlying atmosphere decreases with height. A very rough indication of the pressure decrease is about 34 mb (1 in of mercury) for every 270 m (900 ft), but this applies only to the lowest 1000 m, above which the rate becomes less. From sea-level to 600 m (2000 ft) the decrease is about 4 per cent for each 330 m (1000 ft); from 600 m to 1500 m (5000 ft) it is about 3 per cent; and from 1500 m to 3000 m (10,000 ft) it is 2·5 per cent. At a height of 24 km (15 miles) the rate of decrease becomes virtually imperceptible. If the pressure is 1016 mb (30 in) at sea-level, it will be about 540 mb (16 in) at the top of Mont Blanc; an aneroid calibrated in feet and/or metres is sometimes carried by mountaineers. The same principle is used in the altimeter of an aircraft.

The physiological results of reduced pressure with altitude are shown by the difficulties of acclimatization in high mountaineering. Indeed, it is unlikely that Everest will be climbed without the help of oxygen from cylinders. Sir Edmund Hillary and Tensing Norkey, who reached the summit of Everest on 29 May 1953, each used an open-circuit oxygen apparatus weighing over 14 kilograms, from which they inhaled 3 litres of oxygen per minute during the final stages of the ascent.

Pressure maps Pressure maps are constructed on the same

principles as temperature maps. Stations are located, the mean pressure values for them are plotted, and lines of equal barometric pressure, known as *isobars*, are interpolated. These values are reduced to sea-level; as the fall in pressure with altitude is so rapid, isobars (even more than isotherms) would simply follow the contours. Sea-level variation of pressure is relatively small, ranging (as far as mean values are concerned) only from about 980 mb (29 in) to 1070 mb (31·5 in), whereas it can fall to 540 mb (16 in) in the French Alps, and still lower in the higher ranges. The lowest barometric reading for a place near sea-level in Britain was 925·5 mb (27·33 in) in Perthshire on 26 January 1884; this was associated with a severe storm which caused widespread damage. The world's highest recorded sea-level barometric reading was 1079 mb in Russia in 1900, the lowest 877 mb in the Pacific Ocean in 1958.

There are various kinds of pressure map. One is the *synoptic chart* or daily weather map, which depicts the isobars at a moment in time. Most countries have a central office to which observers send periodic reports, in an internationally accepted code, by teleprinter. The Central Forecasting Office in Great Britain is at Bracknell, where the weather map is prepared, and from which information is sent to airfields and to international meteorological offices.

These synoptic charts are not merely drawn for surface conditions. In recent years high-altitude meteorological research has made it possible to construct upper-air synoptic charts, using radiosonde and rawinsonde ascents (p. 375). These, known as *contourcharts*, show the height of standard pressure surfaces. They are of great value in high-altitude aviation, as well as helping the forecaster to predict surface weather conditions.

Another type is the *average pressure map* on a continental or world scale, which portrays mean pressures calculated over a long period of time. While such a map represents only a broad generalization of conditions, it is helpful because it brings out the patterns and the salient features of pressure distribution over continent and ocean, season by season.

A useful type of chart shows the alteration of pressure over a period of time by means of *isallobars*, lines of equal pressure-change. When these are plotted, their patterns may reveal how a pressure system is developing and moving, thus assisting weather forecasting.

WORLD DISTRIBUTION OF PRESSURE

The planetary system The theoretical concept of pressure distribution on a homogeneous globe at sea-level (an 'ideal' or 'planetary' system) is useful as an elementary basis (fig. 183), but

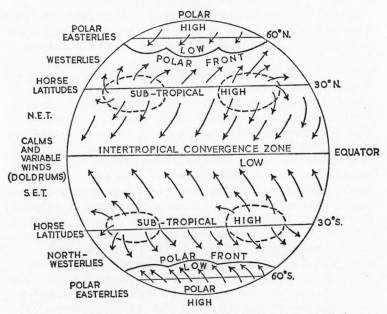

Fig. 183 *The planetary system of pressure and winds near sea-level at the Equinoxes*

it must be realized that this is actually subject to immense modifications.

Within a few degrees of the Equator is a belt of pressure generally less than 1013 mb (29·9 in), the *Equatorial Trough* or *low-pressure belt*. This is an area of high temperature and high humidity, commonly known as the Doldrums, where the air near sea-level is stagnant or sluggish. The low pressure is due to heating, for the pressure of a volume of air decreases as its temperature rises. Increasing numbers of readings from high-altitude observations show that these sluggish conditions rapidly disappear with increasing height, and air-stream flows of considerable velocity can be discerned. It must be emphasized that this is a generalization on a planetary scale, and there are considerable regional modifications (p. 409).

At about latitudes 30° N and S occur the *Subtropical high-pressure belts*, sometimes known as the 'Horse Latitudes', zones of calms and of descending air currents. It is not easy to explain the existence of these belts; they may in part be due to a general poleward movement from the Equator of air in the upper part of the troposphere, which comes under the influence of the earth's rotation so that its centrifugal force throws it back, causing an accumulation in these latitudes (fig. 186). There seems also to be an equatorward movement of air from high latitudes in the upper part of the troposphere, which descends in the Horse Latitudes, and so increases the air accumulation.

Nearer the Poles occur the *Subpolar low-pressure belts*. One reason for this pressure distribution is that the rotation of the earth causes a polar whirl and therefore a tendency towards low pressure at the Poles. But the intense cold around the Poles causes the thermal effect to overcome the dynamical one, with the result that the low-pressure belts tend to be on and just outside the polar circles.

The Antarctic ice-cap, above which lies intensely cold dense air, has therefore a shallow *Polar high-pressure system* over it. The Arctic region, however, is more complicated, for it is an ocean basin, with land-masses grouped round two-thirds of its perimeter; low pressure over the North Pole is not uncommon, even on mean monthly charts.

Fig. 183 portrays the planetary pressure systems at the Equinoxes, but the various pressure belts are displaced some degrees north and south with the seasons. Owing to the apparent movement of the overhead sun through 47° of latitude, the Thermal Equator also moves (p. 456), but not so much (through about 10°), and the Equatorial low-pressure belt moves with it through about 7°. The other pressure belts more or less move likewise.

Pressure 'cells' The main modification to this simple planetary concept is imposed by the irregular distribution of land and water, particularly in the northern hemisphere, which causes marked seasonal temperature changes (p. 385). The result is the interruption of the latitudinal 'belts' by the creation of high pressure over the cold continental interiors in winter and, conversely, of low pressure over the heated continents in summer (figs. 184, 185). This effect is not so apparent in the southern hemisphere, since the areas of land are small; the Subtropical high-pressure belt is not markedly interrupted, except in summer by slight southward extensions of the Equatorial low over the southern continents, while the Subpolar low-pressure belt in the Southern Ocean is continuous throughout the year.

In the northern hemisphere the pressure belts are so interrupted that they form a series of great 'cells', which may be thought of as large-scale eddies in the atmosphere, often of considerable depth and extending up to the tropopause. In winter the intense *Eurasiatic (Siberian)* and *American highs* dominate the continents, and the *North Atlantic* and *North Pacific lows* (sometimes called the *Icelandic* and *Aleutian lows* respectively) are over the northern parts of these oceans. In summer, however, lows form over Asia and North America

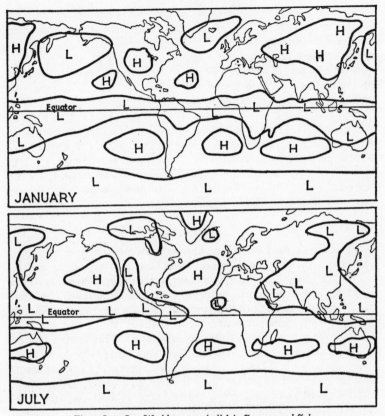

Figs. 184, 185 *World pressure 'cells' in January and July*

The main 'cells' of relatively high pressure and low pressure near sea-level are indicated diagrammatically.

as a result of the heating of the land; the centre of the Asiatic low is in north-western India, of the American low in the south-west of the U.S.A. and northern Mexico. Each is really a northerly extension of the Equatorial low. These lows interrupt the Subtropical high-pressure belt, but the North Atlantic and North Pacific are dominated by the high-pressure cells of the Horse Latitudes, sometimes called the *Azorean* and *Hawaiian highs*. The Icelandic and Aleutian lows are less intense, and are situated farther north than in winter; the Aleutian, in fact, virtually vanishes.

Upper-air pressure In recent years there has been much research into conditions in the upper air, and pressure maps showing these conditions are constructed and published as a routine procedure. However, the question of upper-air pressure is extremely complicated, although it is increasingly realized that developments in meteorology (particularly in long-range weather forecasting) are largely connected with the upper air.

A comparison of a series of high-altitude pressure-maps with the corresponding ones for sea-level shows distinct differences in pressure distribution, sometimes actual reversals of conditions. Over a large anticyclone at sea-level there may be a low at 3000 m (10,000 ft). At no great height, however, in the middle troposphere, the pressure distribution appears to be dominated by a vast low-pressure system in each hemisphere, approximately centred over each Pole, with a series of high-pressure cells more or less over latitudes 15° N and S (fluctuating with the seasons, though less markedly than near sea-level).

WIND

Wind force and velocity The measurement and recording of wind involves both the direction from which it is blowing and its velocity. Until 1949 a variety of codes was in use; in British returns direction was expressed in terms of the cardinal points of the compass, force in terms of the *Beaufort Scale* used for a century and a half and based on personal observation, both on shore and on ship. Thirteen 'forces' are used, from 0 (Calm) to 12 (Hurricane), related to descriptions of their effects and their estimated velocity at 10 m above the ground. Thus Force 4 (moderate breeze) of 13 to 18 knots was distinguished at sea by 'smacks carry all canvas', and on land by 'raises dust and paper; small branches move.' Force 8 (fresh gale) of 39–46 knots was similarly described as 'smacks make for harbour' and 'breaks twigs and branches off trees'. Velocities of over 160 km (100 miles) ph are occasionally reached on the west coast of Britain. During the gales of 16 February 1962 a gust of 285 km (177·2 miles) per hour was recorded at an R.A.F. station on the Shetland island of Unst, a British record.[1] Since 1949 most countries have adopted the recommendations of the International Meteorological Organization, using a standard code of velocity and giving wind directions as true bearings. It must be remembered that velocity and direction are given in average terms, though usually there is a sequence of lulls and gusts (i.e. *turbulence*) of great importance in aviation.

Instruments Apart from direct observation and estimation, several wind-recording instruments can be employed. A commonly

---

[1] Doubt has been cast on the authenticity of this record, since the recording instrument was damaged. The definitely accepted British record is 144 miles per hour (232 km ph) in the Cairngorms in 1967.

used instrument is the *cup-anemometer*, consisting of an upright, at the top of which are three horizontal spokes at an angle to each other, with cups fixed at their ends. These arms rotate horizontally as the wind blows, so moving a central rod which transmits the movement to a calibrated dial.

The *Dines pressure tube* has the merit of being a self-recording instrument which provides a continuous record of both direction and velocity. It consists of a vane placed in a prominent position, which has a shaft going down to the recording unit. The front end of the vane has an opening facing into the wind, and communicates the resulting increase of pressure to the recording unit. This provides a precise indication of velocity, while the movement of the vane records changes of direction. Both are recorded as ink lines traced on a revolving drum.

Records of wind in the upper air were formerly made by releasing small balloons with a fixed rate of climb, and observing their trends by means of a theodolite. The horizontal drift was measured, and some estimate made of the wind force at different levels. This method is slow and inaccurate, and has been superseded at modern meteorological stations by a balloon with a radar target, which can be followed accurately by the observer even when it is cloudy.

Presentation of wind records Observations of wind direction and velocity are usually calculated on a percentage basis. Detailed tables give this information for winds from different directions, for the various velocities and also for calms.

Wind maps Winds are shown on maps by means of symbols, usually arrows pointing in the direction to which they are blowing. Force is indicated either by a number of 'feathers' on the shaft of the arrow to correspond to the code in use, or by arrows of different thicknesses. The mean movement of air, representative of different seasons, may be shown by 'stream-lines', which take the form of elongated arrows curving to denote changes in direction.

A striking depiction of wind is by means of a *wind-rose*, of which there are many varieties. Basically it consists of radiating rays, drawn proportional in length to the mean percentage frequency of winds from each cardinal direction; calms are stated at the centre of the rose. Sometimes part of each ray may be thickened to indicate the frequency of winds above a certain critical force. In other types, twelve radiating rays represent months for the cardinal directions.

WIND SYSTEMS

The relation of wind and pressure Wind may be defined as the movement of air from high to low atmospheric pressure in a virtually horizontal plane. Its direction and strength is the result of four factors: (i) the barometric gradient; (ii) the Coriolis force; (iii) the centrifugal force; and (iv) friction.

(i) *The barometric* or *pressure gradient* Regions of high pressure form centres from which winds tend to blow outward, that is, they are areas of *divergence*. Conversely, regions of low pressure are the foci of winds, that is, areas of *convergence*. When the difference in pressure between two neighbouring points on the earth's surface is great, there is said to be a steep pressure gradient, and on a pressure map the isobars are close together; isobars and pressure maps are thus analogous to contours and relief maps. In such a case the resulting wind is strong. When, however, the barometric gradient is gentle, winds are light.

The result is that air tends to move from areas of high pressure to those of low. The succeeding factors, however, considerably modify this generalization.

(ii) *Coriolis force* The effect of the Coriolis force—the deflecting component produced by the rotation of the earth—may be most clearly stated in terms of *Ferrel's Law*. A body moving on the surface of the earth is deflected to its own right in the northern hemisphere, and to its left in the southern hemisphere, as a result of the earth's rotation. This has an effect on ocean currents, on tidal movements (p. 367), on driftwood which tends to collect along a right-hand river bank in the northern hemisphere, and on rifle bullets, which are slightly deflected. The force is proportional to the speed of the moving object, and it varies with the latitude, being zero at the Equator and at a maximum at the Poles.

When a balanced condition develops between the forces exerted by the pressure gradient in one direction and the Coriolis force in the opposite direction, a steady wind will blow. This is known as a state of *geostrophic balance*, and the air movement as *geostrophic flow* or as a *geostrophic wind*. This will be in a direction parallel to straight isobars.

(iii) *Centrifugal force* When an air-stream moves on a curved course, as in a pressure system with closed or curved isobars, it is subject to centrifugal force acting outwards from the centre of curvature. A *gradient wind* develops as a result of the balance between the pressure gradient on the one hand, and the Coriolis and centrifugal forces on the other. The air thus tends to travel along the sobars, in the northern hemisphere clockwise around the highs and anti-clockwise around the lows, while in the southern hemisphere it would tend to move anti-clockwise around the highs and clockwise around the lows. This relation of wind to atmospheric pressure was put forward by a Dutch scientist, Buys Ballot, in the middle of the nineteenth century; if you stand with your back to the wind in the northern hemisphere, pressure is lower on your left than on your right, and the reverse applies in the southern hemisphere.

(iv) *The effect of friction* Winds approaching pure geostrophic flow are found in the upper atmosphere. Near the ground, however, friction tends to reduce the power of the deflection due to rotation, so the net result is that the air crosses the isobars at a slight angle from high pressure to low pressure. This effect is naturally greater over the land than over the oceans.

With these concepts in mind, the major wind systems of the earth can be described. Just as the theoretical distribution of pressure on a homogeneous globe was a useful initial step in understanding, so it is of considerable help to describe the wind systems which would be associated with this idealized arrangement of pressure belts (fig. 183), and then examine the actual modifications to be found.

(i) **The Trades** The Trade Winds blow from the subtropical highs towards the Equatorial low, but as they are deflected by the rotating earth they blow from a north-easterly direction in the northern hemisphere (the North-east Trades) and from a south-easterly direction in the southern hemisphere (the South-east Trades). The Trades system moves with the other pressure and wind belts north and south with the seasons through about 7° of latitude. In the northern hemisphere winter the North-east Trades cover roughly the zone between Cancer and a few degrees north of the Equator, whereas in summer they occupy a belt farther north, roughly from about 30° N to about 10° N, and the system of the South-east Trades moves accordingly northward.

The Trades are noted for their constancy of force and direction, particularly over the eastern sides of the oceans, though they are markedly stronger in winter than in summer. But many interferences by local pressure disturbances occur on the western sides, and also near the Equator; here the complicating effects of their convergence, together with increased vertical convection lifting, are experienced (p. 412). P. R. Crowe computed the percentage constancy of these winds for each of the 990 sea areas (of 5° of latitude and longitude) between 45° N and S; over much of the oceans they have a constancy of 70 per cent, in the east–centres of the oceans of over 90 per cent. Their force is for the most part from 16 to 30 km (10 to 20 miles) per hour, in places 24 to 40 km (15–25 miles) per hour, while calms are extremely rare, and clear sunny conditions are usual.

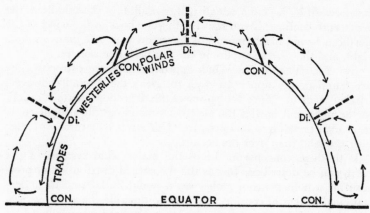

Fig. 186 *The air circulation in the troposphere*

This simplified diagram shows how divergences (shown by pecked lines), regions of descending and outflowing air, occur at the Poles and in the Horse Latitudes, while convergences (shown by solid lines), regions of inflowing and ascending air, occur at the Equator, where air rises to over 6000 m (20,000 ft), and in middle latitudes. The thermally driven circulation from the ascending air at the Equator to the descending air at about 30°, is known as a *Hadley Cell*, after G. Hadley who put forward the concept in 1735. The 'air age' has confirmed the presence of the Hadley Cell in tropical latitudes, what is known as a 'thermally direct cell'. Probably the Polar cell does not actually exist; the tropopause is much lower, the air streams are cold and dense, and tend to move and mix in a more horizontal plane. The mid-latitude cell (sometimes called a 'thermally indirect cell' because it is not directly driven like the tropical cell) consists of two upper vortices moving in an easterly direction around the globe; these, in fact, are the Polar Front Jet-Stream and the Subtropical Jet-Stream described below.

The name Trade Wind comes from the phrase 'to blow trade', i.e. to blow steadily in a constant direction. The name originally had nothing to do with trading, although in sailing-ship days these winds certainly furthered trade.

Between the two Trade Wind systems is found what is known as the Inter-Tropical Convergence Zone (p. 412); part of this is occupied by an area of calms or light winds, the Doldrums (p. 399). However, it has recently been discovered that in the 'summer' hemisphere, and especially over the continents (where thermal heating is pronounced), a zone of westerly winds intervenes between the Trades; these are sometimes called the *'Equatorial Westerlies'* (p. 412).

(ii) **The Mid-latitude Westerlies** The Westerlies comprise the air-flow from the Subtropical highs to the Subpolar lows, from about 35° to 60°; these belts move north and south with the seasonal change of the pressure belts. They consist essentially of two separate circum-polar vortices, increasing in both speed and constancy with altitude, and complicated by jet streams (below) in the upper troposphere. Within the belts both wind direction and velocity are far from constant, the result of local pressure systems, but a westerly component is predominant; the most frequent winds in the northern hemisphere are from the south-west, hence the name 'South-westerly Variables' or 'disturbed Westerlies'.

In the southern hemisphere, where there is continuous open ocean from 40° to 60° S, the Westerlies blow with strength and constancy and are characterized by gales, stormy seas, overcast skies and damp, raw weather. Seamen call them 'the Brave West Winds', and the latitudes in which they blow 'the Roaring Forties'.

(iii) **The Polar Easterlies** Cold air tends to move equatorward from the polar highs. This movement is pronounced in the southern hemisphere, where both the high over the Antarctic plateau and the Subpolar low over the Southern Ocean are clearly defined. As the air moves equatorward it is deflected to the left, and so a belt of winds results, spiralling from the polar region in an easterly direction, forming a polar vortex.

In the northern hemisphere, pressure and wind conditions are so complicated that these polar winds are extremely irregular. Sometimes a large part of North America or Asia is affected by the polar winds which spread southward.

Upper-air movement The winds so far described are largely surface currents, for it is beyond the scope of this work to discuss air currents at high altitudes in any detail (though very relevant to modern meteorological research). At high levels in both hemispheres there seems to be a general movement of air eastward, mainly a result of the fact that the upper-air pressure distribution is lowest at the Poles, highest at the Equator (pp. 401–2). There is therefore a kind of vast low in each hemisphere, centred over each Pole and extending toward the Equator; the air moves in a series of waves or loops at a height of about 10 km (6 miles), in an anti-clockwise direction round the Pole in the northern hemisphere, clockwise in the southern. In these loops originate the depressions of mid-latitudes (p. 414).

The Jet Streams Towards the end of the War of 1939–45, American bombers in the Pacific area encountered powerful currents of air at high altitudes flowing from west to east; an aircraft flying westward could at times hardly hold its own. The term 'jet stream' was coined, rather a misnomer since it is neither a jet nor a stream, but it has been accepted. It is known that there are at least four global jet streams in the northern hemisphere and at least one (almost certainly more) in the southern. The *Polar Front Jet* occurs in middle latitudes at about 9–12,000 m (30–40,000 ft), more or less at the tropopause; it fluctuates, however, both north and south and as low as 3000 m (10,000 ft). Others include the more continuous and persistent *Subtropical Jet* at about 12,000 m (40,000 ft), the *Arctic Jet* which has been traced across Alaska and Canada at 7600 m (25,000 ft), and the recently discovered *Polar Night Jet* above the Arctic Circle in the lower stratosphere. An increasing number of observations indicate that in summer the mean speed of the jets is about 90–110 km per hour (50–60 knots), but in winter it is considerably higher (170–220 km per hour, 95–120 knots), and on occasions an exceptionally strong jet has been experienced, once as much as 460 km per hour (250 knots). These are of such importance to aviation, especially over the expanses of the Pacific, that every day a weather reconnaissance plane flies a 3700 km (2300 miles) route from Honolulu, another from Fairbanks, Alaska, and a third from San Francisco; their reports enable the day's jets to be plotted.

In addition to these global jet streams, there are others of a much more local and transient character, a kind of 'fast river of air' of almost tubular cross-section about 160 km (100 miles) in width and 3–8 km (2–5 miles) deep, extremely turbulent at its margins though smoothly flowing within; east-bound aircraft where possible 'ride the Big Wind' across the Atlantic and Pacific to save time and fuel.

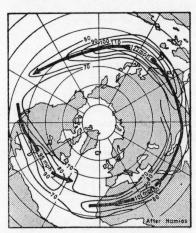

Fig. 187 The Subtropical Jet Stream

The jet streams have considerable significance in connection with surface weather conditions, though this is by no means understood. In general, the more wavy and 'scalloped' the path of the jet, the more numerous and active are local pressure systems near the surface of the earth. In the U.S.A. a distinct relationship has been discovered between jet activity and the incidence of tornadoes (p. 420); when the Polar Front Jet bends south over the Mississippi valley storms may be expected, and the Severe Local Storm Forecast Center at Kansas City, run by the U.S. Weather Bureau, may issue a warning. A dip of the jet toward the ground may mean torrential rain or hail-

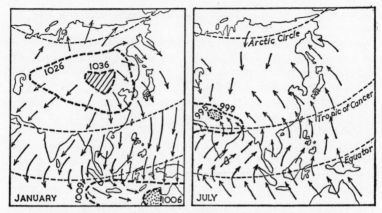

Figs 188, 189 *The Asiatic monsoonal airstreams, January and July*

storms. The position and nature of the jets appear to have important meteorological results, especially on frontogenesis (p. 412) and
on the 'bursting' of the Monsoon.

Interruptions to the planetary wind systems The planetary system of winds holds good to a large extent for most of the
southern hemisphere, where the interrupting effects of the land-
masses are not serious. But in the northern hemisphere and in that
part of the southern hemisphere which lies north of 30° S, the planetary wind systems are considerably modified, as are those of pressure already described. These modifications include: (i) the seasonal
reversal of pressure and winds over the land-masses and neighbouring oceans, usually called the *monsoons*; (ii) the formation of '*air
circulation wheels*', moving around the 'cells' of both high and low
pressure formed by the interruption of the planetary pressure belts
(pp. 400–1); and (iii) the movement of air around more local
small-scale pressure systems, particularly in middle latitudes.

The Monsoons The word 'monsoon' is derived from an Arabic
word *mausim*, meaning season. It has already been seen that the
great differences in seasonal temperature over the continents produce
striking reversals in pressure conditions, particularly in central Asia.
In summer a low develops over north-western India, which is so
intense that it supersedes the Equatorial low. As a result an uninterrupted pressure gradient develops from the subtropical high in the
southern hemisphere across the Equator to north-western India.
The South-east Trades cross the Equator, and, as a result of the
influence of the rotation of the earth, become the South-west
Monsoon over peninsular India, and Burma, Indonesia, China and

Japan experience winds from a south-easterly direction, blowing towards the heated continent (fig. 189).

In winter the great cold of central Asia results in the formation of an intense high, from which cold, dry winds blow outward from the north-west over Japan and China, and from the north or north-east over south-eastern Asia (fig. 188). This outward flow is strengthened by great air-masses subsiding from higher altitudes. The implications of this seasonal reversal are enormous; in Chapter 18 six monsoonal climatic types are distinguished.

This simple and traditional explanation of the Asiatic monsoons, merely in terms of the marked seasonal differential heating and cooling of land and sea, is, however, far from adequate. In addition, the poleward shift of the hemispheric wind-belts in summer, both near the surface and in the upper troposphere, is involved. The Asiatic monsoon is considerably affected by the interrupting effect of the lofty and extensive Plateau of Tibet on the Upper Westerlies. In summer these shift north of the Plateau, emphasized by the northward migration of the Subtropical Jet (p. 408), thus permitting the northward surge of the South-west Monsoon.

The monsoonal effect is felt most markedly in Asia because of its great size, and because of its vast transverse relief barrier. To a less extent Australia, North America, and the coasts of western South America and West Africa, experience the same reversal.

'Air circulation wheels' It has been shown (p. 400) that the planetary pressure belts are in fact replaced by 'cells' of pressure, both highs and lows, fluctuating with the seasons. From the Azorean and Hawaiian highs, air flows outward in a clockwise direction, so moving on their equatorward sides into the Trade Wind belt, on their poleward sides into the Westerly Wind system. Similarly, from the Subtropical highs in the South Pacific and Atlantic air moves outward in an anti-clockwise direction.

AIR-MASSES

The concept of the air-mass is perhaps the most fundamental in modern meteorology. An air-mass consists of a large body of air with its temperature and moisture content horizontally uniform over a considerable distance. If in a vertical cross-section of such an air-mass the isobars and isotherms (that is, the horizontal surfaces of constant pressure and temperature) are parallel, the atmosphere is said to be *barotropic*, and tends to be stable (p. 431) in character.

Air-masses can be divided into groups according to their origin, trajectory and characteristics, both of temperature and humidity.

On a basis of temperature they are known as Polar or Tropical, on a basis of humidity as Maritime (having crossed the oceans, and so moist) or Continental (originating over the continents, and so dry). Their combinations allow four main categories to be distinguished— Polar Maritime (*Pm*), Polar Continental (*Pc*), Tropical Maritime (*Tm*) and Tropical Continental (*Tc*). The name Equatorial Maritime (*Em*) is sometimes applied to the very moist and warm air-masses originating within a few degrees of the Equator; these are a more pronounced form of *Tm* air.

The *Pc* air originates over the continental interiors, the northern tundra lands of North America and Asia, and the Greenland ice-caps, forming masses of cold, dry air. As these masses spread outward toward the coasts and over the North Atlantic, they are warmed from below, and their water-content increases. Thus *Pm* air is formed. From the sub-tropical Horse Latitudes high-pressure cells come the *Tc* air-masses, moving either equatorward or poleward. In the former case the air-mass is known as *TcK* (from the German *kalt* = cold), because it tends to be cooler, especially in winter, than the equatorial air with which it comes in contact. When it moves poleward, it is *TcW* (warm).

The major air-masses from tropical latitudes are of *Tm* quality. They too may move towards either cooler or warmer latitudes, and are denoted by *TmW* and *TmK* respectively, although, paradoxically, as they diverge from the same air-cell reservoirs, *TmK* air is warmed on its equatorward progress and is warmer than *TmW* air which is cooled as it progresses poleward.

A further useful distinction is the addition of (*M*), where the *TmW* air-mass movement is monsoonal in character, signified by the abbreviation *TmW(M)*. Polar continental air may also have a monsoonal effect (*PcM*)—the cold outblowing winter monsoons.

It may be desired to indicate that an air-mass has undergone considerable modification after it has left its source region. In that case a prefix *N* may be added; thus *NTm*. Some meteorologists go further still and add a suffix initial denoting the source region; thus *NP* for North Pacific, *SI* for South Indian Ocean, *A* for Australia, and so on. Others add a suffix *S* or *U* to indicate stable or unstable conditions. Special categories of air-mass may be denoted by *A* for those originating over the Arctic and *AA* over the Antarctic, and a very moist and unstable air-mass near the Equator is labelled *E*. There is a possible source of confusion between Polar and Arctic air-masses. The former term was coined merely to denote a temperature contrast with a Tropical air-mass. The term Arctic was introduced later, after the name Polar had become established, and not all authorities distinguish between them.

Frontal zones As air-masses of differing temperature and humidity move outward from the various major 'high-pressure cells', they come into conflict. A vertical section taken across such a zone of conflict will show that the surfaces of constant temperature and those of constant pressure will intersect at an angle; in other words, such an air-mass has become *baroclinic*, which will result in a condition of large-scale atmospheric instability (p. 430) in the frontal zones. These become the scene of the formation of numerous small pressure systems, a process known as *frontogenesis*.

The most striking frontal zones are in the North Pacific and North Atlantic, where the *Pm* and *Tm* air meet. These are the *Polar Fronts*, the *Pacific PF* and the *Atlantic PF* respectively, and also the *Mediterranean Front*, a winter only frontal zone, lying along the tongue of sea separating the *Tc* air-mass over the Sahara from the *Pc* air-mass over Europe. A *Secondary Polar Front* sometimes develops in the central North Pacific area, if the main Subtropical High Pressure 'cell' has divided into two. The Arctic Fronts (*AF*) lie around the northern margins of Eurasia and America.

The *Inter-Tropical Convergence Zone (ITCZ)* lies more or less around the Equator, where *Tm* air-masses converge. The term Inter-Tropical is better than Equatorial, as it has sometimes been called, since the movement of the Thermal Equator (p. 456) over the land-masses at the Solstices is quite appreciable. It is not a line of true air-mass discontinuity but a broad and complex zone, involving at least two air-mass boundaries, broadly parallel and sometimes distantly separated. Indeed, it has been shown that between the boundaries of the *TmK* air-masses over Asia and Australia respectively, lying some hundreds of kilometres apart, a distinct semi-permanent air-stream (referred to as the *Equatorial Westerlies*) can be discerned in the Indian Ocean, Indonesia and the eastern Pacific Ocean, and also over a small area in Ecuador and Colombia.

Other suggestions for names have been made, such as the 'Equatorial Air-stream Boundaries', and the term 'Equatorial Trough' is also sometimes used, since it implies no causal connection.

Where the ITCZ is only weakly defined, the air-masses may be virtually stagnant; this is in fact the low-pressure area of the Doldrums. Over the ocean the ITCZ is a zone of convergence where the *TmK* air-masses (in effect, the Trade Winds) have more or less similar characteristics of temperature and humidity. Over the landmasses it may be a front in the correct sense, where a *Tm* air-mass comes into contact with a markedly different *Tc* air-mass, forming inter-tropical frontal waves or weak depressions. This is clearly shown, for example, in West Africa. Here *Tc* air from the Sahara high-pressure cell and *Tm* air from the Gulf of Guinea come into contact; these air-streams are recognizable as the excessively dry Harmattan (p. 463) and the moist 'monsoonal' winds respectively.

The importance of air-masses cannot be over-estimated; modern climatology is based on a study of their origin, dispersal and resulting contacts (p. 455). Broadly, Equatorial climates are controlled by Equatorial and Tropical air-masses, mid-latitude climates by both Tropical and Polar air-masses, and high-latitude climates by both Polar and Arctic air-masses. Consider the following two examples. The North American continent is subject to different degrees, at different seasons, to (i) cold, dry stable *Pc* air from the north; (ii) cool, moist unstable *Pm* air from the north-west (Pacific) and north-east (Atlantic); (iii) warm, moist and fairly unstable *Tm* air from the south-west (Pacific); (iv) warm, moist and extremely unstable *Tm* air from the south-east (Gulf of Mexico and Atlantic); and (v) in summer only, hot, dry and unstable *Tc* air forming over the south-western desert-lands. The climates of North America are the sum-total of the effects and interactions of these air-masses. On a different scale, the British Isles and western Europe, essentially zones of transition, are subject to *Pm* air from the west, north-west and north, *Tm* air from the south-west, *Pc* air from the north-east and east, and *Tc* air from the south-east. Most critical is the zone between the *Pm* and *Tm* air to the west.

LOCAL PRESSURE AND WIND SYSTEMS IN MID-LATITUDES

In the North Atlantic and North Pacific, land and sea are inter-mingled in a complicated manner. Here, and particularly in the case of the North Atlantic, are busy oceans, constantly crossed by shipping and air-liners, and flanked by technically advanced countries. In this zone of fluctuating pressure systems and winds, and of changeable weather, meteorology has made most advances. The local pressure systems which so markedly affect these latitudes are described as for the northern hemisphere, though similar systems affect the southern hemisphere.

The main problem of forecasting weather in middle latitudes is concerned with the prediction of future developments and changes in the local distribution of pressure, and therefore of the air-masses concerned, of differing temperature and humidity, and moving in various directions. The synoptic chart, the chief tool of the fore-caster, reveals that the isobars take the form of closed curves, sometimes with prominent bulges. Just as in world pressure distribution the basic distinction is between areas of relatively high and low pressure, so in middle latitudes a system of closed isobars may enclose either high pressure, diminishing outward from the centre (an *anticyclone*), or one of low pressure increasing outward (a *depression* or *low*).

These are the fundamental types of pressure distribution, although varieties can be recognized. A *wedge of high pressure* lies between two neighbouring depressions, and a *ridge of high pressure* is a broader region between two more widely separated depressions; these highs tend to bring very fine, but short-lived weather, the 'borrowed day' that is 'too good to last'.

A *secondary depression* is a small unit on the outskirts of a main depression. It may be just a bulge in the isobars or it may have a separate centre (with closed isobars) of its own, and it may be more intense (with lower pressure) than the primary depression. Most secondaries seem to travel around the primary one, in the northern hemisphere in an anti-clockwise direction. A *trough* is a narrow elongated area of low pressure between two areas of higher pressure, bringing squally winds, with heavy rain or hail.

A *col* lies between either two depressions or two anticyclones, but is not definite enough to be either a trough or a wedge. The weather is extremely variable and a col is often a difficult problem for a forecaster.

Sometimes the chart reveals a pattern of *straight isobars*, when the resulting weather depends on their orientation; if they run north–south, with decreasing pressure eastward, the air-flow will be from a northerly direction, bringing cold weather and probably showers.

The weather in middle latitudes is dominated by an endless procession of these pressure systems moving generally eastward. Each different system brings a specific sequence of weather, which the forecaster tries to interpret by estimating how each successive system will affect the region with which he is concerned.

In a depression it is possible to distinguish lines of separation between the air-masses of which it is comprised. These are *fronts*, and their definition and recognition is a vital factor for a forecaster, since their passage is marked by distinctive changes of weather.

On the old style of weather-map, isobars, wind-arrows (indicating direction and force), and symbols signifying the weather and amount of cloud were shown at each station, and the recorded temperatures added. On the modern type of weather-map, in addition, the various fronts are marked by distinctive lines, against which the isobars often change direction abruptly (fig. 190).

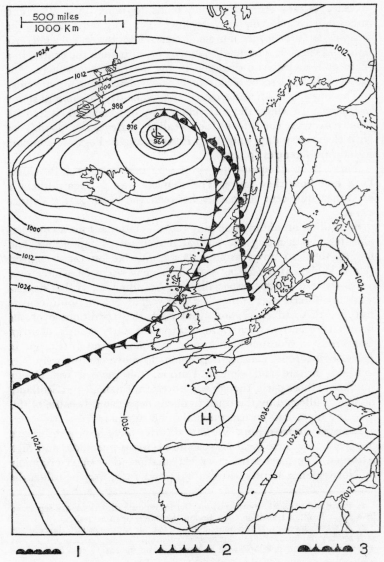

Fig. 190 *Pressure systems in the north-eastern Atlantic*

Redrawn from *The Daily Weather Report* for 6 February, 1952, by permission of H.M. Stationery Office. Details of winds, weather conditions, etc., have been omitted. Note the low-pressure system north-east of Iceland and the high over the Bay of Biscay. **1** Warm front; **2** Cold front; **3** Warm occlusion.

Anticyclones An anticyclone tends to move in a slow and somewhat irregular manner, its slowness usually meaning that the weather is fairly settled for some time. The associated winds are light and variable, and tend to move outwards from the centre, in a more or less clockwise direction in the northern hemisphere. The anticyclone is the scene of extremely slowly descending air currents.

The weather in an anticyclone tends to be dry, warm and sunny in summer. In winter the anticyclone may be associated either with cloudy overcast conditions (the so-called 'anticyclonic gloom'), or with clear, crisp days and sharp frosts by night. Fogs are common in industrial centres (p. 437).

There is no adequate theory of why mid-latitude anticyclones form and how they move. The simplest way to regard them is as rather quiescent air-masses between the more active depressions.

Sometimes an anticyclone may remain stationary for a considerable time, preventing the normal movement of depressions, and forcing them to travel around its margins on less usual courses; this is known as a *blocking high*. This may occur in winter over north-western Europe, especially over Scandinavia, so that depressions are forced over either the Norwegian Sea or south-central Europe; this can cause prolonged cold, dry conditions over the British Isles and western Europe, as in the early months of 1947. Conversely, a blocking high formed over north-western Europe in July–August 1968, causing depressions to move continuously over southern Britain and western Europe; this gave a wet cloudy summer to those parts, though north-western England and Scotland, no longer affected by the usual frequent depressions because of this blocking high, had an exceptionally fine summer.

Frontal depressions Various theories have been put forward to account for these local low-pressure systems in middle and high latitudes. The most satisfactory explanation, developed originally by Norwegian scientists between 1914 and 1918, may be called the

---

1. Polar and tropical air-streams, moving in opposite directions, are separated by a Polar Front, a transition zone, 80 km (50 miles) or so across.

2. A 'wave' develops, as the 'tongue' of tropical air protrudes into the cold air, with the beginnings of an anti-clockwise circulation.

3. A warm sector develops, with a warm front on the east, a cold front on the west.

4. The cold front has overtaken part of the warm front, and part of the warm sector has been lifted off the ground. This occluded front then dissolves, and the depression becomes a mass of homogeneous air, thus 'filling-up' and disappearing.

5. Section through a warm front, showing the sequence of cloud (p. 441).

6. Section through a cold front.   7. Section through a cold occlusion.

8. Section through a warm occlusion.

wave theory. This assumes that two adjacent masses, one of cold (polar), the other of warm (tropical) air, are separated by a frontal zone, sometimes as much as 80 km (50 miles) across. The cold air is flowing south-westward, the warm air north-eastward, more or less parallel (fig. 191). 'Waves' are formed as the warm air-mass bulges into the cold mass, forming a salient, at the most northerly extension of which a low-pressure centre develops. When the pressure at the centre is markedly lower than that at the margin, it is said to be a 'deep depression'; if the difference is relatively small, it is a 'shallow depression'.

The bulge or 'bay' of warm air gradually becomes a much more defined 'tongue', the *warm sector* of the depression. Its eastward edge, advancing in the direction of general movement (that is, eastward or north-eastward), is the *warm front*, along which the warm air is rising above the cold air which it is overtaking. As the depression develops, the cold air undercuts the salient of warm air from the rear, forcing the warm air upward as a *cold front*. It is now realized that conditions at a cold front vary enormously and are much more complex than was thought. If the warm air-mass is for the most part rising over the cold wedge, it is known as an *anafront*, with heavy rain and appreciable temperature change. If the warm air is descending over the cold wedge, it is a *katafront*, with a slight backing of the wind followed by clearing and improvement in the weather.

As the depression advances, the passage of each front is associated with a distinctive sequence of weather—of cloud (fig. 191), rain (p. 448) and temperature change. Ahead of the warm front a broad belt of continuous rain falls from a heavily overcast sky, while the wind backs (changes in an anti-clockwise direction) before the front arrives, and then veers (changes in a clockwise direction) as it arrives. At the cold front there is a marked drop in temperature, rain falls in heavy showers sometimes accompanied by thunder, and the wind freshens from a north or north-westerly direction. Sometimes a cold front may be associated with exceptionally stormy conditions, known as a *line-squall*, moving some distance ahead of the front, a phenomenon especially marked in the Mississippi valley in the U.S.A. This is accompanied by a long cloud-roll, strong squalls of wind sometimes of gale force, occasionally thunderstorms and a downpour of hail.

Gradually the warm front is overtaken by the cold front, which ultimately lifts the warm sector completely off the surface; this is known as an *occlusion*, the centre line of which is an *occluded front*. The cold air in the rear of the depression has now come up against

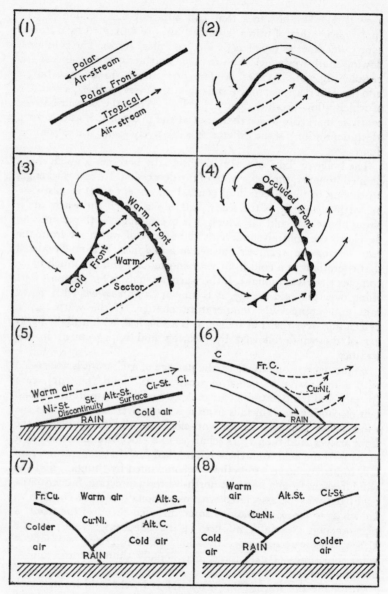

Fig. 191 *The life-history of a depression*

These diagrams are based on the classic originals by the Norwegian meteorologist J. Bjerknes.

the cold air against which the warm front was originally formed. If the overtaking cold air is colder than the air in front, it is a *cold occlusion*; if it is not as cold, it is a *warm occlusion* (fig. 191). After occlusion, the depression tends to 'fill up', that is, die away; most depressions arrive over the British Isles in an occluded state.

Such is the life-history of a depression. As it passes eastward over the British Isles, the weather at any place depends on where this is situated relative to the various fronts. The analysis of large numbers of synoptic charts shows that depressions seem to move more or less parallel to the isobars in the warm sector, and this 'rule of thumb' is one which helps the meteorologist in predicting the future path of the depression.

In any depression there may be more than two fronts. Thus within the cold sector of a depression a temperature discontinuity may form between Polar air which has been incorporated in the depression and accordingly modified, and fresh inbursts of colder Polar air. Such a *secondary front* may complicate greatly the sequence of weather experienced in an 'ideal' depression, and also the problems of a forecaster. Canadian meteorologists believe that depressions over North America frequently have a dual frontal structure involving three or four air-masses, and it seems that frontal waves generally develop in 'families'.

Thermal depressions Intense conditions of low pressure may be caused by the concentrated day-time heating of masses of air in summer. On a large scale this is the cause of the low-pressure conditions which provide a focus for indrawn monsoonal winds (p. 409). On a small scale, thermal depressions form over the hot deserts, resulting in hot, dry 'convection winds' (p. 424). In middle latitudes they may be associated with thunderstorms; small thermal depressions may move northward across the English Channel, bringing short-lived but heavy rain to localities in southern England (p. 449).

Lee depressions A special type of depression seems to form in the lee of a mountain range, probably the result of a dynamically created eddy as an air-stream crosses it. These depressions occur in the western Mediterranean basin to the south of the Pyrenees-Cévennes-Alpine barrier, along the eastern flanks of the Canadian Rockies, and to the east of the New Zealand Alps in the South Island. This phenomenon is particularly important in the Mediterranean Sea; indeed, it has been estimated that 60 per cent of the depressions affecting that area originate in this way. A *PmK* air-stream, associated with a cold front, pours southward through the Rhône valley (in its most pronounced form it is known as the *mistral*

(p. 422)), and bursts into the Mediterranean. The general fall of pressure with the cold front is intensified to form a low-pressure system over the Gulf of Lions, which may then move off southeastward through the Mediterranean.

Tornadoes More intense and localized low-pressure whirls may occur over both land and sea. The land storms, or tornadoes, are common in the Mississippi Basin, averaging about 600 a year. The winds are usually anti-clockwise in the northern hemisphere and of immense velocity; it has been estimated that they may attain 800 km (500 miles) per hour, but no measured records can possibly be taken, for they are the most destructive storms in the world.

They are associated with troughs of low pressure, where cold air from the north and warm damp air from the Gulf of Mexico come into contact along a front. Local heating also causes an uprush of powerful convection currents, like a vortex. They are short-lived, rarely existing for more than an hour, and are usually only two or three hundred metres across, or even less, moving across country at about 65 km ph (40 mph). Their destructive effects are fortunately limited in area, but the centre of the whirl may cause immense chaos. Not only are buildings destroyed by the terrific winds, but the excessively low pressure at the centre causes buildings to collapse outward because pressure within them is so much greater than that outside. Widespread devastation and loss of life was caused by a series of tornadoes in the Midwestern States of the U.S.A. in April 1965; here they are popularly called 'twisters'.

Occasional tornadoes are experienced in higher latitudes, and there are a few records of their occurrence in Britain. On 21 May 1950, for example, a small tornado which had formed on a line-squall overturned a double-decker bus at Ely, and on 8 December 1954 another damaged Gunnersbury tube-station. The worst ever in Britain was probably in 1703, when 8000 people were killed. These 'temperate tornadoes' are fortunately rare.

Water-spouts An intense pressure system similar to a tornado, but over the sea, results in a water-spout. From the low-lying base of a cumulonimbus cloud, a whirling cone of dark grey cloud projects downward, gradually elongating until it touches the surface of the water. The rapidly whirling water-drops are derived both from condensation because lowering of pressure in the centre of the vortex causes cooling, and also from water picked up from the intensely agitated surface of the sea.

TROPICAL LOW-PRESSURE SYSTEMS

So far attention has been concentrated on low-pressure systems in middle and high latitudes. In the Tropics they are also experienced, and they may be divided into two groups: weak and intense.

Weak tropical low-pressure systems These weak shallow lows are common near the Equator and in the more humid parts of tropical latitudes, especially in the area of the Inter-Tropical Convergence Zone. Their movement is slow and vague, though generally westward, and winds are generally light, but they are responsible for long spells of rainfall, as a result of their humid unstable characteristics.

There are several theories concerning their formation, none wholly satisfactory; one envisages their formation on the boundary between two air-streams moving in different directions, another postulates the formation of the centre at the meeting-point of three air-streams, a third the convergence of air to the rear of a wave-trough in the North-east or South-east Trades. Certainly there must be conditions of high temperature and humidity.

Intense tropical low-pressure systems Small intense low-pressure systems, with a diameter of between 80 and 400 km (50–250 miles), originate in warm oceans within tropical latitudes, mostly between 6° and 20° latitude on either side of the Equator. These appear to develop from a local heat source (or 'warm core'), causing the formation of a wave disturbance. This process, known as *cyclogenesis*, is indicated by the development of a whorl of towering cumulus clouds (p. 441). The atmospheric conditions leading to the formation of such a vortex are not yet fully understood, but satellites can now be used to detect incipient storms by their developing cloud patterns (plate 99). They occur all over the oceans except in the South Atlantic, and are known as *hurricanes* in the West Indies and the Gulf of Mexico, as *typhoons* in the China Seas, as *cyclones* in the Indian Ocean and as *willy-willies* off the coast of north-western Australia. The barometric gradient in these lows is extremely steep; indeed, as the storm approaches the pressure may drop by as much as 40 millibars in a few hours, and 950 mb. is commonly recorded. The centre of the storm (or the 'eye', a small region about 20 km (12 miles) across) is an area of calm or at most of light, variable winds, but round it may whirl winds of terrific force, sometimes as much as 270 km (170 miles) per hour. Torrential rain, associated with thunderstorms, also occurs. These tropical storms are a menace to shipping, although when warned by radio a ship may sometimes be able to steam out of the probable path, since they move at a rate of only about 16–24 km ph (10–15 mph). All hurricanes are anxiously tracked by America's National Hurricane Center at Miami, Florida, and their causes and effects are closely studied at the National Hurricane Research Laboratory. Probably the worst hurricane in living memory was that code-named *Camille*, which swept across the Gulf states in August 1969, causing immense destruction and loss of life.

Occasionally hurricanes develop in high latitudes. A recent example occurred on the night of 14–15 January 1968, when a hurricane (code-name *Low Q*) crossed Scotland, leaving a trail of damage across Glasgow, killing 20 people and damaging 30,000 houses. Hurricanes from the Caribbean area may penetrate as far north as New York, occasionally into New England and Canada.

SPECIAL WINDS

Named winds In many parts of the world winds of regular or periodic occurrence are sufficiently distinctive, as a result of their characteristics of temperature and humidity, to be given proper names. They often produce important results on the climate of the areas they affect.

Depression winds The position and relief of certain regions gives distinctive characters to winds associated with depressions. A moving depression involves air-masses originating both on its poleward and equatorward side, and therefore both cold and warm winds may result. A depression moving eastward through the Mediterranean Sea often results in warm, dry winds from the Sahara in the warm sector (fig. 192). The *sirocco* in Italy and North Africa, the *leveche* in Spain, and the *khamsin* in Egypt are examples. The *gibli* in Tunisia is like the hot blast which comes from an opened oven-door. They are often hot, dusty and excessively dry, and may do much damage to the blossoms of the vine, leaving them as if frost-nipped. Sometimes they are humid, as a result of crossing the sea, and the sticky heat is even more unpleasant. In other parts of the world the *brickfielders* of Victoria, dusty winds with temperatures exceeding 38° C (100° F), and the humid *zonda* of the Argentine, are comparable in character.

By contrast, the polar air-masses involved in depressions may cause strong, cold winds; 'bursting' into normally milder areas, they are sometimes called 'polar outbreaks'. In the Mediterranean the *bora* and the *mistral* blow from the 'winter high' over Europe. These bitterly cold winds are felt especially when they blow down a 'funnel' between uplands—the mistral down the Rhône valley, the bora south-westward into the Adriatic. Similar to these are the *southerly-*

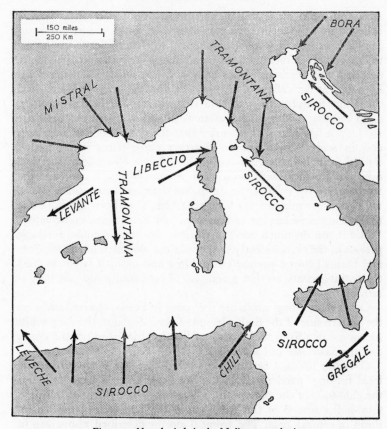

Fig. 192 *Named winds in the Mediterranean basin*

Winds with definite characteristics of temperature and humidity, associated with the passage of depressions, occur with such regularity that they are given special names. The *tramontana* is cold and raw, the *gregale* blustery and accompanied by showers, the *sirocco* hot (sometimes dry, as in northern Sicily, or oppressively humid, as in southern Sicily), the *levante* mild and humid, the *mistral* and *bora* cold and blustery, the *libeccio* boisterous and invigorating, and the *leveche* and the *chili* exceptionally hot and dry.

burster in New South Wales, the *pampero* in the Argentine, the *friagem* or *surazo* in Brazil, the *norther* in Texas, and the *norte* and *papagayo* in Mexico.

Descending winds When an air current crosses a mountain range it is warmed adiabatically (p. 429); the best-known of these winds are the *foehn* in the Alps and the *chinook* in the Rockies. The foehn blows when a depression lies to the north of the main ranges of the

Alps and moist air is drawn northward over the crest. It is now believed that the foehn is in part caused by air-mass turbulence on the leeward side of a mountain range, so that eddies force the air to descend. It is then warmed by adiabatic compression. The foehn descends the northern slopes as a dry, hot wind, following relief lines such as the valleys of the upper Rhine, the Aar, the Reuss, and the Rhône from Martigny to Lake Geneva. It melts the snow rapidly, often causing widespread avalanches, and temperature can rise 10° C in a few hours. It is so dry that the danger of fire is serious, and in some Alpine villages smoking out-of-doors is prohibited during foehn weather. Similarly, the chinook blows from the west over the Rockies, with even more drastic effects; it can raise the temperature by 15° C or more in less than an hour, and a rise of 15° C in three minutes has been recorded. The effects are the same as those of the foehn; the Indian word 'chinook' means 'snow-eater'. While it can do much good in clearing the land of snow, sporadic periods of the chinook early in the year can do much harm, for trees and plants begin a premature budding and animals begin to shed their winter coats, so that a renewal of cold conditions can be disastrous.

Other descending winds are the *samun* in Persia, the *nor'-wester* on the eastern side of the South Island of New Zealand, the *Berg* winds blowing down from the plateau of South Africa towards the coast, and the *Santa Ana* in California.

Convection winds Certain desert winds are the result of intense local heating, particularly in the hot deserts. On a local scale are the *dust-devils* of the Sahara, on a large scale the swirling, burning, sand-laden *simoom* of the Sahara and the *karaburan* of the Tarim Basin in central Asia. The tropical storms described above are in part an intensely exaggerated form of these winds.

Land and sea breezes Land and sea breezes are local winds on a diurnal rather than on a seasonal basis. They result when differential heating takes place within a relatively short distance, a condition which occurs most commonly near the coast. During a summer day heating occurs over the land, forming local low-pressure areas into which moves cool air from the sea, although it does not penetrate far inland. This sets in during the afternoon, when heating has reached its maximum. At night the land cools more rapidly than the sea, forming slightly higher pressure. The cooled heavier air sinks down-hill, and moves out to sea as a land breeze during the night.

These winds are most marked in normally calm areas, for on the coasts of Trade-wind islands, for example, their effect is masked by the dominant winds. On the islands and along the coasts near the Equator, however, the sea breeze brings a welcome relief from the stagnant humid heat. In some islands these winds are so regular that fishing-boats go out at night with the land breeze and return the next afternoon with the sea breeze.

Mountain and valley winds Mountain winds (sometimes called *anabatic* winds) blow up a valley during the day, while valley winds (or *katabatic* winds) blow down a valley during the night, where the general circulation of winds is not strong enough to obscure these local influences.

The down-valley winds are fairly easy to explain. The upper slopes are at night chilled rapidly by radiation, and the cool, dense air flows down-hill by gravity, following the valley since this is an obvious line of least resistance. A very pronounced form of this wind is where cold air blows off an ice-cap, over which chilling is intense. This can be felt in Greenland and the Arctic islands, and in Ecuador the *nevados* blows down from the Andean snowfields into the high valleys.

The mountain or up-valley winds are less easy to understand. They blow during the daytime, and are the result of differential heating. A partial explanation is that the air near the mountain slopes is heated by conduction to a greater extent than air at the same level above the valley floors. This causes convectional rising of air from above the slopes, hence air moves up from the valleys to take its place.

The effect of relief barriers In recent years, much attention has been paid to air-flow across transverse relief barriers. One obvious result is the production of relief rainfall (p. 448) on the windward side of the barrier. More complex are the aerodynamic results on the leeward side.

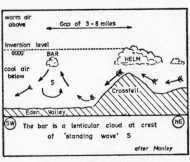

Fig. 193 The Helm Wind

When a light wind crosses the barrier, the effect is limited to simple laminar streaming. With a stronger wind, a standing eddy forms, and with still stronger winds *lee-waves* are formed which may develop into complicated turbulence, what is sometimes termed *rotor-streaming*. This may produce striking cloud forms. For example, an easterly or north-easterly wind may blow strongly and gustily down the Crossfell escarpment in the northern Pennines into the Eden valley to the west. This results in the lenticular *helm-cloud* resting on the ridge, a type of banner-cloud, while a few miles to the west is a parallel line of cloud known as the *helm-bar*, at the crest of the standing lee-wave (fig. 193).

CHAPTER 17

Humidity and Precipitation

WATER-VAPOUR IN THE ATMOSPHERE

ONE of the most important ingredients in the atmosphere is water-vapour, that is, the invisible, gaseous form of water. At least half of the total vapour in the atmosphere occurs in the lowest 2300 m (7500 ft). Its importance in meteorology lies in the fact that most of the features of the weather are closely related to its ever-varying presence in the atmosphere, from place to place and from season to season. Water is turned into vapour by evaporation; vapour returns to the liquid form by condensation. Hence the hydrological cycle (p. 110) begins with evaporation to form water-vapour; something of the order of 400,000 cubic km (100,000 cubic miles) of water are evaporated into the atmosphere each year from the earth's surface, to be returned as precipitation in the hydrological cycle. Next condensation forms clouds and then water is precipitated on to the surface of the earth, from which a large part returns to the sea by way of rivers.

The term which is applied to the presence of moisture in the atmosphere is *humidity*; this may be expressed either absolutely or relatively.

Absolute humidity The absolute humidity is the actual amount of water-vapour present in a certain quantity of air, expressed in grammes per cu. m or grains per cu. ft. A mass of air, of a given temperature and pressure, can hold vapour up to a certain limited amount; when this limit is reached (that is, when as many molecules of water enter the air as are leaving it), the air is *saturated*, that is, the *saturation vapour pressure* has been reached. Saturated air at 10° C contains 9·41 grammes of vapour per cu. m, at 20° C it contains 17·117 grammes, and at 30° C it contains 30·036 grammes. If air is not saturated, the mass of vapour required to bring it to the point of saturation is called the *saturation vapour-pressure deficit*.

The absolute humidity over land areas tends to be highest near the Equator throughout the year, and to be lowest in the great high-pressure systems covering central Asia in winter and in the Antarctic. Elsewhere it is extremely variable. Paradoxically, it is often extremely high in the hot deserts; much of the trying quality of Aden or the Persian Gulf in August is the result of the 'sticky heat', the combination of high humidity and high temperature.

Vapour-pressure Water-vapour exerts a definite pressure. The maximum vapour-pressure at any temperature occurs when the air is saturated; tables are available giving the vapour-pressure of saturated air at different temperatures. The average vapour-pressure in England on a summer afternoon amounts to about 15 millibars (approximately equivalent to 11·4 mm, 0·45 in of mercury), at the Equator about 30 millibars. Absolute humidity can, therefore, be expressed either in terms of the mass of water-vapour per cubic unit or in terms of the pressure it exerts.

Other practical expressions of moisture content are *specific humidity* (the ratio of the weight of water-vapour present to the total weight of the air with its moisture content, usually expressed as grammes per kilogram) and the *mixing ratio* (the ratio of the weight of water-vapour present to the weight of the air less its moisture content).

Relative humidity There is no change in absolute humidity unless vapour is actually added to or removed from a body of air. If the amount of vapour present is expressed as a percentage of the total amount that would be present were the air saturated, the relative humidity is obtained. Saturated air at 20° C contains 17·117 grammes of vapour per cu. m; if, for example, a mass of air at 20° C contained only 8·262 grammes, the relative humidity would be (8·262 × 100) ÷ 17·117, approximately 48 per cent.

Relative humidity varies not only with the absolute humidity, but with the temperature of the air, for as the temperature rises so the relative humidity falls. If a mass of air, saturated at 4° C (that is, with relative humidity = 100 per cent), is warmed, at 10° C the relative humidity falls to 71 per cent, at 15° C to 51 per cent, and at 32° C to 19 per cent. Conversely, when a mass of unsaturated air is cooled, the relative humidity rises until it reaches the point of saturation at 100 per cent. Beyond this, any further cooling normally causes condensation of the excess vapour which the air is no longer able to hold, in the form of minute drops of water. This critical temperature is known as the *dew-point*.

Measurement of relative humidity The most commonly used instrument for determining relative humidity is a 'wet- and dry-bulb thermometer'. This consists of a pair of mercury thermometers, mounted side by side on a stand; round one of the bulbs is tied a cotton or muslin bag, kept moist by a wick leading down into water. If the air is saturated, both the thermometers show the same reading. If the air is not saturated, evaporation will take place from the moist cloth and so cool the wet-bulb, since latent heat (that is, the heat expended or consumed in doing work by changing the physical state of a body) is used up in the process of vaporization. The difference in the temperatures recorded by the two thermometers is noted, and the relative humidity is obtained from tables which are available. Thus, if the dry-bulb thermometer reads 21° C and the wet-bulb 15° C, the relative humidity can be seen from tables to be 54 per cent.

To ensure that maximum evaporation is taking place from the wet-bulb, various refinements have been devised. One method is the mounting of the thermometers in a sling or in a kind of rattle, so that they can be swung around; these are *whirling psychrometers*. The *Assman psychrometer* incorporates a small electric fan.

Continuous records of relative humidity, although not particularly accurate, are made by a *hygrometer*. This consists of strands of human hair, which lengthen and shorten according to the humidity; these minute changes are amplified and transferred to an inked pen, which traces a record (a *hygrograph*) upon a chart fixed to a rotating drum.

THE HUMIDITY OF AIR-MASSES

An important effect of the presence of water-vapour in a body of air is the reduction of its density. A quantity of water-vapour is lighter than the same volume of dry air, in the proportion of about 5 to 8. When dry air takes up water-vapour by means of evaporation, this is not just a net addition, but it replaces an equivalent volume of air. The important result follows that moist air is less dense, that is, lighter, than an equal volume of dry air at the same temperature and pressure.

One other physical principle must be stressed at this point. If air is compressed, not only does its density change, but its temperature also; if a bicycle tyre is blown up, the valve becomes uncomfortably warm because compression causes dynamic heating. Expansion, on the other hand, results in dynamic cooling. The usual way in which an air-mass can expand is when it ascends bodily into the upper atmosphere, since there will be a smaller amount of air above it and therefore lower pressure. When an air-mass undergoes a change of temperature without any heat being lost or gained by the air-mass from outside, it experiences *adiabatic warming* (on compression) or *adiabatic cooling* (on expansion). This is in contrast to *diabatic* change, when a temperature change at the earth's surface involves the mixing of air, with a definite gain or loss of heat from its surroundings.

Lapse-rates The average decrease of temperature in still air with altitude (the *environmental lapse-rate* or *vertical temperature* *gradient*) is, though extremely variable, about 0·6° C per 100 metres (3·3° F per thousand feet) (p. 390).

When an unsaturated air-mass ascends vertically it expands, and therefore cools adiabatically. This loss of temperature is known as the *dry adiabatic lapse-rate*, a meteorological constant, which is a temperature decrease of 1° C for every 100 m. of ascent (5·4° F for every thousand feet).

If a saturated air-mass ascends vertically, it also expands and cools, and as it is saturated to start with, some of its water-vapour is immediately condensed. This means that a certain amount of latent heat is liberated, which reduces the rate at which the ascending air-mass cools. This *saturated* (or *wet*) *adiabatic lapse-rate* can lie anywhere between 0·3° C and 0·9° C per 100 m. of ascent, according to the amount of water vapour present and to the temperature of the air. Thus an air-mass of about 30° C may contain so much water-vapour, and therefore release so much latent heat, that the saturated adiabatic lapse-rate is only 0·4° C per 100 m., while in a very cold air-mass or at high altitudes there may be so little water-vapour that the saturated adiabatic lapse-rate will not differ materially from the dry adiabatic lapse-rate. The following table gives conditions for various lapse-rates.

Metres	(*In degrees Centigrade*)		
1200	12·8	8·0	15·2
900	14·6	11·0	16·4
600	16·4	14·0	17·6
300	18·2	17·0	18·8
Sea-level	20·0	20·0	20·0
	Environmental atmospheric conditions in still air (average)	Column of rising unsaturated air	Column of rising saturated air (at 0·4° C per 100 m)

Fig. 194 *Lapse-rates*

Instability 'Pockets' of air are forced to rise in several ways. One is when local heating of the earth's surface takes place; the overlying air-mass is warmed by conduction (p. 382), and a vertical convection current is set up. Another cause is when a wind blows against a mountain-side, mechanically forcing the 'pocket' up the slope. A third involves the rising of one mass of air above another at a frontal surface (p. 417). Finally, vortical ascent may take place, as in a tropical storm.

When a 'pocket' of air is warmer, and therefore lighter, than its surroundings, it will rise bodily; such a mass is said to be *absolutely unstable*, or in *unstable equilibrium*. Instability is much more common in the case of saturated than of unsaturated air, since it cools much less rapidly, and so remains warmer than its surroundings. A warm, damp air-mass may therefore rise to great heights, and cause very unstable atmospheric conditions, building up great clouds, and possibly causing heavy rainfall, hail and thunderstorms (fig. 195). When ultimately the air-mass reaches a height at which it has the same temperature as the surrounding air, vertical ascent ceases; it is now in *neutral* (or *indifferent*) *equilibrium*.

If a mass of dry, or unsaturated, air rises as a wind up a mountain slope, it cools at the dry adiabatic lapse-rate until it reaches its dew-point. Then condensation begins, and it now cools at the saturated adiabatic lapse-rate. This may mean that even though the initial cause of the uplift was mechanical, the lower rate of cooling results in the air-mass remaining relatively warmer than its surroundings, and so it will continue ascending of its own accord. This state of affairs, known as *conditional instability*, is a common cause of storms and heavy rain; the term is derived from the fact that instability is conditional upon the presence of sufficient water-vapour. It is not necessary to have a forced orographic ascent for conditional instability to develop, although this is the commonest cause; it may develop whenever the general environmental lapse-rate lies between the dry and saturated adiabatic rates.

If, however, the mass of dry air which is rising in the form of a wind has a lapse-rate greater than that of the surrounding air, and does not reach dew-point, it will in due course become cooler than its surroundings. As soon as the mechanical force ceases (that is, when the wind drops) it tends to sink back to lower levels, since it is heavier, that is, denser, than its surroundings. This is known as *stable equilibrium* or simply as *stability*. A mass of air, which becomes conditionally unstable when forced to rise over a relief barrier or over a cold air-mass at a front, is referred to as being in a state of *potential instability*.

Stability and instability have been discussed in some detail, since the vertical movement of air-masses of differing temperature and humidity is intimately associated with atmospheric disturbances. This is, moreover, the most potent cause of precipitation.

EVAPORATION

Evaporation is the process by which water is changed from the liquid to the gaseous form, a process of molecular transfer, and so incorporated into the atmosphere. The rate and amount of evaporation in many ways is as important to the climatologist as is the amount of rainfall. A high evaporation rate reduces the effectiveness of the rainfall; in northern Ceylon, for example, the 'Dry Zone' receives about 130 cm (50 in) of rain per annum, an amount which would be regarded as distinctly heavy in middle latitudes. The *effectiveness of precipitation* is taken to be the total rainfall minus the total possible evaporation.

Evaporation is measured in a variety of ways, though none of them is entirely satisfactory. The *Piche evaporimeter* is commonly used, in which water in a tube is allowed to evaporate from a piece

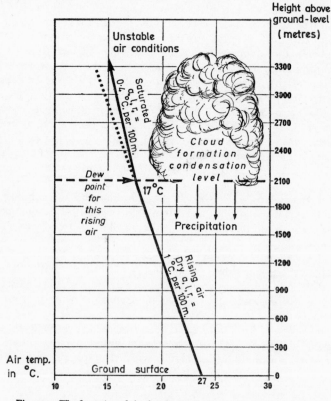

Fig. 195 *The formation of cloud and precipitation in an ascending air-mass*

of porous paper, and the loss in a certain time is measured on a graduated scale along the tube. Some evaporation statistics are calculated from measurements taken of the water-level in large open tanks. This is more correctly known as *potential evaporation*, since it depends on the availability of a constant supply of water.

The *rate of evaporation* is mainly a function of relative humidity, of absorbed radiation and of air movement, but it depends, too, on the nature of the surface. The loss from bare soil is very rapid, but where the surface is covered with a loose tilth (as in dry-farming practice) it is much reduced. A plant cover may protect the ground itself against direct evaporation by the shade it provides, but the loss by transpiration (p. 514), plus direct evaporation (a joint effect sometimes called *evapotranspiration*), may be very great. Experiments were carried out by the Fylde Water Board at the Stocks Reservoir, Slaidburn, Yorkshire, where the close planting of spruce has taken place. A total of 98·4 cm (38·75 in) of rain fell there in one year, but as a result of the interception of rainfall by the trees and of evaporation from the foliage, only 61 cm (24 in) reached the ground, and of that only 27·3 cm (10·75 in) became available for water supply. In other terms, on 600 hectares (1500 acres) of catchment the planting of spruce reduced the available water supply by 4·5 million litres (1 million gallons) a day. It must be admitted, however, that not all authorities accept the validity of this experiment and the deductions from it.

The highest potential evaporation records come from the Trade-wind deserts, where the effects of high temperatures, strong fresh winds and a bare sandy or rocky surface are combined. Atbara and Khartoum in the Sudan have mean annual potential evaporation figures of 625 cm (246 in) and 541 cm (213 in) respectively, Helwan in Egypt of 239 cm (94 in). There are considerable seasonal differences; the mean evaporation at Helwan in June is 33 cm (13 in), but in January and December it is only 8·9 cm (3·5 in). Vast quantities of water are evaporated from the oceans in subtropical latitudes, are injected into the atmosphere, condense as clouds, and move Equatorwards.

Evaporation rates are low in the equatorial belt, where they only amount to 5–8 cm (2–3 in) a month. They are low, too, in cool middle latitudes; the mean annual rate of evaporation for London is only 46 cm (18 in) (fig. 196).

Causes of condensation Condensation has been defined as the formation of water-droplets when air has been cooled to and

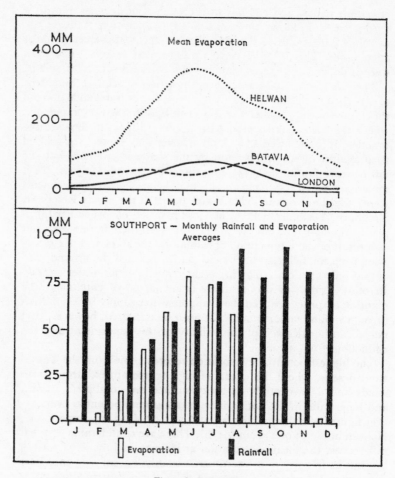

Fig. 196 *Evaporation*

beyond its dew-point. There are several ways in which air may be so cooled, including direct radiation from the surface of the earth during a clear night (p. 382), the horizontal movement of warm air over a cold surface, the mixing along the margins of two air currents of markedly different temperature, the movement of air from warmer to cooler latitudes, and, by far the most important, by ascent. Each form of cooling may produce condensation of different degrees and with different results.

434

It has been shown by experiment that completely pure air can be cooled, under laboratory conditions, to temperatures well below the dew-point without condensation taking place. It is necessary for some kind of nuclei to be present on which the droplets can form. These nuclei include particles of dust and smoke, salt from the ocean, pollen, and even negative ions (atoms carrying a negative electrical charge produced by the passage of radiation through the atmosphere).

The condensed droplets are only about 0·05 mm in diameter when they form, and are so minute that they float in the air as fog or clouds. Larger drops form on leaves and grass as dew, or if the temperature is below freezing-point as hoar-frost. When droplets coalesce in the air to a certain critical size, they may fall to earth as one or other of the forms of precipitation (rain, snow, hail or sleet). The formation of raindrops is, however, much more complex than this simple account would imply. Indeed, it is not known how droplets actually coalesce, though various theories propose electrical attraction, supercooling followed by freezing into ice particles (which form nuclei for condensation), and turbulence causing coalescence by collision.

It is sometimes possible for drops to form and then continue to exist in the liquid state, even when the temperature is well below freezing, as long as the air is undisturbed. This phenomenon is known as *supercooling*, and has a number of important meteorological results. One practical effect of great importance is the accretion of ice on aircraft. If an aircraft passes into a cloud consisting of large drops of supercooled water, with the air temperature at or below freezing-point, a considerable thickness of clear ice may form as each drop freezes on impact with the leading edges of the wings. A similar phenomenon is known as *glazed frost* or as 'silver thaw' in America. The super-cooled water freezes on branches, telegraph wires (which may be brought down by the weight), and road surfaces (which become treacherous). Rime (p. 439) can also form on aircraft. Incidentally, another cause of icing on an aircraft is when it is flying in a layer of air of temperature below freezing-point, but not in a cloud; if heavy rain falls from a cloud above, such as might occur in a warm front over a cold air-mass, it will immediately freeze into clear ice on the aircraft. The provision of icing warnings has become an important function of forecasting at the meteorological stations attached to airports and Service airfields.

Dew Dew occurs when condensation takes place on the surface of bodies which have cooled by nocturnal radiation to a temperature below the dew-point of the layer of air resting on the earth's surface.

This is particularly likely to occur after a warm day, during which there has been much evaporation to increase the absolute humidity of the air. The evening must be calm, since wind mixes the layers of air and does not allow one layer to remain sufficiently long in contact with the earth for cooling to take place, and the sky must be clear since absence of cloud allows radiation of heat to take place readily (p. 382).

The dews of spring and early summer are probably mainly derived from water-vapour in the atmosphere. However, experiments have shown that, particularly in autumn when the earth is warm, the moisture deposited as dew can come from the ground itself, both directly and transpired by plants (*guttation dew*). It is condensed on grass and other bodies which have cooled by radiation —undersides of objects left overnight on a lawn are often wet with dew in the morning.

A practical application of dew formation is shown by the *dew-mounds* which are a common feature in the drier parts of Israel. A mound of earth is covered with flat stones on which copious dew condenses, trickling between the stones into the earth, which is kept moist. Citrus fruit trees grow out of the mounds, and are thus supplied with water.

Mention may be made of a *dew-pond*, although this seems to be something of a misnomer. Hollows were dug in the South Downs and lined with puddled clay and straw, in order to provide sheep with water. The problem is, where does the water which accumulates therein originate? Little can be derived from direct rainfall, usually the dew-pond has no catchment area, and dew condensation has been shown to account for only an inch or so in the year. Perhaps sea-mists blowing off the Channel are contributory sources, but this is unlikely.

Hoar-frost forms when the dew-point is below freezing-point, so that the vapour is deposited directly in the form of tiny ice spicules, without passing first through the liquid state.

Mist and fog Fog, another result of condensation, is notable mainly for the resulting reduction in visibility, and in this respect it is an important meteorological element which forms a major handicap to transport by land, sea and air. The term *fog* is used (under the International Meteorological Organization Code) when the visibility is less than 1 km, *mist* when the visibility extends from 1 to 2 km. In Great Britain, however, 'thick fog' implies a visibility of less than 220 yards. The term *haze* is reserved for impaired visibility of 1 to 2 km as a result of dust or smoke particles. The categories of 'poor', 'moderate', 'good' and 'very good' visibility are defined on an international scale in terms of the range of vision from an observing station.

The main types of fog, according to the causes responsible for them, are (i) radiation fog; (ii) advection fog; (iii) frontal fog; (iv) steam fog; and (v) hill fog.

(i) *Radiation fog* This type owes its formation to an amplification of the conditions which produce dew. During a period of calm, clear weather in Britain, particularly during spring and autumn, the surface of the earth is rapidly cooled by radiation, and so the layer of air resting upon it is also cooled. In hilly country the cooled dense air flows by gravity into hollows, and as it comes into contact with the cold earth, condensation produces a shallow horizontal layer of white radiation fog. Often the layer of fog survives well after sunrise; in the English Lake District, particularly in early summer, one can stand on a hillside in the morning sunlight while the valley below is still mist-filled, but as the sun rises in the sky the fog is dissipated.

Under cold anticyclonic conditions in late autumn and winter, the radiation fog may be thicker and more persistent. The sun's rays are then less powerful, and the fog may linger until the calm weather of the anticyclone is interrupted. It tends to form over valley floors, where the air is moist (particularly when there is a river), and where the chimneys of a large town pour into the atmosphere quantities of soot which act as nuclei. Such are the 'pea-soup fogs' or 'smogs' formerly common in London, Merseyside and most big urban areas. Sulphur dioxide poured into the atmosphere from burning coal combines with the moisture present to form sulphuric acid, thus giving these urban fogs their acrid flavour.

One of the worst fogs ever experienced was in London in December 1952. An anticyclone over the Thames valley formed a deep pool of cold stagnant air, with a much warmer layer above, creating an impenetrable inversion and resulting in an enormous accumulation of soot. Between 6 and 8 December visibility was down to a few yards. The smog resulted in 4000 deaths from bronchitis and pneumonia, and a vast bill for cleaning curtains and clothing. This state of affairs has led to campaigns for smoke abatement, the creation of smokeless zones, and the use of smokeless fuels. The National Smoke Abatement Society publishes a journal, *Smokeless Air*. Improvements can be effected, as shown by Pittsburgh, one of the dirtiest towns in the U.S.A. for many years, which achieved a reduction of atmospheric pollution of over 90 per cent in six years.

(ii) *Advection fog* When a warm, moist air current is cooled as it moves horizontally over a cold land or sea surface, an advection fog may be formed. The term advection is used of a horizontal transfer of air, as compared with convection when the change is vertical.

Where the hot deserts reach the west coasts of continents occur what are called 'cold-water coasts'. These are due to the presence near the coast of cool currents flowing equatorward, accentuated by the upwelling of cold deep water (p. 353). Offshore fogs, formed by condensation over the cool water, are blown for a few km inland by the sea breezes, but they gradually dissipate over the much warmer land. The Golden Gate at the entrance to San Francisco harbour experiences dense fogs of this nature on an average of forty days in the year.

A striking example of an advection fog forming over the sea is in the neighbourhood of the Grand Banks of Newfoundland. When air from over the Gulf Stream, moving in a northerly direction, crosses the line of the 'cold wall' (p. 353), it passes over the waters of the Labrador Current, which are some 8° to 11° C cooler, since the Labrador is bringing melt-water from disintegrating pack-ice, while the Gulf Stream 'transports' high temperatures from much farther south. A thin layer of dense fog persist for many days; in the Strait of Belle Isle it occurs on an average on more than 100 days in the year, and off Newfoundland on more than 70 days. These fogs are common along the coast of north-eastern America and cause much delay to shipping. In July 1959, for example, the *Queen Elizabeth* was in collision with a freighter in thick fog soon after leaving New York.

(iii) *Frontal fog* Short-lived fog, really a very thick, fine drizzle, is sometimes associated with the passage of the warm front of a depression (p. 417). The warm rain falls into the cold underlying air near the ground, and if there is a marked temperature difference gloomy foggy conditions may prevail for a time.

(iv) *'Steam fog'* This type is comparatively rare, but it is included for its meteorological interest. It is formed when cold air passes over a much warmer water surface, so that the water appears to 'steam'. In high latitudes it is known as *ice-fog*, where the moisture in the air is converted into ice crystals, but it rarely lasts for any length of time. A distinction is sometimes made between steam fog, which develops over fresh water, and 'Arctic Smoke' over salt water.

(v) *Hill fog* This is simply low sheet-cloud, which may envelop, for example, the hills of western Britain when a moist air-stream is moving inland. All mountainous districts in middle latitudes are liable to spells of this hill fog at any time of year during unsettled weather.

96 An ice-berg in the Antarctic, which has just 'calved' from the tide-water
glacier in the right middleground
(Aerofilms Ltd)

97 The Severn bore near Gloucester
(Topix)

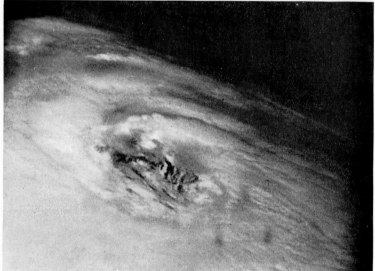

98 Altocumulus clouds
(Clarke Collection, Royal Meteorological Society)

99 The cloud-pattern in Hurricane 'Debbie', September 1961, photographed from over 160 km above the earth by the 'Mercury' spacecraft, while travelling at over 28,000 km per hour. The 'Eye' of the hurricane, the comparatively calm centre within the vortex of spiralling winds, can be clearly seen.

(NASA/Camera Press)

Rime When a fog composed of supercooled droplets (p. 435) is driven by a slight wind against prominent objects such as telegraph poles, wires and trees, the drops freeze on to them as rime. The temperature both of the droplets and of the objects is below freezing-point. The ice particles imprison some air, so that the rime has a white, opaque appearance. This is a common feature on the mountains of Britain, and the climber in winter may find his progress impeded by the covering of 'frost-feathers' on the rocks.

CLOUDS

Clouds consist of tiny particles either of water or ice, which float in masses at various heights ranging from ground-level (where they occur in the form of fog) to the highest wisps at 12,000 m (40,000 ft).

Measurement and recording The amount and nature of the cloud cover is recorded at meteorological stations, and appears on the weather-maps. The amount is recorded in terms of the proportion of the sky which is covered, expressed in eighths (*oktas*), and indicated on the weather-map by a shaded disc. The height and nature of the clouds is shown by means of symbols and figures, according to an international code. If statistics are taken in sufficient detail over a long enough period of time, the mean annual or monthly degree of cloudiness for each station can be plotted, and lines of equal cloudiness (*isonephs*) can be interpolated; these are not particularly useful, however, since cloud is such a variable feature.

Classification of clouds Clouds may be classified either according to their *height* into three groups—high, medium and low cloud, or according to their *general form and appearance* also into three groups—the feathery or fibrous types (*cirrus*), the globular or heaped types (*cumulus*), and the sheet or layer types (*stratus*).

These simple classifications are, however, inadequate to describe the variety of cloud forms. The International Cloud Code lists twenty-eight different types, but ten fundamental forms can be recognized. These are distinguished partly by combinations of the three form-names mentioned, partly by adding the suffix '*alto*' to indicate height, and partly by adding '*nimbus*' to signify rain. Other variations (not included in this limited group of ten) are the lens-shaped *lenticularis* cloud (often associated with air-streams crossing relief barriers and producing eddies or standing waves on the lee-side (p. 425)), the turret-shaped *castellatus*, and the breast-shaped *mammatus* clouds (the two latter are often associated with towering thunder clouds). The prefix '*fracto*' indicates tattered, ragged clouds, such as *fractostratus* or scud, which is a bad-weather cloud.

Leaving these variants aside, the ten fundamental forms may be summarized:

Relative position	Height of base (m)	Name
High clouds	6–12,000	Cirrus (*Ci*)
		Cirrocumulus (*Ci-Cu*)
		Cirrostratus (*Ci-St*)
Middle clouds	2400–6,000	Altocumulus (*Alt-Cu*)
		Altostratus (*Alt-St*)
Low clouds	Up to 2400, usually less	Stratocumulus (*St-Cu*)
		Stratus (*St*)
		Nimbostratus (*Ni-St*)
With considerable vertical development	—	Cumulus (*Cu*)
		Cumulonimbus (*Cu-Ni*)

Description of cloud forms The student is referred to the *International Atlas of Clouds*, which contains illustrations, definitions and descriptions of the ten fundamental types and their variants.

Cirrus is a delicate fibrous or wispy cloud, consisting of tiny spicules of ice. It is often a fair-weather cloud, though if it is succeeded by cirrostratus it may be an indication of an approaching depression. Where the cirrus is drawn out in a 'mare's tail', it signifies strong winds in the upper atmosphere. *Cirrostratus* is a more complete milky layer of high cloud through which the sun shines with a distinct halo, while *cirrocumulus* forms a series of lines of small globular clouds with a rippled appearance, sometimes called a 'mackerel sky'.

Altostratus is a greyish sheet-cloud, much denser than cirrostratus, but through which it is still possible to see the sun, in the words of the official simile, 'as through ground-glass'. The sky often has a 'watery look', and as this cloud so often heralds the approach of a warm front the appearance is usually not belied. *Altocumulus*, a fleecy cellular cloud in bands of globular patches separated by blue sky, is usually, though not always, a sign of fair weather (plate 98). If, however, the masses develop into either castellatus or lenticularis, it may mean heavy rain or thunder.

The low varieties of stratus consist of grey uniform sheets of cloud. *Stratus* itself forms a depressing, heavy grey pall; if continuous rain is falling from it, as at the warm front of a depression, it is termed *nimbostratus*. *Stratocumulus* is a darker, lower and heavier type of altocumulus, which often covers the sky for long dreary periods in winter. Occasionally clearly defined rays of sunshine, known as *crepuscular rays*, appear to fall to earth through chinks in the cover.

Cumulus is the convection cloud, which grows when rising air-masses reach the level at which condensation takes place, hence the horizontal base of the cloud (fig. 195). Many of these cumuli develop into large white globular masses, but usually they are fair-weather clouds and die away in the evening. If, however, the rising air currents are sufficiently strong, the cumulus cloud continues to grow to an immense vertical height, sometimes upward from a base at about 450 m (1500 ft) for 10–11 km (6–7 miles). These towering clouds are *cumulonimbus*, and from them heavy showers of rain or hail may fall, accompanied by thunderstorms, hence the name 'thunderhead'. Although from the side the cloud is dazzling white, from below its base may be almost black. The upper parts of a large cumulonimbus cloud may spread out in the form of an anvil in the direction of the high-level winds. Ice and snow crystals falling below the spreading layer produce the characteristic wedge-shaped cloud.

The meteorologist pays close attention to the nature of the clouds, for this affords evidence of the trend of the weather, particularly when the sequence of developing cloud forms is studied. For example, a definite sequence of cloud is visible during the passage of a depression (fig. 191). Heralded by cirrus and cirrocumulus, a milky pall of cirrostratus then veils the sky, to thicken into altostratus. As the warm front appears, low stratus and nimbostratus from which rain is falling cover the sky. As the cold front arrives, the sharply under-cutting cold air may cause cumulonimbus to develop, with heavy showers of rain, to be succeeded by more ragged fractocumulus in a rain-washed hard blue sky as the weather gradually clears.

World distribution of cloud The rainy parts of the world, such as the Equatorial and the Cool Temperate West Marginal types (Chapter 18), experience a considerable amount of cloud. There is a diurnal cycle of cloudiness in equatorial latitudes—a clear morning sky is slowly obscured by cumulus and cumulonimbus, though by evening the sky begins once again to clear. The cool mid-latitude zone in the southern hemisphere, the latitudes of the 'Roaring Forties' (p. 407), is one of the cloudiest parts of the world. So, too, are the mountainous areas in the northern hemisphere—the coasts of Norway, western Britain, Ireland and British Columbia. The continental interiors in high latitudes have clear skies in summer, but are subject to long periods of anticyclonic 'gloom', with much low stratus cloud, in winter.

The least cloudy parts of the world are the hot deserts, which are also the most sunny (p. 482).

In 1960 the U.S. 'Tiros I' satellite produced the first high-altitude photographs of cloud-patterns. In August 1964 'Nimbus', the

U.S. weather satellite, was launched into orbit, and its television cameras cover every part of the earth's surface in 24 hours, photographing cloud patterns. The first complete view of the world's cloud cover was made for 13 February 1965. This was obtained by the U.S. Weather Bureau by assembling 450 individual photographs taken by the cameras carried by another weather satellite, 'Tiros IX', during a period of 24 hours.

RAINFALL

Measurement and recording The amount of rainfall, in units of inches or millimetres, is the theoretical layer of water that would cover the level ground, assuming that none is lost by evaporations runoff or percolation. One inch of rain is equivalent to 100·9 tons of water per acre, 14,460,000 gallons per sq. mile, or $4·2 \times 10^6$ litres per km^2.

A *rain-gauge* consists of a metal cylinder within which a funnel leads into a collecting vessel. This vessel is emptied periodically into a measuring cylinder, graduated so that the depth of rainfall can be read off directly. Should an ordinary measuring cylinder graduated in cubic cm be used, it is necessary to divide this reading by the funnel area in sq. cm. The gauge must be very carefully sited; the standard height of the funnel rim is 0·3 m (1 ft) above the ground, and the instrument should be placed well away from trees, high rocks or buildings; it should if possible be twice as far from the nearest building as the latter is high. The gauges are emptied and measured at regular intervals, at an observatory once a day or more frequently. In recent years many gauges have been placed among the British mountains; these are much less accessible and many are visited only monthly.

Self-recording (or *tipping-bucket*) gauges are now used more commonly. The water accumulates in a container which is carefully balanced, and when 5 mm have been collected the water is automatically tipped off. This movement is connected to a pen which traces an ink line on a rotating drum; it shows the rise to 5 mm, then there is a vertical drop, and the rise begins again.

Compilation of rainfall records From rain-gauge measurements all over the world a wide variety of records is compiled of interest to the geographer. For general purposes the *monthly means* of rainfall are either simple arithmetic means of the total rainfall for each month over a period of years, or are weighted figures in which all months are reduced to an equal length to avoid false deductions (as, e.g., if a comparison has to be made between February and August). Monthly means are the basis of the graphs used in

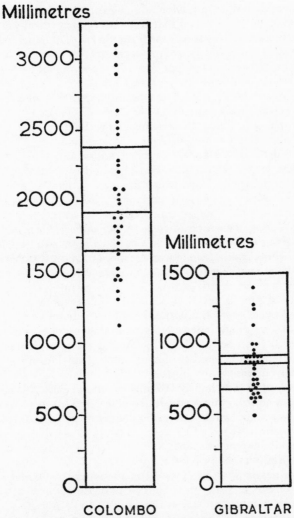

Fig. 197 *Annual rainfall dispersion diagrams for Colombo and Gibraltar*

Chapter 18, and they enable seasonal rainfall to be examined. Annual figures, both means and for actual years, are also useful.

Dispersion diagrams (fig. 197), in which each year's rainfall is shown by placing a dot on a vertical scale, enable one to see at a glance the whole range of wet years, average years and dry years over a long period of time. This is often much more informative than an arithmetic mean, which disguises these differences, of such importance to the people of that area. On fig. 197 'flood years' and 'famine years' can be distinguished. Another way of indicating these annual variations is in terms of the departure from the annual mean. In India a deficiency of 25 per cent will injure the crops, while one of 40 per cent will cause widespread famine.

The number of *rain-days* (days with more than 0·25 mm (0·01 in), and of *wet-days* (days with more than 1·0 mm (0·04 in)) are both useful[1]; Hassness (near Buttermere in the English Lake District) has 228 rain-days per annum. *Rain-spells, wet-spells, droughts* and *dry-spells*, all defined by standard practice, are recorded in some countries. The longest consecutive run of rain-days in Britain was at Eallbus in the Isle of Islay, which had no less than 80 in 1923. An *absolute drought* is defined as a period of at least fifteen consecutive days each with less than 0·2 mm; the British record was 73 days in London during the spring of 1893. A *partial drought* is a spell of 29 consecutive days, some of which may have slight rain, but during which the total rain does not average more than 0·2 mm per day.

Intensity of rainfall is important, since the rate at which rain falls (ranging from a fine drizzle to a torrential downpour), is related to problems of runoff, percolation into the soil, evaporation, soil erosion and flood control. Information about intensity of rainfall is almost as vital to an understanding of rainfall regimes as are totals. The intensity can be calculated by dividing the total rainfall over a given period by the number of hours during which rain fell (giving the hourly intensity), or by the number of rain-days (giving the daily intensity). The hourly intensity of rain at Boston (U.S.A.) is 0·91 cm (0·36 in), compared with 10·6 cm (4·17 in) at Cherrapunji in north-eastern India. Cherrapunji once experienced 103·6 cm (40·79 in) in a period of 24 hours; Baguio in the Philippines has recorded 116·8 cm (46·0 in); and the world record for 24 hours' precipitation was on 16 March 1952, when a station situated at a height of nearly 1200 m (4000 ft) in the island of Réunion in the Indian Ocean had 187 cm (73·6 in). At Unionville (Maryland, U.S.A.) on 4 July 1956 3·12 cm (1·23 in) fell in 1 minute.

Presentation of rainfall records The range of rainfall records may be depicted, graphically and cartographically, in a variety of ways. The most common graphical device to show mean monthly rainfall uses vertical columns, one for each month; they may be shaded or filled black for effect. These columnar diagrams may be strikingly located on a map in the approximate position of

---

[1] These definitions apply specifically to Great Britain.

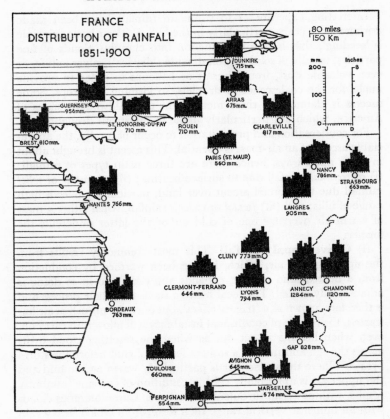

Fig. 198 *Rainfall map of France*

The columnar diagrams enable the monthly distribution of rainfall to be seen at a glance. Note the summer drought in the south and the continental summer maximum in the east.

the station to which they refer (fig. 198). *Isohyets*—lines on a map to indicate the mean annual, seasonal or monthly rainfall total—are interpolated after values for available stations have been plotted.

TYPES OF RAINFALL

When minute droplets of water are condensed from water-vapour in the atmosphere on to nuclei, they may float as clouds. If, however, the droplets coalesce, they form larger drops, which, when heavy enough to overcome by gravity ascending air currents within a cloud, fall as rain (p. 441). One theory maintains that the presence of ice spicules acting as nuclei is necessary for precipitation to occur.

445

Interesting experiments to stimulate rainfall have been made during recent years, particularly in the U.S.A. Their basic principle is 'seeding', that is, the introduction into cumulus clouds of fine particles, usually of solid carbon dioxide ('dry ice'), silver iodide, or even volcanic dust, from aircraft or balloons in order to provide nuclei for the coalescence of droplets, so forming raindrops. Some success is claimed for these methods in the Midwestern states of America, which are particularly liable to disastrous droughts.

For condensation and precipitation to occur naturally, the appreciable ascent of an air-mass is essential. This ascent is brought about in three main ways, hence there are three main types of rainfall: (i) *convectional* rainfall due to surface heating; (ii) *orographic* or *relief* rainfall due to a forced ascent over land, particularly over a high range of hills; and (iii) *frontal* or *cyclonic* rainfall, when either a mass of warm air overruns one of cold air or the latter undercuts the former.

(i) **Convectional rainfall** This most commonly results from the updraught of air, which, having been warmed by conduction from a heated land surface, expands and rises, and in so doing is adiabatically cooled. The local heating starts the whole process and is therefore known as a 'trigger effect', but once the uprush of air has started, the release of conditional instability will allow it to carry on, even when heating is cut off, as when the resulting cloud drifts across the sun and there is a sudden feeling of chill before the storm breaks. Where the surface air is particularly warm and humid and the upper air is abnormally cold, a condition of extreme instability results, with powerful turbulent up-currents. Cumulonimbus clouds with an immense vertical range then form, from which heavy rain may fall. This is especially marked in the 'hot tower' of a tropical low pressure system.

Convectional rainfall occurs throughout the year near the Equator, where constant high temperatures and humidity produce this type of rainfall almost daily in the afternoon. With increasing distance from the Equator the rainfall is associated more markedly with summer, and both the total amount and the duration of the rainy season decrease as the hot deserts are approached.

In middle latitudes convectional rainfall occurs in early summer, when the upper atmosphere is still cool following winter, but when heating of the earth's surface is becoming active. The rainfall in the continental interiors in early summer is mainly of this type.

Thunderstorms A development of convectional overturning under extreme conditions of instability may produce thunderstorms. As the towering cumulonimbus cloud advances across the sky, the barometer falls markedly, and a strong fresh wind relieves the sultriness which usually precedes a storm. Heavy rain or hail falls as the cloud passes overhead, accompanied by lightning, either 'fork' or the common 'sheet' lightning of summer.

When condensation takes place at the level of saturation, the uprush of air is still so great that the droplets are carried upward; cases have even been known whereby pilots of aircraft baling out by parachute have been carried upward for some distance by these air currents. The largest possible droplet that can form is 5·5 mm in diameter, for beyond that size it becomes unstable and breaks into one or more droplets; this may go on repeatedly, but if the rising air currents decrease in strength, the drops will fall to earth.

One theory assumes that each time a drop breaks up, the resulting droplets assume a positive charge of static electricity, the surrounding air taking a negative charge. The upper part of the cloud becomes positively charged, the lower part and the surrounding air negatively charged, with a further small subsidiary positive charge near the base of the cloud. A difference of as much as 100 million volts may develop, and a lightning flash represents the re-uniting of the separate charges along a direct channel through the atmosphere, either within the cloud, or from cloud to ground. The thunder which follows the discharge is explained in terms of the disturbance of the air particles in the atmosphere, producing a noise reflected to earth from the cloud surfaces. But this explanation of thunderstorms is much too simple; scientists who have spent many years studying these features are still not certain of what really happens. Much work has been done particularly by B. J. Mason; two important concepts have been suggested by him. The first involves the creation of vertical 'cells' within the thunder cloud, extending throughout its full height, with a very strong updraught and the formation of ice crystals in the upper part, and a corresponding downdraught. The other is that when ice particles of initially different temperatures collide, the 'warmer' piece acquires a positive charge, the 'colder' piece an equal negative charge. A collision between super-cooled water droplets and hailstones, both present in a cell, will also cause a separation of positive and negative charges.

Other types of thunderstorms occur in addition to the convectional or thermal varieties, though these are the most common. *Frontal thunderstorms* in middle latitudes are associated with the movement of a cold front, particularly with an accompanying line-squall (p. 417). Occasionally even, they are associated with a warm front when the warm air-stream is particularly unstable. Thunderstorms frequently accompany orographic rainfall, particularly in the Tropics

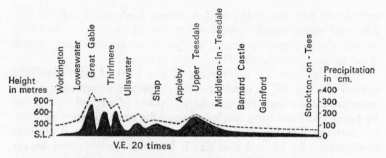

Fig. 199 *Relief and rainfall profile across northern England*

The close association between the relief and the mean annual rainfall totals will be noted. The solid area indicates the relief profile. The distance from Workington to Stockton is 145 km (90 miles).

where markedly moist and warm air-masses rise sharply over steep mountain ranges.

(ii) **Orographic rainfall** This type of rainfall occurs when air is forced to ascend the side of a mountain range. This may 'trigger off' conditional instability (p. 431); it may cause convergence and uplift; it may increase precipitation by retarding the rate at which a depression moves; and it may steepen a cold front by friction. It is found wherever hills lie parallel to the coast over which blow moist winds from the sea. There is a pronounced difference between the windward and leeward sides of the mountains; the markedly drier leeward side is the *rain-shadow* (fig. 199). The orographic factor usually increases precipitation produced by other causes. Thus the occurrence of high relief along the margins of western Britain causes an intensification of frontal (depressional) precipitation.

(iii) **Frontal rainfall** This type of rainfall occurs along the frontal zones of convergence, notably at the Inter-Tropical Convergence Zone (p. 412) and at the Polar Fronts. In the case of the ITCZ the diurnal convectional rhythm is strongly marked, superimposed upon frontal influences. Patterns of air convergence in the ITCZ waves may, under the prevailing warm, humid conditions, cause the build-up of massive cumulonimbus clouds, with torrential rain and thunderstorms.

In the case of fronts associated with local low-pressure systems which usually travel from west to east in middle and high latitudes, continuous drizzling rain falls in a broad belt along the warm front and in the warm sector, while more concentrated squally showers fall as the cold front passes (p. 417), The rainfall is intensified by the effect of relief as the depression crosses the coasts, as in western Britain, Norway and British Columbia. Occasionally small depressions of great intensity, with a warm exceptionally humid air-stream in the warm sector, may produce torrential downpours rare in middle latitudes; 28 cm (11 in) of rain fell at Martinstown near Dorchester (on 18–19 July 1955) during a period of only 9·5 hours. The actual days of heaviest rain recorded in England were 26·67 cm (9·56 in) at Bruton in Somerset (28 June 1917), 23·88 cm (9·40 in) at Cannington in Somerset (18 August 1924), and 23·11 cm (9·1 in) at Simonsbath in Somerset (15 August 1952).

Aridity The causes and distribution of aridity are in effect complementary to those of rainfall. Although in a sense it is a nega-tive feature, aridity (both seasonal and annual) is a major climatic fact, and its importance is seen in the delimitation of climatic types in the following chapter.

Aridity occurs on leeward slopes, that is, in rain-shadow areas. It is also found where an area is subject to dry land-winds, or to winds blowing from cooler to warmer latitudes, which exercise a drying effect. Again, it occurs where stationary anticyclones appear either on a small scale in middle latitudes or on a large scale in the sub-tropical land areas and mid-latitude continental interiors, the great air-mass source-regions. Aridity also occurs where the atmosphere is at a constantly low temperature and can contain little vapour, as in the Tundra (p. 479) and the polar regions.

It is important to remember that aridity depends not merely on the amount of precipitation, but on its effectiveness, hence the term *precipitation-effectiveness* (PE). Various empirical formulae, involving both precipitation and temperature, have been devised to obtain values, especially for use in climatic classification.

OTHER FORMS OF PRECIPITATION

Snow Snow is formed when water-vapour condenses at a tem-perature below freezing-point, passing directly from the gaseous to the solid state and forming minute spicules of ice. These particles unite into crystals which basically are either flat hexagonal plates or prisms; both reveal infinite variations of great beauty in their symmetrical patterns. If condensation continues, these crystals unite into snow-flakes and where the lower atmosphere is sufficiently cool they will reach the ground without melting. For snow to fall there must be both plentiful vapour in the atmosphere and a sufficiently low temperature. Its geographical occurrence and the position of the permanent and winter snow-lines in various latitudes have been discussed in Chapter 8.

Snow which settles on the ground may be very dry and powdery under low-temperature conditions, as in Antarctica, or it may be wet and compact. In the former case about 30 cm of fresh snow are the equivalent of 1 cm of rain but in the latter 4 or 6 cm of snow will melt to form 1 cm of water.

For record purposes in Britain, snow is melted and included in the total precipitation figure. A tall cylinder is added to gauges in mountain districts, so that the snow can accumulate and then melt into the container; considerable inaccuracies are inevitable, since often the gauge becomes choked with snow, which then may freeze solid and prevent any further addition. In Canada, however, the actual depth of fresh snow is directly measured with a rule.

Sleet Sleet is an intermediate form of precipitation, though there is some difference between the British and American definition. The former regards it as a mixture of snow and rain, or of partially melted snow, the latter as raindrops which have frozen and partially melted again.

Soft hail This consists of small spherical grains formed by the aggregation of tiny ice particles, deposited by direct freezing from water vapour. The hail is white and opaque in character.

Hail True hail is a form of precipitation associated with extreme instability. It falls from lofty cumulonimbus clouds, frequently at the passage of a cold front, and after exceptional local heating and convectional overturning. When droplets form in the lower part of a cloud, they may coalesce with others to form large drops. If the ascending air currents are particularly strong, the drops may be carried upward to a point at which they freeze into ice-pellets. In addition, drops of super-cooled water (p. 435) on colliding with tiny ice particles immediately freeze around them as a layer of clear ice. These pellets grow by being carried still higher, so that vapour freezes directly on to them as ice crystals. As the strength of the vertical uprush in a turbulent cloud is extremely variable, the pellets may fall for some distance, partially melt at lower levels, and then be carried upward once again. This may happen several times until the weight of each hailstone is sufficiently great to overcome any up-rising current, and they then fall to earth. This procedure helps to explain the fact that if a large hailstone is cut across, it will probably be found to consist of alternate concentric layers of the clear ice of directly frozen water, alternating with the white opaque ice of frozen vapour; as many as twenty-four layers have been counted.

A more recent and indeed more acceptable theory involves the concept of the 'vertical cell' in an atmospheric section within a thundercloud (pp. 446-7). This postulates that ice-pellets fall out ahead of the updraught, the result of strong upper winds, and are swept up again by the advancing storm before reaching the ground. In other words, the formation of hailstones does not depend on an irregular updraught.

Hailstones may attain very large sizes, and there have been some fantastic records claimed. The largest hailstone recorded in Britain had a diameter of 5 cm. Elsewhere stones of a diameter of 10 cm and of 1 kg in weight have definitely been authenticated, although the unconfirmed claims of hailstones in China 30 cm in diameter and 5 kg in weight seem hard to believe. Hailstones can do great damage to orchards and glass-houses, and in India and the U.S.A. animals and men are occasionally killed.

Hail occurs widely in the world, except in the polar regions where thunderstorms are rare, in the equatorial zone where the air temperature is so high that they melt before reaching the ground, and in the hot deserts. Hailstorms are especially common in spring and early summer in mid-latitudes, such as in the U.S.A. and China, but in Britain they are mainly a winter phenomenon. Other areas of frequent occurrence are northern India and the plateau of South Africa.

CHAPTER 18

Climatic Types

THE classification of climatic types represents an effort to distinguish various combinations of climatic elements throughout the world, and thereby to delimit areas in which these occur. A regional classification of climates, as of land-forms, soils and vegetation, is a classification of convenience, in that it helps to systematize and to marshal the large amount of available data.

METHODS OF CLASSIFICATION

Probably the earliest classification of climatic types was made by the Greek philosophers, who distinguished three latitudinal temperature zones—*torrid* (within the Tropics), *frigid* (within the polar circles) and *temperate* (between the two). Medieval scholars, and in fact text-books as late as the end of the nineteenth century, followed this simple arbitrary division.

Various German climatologists have produced classifications based on more complicated criteria. One system, put forward in 1896, used two significant isotherms—the 20° C (68° F) annual isotherm and the 10° C (50° F) summer isotherm. These values were not selected arbitrarily, but because they indicate certain biological responses—20° C was taken to be the limiting temperature for palms (indicative of hot climates), while the isotherm of 10° C for the warmest month forms a boundary between the tundra and the coniferous forest. Using these isotherms, the world was divided into a series of 'temperature provinces'.

A few years later this classification was elaborated, using the same critical mean figures, but based on the number of months experiencing mean temperatures above or below them. Thus the tropical zone has twelve months with mean temperatures above 20° C, the polar zone twelve months below 10° C, and there are various intermediate zones. While this classification has the virtue of symmetrical simplicity, it groups areas together which on investigation are clearly dissimilar; on this basis the British Isles, California and the Plate estuary are all in the same climatic zone.

Other classifications have sought to introduce refinements in the form of more varied criteria, particularly using rainfall in conjunction with temperature. A French geographer, for example, defined a dry climate as 'one in which the mean annual rainfall (measured in centimetres) is less than double the mean annual temperature (measured in degrees Centigrade)'. Many other Temperature/Precipitation ratios of varying degrees of complexity have been determined, in order to arrive at indices which can be plotted on a map and so enable lines bounding climatic regions to be interpolated. Many of these refinements have only been attained at the sacrifice of simplicity, and when working on a map of world or continental scale, where the actual thickness of a line may represent many km, such refinements are often unnecessary.

Apart from producing some system of classification which is free from anomalies, the chief difficulty is that climatic conditions change gradually, one type shading into another, quite apart from infinite local complications. There are, in fact, no boundaries, but instead 'climatic transitional zones'. The true centres of climatic regions, the type regions proper, are rarely in doubt as long as there are sufficient statistical data; the problem is where to draw the line.

A. A. Miller's classification The late A. A. Miller produced an eminently workable classification, of great value to geographers because of its applicability to regional description, by combining a map of temperature zones with one showing the seasonal distribution of rainfall. His map of temperature zones used the mean annual isotherm for 70° F, the 50° F mean isotherm of the warmest month, the 43° F mean isotherm of the coldest month, and an isopleth defining areas with six months or more below 43° F. These figures were selected because of the significance of their biological responses (pp. 513–16). He thus worked out a system of five temperature zones: *hot, warm temperate* (or subtropical), *cool temperate, cold* and *arctic*. To these zones based on temperature Miller added two others —*mountain* and *desert* climates. Mountain climates are found where altitude produces marked modifications of the 'base-level climate', and desert climates are defined broadly by a maximum annual rainfall of 25 cm (10 in).

Then he superimposed a seasonal rainfall map in order to obtain further subdivisions. This map distinguished (i) areas with rain at all seasons (including regimes of evenly distributed rainfall, of double maxima and of a summer maximum); (ii) areas of constant drought; and (iii) areas of markedly periodic rain, that is, including a season of definite drought.

The final classification involved seven main climatic groups, with nineteen subdivisions, reproduced below, together with a map

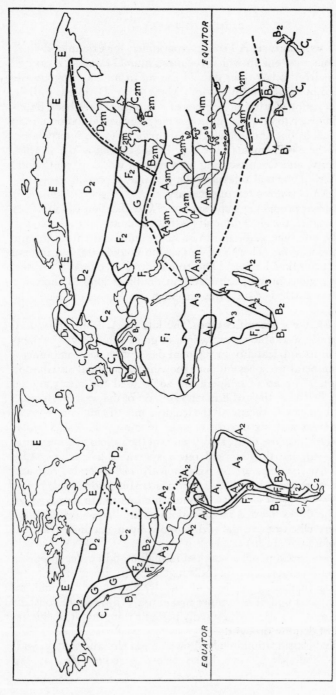

Fig. 200 The classification of world climatic types

Based, by permission, on A. A. Miller's classification. The lettered and numbered types are listed on pp. 455–6. The limit of the monsoonal types is shown by a pecked line. The dotted line in eastern North America delimits the variants mentioned on p. 476. Many relatively small areas of Mountain Climate (G) are omitted owing to the small scale of the map.

100 Soil profile developed on Millstone Grit, Goyt's Moss, Derbyshire
(R. Kay Gresswell)

101 Cave-dwellings in loess, Shansi Province, north-western China
(Paul Popper)

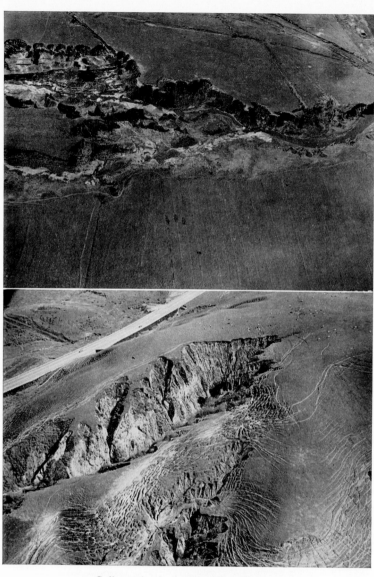

102 Gully erosion in the Republic of South Africa
(Aerofilms Ltd)
103 Soil erosion in southern California
(Soil Conservation Service)

showing the world distribution of these climatic types (fig. 200). Each type is described in turn and illustrated by more detailed regional descriptions of specific areas. As these descriptions are based on mean temperatures and rainfall totals, to avoid constant repetition the word 'mean' is usually omitted. Unless the text definitely states that actual figures are used, the reader must infer that mean values are involved. A series of temperature and rainfall graphs has been drawn to illustrate each climatic type and any interesting variants.

Once again the importance of air-masses in climatic study must be emphasized. As Miller so clearly expresses it, 'The weather at any particular time and place is the weather associated with the air mass in temporary occupation, or is the result of conflict between them for possession of the territory.' Following on from this is another statement: 'Much will have been gained if it can be shown that the climate of a place is the statistical frequency with which the weather associated with each air mass (or front) affects the area.'

Synoptic climatology A similar concept—that of synoptic climatology—has been developed by H. H. Lamb for a classification of 'weather-types' over the British Isles. This recognizes five major types named after the direction from which the air-flow comes (westerly, north-westerly, northerly, easterly and southerly), and two further types, cyclonic and anticyclonic, which refer to the dominance of either low or high pressure. This recognition of synoptic climatic types is a lengthy business, and the analysis of temperature, precipitation and other elements in relation to such types of air-flow necessitates the use of a computer.

CLIMATIC TYPES

A. Hot Climates

(Mean annual temperature exceeding 70° F (21° C))

1   Equatorial: double maxima of rain
1m Equatorial (monsoon variety)
2   Tropical, marine: no marked dry season
2m Tropical, marine (monsoon variety)
3   Tropical, continental: summer rain
3m Tropical, continental (monsoon variety)

B. Warm Temperate Climates

(No cold season, i.e. no month below 43° F (6° C))

1   Western Margin (Mediterranean): winter rain
2   Eastern Margin: uniform rain
2m Eastern Margin (monsoon variety): marked summer maximum of rain

C. Cool Temperate Climates

(Cold season of one to five months below 43° F (6° C))

1 Marine: uniform rain or winter maximum
2 Continental: summer maximum of rain
2m Continental (monsoon variety): strong summer maximum of rain

D. Cold Climates

(Long cold season, of six or more months below 43° F (6° C))

1 Marine: uniform rain or winter maximum
2 Continental: summer maximum of rain
2m Continental (monsoon variety): strong summer maximum of rain

E. Arctic Climates

(No warm season, twelve months below 50° F (10° C))

F. Desert Climates

(Less than 25 cm (10 in) of rain annually)

1 Hot Deserts: no cold season, no month below 43° F (6° C)
2 Mid-Latitude Deserts: with cold season, one or more months below
   43° F (6° C))

G. Mountain Climates

A. HOT CLIMATES

The group of Hot Climates (fig. 201) is bounded on the poleward sides by the 21° C (70° F) annual isotherm, enclosing a zone which extends in parts beyond the Tropics. Much consists of ocean, but large parts of the three southern continents and of the peninsulas and islands of southern Asia are included.

A line drawn around the earth, joining the point with the highest average temperature on each meridian for each month, may be called the *Thermal Equator*. This does not correspond to the true Equator, but moves north and south during the year with the apparent motion of the sun (p. 383). As the greater land-masses lie more to the north of the true Equator and form the areas of greatest heating and highest temperatures, the Thermal Equator moves farther north in July (as far as 28° N) than it does south in January (as far as 23½° S). In fact, if a mean annual Thermal Equator is plotted by drawing a line to join the highest mean annual temperature on each meridian, it will lie for the most part well north of the true Equator.

This apparent movement of the sun brings its midday rays vertically overhead at the Equator twice in the year, in March and September, resulting in two delayed maxima of temperature in

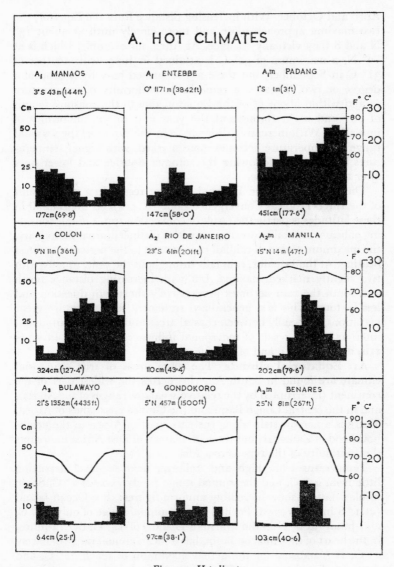

Fig. 201 *Hot climates*

The latitude, height above sea-level, and mean annual total rainfall are stated.

April and October. With increasing distance from the Equator, the two maxima approach more and more closely until at about 13° N and S they virtually coalesce. At Aden, for example, which is at 13° N, a maximum of 32° C is attained in June, with another of 31° C in September, but these are separated only by a drop of a degree or two. There is a remarkable uniformity of temperature at and within about 5° of the Equator, due to the constant length of day and night throughout the year and to the uniformity of insolation. With increasing distance from the Equator, the seasonal range of temperature becomes more marked, with higher temperatures in the weeks following the summer Solstice and lower ones after the winter Solstice.

The movement of the Thermal Equator results in a general shift of the Inter-Tropical Convergence Zone (p. 412). As the rainfall in these latitudes is partly the result of convection (pp. 446-7), consisting of heavy instability downpours with a distinct diurnal periodicity, the maximum season of rainfall corresponds to the period of greatest heating. At the Equator, rain falls throughout the year, usually with two slightly marked maxima, but with increasing distance north and south the rain becomes progressively shorter in duration and less in total. This is a generalized statement and there are great variations, especially between coastal areas and those in continental interiors, as the result of monsoonal influences (as shown in types **1m, 2m** and **3m**), and of relief features.

A1. Equatorial climate The main areas of true equatorial climate are found in South America across the widest part of the continent (interrupted by the relatively narrow ranges of the Andes), and in the central Congo Basin and the Guinea coastlands of Africa. There is a narrow strip along the east coast of Africa in the neighbourhood of Zanzibar, but the high plateau of East Africa interrupts the continuity of the type across Africa.

Temperatures are high and uniform, near sea-level departing little from 27° C, and the annual range rarely exceeds 2° C. Some insular stations show incredibly uniform figures; thus Ocean Island (at 1° S in the western Pacific) has an annual range of only 0·1° C, and Jaluit (at 6° N in the Marshall Islands) of 0·4° C. Even Iquitos, in the heart of the Amazon Basin, has a mean annual range of barely 2° C. It is unusual for the actual maximum to reach 32° C, and very rare for it to attain 38° C. The diurnal range is much more marked than the seasonal one, sometimes attaining 8° C.

The humidity and the amount of cloud are high throughout the year. The constantly high humidity, producing a monotonous 'sticky heat', is extremely oppressive and enervating, and forms perhaps the most trying feature of the climate for Europeans. Moreover, the equatorial belt is a region of stagnant air, of calms or light winds, except at coastal stations where the alternation of land and sea breezes may provide some slight relief.

As a rule, a clear morning sky is gradually obscured by a rapidly increasing amount of cloud; cumulus develops into cumulonimbus, and in the mid-afternoon (following the maximum heating and convectional rising) rain falls in the form of torrential showers, often with thunderstorms. A fine evening commonly succeeds the downpour.

The true equatorial climate has a well distributed rainfall of some 150–200 cm (60–80 in) per annum, with no marked dry season but with two maxima shortly after the time of the overhead sun, that is, at about April and October. When rainfall figures are examined for a number of equatorial stations, however, various modifications of this typical regime can be found. Sometimes one maximum is manifestly greater than the other. Elsewhere, as in the case of Belém and Manaus, there is only one maximum and one minimum; in the case of the former station there is a marked minimum from August to November, in the case of the latter it occurs from June to September. No adequate explanation of these phenomena can be offered. The relief factor can strikingly affect the total; Iquitos in the west of the Amazon Basin, with the Andes beyond, has 262 cm (103 in) per annum. Some of the equatorial islands experience tremendous totals; thus Jaluit has 450 cm (177 in). On the other hand, island groups situated within areas of divergence in the Doldrums, on the eastern sides of the oceans, may show a remarkable aridity, for example, the Galápagos Islands.

Aim. Equatorial (monsoon) climate A marked monsoonal effect is experienced in many coastal areas near the Equator, particularly where mountain ranges near the coast cause moist inflowing air-streams $(TmK(M))$ to rise sharply. On to the convectional rainfall total are superimposed monsoonal and orographic effects; this is shown by the west coast of Colombia which has totals approaching 750 cm (300 in) (Buenaventura, 714 cm (281 in)), and the Guinea coast of Africa where in July the South-east Trades are drawn across the Equator as a south-west monsoon (Freetown, 445 cm (175 in)); Cameroon Peak, probably more than 1000 cm (400 in).

Indonesia The most striking area with an equatorial monsoon climate is Indonesia, which lies across the Equator between the two monsoon centres of south-eastern Asia and northern Australia (figs. 188, 189). The air-flow during the period from December to March is from the Asiatic high-pressure system toward the low-pressure centre over northern Australia; over Java and Sumatra the wind direction is from between west and north-west, particularly in January–February, and so it is locally known as the 'west monsoon'. From May to September the direction of air-flow is almost reversed, so that the winds blow more or less from the east or south-east across the Equator towards the Asiatic low-pressure area. During the change-over period in April–May, and again in October–November, light variable breezes and calms are experienced, for the pressure gradient is then slight.

The force and direction of the monsoonal air-streams over Indonesia are much affected by the arrangement of land and sea, and by the alignment of the mountain ranges. In many parts a distinct foehn effect (p. 423) is experienced. Sudden squalls are frequent, as in the Malacca Straits where the *Sumatras* occur, usually accompanied by thunderstorms; these are linear disturbances which cross Malaya from west to east during the monsoon.

A variety of factors has combined to make the Indonesian islands one of the wettest parts of the world. They experience both the convectional rainfall of equatorial latitudes and the heavy relief rain brought by the moist air currents of alternating monsoons crossing steep mountainous islands. The fact that they lie in a warm ocean accounts for the constantly high humidity. Djakarta has 180 cm (71 in), Padang in western Sumatra 452 cm (178 in), and Pontianak in Borneo 320 cm (126 in) per annum. All these are coastal stations, and at higher altitudes the totals increase still more; Craggan in the mountains of western Java has no less than 732 cm (288 in).

The several factors accounting for the rainfall produce great variety in both totals and seasonal distribution. The normal equatorial double maxima can be distinguished at Padang, while farther from the Equator a dominant rainy season is associated with the west monsoon. The period of the east monsoon tends to be drier, since the air-streams from the Australian high-pressure area, crossing but a little subsequent stretch of ocean, do not carry as much water-vapour. The rainfall, as might be expected from the high vapour content of the atmosphere and the powerful ascending air currents, occurs in showers of short duration but of considerable intensity. Most stations in Java have experienced 30 cm (12 in) in 24 hours, and 50 cm (20 in) has been recorded in the same period. Indonesia is one of the most thundery regions in the world as a result of this atmospheric instability. At Bogor in the hills above Djakarta thunder is heard on an average on 322 days in the year.

In spite of the constantly high humidity and considerable rainfall,

Indonesia was one of the areas near the Equator where Europeans had settled successfully. The main reason was the ready accessibility of the numerous hill-stations from the enervating coastal districts. Bandung at a height of 730 m (2395 ft) has a mean annual temperature of 22° C, while Tosari at 1735 m (5692 ft) has one of 16° C. On the high plateaus night-frosts are experienced, and a stimulating contrast is provided between the hot sunny days and the need for blankets and a fire at night. These hill-stations largely accounted for the success of the former Dutch colonial empire.

A2. Tropical marine climate A marine type of tropical climate is experienced on the eastern sides of land-masses, especially where bounded by steep highland edges, and on islands within tropical latitudes. These areas experience not only convectional rainfall for part of the year, the result of the seasonal migration of the Inter-Tropical Convergence Zone, but also relief rainfall brought by *TmK* air-masses (in effect the Trade Winds) for part of the year. They are, in fact, sometimes known as 'Trade-wind coasts'. The coasts of the Guianas in South America, the east coast of Brazil, the narrow isthmus of Central America, the West Indian islands, the coasts of Portuguese East Africa and of eastern Madagascar, and the Hawaiian Islands (where Mount Waialeale has recorded a short-term mean precipitation of 1209 cm (476 in) are all included.

One feature of these Tropical Marine coasts (and also of the Tropical Monsoon coasts in similar latitudes) is the occurrence of violent storms (p. 421), known as *cyclones* (in the Indian Ocean), as *hurricanes* (near the West Indies and the Gulf of Mexico), and as *typhoons* (in the China Seas and the western Pacific margins) (plate 99).

Jamaica The details of rainfall totals and distributions in these Tropical Marine areas vary considerably with location, since the orographic effect is so important. Jamaica, just north of 18° N, is an island orientated more or less from west to east, with a mountain backbone in the same direction culminating in the Blue Mountains rising to over 2100 m (7000 ft). The combination of convection and the Trade Winds makes summer the rainy season, but even in winter the Trades bring some rainfall. The effect of aspect is shown by the contrasting totals of Port Antonio on the north-east coast (353 cm, 139 in), Blue Mountain Peak (445 cm, 175 in), and Kingston in the rain-shadow on the south coast (86 cm, 34 in). Kingston has a pronounced dry season from January to April, during which less than 10 cm (4 in) fall, for both convection and Trade Wind activity are at a minimum. However, at Port Antonio even the driest month (March) receives an average of nearly 13 cm (5 in).

Little need be said about temperatures, since they resemble the equatorial figures; Kingston has a mean annual temperature of about 26°C and a seasonal range of under 3° C. The maritime location is responsible for these equable temperatures. Conditions are, however, far less enervating for Europeans than the true equatorial regime, because of the strong 'fresh' quality of the on-shore Trades.

A2m. Tropical marine (monsoon) climate Certain areas on the east coasts of tropical lands experience Tropical Marine conditions, but because they lie near land-masses strongly affected by monsoonal reversals, a distinct monsoonal regime is superimposed. These areas include the coast of Indo-China, the Philippine Islands, north-eastern Queensland and possibly Ceylon. The last-named in many ways might be included in Type **A1m** in common with Indonesia, but as the seasonal temperature range is rather more marked in many parts of Ceylon (the northern plains have monthly temperatures varying between about 24° C and 32° C), it is better to include the island in this present category. The equable effects of its insular position are, however, shown by the monthly temperature figures for Colombo, which vary only between 27° C and 28° C.

This climatic type resembles the Tropical Marine type (**A2**) in that no part of the year is rainless, but a marked monsoonal effect accentuates the rainy season. Mon Cai in N. Vietnam has a total of 269 cm (106 in), with the period of heaviest rainfall between May and September (maximum, August 61 cm (24 in)), but it receives only 25 cm (10 in) between November and March. Similarly, Cairns on the coast of Queensland has a total of 226 cm (90 in), with a rainy season from January to April, when the Trades are drawn in towards the low pressure of northern Australia as an easterly monsoon, and cross the Great Dividing Range more or less at right angles.

Ceylon The rainfall regime in Ceylon is more complicated, because three distinct factors are involved. In the first place, a monsoonal reversal affects various parts of the island at different times. The South-west Monsoon brings rain to western Ceylon in June–July, the retreating South-west Monsoon brings rain to eastern Ceylon in October–November, and north-easterly winds from the Bay of Bengal bring rain to eastern Ceylon in December–January. The second factor is the mass of uplands rising to 2500 cm (8300 ft) in the south-centre of the island, which accentuates the monsoonal effect on either coast according to the direction of the particular airstream. The third factor is that Ceylon is an island located on the northern edge of the Doldrums, subject to high temperatures and convectional effects and surrounded by a warm ocean over which originate numerous moist air streams.

Colombo, therefore, on the heated western coastal plain yet backed by mountains, receives rain every month, but with two well-defined maxima, one in April (25 cm, 10 in), May (28 cm, 11 in) and June (18 cm, 7 in), and another in October (33 cm, 13 in) and November (30 cm, 12 in). At Trincomalee on the north-east coast there is a more marked dry season between February and July (these six months receive only about 5 cm (2 in) each), and the main period of rains occurs in November (36 cm, 14 in), December (36 cm, 14 in) and January (18 cm, 7 in).

The effect of the relief is evidenced by the two 'dry belts', one in the northern plains, the other along the south coast, where the total is about 130 cm (50 in). Admittedly this would be distinctly wet for middle latitudes, but its effectiveness (p. 432) is reduced by the rapid evaporation and the torrential short-lived showers.

A3. Tropical continental climate Between the equatorial belt of rainfall and the hot deserts lies a transition zone which experiences convectional rainfall within the Inter-Tropical Convergence Zone during the summer months, but is under the influence of *TmK* air-masses (the Trade Winds) for the rest of the year. The Trades are rain-bringers only to certain eastern margins; in the continental interiors they are dry winds. The North-east Trades, for example, blow from the Sahara as far as West Africa; near the coast they are known as the *Harmattan*, sometimes as 'the Doctor' because the dryness of the wind forms a healthy contrast to the humidity of summer. Inland, however, these winds are extremely dry and dust-laden, and are very trying.

As a result of this seasonal change, areas with a Tropical Continental climate experience a period of summer convectional rain, alternating with a dry winter during which either the Trades or stable masses of *Tc* air are dominant. As the distance increases from the Equator, both the rainfall total and the length of the rainy season decrease. Thus while Akassa in the equatorial region at the mouth of the Niger delta receives 366 cm (144 in), Kano farther north has 81 cm (32 in), and the south of the Niger Republic only about 50 cm (20 in). Kano has no rain at all between November and February, and very little in the four succeeding months until the rainy season of July (18 cm, 7 in), August (28 cm, 11 in) and September (13 cm, 5 in). Most of this rain is short-lived and torrential, a disadvantage for agriculture which is emphasized by the high rate of evaporation, and moreover the rainfall total is also extremely unreliable from year to year.

The seasonal range of temperature gradually increases as the difference between summer and winter insolation becomes more pronounced with increasing latitude. Maximum temperatures are attained before the onset of the rains as a result of the dry air and cloudless skies. Monthly mean maxima exceeding 32° C are common as the desert borders are approached, and shade temperatures of 43° C are experienced. The diurnal range of temperature is marked, since the clear skies which promote daytime heating also allow night-time radiation, so that even within the Tropics night-frosts are not unknown.

The main areas experiencing a Tropical Continental climate occur in Africa to the north, east and south of the Guinea Coast–Congo Basin equatorial area, particularly as a continuous 1000 km (600 mile) wide strip across the continent from Cape Verde almost to the Eastern Horn, and extending through East Africa almost as far south as Cape Province. In South America the type is found on the Guiana and Brazilian Highlands, and on the Mexican plateaus. This climate is responsible for the widespread Tropical Grasslands or Savanna (pp. 528–9).

A3m. Tropical continental monsoon climate The main causes and features of the monsoonal reversals of air-flow have been discussed, and the results have already been appreciated in the description of types **A1m** and **A2m**; in these cases the monsoonal effect accentuates the rainfall maximum, but the rest of the year is far from rainless.

However, in India, Burma, Thailand, southern China, northern Australia and the lands near the Eastern Horn of Africa, there is a complete seasonal contrast between the inblowing rain-bringing monsoon ($TmK(M)$) and the outblowing dry monsoon ($Pc(M)$). The dry season is so marked that it justifies the delimitation of a Tropical Continental Monsoon type.

India The subcontinent of India exhibits this monsoonal climate very strikingly. The climatic year is usually divided into four seasons —the 'Cold Season', the 'Hot Season', the 'Season of Rains' and the 'Season of the Retreating Monsoon'.

The *Cold Season* in January and February is characterized by clear skies and dry, sunny weather. The mean temperature in the Punjab is about 10° C, in southern India about 21° C. The diurnal range is appreciable; in New Delhi mid-afternoon temperatures in February sometimes reach 29° C or 32° C, while there is a sharp fall at evening, often to temperatures below 10° C, and frosts are not unknown. During this season the air is cool and stable, and what winds do blow are offshore, so that much of India is dry. The only areas with rain are the Punjab and the extreme south-east of the peninsula. The former receives its rain from shallow depressions which move eastward; their origin is obscure, but they either form over the mountains of Baluchistan or penetrate all the way from the Mediterranean Basin. The amount of rain is not much, 2·5 to 5 cm (1 to 2 in) per month, although in the Himalayan foothills it is greater; Simla has about 18 cm (7 in) during each of January and February. The south-east coast, south of Madras, receives a few cm of rain from north-easterly air-streams associated with the retreating Inter-Tropical Convergence Zone (p. 412).

From March to mid-June is the *Hot Season*, and the heat steadily increases as the overhead sun appears to move north until maximum daily temperatures of well over 38° C are attained in May and early June; shade temperatures of 49° C have been recorded in Sind. Inland this season is characterized by heat, glare and excessive dryness, sometimes with leaden, hazy skies, but near the coast the humidity may be higher and conditions even less pleasant. Meanwhile, as the temperature increases so does the low pressure gradually build up in the Punjab. Storms are common, usually accompanied by dust-storms, and there may be occasional heavy downpours of rain or hail near the coasts, where moist air from over the sea is involved and atmospheric conditions are highly unstable.

From mid-June to mid-September is the *Season of Rains*, associated with the South-west Monsoon. As the low pressure over north-western India reaches its greatest development, the light air circulation of the Indian Ocean is overcome. The South-east Trades then sweep across the Equator, are deflected to the right in the northern hemisphere, and so form the South-west Monsoon over peninsular India. This northward penetration is facilitated as the Westerlies in the upper troposphere retreat to the north of the Plateau of Tibet with the seasonal readjustment of the general circulation (p. 410). That part of the monsoonal air-stream which crosses the Bay of Bengal turns north-westward up the Ganges valley towards the low-pressure focus in Sind, guided partly by the Himalayan mountain barrier.

As the result of their 6400 km (4000 miles) fetch over a warm ocean, the winds are laden with immense amounts of water-vapour. Their arrival in India and Burma is heralded by a few days of light rain and then a striking 'burst' occurs, associated with torrential rain and thunderstorms. The heaviest rain occurs along the ranges of hills lying at right angles to the wind direction—the Western Ghats, the Arakan Yoma and the Khasi Hills. Bombay has 50 cm (20 in) in June, 61 cm (24 in) in July, 38 cm (15 in) in August and

28 cm (11 in) in September. Akyab on the west of the Arakan Yoma in Burma has 518 cm (204 in) per annum, and Cherrapunji at an altitude of 1358 m (4455 ft) in the angle of the Khasi Hills in north-eastern India has 1161 cm (457 in), one of the wettest places in the world. In fact, 300 cm (905 in) was recorded here in one year, 930 cm (366 in) in one month, and 103·6 cm (40·79 in) in a single 24 hours. The lowland areas have much less and totals gradually decrease towards north-western India— Calcutta (163 cm, 64 in), Benares (102 cm, 40 in), Delhi (66 cm, 26 in) and Lahore (46 cm, 18 in). Moreover, the rain-shadow areas stand out strikingly; much of the Deccan has between 50 and 75 cm (20–30 in), Poona in the rain-shadow of the Western Ghats has only 74 cm (29 in) and Mandalay in the 'dry belt' of Burma behind the Arakan Yoma has 84 cm (33 in). Shillong, less than 40 km (25 miles) from Cherrapunji on the north-eastern leeward side of the mountains, has 89 cm (35 in).

In the north-west of India is the Thar Desert, where Jacobabad, 80 km (50 miles) north-west of the Lloyd Barrage, has a mere 10 cm (4 in) of rain. This extreme aridity is partly the result of its location near the low-pressure focus where the winds are practically destitute of moisture at the end of their long overland journey. It is also partly the result of a current of hot, dry air flowing at high altitudes from Baluchistan, thus precluding vertical ascent and the formation of cumulus clouds with possible convection rain.

Finally, from mid-September to December is the *Season of the Retreating Monsoon*. The effects of the rain-bearing winds are experienced farther and farther south as they leave the northern areas, to be replaced by light, variable northerly winds. In October and November these retreating monsoonal winds bring some rain to the south-east coast, for they still blow towards the land; thus Madras has its maximum in October (28 cm, 11 in) and November (36 cm, 14 in). Over most of India the sky clears, the sun shines continuously, the relative humidity falls and the lower temperatures of the 'Cold Season' approach.

Australia The monsoonal conditions experienced in northern Australia are complementary to those of Asia, though less potently developed. The focus of the summer monsoon is the low-pressure system which develops over north-central Australia, where mean monthly temperatures exceeding 32° C are experienced during December. The rain-bearing winds blow from the north-west and north across the Indonesian archipelago on to the northern coast. In January the area of rains almost reaches Capricorn. Wyndham, on the north coast of Western Australia, has a total of 71 cm (28 in), of which 56 cm (22 in) fall between December and March, while Darwin, nearer the Equator and backed by low hills, has 157 cm (62 in) with 42 cm (16 in) in January.

During the rest of the year Australia south of Capricorn experiences high-pressure conditions, and dry air moves northward across the coast towards Indonesia. Skies are cloudless, the relative humidity is low, and the winds are strong and steady, while virtually no rain falls in northern Australia.

B. WARM TEMPERATE CLIMATES

The group of Warm Temperate climates (fig. 202), characterized by hot summers and mild winters (that is, with no cold season), is transitional between the tropical areas and the Cool Temperate latitudes. There is a definite distinction between the western and eastern margins of the continents. The western margins receive winter rain from *Tm* air-masses and summer drought resulting either from the presence of dry Trade Winds or the effects of the calms of the Sub-tropical high-pressure area (*Tc* air-masses). The eastern margins tend to have rain throughout the year, both from the Trades and various local onshore winds, and from depressions which have penetrated eastward.

One further complexity is the monsoonal effect felt on the margins of the Asiatic land-mass in southern China, which is sufficiently distinctive to merit an individual climatic type.

B1. Western margin (Mediterranean) warm temperate climate This type occurs on western continental margins between latitudes 30° and 40°. In the New World the area of occurrence is sharply defined inland by the mountains in both central California and central Chile. During the winter depressions pass eastward from the Pacific Ocean, resulting in variations in the directions of the winds, but providing considerable periods with winds from the west. Between October and March, therefore, San Francisco has 48 cm (19 in) and Los Angeles 38 cm (15 in); similarly in the southern hemisphere winter, Valparaiso has 43 cm (17 in) between May and August. The rainfall totals increase poleward and decrease equatorward, while the duration of the rainy season alters similarly.

In summer the Sub-tropical high-pressure belt forms a distinct anticyclone off the Pacific coast in both the northern and southern hemispheres; winds are generally weak, with long periods of calm. This, coupled with the fact that along both coasts there are cool surface ocean currents (pp. 357-8) so that any air-masses moving from over the sea are cool compared with the heated land, results in

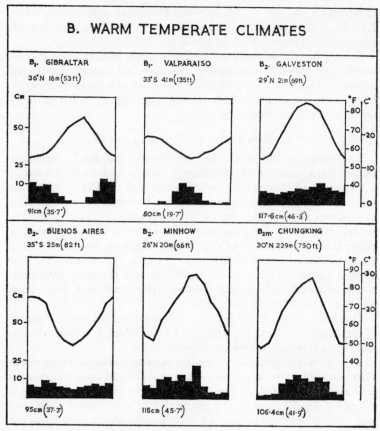

Fig. 202 *Warm temperate climates*

a virtually rainless summer. Fog is, however, very common near the coast, notably in the vicinity of the Golden Gate in California.

Even in the summer months more than 90 per cent of the winds in both California and Chile come from the sea. The mountain ranges virtually cut off these coastal margins from the rest of the continent; the common explanation of the summer drought as being the result of dry land-winds is incorrect.

One other effect of the cool ocean currents is that they ameliorate the summer temperatures of coastal stations; thus San Francisco has a mean July temperature of only 14° C. Inland, however, the cooling effect rapidly disappears, particularly in the Central Valley of California, where, for example, Red Bluff has a July mean of 27° C.

This Western Marginal regime is also experienced in the south-west of Cape Province in the Republic of South Africa and in south-western and southern Australia. The absence of any marked mountain barriers and the narrowness of the continents cause the climate to change gradually eastward in both Cape Province and southern Australia into the Eastern Marginal type (**B2**), while in Western Australia it passes into semi-desert and desert.

The Mediterranean basin The largest extension of this climatic type is round the Mediterranean Sea, which is so regarded as the type area that the word 'Mediterranean' has become synonymous both for this winter rain–summer drought regime and also for the characteristic type of associated vegetation cover (pp. 524–6). In point of fact, the Mediterranean Basin, with its alternation of seas with mountainous islands and peninsulas, and its lengthy eastward extension into the heart of the Old World land-mass, provides complex variations to the general regime. The broad features of the type are indeed present—winter rainfall and more or less complete summer drought, mild winters above 6° C, and hot sunny summers usually above 21° C. The rainfall total decreases generally eastward— Gibraltar (91 cm, 36 in), Palermo (64 cm, 25 in) and Athens (41 cm, 16 in), but aspect (such as the difference between the western and eastern sides of the Italian peninsula) is extremely important.

The Mediterranean Sea is an area of seasonal change of pressure and winds. It has been called 'a lake of winter low pressure', but more accurately it is a distinct frontal zone (the Mediterranean Front, or MF) (p. 412), characterized by a succession of depressions moving eastward between the high-pressure areas over central Europe and northern Africa. These depressions move slowly, following more or less well-marked tracks, and the passage of each results in successive changes in wind direction. Some depressions penetrate from the Atlantic through the Straits of Gibraltar or through the Gate of Carcassonne, while others form within the Mediterranean itself, the so-called *lee-depressions* (p. 419). The particular pressure distributions are therefore associated with various dominant wind directions, each of which has distinct characteristics of temperature and humidity (fig. 192).

In summer the mid-latitude high-pressure belt of Tc air-masses extends over southern Europe and the Mediterranean, and the basin is characterized by calms or light northerly winds. On a pressure map isobars run generally north–south, and pressure gradients are slight. An occasional depression may penetrate as far as Corsica, but this is uncommon.

The climate of Corsica shows clearly both the main features of the general climatic type and striking local variations, for which several factors are responsible. The island is situated in the centre of the western Mediterranean Basin in latitude 43° N, it experiences the modifying effects of an insular position, and much of its surface is mountainous. The insularity ameliorates summer temperatures round the coast, while local diurnal temperature and pressure changes cause pronounced land and sea breezes. The altitude results in a decrease of temperature (hence one can distinguish a series of altitudinal climatic zones), it emphasizes contrasts due to aspect, and it increases precipitation on windward slopes.

There is a well-marked autumn and winter maximum of precipitation. Ajaccio receives 74 cm (29 in) per annum and Bastia 91 cm (36 in), but totals increase considerably in the interior. Vizzavone, at a height of 1160 m (3800 ft) has 165 cm (65 in); probably the heads of westerly-facing funnel-shaped valleys have even more, but there are no rainfall records. An isohyetal map would be very similar to a relief map.

These precipitation statistics include snowfall. While this is rare at sea-level, it covers the island above 1500 m (5000 ft) from mid-December to April, and lingers even longer in north-facing gullies.

B2. Eastern margin warm temperate climate Eastern margins in warm temperate latitudes share the mild winters and hot summers of the western margins, but both rainfall amount and its seasonal distribution differ markedly. This Eastern Marginal type is found along the south-eastern Atlantic and Gulf coasts of the U.S.A., around the Plate estuary, in the extreme south-east of Africa, and in south-eastern Australia. The North Island of New Zealand is in a sense intermediate between the Western and Eastern Marginal types, since the width of the island is so small that it is exposed to maritime influences both from west and east.

The main features of this climatic type include mild winters and warm summers (Buenos Aires, July 9° C, January 23° C; New Orleans, January 12° C, July 28° C). Rainfall varies considerably from one station to another, depending on situation or aspect, but it is fairly well distributed throughout the year, since it is affected by *Tm* air-masses throughout the year; depressions moving eastward bring cyclonic rain in winter, and onshore winds from an easterly direction together with convectional influences result in summer rain. Sydney has 56 cm (22 in) during the six summer months and 66 cm (26 in) during the six winter months, while New Orleans has 64 cm (25 in) and 79 cm (31 in) respectively. The summer rain tends to fall in heavy short-lived showers, the winter rain more in the form of prolonged lighter drizzle. The rainfall total, however, is subject to great variations from year to year; Buenos Aires has a mean annual total of 94 cm (37 in), but has recorded as much as 203 cm (80 in) and as little as 53 cm (21 in).

The Plate Estuary This example includes the extreme south of Brazil, Uruguay and Argentina as far south as Bahia Blanca, extending inland until increasing aridity produces a semi-desert climate.

During summer, high pressure lies over the South Atlantic and the South Pacific in subtropical latitudes (p. 399). The winds at the east coast, although irregular, tend to blow from an easterly direction, that is, onshore. In winter the high-pressure belt extends across South America just to the south of Capricorn. A procession of depressions, forming over Patagonia and the adjacent sea area, moves eastward; these depressions have associated moisture-bearing south-easterly winds and so bring rain to the coastal areas.

In the coastal areas, therefore, the rainfall is fairly well distributed throughout the year, but farther inland both the total and the percentage received in winter decrease; Cordoba, with a total of 71 cm (28 in), receives less than 2·5 cm during the three winter months.

B2m. Eastern margin (monsoon) warm temperate climate None of the coastlands of eastern Asia in warm temperate latitudes is included in the Eastern Marginal type because the Asiatic coasts are dominated by the monsoonal air-stream reversal, with $TmK(M)$ air-masses in summer, $Pc(M)$ in winter. It is therefore necessary to distinguish a monsoonal variety of the Eastern Marginal type. This is found in China between the valleys of the Yangtse- and Si-kiang (it is, in fact, often known as the 'China Type'), and it also occurs in southern Japan.

The rainfall regime does not differ materially from that of the Eastern Marginal type, except that the strong South-east Monsoon makes the summer maximum much more clearly defined, especially in the interior. Thus at Minhow 74 cm (29 in) falls in the five months from April to August out of a total of 117 cm (46 in), while at Chungking in Szechwan only 15 cm (6 in) comes between November and March out of a total of 107 cm (42 in). What winter rain there is results from depressions which pass eastward along the Yangtse- and Si-kiang valleys, temporarily interrupting the cold outblowing monsoonal winds. Like the depressions which affect northern India (pp. 465-6), their origin is not very clear, since it is improbable that they could travel eastward through Europe and Asia from the Atlantic and still retain appreciable moisture. They must originate within central Asia itself.

Another distinction between the monsoonal variety and the Eastern Marginal type is the greater seasonal contrast in temperature. Southern China is exceptionally cold for its latitude, because of its exposure to the cold North-west Monsoon blowing out from the Asiatic high-pressure area. Thus the mean January temperature at Shanghai (32° N) is only 3° C. Chungking, however, in the Szechwan Basin sheltered by a mountain rim, has a mean January temperature of 9° C, an example of an inland station experiencing higher winter temperatures than a place on the coast in more or less the same latitude.

Southern Japan has its winter temperatures modified by insularity, with January figures around 7° C. Summer temperatures are in the neighbourhood of 27° to 29° C.

C. COOL TEMPERATE CLIMATES

The distinguishing criterion between the Warm and Cool Temperate climates is the fact that the latter have a clearly marked cold season, experiencing from one to five months with mean temperatures near or below 43° F (6° C). The Cool Temperate group can be divided into three (fig. 203).

Those areas which lie on the west coasts of continents between latitudes 40° and 55° are subject to marine influences, with mild equable temperatures and with depressions and associated westerly winds throughout the year (Marine type). Farther inland is a Continental variety, with increasingly extreme seasonal temperatures and a decrease in the rainfall total. Thirdly, on the Asiatic margin the strongly marked seasonal periodicity of the monsoonal regime produces a Continental Monsoon variety.

C1. Cool temperate marine climate The Marine variety occurs in North America along the coasts of British Columbia, Oregon and Washington, where it is limited by the Coast Ranges to a narrow strip. It extends as a broader zone in Europe, including the British Isles, southern Scandinavia and western Europe, but the absence of high mountain barriers allows the Marine regime to grade eastward into the Continental variety, and still farther eastward into the Cold Continental type. In the southern hemisphere the Marine regime occurs along the narrow coastline of southern Chile, in Tasmania and in the South Island of New Zealand.

Their main characteristic is the non-periodic passage of different pressure systems, with associated air-masses of varying temperature and humidity (pp. 413–19). All the main air-mass types can be distinguished at some time of the year, though Pm types are the

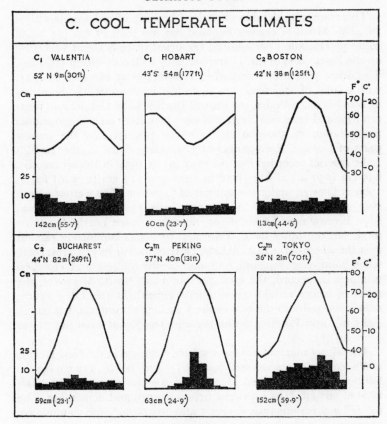

Fig. 203 Cool temperate climates

most frequent. In the words of A. A. Miller, 'by the greater or lesser proportion of these two qualities of purity and frequency of occurrence of each air-mass and its weather type are explained the climatic contrasts of maritime and continental varieties'.

Some generalizations are perhaps possible. There is a small annual range of temperature (Valentia, January 7° C, July 15° C; Hobart, July 8° C, January 17° C), and in true maritime situations such as the Scilly Isles frosts are extremely rare; the lowest temperature ever recorded in the Scillies was —3° C. Rainfall is evenly distributed throughout the year, but there is a tendency to a winter maximum and a spring minimum. Thus Valentia has 142 cm (56 in), of which only 25 cm (10 in) fall in the period of April to June, and the wettest month is December with 15 cm (6 in).

There are considerable variations because of orographic effects (p. 488). Much of eastern England has less than 61 cm (24 in) of mean precipitation per annum; the driest place is Great Wakering on the Essex coast of the Thames estuary with only 46 cm (18·1 in). The driest year ever recorded in Britain was at Margate in 1921, with a total of only 23·60 cm (9·29 in). By contrast, the Snowdon area in northern Wales, the central English Lake District and western Scotland near Ben Nevis and elsewhere have an average annual precipitation of 380–500 cm (150–200 in), forming the wettest parts of Europe. Rain-gauges at Sprinkling Tarn in the English Lake District recorded 653 cm (257 in) in 1954 (a British record), 635 cm (250 in) at Sty Head in 1928 and 617 cm (243 in) fell in 1909 at Glaslyn under the summit of Snowdon. The wettest month ever recorded was October 1909 with 144 cm (56·54 in) at Llyn Llydaw near Snowdon, or more than twice London's annual total. On the other hand, February 1932 was completely rainless over the whole of the English Lake District, and in July–August 1968, when south-east England had one of the wettest and cloudiest summers on record, the area had one of its finest. Moreover, the areas of heavy rainfall are limited; Penrith in the Eden valley, only a few miles to the east of the Lakeland mountains, has 92 cm (36·4 in), and Workington on the windward west coast has 94 cm (36·9 in) (fig. 199).

The winter maximum is more strongly marked in the New World; of Vancouver's mean annual total of 150 cm (60 in), 112 cm (44 in) fall between October and March, and 200 cm (79 in) of Valdivia's total of 267 cm (105 in) occur between April and September. The rainfall is predominantly cyclonic and orographic, and so totals are heaviest on western margins and decrease eastward.

This Marine variation of the Cool Temperate climate changes gradually eastward into the Continental variety and equatorward into the Warm Temperate type. Fig. 198 plots the monthly rainfall for stations in France, revealing strikingly the transitional nature of its position.

C2. Cool temperate continental climates The Continental variety lies inland of the Marine type, grading into it in Europe since there is no major mountain barrier, and extending eastward until increasing continentality results in the long cold season which defines the cold climates. In North America, however, the Cordillera forms a distinct climatic boundary, and a zone of 'Mountain climates' (**G**) separates the two types.

The Cool Temperate Continental climate in Europe is characterized by a gradual decrease eastward in rainfall total, with an increasing tendency toward a summer maximum (the result of convection influences), and a much more marked seasonal range of temperature, particularly an increasingly severe winter. Precipitation in winter is usually in the form of snow, which lies long but rarely attains much depth because of the low humidity.

Cyclonic influences are less marked, especially in winter when the more stable anticyclonic conditions, the result of the cooling of the Eurasiatic land-mass, prevent the eastward penetration of depressions. Mild spells in winter become progressively rarer, and sometimes a westward extension and intensification of the winter Asiatic anticyclone may cause a long spell of exceptional cold. The air is on the whole drier, and there is an increasing tendency to long periods of drought, clear skies and settled fine weather.

Germany Germany (including both West and East) lies athwart the transition zone between the Marine and Continental types and it is possible to distinguish four main climatic sub-regions illustrating the features of the Cool Temperate climates.

A relatively narrow area bordering the North Sea coasts has a 'modified Marine' type of climate. Hamburg has a January mean of 0° C (the Elbe sometimes freezes) and a July mean of 17° C. Rainfall of about 71–76 cm (28–30 in) total, is fairly well distributed, but there is a slight autumn maximum and a more marked spring minimum.

A 'modified Continental' type of climate lies farther inland, extending eastward to the middle Elbe with a slightly more marked temperature range; Magdeburg has a January mean of −0·5 C. and a July mean of 18° C. There is a slight decrease in precipitation, with totals about 58–66 cm (23–26 in), a definite summer maximum as convection becomes more effective, and a pronounced spring minimum. The effect of relief is shown by the isolated mass of the Harz Mountains rising from the North European plain, with a total precipitation exceeding 150 cm (60 in).

East of the Elbe is the true Cool Temperate Continental variety, which is still more extreme. Berlin has a January mean of − 1° C and a July mean of 19° C, with 58 cm (23 in) of rainfall, the wettest month being July (8 cm, 3 in), the driest February (3·6 cm, 1·4 in). The variability in the rainfall is remarkably small; the figures available for Berlin for seventy-three years show that the annual total is rarely much below 50 cm (20 in), and rarely above 71 cm (28 in). The wettest year had 76 cm (30 in) (in 1882), the driest had 36 cm (14 in) (in 1857).

Southern Germany exhibits all the local variations which result from a diversified relief. Temperatures are to a large extent the result of altitude; the Zugspitze observatory (at 2963 m, 9720 ft) has a January mean of − 12° C and an August mean of 2° C. Considerable areas in the Black Forest receive more than 130 cm (50 in) of rain and the Alpine foothills have 150–200 cm (60–80 in), though by contrast the Rhine valley has only about 50 cm (20 in) as it lies in the rain-shadow of the Vosges. Much of the winter precipitation is in the form of snow; together with the considerable sunshine which is the result of frequent anticyclonic conditions, this is responsible for the attractiveness of winter sports in the Bavarian Alps; the winter Olympic Games of 1936 were held at Garmisch-Partenkirchen.

North America The Cool Temperate Continental type in the New World extends eastwards across southern Canada and the U.S.A. as far as about 35° N. This area has a complicated regime, since it is affected both by depressions which form over the continent itself and by winds from the Atlantic. The absence of a transverse barrier enables air-masses both from the Gulf of Mexico and from beyond the Arctic Circle to reach the area; New York and Chicago are sometimes hit by devastating winter blizzards, but a summer heat-wave may push the thermometer above 38° C.

Omaha provides a typical example of the true Continental type. Its mean January temperature is − 6° C, its July temperature 25° C, while 56 cm (22 in) of its annual total of 74 cm (29 in) fall between April and September.

Eastern U.S.A. The eastern coast of the U.S.A. is included in this climatic type because of the seasonal temperature range, although this is not as marked as in the continental interior (fig. 185). Boston has a January mean of − 3° C and a July mean of 22° C, while Washington has respective means of 1° C and 25° C. However, maritime influences, brought by south-easterly winds from over the Atlantic in front of easterly-moving depressions, cause quite an even distribution of rainfall. Washington has only one month with a mean rainfall figure below 8 cm (3 in) out of its annual total of 104 cm (41 in).

C2m. Cool temperate continental (monsoon) climate A small part of northern China, Korea, southern Manchuria and central Japan falls into the Cool Temperate type because of its cool season of one to five months' duration. These, like the other eastern margins of the Asiatic land-mass, experience the strongly marked periodicity of the monsoons. Rainfall is mainly in summer from the south-easterly sea-winds; Peking has 64 cm (25 in) per annum, with 53 cm (21 in) falling between June and September. The cold out-blowing winds result in very low winter temperatures

D. COLD CLIMATES

D₁. BODÓ
67°N 2m(7ft)

D₂. WINNIPEG
50°N 232m (760ft)

D₂ₘ. HARBIN
46°N 160m (525ft)

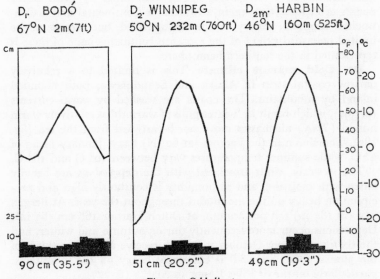

90 cm (35·5") 51 cm (20·2") 49 cm (19·3")

Fig. 204 *Cold climates*

for the latitude; Peking has a January mean of −5° C and Mukden of only −13° C. Summer temperatures are, however, high, and even at Mukden (at latitude 42° N) the July mean is 25° C.

Central Japan has a definite monsoon climate, but its insular position, particularly in relation to the warm Kuroshio (p. 357), helps to modify somewhat the extreme features experienced on the mainland. The mountainous islands are affected by depressions and by both monsoons; western Japan has more rain in winter, eastern Japan more in summer, but most parts have some rain throughout the year. Tokyo has 150 cm (60 in), of which almost exactly two-thirds falls between May and October. Temperatures are less extreme than on the mainland; Tokyo, in practically the same latitude as Peking, has a January mean 8° C higher.

D. COLD CLIMATES

Much of what has been said concerning the Cool Temperate climates applies in greater degree to Cold climates (fig. 204). They lie within the Westerly wind-belt, and show the same transition inland between the Marine and Continental types, with decreasing precipitation, increasing range of temperature and accentuation of the winter cold. But winters are much longer and much more severe than in the Cool Temperate zones, dominated by great stable masses of *Pc* air. Once again a Marine, a Continental and a Continental Monsoon variety can be distinguished, but because of the limited poleward extent of the southern continents none of these is represented in the southern hemisphere.

D1. Cold marine climate This is limited to a relatively narrow coastal strip in Alaska and Scandinavia, both bounded inland by mountains. The coasts are washed by warm currents (p. 392), which result in 'winter gulfs of warmth' as relatively warm humid (*TmK*) air-masses move north-eastward from the sea (fig. 182). Thorshavn in the Faeroes (at 62° N) has a January mean of 3° C, while summer temperatures vary between 10° C and 16° C.

The westerly winds associated with the depressions are heavily laden with moisture, and so humidity is constantly high and precipitation heavy on the mountains throughout the year. At Bergen 213 cm (84 in) fall per annum, at Dutch Harbor 160 cm (63 in). Depressions occur more frequently during autumn and winter, and the six months from October to March receive about two-thirds of the annual total. Snowfall is extremely heavy on the mountains, particularly on the St Elias Range in Alaska (p. 227).

D2. Cold continental climate This type extends across North America from Alaska to the Gulf of the St Lawrence, and across Eurasia from the Baltic almost to the Pacific. Temperatures rapidly become more and more extreme as the continental interiors are approached; in central Canada Winnipeg has a January mean of —20° C and Dawson City (at 14° of latitude farther north) has —31° C. In Asia the size of the land-mass makes the winter cold even more intense. The January temperature at Bergen is 1° C, at Oslo —4° C, at Leningrad —9° C, at Tobolsk —19° C, at Olekminsk —68° C, and at Verkhoyansk (which is virtually at the 'cold pole' of the globe) it is —51° C; the phenomenally low temperature of —68° C has been recorded at the last-named place. Yet summer temperatures are quite high; Verkhoyansk has a July mean of 16° C and Winnipeg of 19° C.

Precipitation is sparse, due to the distance of these continental interiors from the sea and (particularly in North America) the presence of mountain barriers in the west. Many parts receive less than 50 cm (20 in); Edmonton has 43 cm (17 in), Verkhoyansk only 13 cm (5 in). What precipitation does occur in winter is in the form of dry blizzard-snow of negligible amount, and more than half of the precipitation falls during the three months of summer.

Eastern Canada The east coastal margins of Canada qualify for inclusion in this Cold Continental climatic type because of the severe winters and the considerable annual temperature range. Quebec has a January mean of −12° C, a July mean of 19° C. However, precipitation is considerably heavier than in the continental interior. Quebec has an annual precipitation total of 104 cm (41 in), and only one month has less than 8 cm (3 in). This is the result of moisture-laden sea-winds, associated with the frontal areas of depressions moving eastward over the Great Lakes and the St Lawrence valley. Most winter precipitation is in the form of heavy snow.

D2m. Cold continental (monsoon) climate There is little need to say much about the monsoonal variety of the Cold Continental type, which affects central and northern Manchuria, the Pacific coastlands of the U.S.S.R., the island of Sakhalin and the Kamchatka peninsula. In winter conditions are virtually identical with those of the Continental type, with bitter cold (Vladivostock, January −15° C, Okhotsk, −24° C, and the icy winds of the out-blowing North-west Monsoon. In summer temperatures rise considerably (Vladivostock, August 21° C), and the South-east Monsoon, blowing from the sea, brings much more rainfall than is experienced in the interior of Siberia; Vladivostock receives nearly 36 cm (14 in) between April and October.

E. ARCTIC CLIMATES

Arctic climates (fig. 205), which have no month with a mean temperature as high as 10° C, cover the extreme north of Alaska, Canada and Labrador, the whole of Greenland, the interiors of islands within the Arctic Circle, and the northern coast of the U.S.S.R. In the southern hemisphere there is the Antarctic continent. Meteorological knowledge of Arctic climates until the last two decades or so was scanty and sporadic, based on explorers' records which, although of great interest as 'climatic samples', were often of a short-term and unsystematic character. Now, however, a considerable number of permanently staffed meteorological stations has been established during and since the War of 1939–45 in Greenland, Iceland, Jan Mayen, Spitsbergen, Alaska, the Canadian islands, some of the Antarctic island groups and on the Antarctic continent itself.

This climatic type may be subdivided into three. The *Tundra* climate has a brief summer above freezing-point during which there is continuous daylight, and so the characteristic tundra vegetation (pp. 532–3) develops. The sky is usually clear in summer, long hours of sunshine are recorded, and while air temperatures are not high,

E. ARCTIC CLIMATES

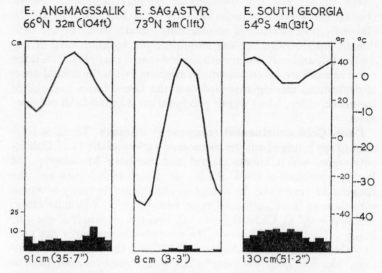

E. ANGMAGSSALIK
66°N 32m (104ft)

E. SAGASTYR
73°N 3m (11ft)

E. SOUTH GEORGIA
54°S 4m (13ft)

91cm (35·7") 8 cm (3·3") 130cm(51·2")

Fig. 205 *Arctic climates*

sun temperatures (as recorded by a black-bulb thermometer) may exceed 38° C. Winter temperatures fall well below freezing (Sagastyr on the Siberian coast has a February mean temperature of −38° C, which is, however, not so low as those in the heart of the Asiatic Cold Continental type). Precipitation is usually in the form of blizzard-snow and rarely amounts to more than about 28 cm (12 in).

The true *Polar* climate is one of perpetual frost, with intense winter cold (Admiral Byrd recorded −51 °C on twenty days during a winter spent alone at 80° S in 1933–4), frequent blizzards, and summer air temperatures also below freezing but with often quite high sun temperatures.

While the Tundra and Polar regions have been described as Arctic climates, they may logically be included also in the group of Desert Climates (F), as 'Cold Deserts'. Precipitation is low, usually in the form of fine blizzard-snow; Kola (near Murmansk) has 20 cm (8 in) and Sagastyr only 8 cm (3 in).

Because of the number of island groups in both the Arctic and Antarctic Oceans, it is possible to distinguish a *Marine Arctic* type. Here winter temperatures are considerably ameliorated. Jan Mayen, for example, situated at 71° N, has a March temperature (the lowest

F. DESERT CLIMATES

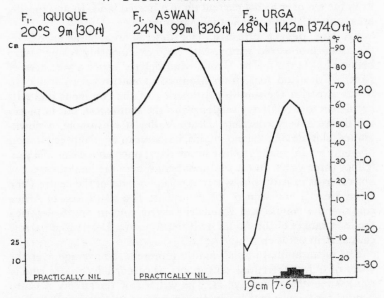

Fig. 206 *Desert climates*

monthly mean) of −4° C, modified by the influence of the North Atlantic Drift, and Spitsbergen at 78° N has a February mean of −19° C. The precipitation figures tend to be higher than those of other Arctic areas, so far as the short-term available records indicate, the result of depressions finding their way into the polar zones. Angmagssalik on the south-east coast of Greenland has 91 cm (36 in) and South Georgia in the Antarctic has over 130 cm (50 in).

F. DESERT CLIMATES

Deserts (fig. 206) are the result of aridity, the implication of which was defined on p. 449. Aridity may result in the first place from a location in a continental interior, far from winds deriving their moisture content from an adjacent ocean, as in central Asia. In the second place it may occur in an area permanently under the influence of a large stable high-pressure system (a *Pc* or *Tc* airmass), with low humidity and descending air currents. In the third place it may be found in an area under the influence of dry or drying winds, such as a local 'rain-shadow', or where winds such as the Trades have blown over an extensive land area, and, moreover, are blowing from cooler to warmer latitudes, or where a wind is blowing from the sea over a cool current on to a heated land. In any specific area, any or all of these factors may operate.

F1. Hot desert climate Hot deserts are characterized by the absence of a cold season, with no mean monthly temperatures below 6° C (43° F). The 'Trade-wind deserts' cover a vast area of North Africa and Asia, from Morocco to north-western India. In North America the southward tapering of the continent limits its hot desert to a relatively small area in the south-west, but in places the desert is very complete. Death Valley, for example, a down-faulted 'slot' among the mountains, has a mean annual precipitation of 4 cm (1·5 in) and 351 clear sunny days; once no measurable rain fell for three years. In the southern hemisphere the largest extent of hot desert is in Australia, where it occupies most of the centre ('the Dead Heart') and west of the continent. The south-west of Africa contains the Namib and Kalahari Deserts, and in South America the long ranges of the Andes limit the Atacama Desert to a narrow coastal strip between 7° and 32° S.

The mean annual rainfall figures represent the average over the years of a few torrential downpours, succeeded by many wholly rainless months or even years. The writer saw the streets of Cairo flooded for twenty-four hours during a day's continuous downpour in February 1945, but the mean February figure is only 0·5 cm (0·2 in) and the mean annual rainfall 3·3 cm (1·3 in).

A distinction can be made on the basis of temperature between coastal and continental locations.

Coastal Hot Deserts Where the hot deserts reach the west coasts of continents, they are affected by equatorward-flowing cool currents and by upwelling cold water. These currents result in surprisingly cool summers, for during this season winds blow from the sea on to the land. The mean temperature of the hottest month at Walvis Bay is 19° C and at Iquique only 22° C; the latter figure is remarkably low for a station well inside the Tropics. This also reduces the seasonal range; the difference between the coolest and warmest month at Walvis Bay is a mere 5° C.

One other effect of these 'cold-water coasts' has been noted in connection with Warm Temperate climates: the high humidity, causing fogs and heavy dews but seldom producing rain (p. 468). The fogs form over the sea near the coast and roll inland for distances varying from 3 to 110 km (2–70 miles), they may cause sufficient condensation to nourish a scanty vegetation, as in the Atacama Desert.

Continental Hot Deserts Those deserts which occur in the continental interiors are far more extreme than those along the coasts. The world's highest temperatures are recorded here, the result of intensely dry air, cloudless skies and continuous daytime insolation. Temperatures above 50° C are recorded; the world record shade temperature is reputed to be 58° C, recorded in 1922 at Azizia, some 40 km (25 miles) south of Tripoli. Probably the highest monthly mean over a long-term period of observations is at In Salah in Algeria, whose July mean is 37° C, while Jacobabad, in the desert of Sind, has a June mean of 36·6° C.

The seasonal range in these Continental Hot Deserts, while higher than in the 'cold-water deserts', is not very great, usually less than 17° C. The diurnal range is, however, very pronounced. The clear skies which allow great insolation during the day allow very rapid terrestrial radiation by night, and there may be a fall of 14° to 17° C in the couple of hours succeeding sunset. Diurnal ranges of 33° C or even 39° C are commonly recorded, and frost is not unknown during winter nights. Azizia in Tripoli once recorded 52° C and −3° C during the same 24 hours, a record diurnal range.

F2. Mid-latitude desert climate Some deserts occur in the interior of the Asiatic and North American land-masses among the high basins and plateaus with their mountain rims. Their continentality and enclosed nature result in high summer temperatures (Kashgar, July 27° C, Luktchun, 32° C), and day maxima exceeding 43° C are commonly recorded. But the same continentality allows cold anticyclonic conditions to dominate in winter, and most of the Asiatic cold deserts have winter temperatures well below freezing; Luktchun, which lies in the Turfan Basin 17 m (56 ft) below sea-level, has a January temperature of only −11° C.

Rainfall is negligible, and what does occur is the result of the penetration of an occasional stray depression in winter or of a rare convectional storm in summer. Kashgar has a mean annual total of only 9 cm (3·5 in).

Patagonia This southern part of the Argentine qualifies on grounds of aridity for inclusion in the Desert type, but the South American continent is so narrow that the temperature range is reduced because of the proximity to the ocean. Santa Cruz on the coast has a range of 16° C, with a July mean of 0·5° C and a January mean of 16° C. The aridity results from its position on the leeward side of the Andes, and only 13–15 cm (5–6 in) of rain falls.

G. MOUNTAIN CLIMATES

It is hardly possible to describe Mountain Climates in general terms, since it is often said that 'Mountains make their own weather', and if it were possible to construct accurate actual temperature and rainfall maps of mountain regions, the isotherms and isohyets would to a large extent run with the contours.

Every mountain range has a series of broad altitudinal climatic zones, the nature of which depend on (i) the 'base level' climate, and (ii) the actual height of the range. The Andes, for example, present a range of climatic zones from Equatorial to Polar in character. Quito, the capital of Ecuador, on the Equator but at an intermediate height of 2850 m (9350 ft), has mean monthly temperatures varying less than a degree from 13° C; it presents both the equable effects of its latitude and the tempering effects of altitude.

Altitudinal zones Two examples must suffice to illustrate mountain climates. In *Mexico and Central America* from sea-level to about 900 m (3000 ft) is the *Tierra Caliente*, the hot tropical coastlands. Between 900 m and 1800 m (3–6000 ft) is the *Tierra Templada*, with temperatures ranging between 18° C and 24° C. Above that again is the *Tierra Fria*, the region most suitable for European settlement. Mexico City lies at a height of 2250 m (7400 ft); its coolest month is January with a mean temperature of 12° C, its warmest month is May, only 6° C higher, and its rainfall of 58 cm (23 in) falls mainly in the summer months.

Corsica presents a series of altitudinal zones which emphasize the variability of the generalized Western Margin Warm Temperate climate. From sea-level to 180 m (600 ft) the climate is 'true Mediterranean', with hot summers, though moderated by sea-breezes, and warm winters. From about 180 m to 900 m (600–3000 ft) is an equable and pleasant climate; the summers are cooler, the precipitation is greater and the dry season shorter than at sea level. From 900 m to 1500 m (3000–5000 ft) there are still warm sunny summers but the winters are quite cold; rain or snow may occur at all seasons, although the summer minimum is still evident. From about 1500 m (5000 ft) to the highest summits are cold, even severe, winters, with some five to six months of snow-cover.

General features of mountain climates Many features in mountainous areas are the result of the local configuration, and each area must be discussed in its own right. The climatologist must examine all the local details of exposure to or shelter from the dominant winds, and the resulting rainy or rain-shadow conditions (p. 448); the incidence of snowfall and the nature of the snow-cover (pp. 210–11); temperature inversion and the incidence of frost and fog in valley bottoms (p. 391); the amount of insolation received in the valley bottoms and on the mountain-sides; contrasts in aspect, as in the case of the shady side (*ubac* or *Schattenseite*) and sunny side (*adret* or *Sonnenseite*) in an Alpine valley; the incidence of winds such as the foehn or chinook (p. 423); the frequency of mountain and valley winds (p. 425); and the occurrence of 'mountain storms', often of great violence and intensity.

CHAPTER 19

The Soil

THE thin surface layer of the earth, known as the regolith (p. 95), is formed by the break-down of the rocks in various ways and by various processes (plate 100). Its nature therefore depends first on this 'parent' rock from which it develops. This is modified and altered by weathering, which involves both physical and chemical processes (Chapter 4). But mere disintegrated rock particles are not soil. In addition, there are the results of the interaction of various biological processes, particularly the growth and decay of the vegetation layer and the action of soil organisms upon it; there is a kind of cycle, for vegetation derives mineral salts from the soil, but when it decays these return to the soil. This cycle is known as a *soil-vegetation system*, which may achieve a certain balance or stability but may be radically altered either by natural processes or by man.

The particular climate affects the type of soil, both directly by means of its weathering effects and indirectly as a result of the vegetation cover for which climate is largely responsible (Chapter 20). One type of soil may form under mid-latitude grasslands, another under coniferous forests; one type may form in a hot, wet climate, another in a cold, tundra climate. There is a tendency for a particular climate to develop a particular climatic soil type; in the U.S.S.R. and North America a broad zonal classification of soils can be made which corresponds equally broadly to the latitudinal climatic zones (fig. 208). But in an area such as the British Isles, where the parent rocks, the relief, the drainage and the climate are all extremely diversified, so too are the types of soil. Moreover, the period of time during which all these factors have been interacting must be taken into account.

A soil therefore consists of mineral particles, a certain proportion of decayed organic material, soil water, a soil atmosphere, and living organisms, which exist in a complicated and dynamic relationship one with another. Particularly important, though microscopically small (less than $2 \cdot 5 \times 10^{-5}$ cm in diameter), are the colloidal particles or *colloids*, of both mineral and organic material, which play a vital though highly complex role in soil chemistry. On the one hand their function is physical, in that they impart to certain soil constituents (notably clay particles) an adhesive quality. On the other hand their function is chemical, in that they are electrically charged and can attract and hold ions of dissolved substances, particularly of bases such as calcium, magnesium and potassium, and also hydrogen.

THE MINERAL CONTENT OF THE SOIL

The parent minerals These include both hard, resistant rocks such as granite and slate, and also less resistant rocks such as recent volcanic lavas and ashes, and most of the sedimentary rocks (sandstone, clay and limestone). In the British Isles the old rocks occur mostly in the west and north, the new rocks in the south and east. The major distinction drawn between 'upland Britain' and 'lowland Britain' is reflected in the nature of the soil-cover; the poor, thin soils of the mountains and moorlands contrast with the deeper lowland soils.

The term 'rock' is strictly applied not only to granite, sandstone and the like, but also to gravel, clay and unconsolidated sand. These materials, laid down haphazardly in geologically recent times, cover very large areas. Many are unconsolidated, so that roots can readily penetrate; some like sands and gravels are so loose that the most essential soil-forming process is to make the shifting, unstable particles more cohesive. Others, such as heavy till, are already cohesive and 'sticky'. These superficial deposits are usually 'young', and the soil-making processes have not as a rule progressed far, but some deposits are themselves the product of disintegration. Loess, for example, is fine-grained yet porous, compact yet not cemented, and will respond rapidly to soil-forming processes. So too will river-borne alluvium; the river flood-plains of south-eastern Asia form some of the most intensively cultivated land in the world.

Limestone Limestone as a soil parent material is really in a class by itself. It is of widespread and varied occurrence (p. 130), but essentially it consists of calcium carbonate which is soluble in rain-water containing carbon dioxide. Varying amounts of insoluble material are also present—much in impure limestone, little in pure chalk—which remains as a residue after long-continued solution has operated. The Clay-with-flints, which lies in patches on the Downs of south-east England, is to some extent a residue from the weathering of chalk.

Most soils formed on limestone are thin and dry, and on steep slopes soil may even be non-existent. But where the residue can accumulate, retaining some of the lime content, a sweet, short turf, characterized by lime-loving plants, will develop. Two major soil types are recognized which develop on limestone, *terra rossa* and *rendzina* (p. 502).

Elements in the soil The minerals which form the bulk of the soil consist of silica (quartz) grains, the commonest product of rock disintegration, of silicates (such as aluminium silicate), and of oxides (notably various iron oxides). Quartz grains are resistant to chemical weathering and so form a 'stable' or 'inert' soil constituent; almost all sandy soils consist largely of it. Most silicates undergo hydration and their end-product is clay.

Soil contains other elements in small and varying amounts, but of the utmost importance because they supply plants with food. They include compounds of calcium, sodium, potassium and magnesium; of nitrogen, sulphur and phosphorus derived partly from organic substances, partly from the parent rocks; and of oxygen, hydrogen and carbon obtained from the air and water. In addition, minute quantities of 'trace elements', such as boron, manganese and iodine, are also present.

One important aspect of farming practice consists of replacing in the soil the elements which successive crops have removed. Rotation helps since different crops have different needs, while leguminous plants such as clover can 'fix' nitrogen from the atmosphere in their root nodules. Farmyard manure, apart from its ideal physical and bacteriological advantages, provides both nitrates and phosphates. The use of 'artificials' enables mineral deficiencies to be accurately restored; nitrogen is provided by ammonium sulphate and sodium nitrate, phosphorus by basic slag and bone-meal, and potash by kainit (crude sulphate of potash) and wood ashes.

Soil acidity In many cool, moist areas, percolating ground water gradually leaches out the soluble bases, particularly calcium, for rain-water, which is a weak solution of carbon dioxide, can readily dissolve these bases. As a result the soils gradually become lime-deficient, in other words increasingly acid and 'sour'. In the U.S.A. the distinction between the soils to the east of the Mississippi, where precipitation exceeds 64 cm (25 in) and leaching is rife, and those to the west, where it is less and leaching is also much less, is so marked that they are classified respectively as *pedalfers* (with an appreciable content of aluminium and iron hydroxide) and *pedocals* (with a high content of calcium carbonate).

The quantitative degrees of acidity and alkalinity are expressed in the *p*H *value*, a scale measured in terms of the hydrogen-ion concentration held by the soil-colloids, for which values are so inconveniently small that the negative index of the logarithm of this concentration is used. In pure water one part in 10 millions is dissociated into hydrogen ions, that is 10^{-7}, and the pH is 7; this is a neutral state on the scale of acidity. If a strong alkali such as caustic soda is dissolved in water, the solution is markedly alkaline, and an infinitesimal part (10^{-14}) is dissociated into hydrogen ions, hence the pH is 14. By contrast, with hydrochloric acid one part in 1000 is dissociated, that is, 10^{-3}, or a pH of 3. A neutral soil has a pH value of about 7·2, an acid soil of less than 7·2, and a strongly alkaline soil of about 8·0 or higher.

Both natural and cultivated vegetation vary enormously in tolerance to acid conditions. Some heathland plants such as ling flourish in an acid soil, and a number of plants such as rhododendrons and azaleas need a wholly lime-free soil. Most cultivated crops, however, suffer if the soil becomes unduly acid, and farmers apply lime in various forms to meet the requirements of the soil; in practice a pH of between 6·0 and 6·5, that is, very slightly acid, is desired. Lime not only helps to neutralize the excess acids and so 'sweeten' the soil, but it encourages bacteria and helps to improve the physical texture of heavy soils.

Soil texture The physical texture of the soil, which depends to a large extent on the size of the individual particles, is a matter of great importance to a farmer or gardener. By definition soil particles do not exceed 2 mm in diameter; anything larger is classed as stones and though of course they are often present in the soil they are not part of it. Some soils are commonly called 'light', others 'heavy', while between the two are the favourable 'loams', but there are many variations; the United States Department of Agriculture specifies no less than twenty textural groups. For practical purposes it is sufficient to distinguish soils of three grades of mineral particles: (i) coarse-grained, that is, of sands; (ii) fine-grained, that is, of clay; and (iii) with grains of an intermediate size, that is, of silt. If one handles a moist sample of each of these they feel gritty, sticky and silky respectively.

Sandy soils A light soil consists mainly of sand, that is, grains of quartz, with considerable air spaces between them. The sand may either be 'coarse' where the particles are between 0·2 and 2·0 mm in diameter, or 'fine' between 0·02 and 0·2 mm where the grains are just visible to the naked eye. These light soils allow water to drain through rapidly, taking soluble plant foods with it. They are therefore often 'hungry' soils, which not only need constant manuring, but may dry out completely during a period of drought so that shallow-rooted crops fail and pastures 'burn'. They are, however, well-aerated soils and warm up quickly in spring, so that provided plant food and water can be readily supplied, they are good for horticulture, particularly for the production of early vegetables and especially root-crops which desire a deep root-run. Elsewhere they may be planted with conifers, or are covered with heath (p. 537). Sometimes they are improved by *marling*, the addition of a calcareous clay which gives a light sandy soil some cohesion and 'body'.

Clay soils Heavy soil consists of a large proportion of clay, that is, of very fine mineral particles with individual grains less than 0·002 mm in diameter. These particles are chemically extremely complex, but consist mainly of hydrated aluminium silicates. A clay contains little air; moreover, it can hold much water, so forming a tenacious sticky mass, but when it dries out completely it forms a hard, concrete-like surface, seamed with numerous cracks. Sometimes matters are made worse by the formation of a *claypan*, a compacted solid layer of clay in the subsoil, which is often hard to dig or plough. Moreover, in Great Britain the water which clay holds from the winter rains makes it a 'cold' soil and usually a 'late' soil, since the farmer cannot get on to it until the spring is well advanced.

Nevertheless, a clay soil is often rich in plant foods, so that if the farmer can cultivate it well and 'keep it in good heart', it will yield heavier crops than will a sandy soil. He tries to maintain a 'crumb-structure', in which the fine particles collect in crumbs with air spaces between. Ridging in the autumn so that the frost can act on the largest possible surface, frequent liming which causes 'flocculation' (that is, the collection of the fine particles into crumbs, thus coarsening the texture), and the addition of as much humus as possible will help to make a clay soil more workable.

In most areas the heavy clay-soils are sown with selected mixtures of perennial grasses and clovers and left down for a long term of years, providing permanent pasture for dairy- and beef-cattle. Efficient drainage methods, modern machinery and careful liming enable clay-soils to grow roots, green crops and cereals.

Silty soils More favourable soils are those intermediate in physical character to the sands and clays. A silty soil has particles considerably larger than those in clay, from 0·02 mm to 0·002 mm in diameter, but much finer than those in sand; its characteristics are therefore likewise intermediate.

Loamy soils The ideal soil for agriculture is a loam, which consists of a mixture of particles of many different sizes. It may be a sandy loam, a silty loam or a clay loam, but its advantage is that while it can retain some moisture and plant food, it is well aerated and drained, and can be readily worked.

THE ORGANIC CONTENT OF THE SOIL

Where the soil is immature and consists almost entirely oɪ mineral particles derived from disintegrated rock, it is described as a *skeletal soil*. Various natural biological processes gradually add a certain amount of organic matter or humus.

Humus Humus consists of the remains of plants and to a lesser extent of animals, which have decomposed through the action of bacteria and other micro-organisms into a darkish, amorphous mass. Where the process of decay has not been completed, it is usually possible to pick out details of the vegetable structure in the humus.

The humus content of the soil provides nitrogen and other elements such as phosphorus, calcium and potassium, which are broken down from the decaying plant tissues by soil bacteria, and so made available in a form which plants can absorb through their roots. It also has the beneficial physical effect of improving the texture; on the one hand it helps a sandy soil to retain moisture, while on the other hand it 'opens' and aerates a clay soil. In order that a cultivator can maintain the humus content of his soil, it is necessary for him to dig in farmyard manure or compost, rotted garden refuse. In the West Riding of Yorkshire shoddy-waste is dug into the soil, and in Cornwall and Brittany loads of seaweed are used. A peat soil consists almost entirely of humus, a chernozem (p. 498) includes about 12 to 16 per cent of humus, but a podzol (p. 493) has only 3 per cent or less.

Humus readily forms in nature under certain conditions. In deciduous woodlands leaf-mould accumulates as a result of the autumnal leaf-fall; this is 'mild humus' or *mull*, with a pH of 4·5 to 6·5, and to dig it into the soil benefits a garden enormously. In virgin prairie or steppe the layer of humus lies under the surface, since the rainfall is inadequate to carry it to deeper levels. The chernozem of the Ukraine owes its dark colour to the humus formed from the decayed mat of grasses; the humus layer may exceed 0·6 m in thickness.

Not all natural humus is of the same agricultural value to the soil. In contrast to mild humus, in some environments 'raw humus' or *mor* can form, particularly on heaths and moors and in pine woods. The whole process of decay is slower because of cooler, moister conditions, the absence of worms and other soil organisms, and the toughness of conifer needles. Usually bases such as calcium are deficient, so that the *mor* is markedly acid, with a pH value of less than 3·8. Nitrogenous compounds are likewise deficient, a fact which slows down decomposition because micro-organisms in the soil need nitrogen to live; when a gardener is building a compost heap of leaves, weeds and grass-mowings, he sprinkles each layer with a nitrogenous 'accelerator' to help the process of decomposition. Where little or no decomposition has been able to take place, the mass of accumulated vegetation is known as *peat*.

THE AIR AND WATER CONTENT OF THE SOIL

Air in the soil The air content of a soil is vital both to the development of the soil itself and to the organic life within it. Except where a soil is waterlogged, a certain amount of air is contained between the individual particles. One immediate effect is the chemical change known as *oxidation*, which converts part of the organic material into nitrogen in a form readily available to the plant. On the other hand, too high a degree of oxidation (sometimes induced in tropical lands by over-frequent ploughing) may consume so much organic material that the soil becomes increasingly sterile.

In addition, most bacteria, present in soil in infinite numbers but of microscopic size, require oxygen and are said to be *aerobic*. As these organisms are partly responsible for breaking down plant remains, absence of air limits their activity. Earth-worms too have an important effect on soil processes.

The air content of the soil is usually related to its water content. A waterlogged soil is virtually airless, that is, *anaerobic*.

Water in the soil Water moves downward by percolation, depending on the texture of the soil or rock (p. 111), and moves upward by capillarity. It becomes, in fact, a weak solution of innumerable compounds, both organic and inorganic. The chemical processes which take place in the soil do so mainly in solution, and plants derive most food from these solutions by way of their roots. The amount of water in the soil varies from almost nil in arid climates which makes life virtually impossible for organisms, to a state of complete waterlogging which excludes all air, causes a reduction of bacteriological activity, and limits decomposition, so allowing the formation of peat.

In damp climates, especially in high latitudes where the evaporation rate is low, water tends to move predominantly downward, particularly in coarse-grained sandy soils. This dissolves the soluble minerals in the soil, together with soluble humus material, and carries both downward, a process called *leaching* or *eluviation*. The surface layers from which soluble materials, humus and the lighter fractions are carried down are in fact known as the *zone of eluviation* (fig. 207). A typical leached soil is a *podzol*, a Russian word meaning 'ash', because the surface layer is often greyish or ash-coloured. Such a podzol can be seen on many heaths and moors in Britain, and in the coniferous forests of Eurasia and North America. Below this leached surface layer often occurs a hard, thin stratum, consisting either of redeposited humus compounds forming a hard, dark impervious band (*moorpan*), or of sand grains or gravel, stained reddish-brown and cemented by ferric salts deposited by percolating solutions (*ironpan*). Other compact layers may be formed by washed-down or synthesized clay (*claypan*), and by redeposited calcium carbonate (*limepan*). The underlying layer of the soil in which redeposition occurs of the materials removed from the upper layers is known as the zone of *illuviation*. Below the pan lies the unaltered subsoil or the solid rock.

In a hot, arid climate, however, evaporation exceeds precipitation for much of the year (p. 432), so that water tends to move upward and the soil dries out; in some areas a thin salty crust is formed on the surface. This process of salinization can produce an extremely saline soil, known as a *solonchak*, as in the Great Basin of Utah, in the Jordan Valley and in the neighbourhood of the Caspian Sea.

Drainage It is clear that water in the soil has a profound effect on the soil-forming processes and on its suitability as a habitat for cultivated plants. A farmer is therefore very concerned about the drainage of his soil. Natural drainage depends on the surface configuration, the degree of permeability of the rocks (p. 111), the amount and seasonal distribution of precipitation, and the nature of the vegetation cover. Drainage may be *excessive* (the soil dries out quickly in time of drought); *satisfactory* (it is porous so that it holds water between the grains but is also permeable so that it allows surplus water to drain away); *deficient* (in time of heavy rain the water does not get away reasonably quickly); or *impeded* (it is nearly always waterlogged).

In situations where a farmer is faced with deficient or impeded drainage, he is obliged to use such expedients as surface trenching, particularly deep trenches into which water can seep horizontally, the insertion of field drain-pipes, and methods of 'mole-ploughing'.

The problems of excessive drainage have to be coped with by adding marl and humus to thin sandy or chalky soils to help their water-holding capacity. The maintenance of a fine crumbly surface tilth so as to prevent the drying-out and cracking of clay soils is an essential part of 'dry-farming' practice, except where wind-erosion threatens to remove this fine top layer. The cultivation of special grass mixtures, which are quick-growing, drought-resistant and possess tenacious binding roots, is practised in areas of reclaimed heathland. But in many arid regions the only solution to the deficiency of ground-water is the introduction of various systems of irrigation.

THE SOIL PROFILE

In a study of soil and its formation careful attention must be given to what is happening at different levels below the surface. If one examines the side of a quarry or a road cutting, or digs a deep hole, there nearly always seem to be distinct horizontal layers, often of different colours and textures, which sometimes grade into each other, at other times change more sharply. Such a section, showing the successive layers, is a *soil profile*, within which each main layer or zone is a *horizon*. Every main soil type has its own profile, and a pedologist determines very carefully its nature. Sometimes he takes away a complete profile in a soil-box for laboratory examination.

A cool, damp climate, particularly in areas of light sands such as heathland, tends to produce a profile with four main horizons. The downward percolation of water leaches out soluble minerals, and develops a podzol soil type. Under the mat of surface vegetation and raw humus the uppermost layer is known as the *A*-horizon, extending down for 0·3 to 0·6 m; through it water constantly passes, producing a greyish soil from which have been leached both the bases and also most of the fine clay and humus particles. Below this is the *B*-horizon, where most of the material washed down has accumulated, with possibly a hardpan. Below that again are the *C*-horizon, the layer of partly weathered rock, though with no humus or other additions from above, and finally the *D*-horizon, the solid underlying parent rock.

Soil scientists can produce much more detailed subdivisions of the various horizons than these three; in a podzol A_0 is sometimes used to indicate the thin surface layer of humus, the partly leached layer is A_1, while the ash-coloured soil below is denoted by A_2. Again, where the subsoil is water-logged through impeded drainage, oxidation cannot take place, so that ferrous salts give the soil a bluish-grey appearance. These are known as *gley* or *glei* soils, and the layer is referred to as the *G*-horizon.

Different soil profiles can be seen under other conditions. Under a deciduous forest a brown-earth or brown forest-soil is produced, under mid-latitude grassland a black-earth (chernozem). In both instances there is an abundant supply of humus, and leaching is only slight. In a brown-earth, for example, the *A*-horizon consists of a dark-coloured mass of well-decomposed humus (*mull*), which shows a gradual transition downward, with hardly any clearly defined visible differentiation between the A- and B-horizons, to the parent rock material.

Such profiles develop only when man does not disturb them; in fact, the whole purpose of ploughing or digging is to provide a deep layer of soil suitable for plant growth. The gardener 'double digs' his soil, breaking up the subsoil and incorporating manure and organic refuse, but replacing the top layer or 'spit'. A farmer usually 'surface ploughs', or occasionally he may 'deep plough' to break up the hardpan in the B-horizon and so improve drainage. A cultivated soil has therefore a different profile from that of a natural

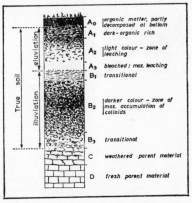

Fig. 207 *A soil profile*

soil; drainage, ploughing, fertilizing and the cultivation of particular crops affect not only the surface layers but also the lower ones, and mix them together.

THE MAJOR SOIL TYPES

Soils may be divided into *zonal*, *intrazonal* and *azonal* orders. *Zonal* soils occur most commonly on gently undulating land where drainage is free and where the parent material is of neither extreme texture nor chemical composition. Since climate exerts such a profound influence on weathering, water content and plant cover, the basis of a broad analysis of major zonal soil groups is climatological, and they occur in latitudinal zones in such extensive land areas as the U.S.S.R. (fig. 208) and the U.S.A.

Intrazonal soils occur where special conditions of relief or parent material exert a stronger influence on the soil than climate or vegetation. They depend on a specific kind of parent rock (such as limestone), on the presence of large amounts of salt (such as solon-chaks), on the presence of much water (such as bog-peat, fen-peat or meadow soils), or on a coastal habitat (salt-marsh soils).

Azonal soils are without well developed characteristics, either because they are 'young' or because the parent material and relief conditions have prevented the development of more definite characteristics.

495

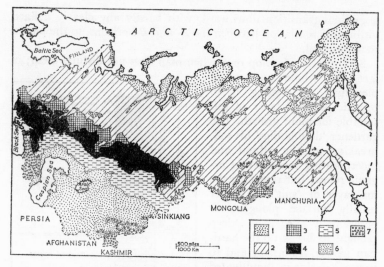

Fig. 208 *Soil map of the U.S.S.R.*

Based on a map in *The Great Soviet World Atlas* (Moscow, 1938).
1 Tundra and Mountain soils; 2 Podzols; 3 Grey-earths; 4 Chernozem;
5. Chestnut-coloured soils; **6.** Arid soils, including sandy soils, saline soils etc.;
7. Various upland soils (upland pasture soils, mountain forest soils, etc.).
(*Note:* the Soviet frontiers are those of pre-1939.)

Defects can be found in all classifications of soil proposed. One difficulty is that definitions are based on virgin soils undisturbed by man, so that cultivated soils do not easily fit into the pattern; man has so altered the characteristics of some that virgin and cultivated examples of the same type may be very dissimilar. Nor are soils easy to define in exact terms, since one may grade into another and only arbitrary divisions between them can be made.

MAJOR SOIL TYPES

I. Zonal soils

   A. Cold Climates

                          tundra soils

   B. Temperate Climates
      Humid   podzols
                          grey-brown podzols
                          brown-earths (or brown forest-soils)
      Seasonal rainfall   chernozem
                          prairie soils
      Semi-arid and   chestnut-coloured soils
        arid   cool desert soils (grey-earths or serozem)

C. *Tropical Climates*
    Humid
                        tropical red-earths
                        lateritic soils
                        tropical black-earths (Regur)
    Arid   red desert soils

II. Intrazonal soils

  A. *Due to saline content*   solonchak
                                          solonetz
  B. *Due to excessive moisture*   meadow soils
                                          fen-peat soils
                                        bog-peat soils
                                        dry-peat soils
  C. *Due to calcareous parent material*   rendzina
                                          terra rossa

III. Azonal soils (immature and skeletal soils)

  A. *Mountain soils*   scree soils
  B. *Alluvial soils*   river-borne material
  C. *Marine soils*   salt-marsh soils, mud-flat soils, marine
                                          clays
  D. *Glacial soils*   till soils, fluvio-glacial soils
  E. *Aeolian (wind-blown) soils* dune soils, loess soils, *limon* soils
  F. *Volcanic soils*   recent lava and ash soils

ZONAL SOILS

(i) **Tundra soils** The main conditions controlling the development of tundra soils comprise low annual temperatures, a permanently frozen subsoil, a slow rate of evaporation, and a brief growing season for plants. Chemical weathering and biological activity are both very limited, and waterlogging is widespread in summer. A typical tundra soil profile includes a surface layer of peat formed from mosses and lichens, then a horizon of mud, frozen in winter, with a bluish tinge due to the absence of oxygen, and below this the permanently frozen horizon.

(ii) **Podzols** This group of soils has already been mentioned (p. 493) since they occur widely in the moist cool temperate climates.

(iii) **Grey-brown podzols** An important group of soils, transitional between the podzols and the brown-earths, deserve separate recognition as grey-brown podzols. They are more acid in quality than the brown-earths, due to leaching under humid conditions, but the grey podzolic layer is replaced by a brownish layer with some organic content. These soils are found widely in Britain and western Europe and over much of the north-eastern corner of the U.S.A.

Many of the freely-drained soils in western England, Wales and Scotland can be included in this category.

Though not as good agriculturally as the brown-earths, when carefully farmed they are quite productive. In Britain they carry good grasslands, and in America they are the scene of mixed farming.

(iv) **Brown-earths or brown forest-soils** These soils, typical of much of the area formerly covered by the deciduous woodlands of western Europe, are rich in organic matter derived from the accumulation and decay of leaves. Agriculturally they are much superior to podzols. The surface soil is usually slightly acid, since the climate is humid and some leaching takes place. Most of these soils have been cultivated for centuries, the original forest cover has been long removed and constant manuring and liming are now necessary.

(v) **Chernozem** These famous steppe-land 'black-earths' are rich in humus because there is insufficient rain to carry the grass-derived humus deeper. The chernozem belt extends (except for interruptions by mountains) across central Asia, from Manchuria, through southern Siberia and central Russia, into the Ukraine (where it attains its maximum development), and on into Rumania, Hungary and Moravia. The Canadian black-soils are of this type, as are those in the U.S.A. extending southward from North Dakota to Texas. Parts of the Argentine pampas and of the downs of south-eastern Australia have a cover of black-earths, which may be akin to the true chernozems.

These soils afford a very striking example of origin due to climate and very little to parent rock material, since they occur widely over a variety of bedrocks. Their loose, crumbly texture and natural richness in plant foods, together with the wide extent of gently rolling land and a suitable climate, make the areas of chernozem the great wheat-lands of the world.

(vi) **Prairie soils** These dark brown soils are transitional between the brown forest-soils, developed under humid conditions, and the chernozems of sub-humid or semi-arid conditions. They form in continental interiors, where rainfall, though markedly seasonal, totals 60 to 100 cm annually; they are therefore transitional between pedalfers and pedocals. The conditions allow leaching, but not so much that a large part of the calcium content is removed, nor so little that calcium salts have accumulated. The tall grass vegetation stimulates a rich humus content. These soils occur in eastern Europe, and in the U.S.A. from southern Minnesota to Oklahoma; the greater part of the Corn Belt has such soils.

(vii) **Chestnut-coloured soils** These soils are essentially a variety of chernozems, modified by greater aridity, for they are characteristic of areas of sparse dry steppe with only about 20 to 25 cm of rainfall. They are loose and friable, containing much humus because the natural vegetation consisted of coarse grass, but hardly at all leached because of the low rainfall. The upper horizon has a distinctive dark-brown colour, becoming paler with depth. These soils have a very wide distribution, covering most of the drier steppe-lands to the south of the chernozem in Russia, and extending westward into Rumania and Hungary. They are also found in the High Plains of the U.S.A., and in the drier parts of the Argentine pampas and the South African veld.

(viii) **Cool desert soils** A group of soils, sometimes known as *serozems*, occurs in the more arid areas of middle latitudes. The low rainfall results in a small organic content, but also in little leaching. The upper horizon is a light grey (they are often called *grey-earths*), grading downward into a browner shade. Much of the extreme south of the U.S.S.R. to the east of the Caspian Sea and parts of western U.S.A. are covered with these soils. As they have suffered little leaching they are rich in plant foods, and if irrigation can be practised are very fertile. A large part of the U.S.S.R.'s cotton is grown on this type of soil.

(ix) **Tropical red-earths** In tropical areas high temperatures and humid conditions with markedly seasonal rainfall result in very intensive chemical weathering, producing a loamy mixture of clay and quartz. This rock disintegration may go on to a considerable depth, sometimes for as much as 15 m (50 ft). The widespread presence of iron compounds causes the characteristic red colour. Leaching tends to remove some of the soluble bases, but when the virgin soil is cultivated it produces rich crops for a while; heavy fertilization is then essential since the soil becomes rapidly exhausted. These red-earths cover extensive areas on the tropical savanna plateaus, as in Brazil and Guyana, East Africa, the southern Deccan and Ceylon (where the dark green of the tea-bushes contrasts with the red hill-sides), and also in the tropical monsoon forest lands, as in Burma and Indo-China.

(x) **Lateritic soils** These are formed under conditions which favour much more potent leaching than in the case of the red-earths, although superficially they are similar in appearance. They consist mainly of sesquioxides of aluminium, a hydrated form of oxide of aluminium, which give it a plastic quality when wet, hence an alternative name of 'brick-earth' (Latin *later*—a brick), but when dry it forms an extremely permeable crumbly earthy soil.

499

Lateritic soils are not usually of any great agricultural value, because they are so porous that irrigation is difficult. They occur widely in Brazil, the West Indies, West and East Africa, southern India and Ceylon. Immense thicknesses occur in northern Ceylon in the 'Dry Zone', covered with scrubby jungle, the scene of shifting cultivation. They are sometimes referred to as *latosols*.

(xi) **Tropical black-earths** A type of tropical soil, formed by weathering under conditions of heavy rainfall and high temperatures, is found in the north-west of the Deccan, where it is termed *Regur*. Its blackness is due to the high content of titanium salts in the basalt bedrock. In the dry season if it has been ploughed it forms a dark grey or black soil which easily crumbles into dust, but if it has not been ploughed it dries out into a hard black sheet, seamed with deep cracks. In the wet season, by contrast, it becomes plastic, even sticky. The soil is characterized by a low organic content and by a zone of calcium carbonate concretions. The chief cotton-growing areas of India are found here. Very similar soils are found in Kenya (where they are also called 'black cotton soils'), in Morocco, in northern Argentine, and in some small areas in the West Indies.

(xii) **Red desert soils** These are the reddish-brown soils of the hot deserts. They are mainly of a sandy texture and contain much salt because of the absence of leaching.

INTRAZONAL SOILS

(i) **Saline soils** A soil in which soluble salts are present in considerable quantity affords an example of an intrazonal category. These are of widespread occurrence wherever there is a sufficient degree of evaporation, both in the hot deserts and in the cooler continental interiors where summer heat allows seasonal evaporation. The strong salt solutions rise by capillarity and form a greyish surface crust, below which is a granular salt-impregnated horizon. Typical saline soils are called *solonchaks*. If there is a rather higher rainfall some of the surface salts may be leached out, so that the highly saline layer occurs lower down in the *B*-horizon; these are *solonetz* soils.

(ii) **Peat soils** When a soil is waterlogged, air is virtually absent and few organisms can live, so that bacteriological activity is reduced. Instead of the chemical processes of oxidation producing nitrates, carbon dioxide, sulphates and ferric oxides, reduction takes place forming ammonia, marsh-gases, sulphides and ferrous oxides. The decay of vegetation, though only partial because of the lack of oxygen, enables a considerable thickness of peat to accumulate; as it increases the lower layers are compressed but retain to some extent their fibrous form. Pollen grains preserved in the peat enable its constituents to be determined after thousands of years; such examination under a microscope is known as pollen analysis (*palynology*), a helpful method of enquiry into bygone climatic and vegetational conditions. Moreover, fragments of plant-tissue in the peat can be dated by the carbon-14 technique (p. 14).

Peat is not a soil, it is a soil-forming material. But when the soil-forming processes are allowed to act on an accumulation of peat, either by natural changes such as increasing aridity or by the work of man, notably drainage and deep-ploughing, peat may form a richly organic soil.

Fen-peat soils Fen-peat accumulates in an area where the water contains much calcareous material derived from limestone or other alkaline salts. This neutralizes the humic acids liberated during plant decay, and so forms a thick black spongy mass of 'mild' peat, which when drained provides rich agricultural soil, as in the case of the English Fenlands. It is noteworthy, however, that in this area repeated ploughing has helped the oxidation of much of the original organic matter, so that the 'crumb' structure has been replaced by a finer particled soil. Not only must it be fertilized and manured, but care must be taken to avoid the erosion of the fine top-soil by the wind, which is causing worry in the Fenlands (p. 507).

Meadow soils Akin to fen-peat soils are those which form in river flood-plains where there is flooding during part of the year, accompanied by a widespread deposition of silt and mud, but with considerable growth of vegetation. The surface horizon is dark, including much organic material, but below that the G-horizon, the result of waterlogging, is clearly marked.

Bog-peat soils Bog-peat forms in moist areas, particularly on high moorlands (p. 538), where the humic acids developed by plant decay are not neutralized. This raw humus or *mor* is therefore highly acid. The development of good agricultural soils from such a parent material is difficult; the farmer has to drain and deep-plough it, mix the peat with sand or silt, and heavily lime it, as has been done in parts of the German moorlands.

Dry-peat soils A type of dry acid-peat may develop on sandy heathlands (p. 537), often only a few cm thick, derived from mosses, lichens and the roots of ling. On such sands leaching is rife, so that below the peat is a typical podzol profile; in fact, the peat layer is the A_0-horizon already mentioned.

(iii) **Calcareous soils** The characteristics of limestone as a parent material have already been discussed (p. 487). Two main soil groups are developed on calcareous material, rendzina and terra rossa, the latter formed under rather more arid conditions.

Rendzina These soils are dark coloured, with a surface horizon of friable, almost granular, loam, lying on a subsoil containing limestone or chalk fragments, which in turn rests on the solid rock. The group is an example of the effect of a specific parent rock; elsewhere, under the same climatic conditions, brown-earths would occur. Soils developed on chalk occupy a considerable part of southeastern England. Apart from the high content of calcium carbonate in chalk, its other characteristic is its extreme permeability (p. 111). These soils for many centuries were under pasture, 'downland grazings' or 'basic grasslands' (p. 536), but large areas now grow specially bred varieties of wheat and barley. In recent years irrigation, in the form of large-scale overhead sprinklers, has been used to improve grasslands and particularly fields of alfalfa.

Terra rossa This name is applied, somewhat loosely, to reddish residual soils which accumulate in limestone depressions, usually under semi-arid conditions, or at least those of prolonged summer drought, as in the Mediterranean area. In parts of Spain, the Causses of southern France, northern Italy, Malta, the Karst of Jugoslavia, and northern Greece, the steep slopes, the concentrated winter rains and the rapid run-off result in the indissoluble limestone residue being washed into crevices, hollows and basins. In some parts of Jugoslavia one can see peasants collecting baskets of this soil to accumulate in small stone-walled enclosures; they are in fact 'making' their fields.

AZONAL SOILS

Weathering and other soil-forming processes need to operate for a considerable time to allow the soil to become mature before a typical profile is developed. New soil-forming material may be deposited by rivers, the sea, glaciers or the wind. Alluvium is not a soil, although it usually forms a medium ready for cultivation (hence *alluvial soils*), as in the case of the Nile valley. It may also give rise to marsh vegetation, so that dark, fibrous fen-peat soils, as on the margins of the Wash, develop from alluvium.

(i) **Mountain soils** On stony unstable surfaces such as scree slopes (p. 98) or on glacial moraines (p. 241), the soil may consist to a large extent of rock fragments, which disintegrate physically, mainly by frost action, but suffer little chemical change. These are known as *scree soils*.

Soil-wash and soil-creep usually move finer material gradually downhill. But if the scree-surfaces become more stable through the growth of mountain vegetation (coarse grasses, bilberries and mosses), they tend to be leached because of the heavy rainfall often experienced in the mountains. The products of the leaching move downhill on to the lower slopes or valley bottoms, where they accumulate, forming soils rich in bases, known as *flushed soils*.

(ii) **Alluvial soils** These soils are derived from a mixture of sand, silt and clay, consisting of well-mixed rock waste transported and redeposited in level beds by running water, and replenished at times of flood. Because of its fine texture and its deposition in level sheets, alluvium forms good agricultural soils. It is a 'young' deposit, and has not been deprived of its mineral content. Moreover, the soils are usually near the river responsible for their deposition and therefore irrigation is possible. Some alluvial plains, particularly those of south-eastern Asia, form great areas of intensive cultivation and of dense population. Similarly, in the British Isles and western Europe, wherever the flood-plains are not marshy or can be drained, good agricultural soils result. The river-clays of the Netherlands, covering the flood-plains of the Maas and the Rhine distributaries, are dyked against floods and drained, providing valuable soils for mixed farming.

(iii) **Marine soils** Materials derived from marine origin are built up along low-lying coasts in the form of mud-banks, sand-banks and dunes (pp. 300–2) by natural processes, stimulated by artificial reclamation. Along the coasts of Belgium, the Netherlands, Germany and Denmark, areas of former sea-floor, known as *polders*, are enclosed by dykes and then drained, producing various soils developed on marine clays (sometimes called 'sea silt'). The nature of the soil depends on the level of the artificially controlled water-table; where it is high peat soils are usual and the land is mainly under pasture, but where it is low heavy clay soils predominate. The removal of salt can be accelerated by applying gypsum (hydrated calcium sulphate) to the soil; this combines with the sodium chloride to form calcium chloride, which is highly soluble and can be easily washed-out. Thus the Wieringermeer, the north-western polder of the Zuiderzee scheme, was enclosed by a dyke in 1930, when pumping began, and produced its first crops four years later.

(iv) **Glacial soils** Soils developed on glacial materials are widespread in the northern hemisphere. The variety of glacial deposits includes various tills, outwash sands and gravels, and re-sorted clays laid down in glacial lakes. A wide range of soils has developed on them.

(v) **Wind-blown soils** The depositional effect of the wind, producing sand-sheets, dunes and *loess*, has been described (pp.

266–72). The loess of north-western China (plate 101), that of the Börde of West and East Germany, and the *limon* of Belgium and north-eastern France may develop fine-textured, easily worked soils, deep and well drained, giving a good tillage bed. As loess is a product of disintegration it responds rapidly to the soil-making processes, and forms the most important agricultural regions of western Europe. Those parts of the limestone plateaus of Beauce and Brie in the Paris Basin which are not limon-covered carry a scanty herbage and are used for sheep grazing, while the limon-covered areas are the main wheat and sugar-beet lands in France.

(vi) **Volcanic soils** Recent volcanic activity deposits lava, ash and pumice, which weather easily and are transported lower down the slopes of volcanoes by rain-wash and torrents. The new higher lava forms bare grey sheets, but the lower weathered materials are immensely fertile. Villages and vineyards cling around the slopes of Vesuvius and Etna, in spite of repeated disasters in the past and the ever-present threat of a recurrence of volcanic activity.

THE SOILS OF BRITAIN

On a small-scale map of the major world groups of soils, much of Britain and neighbouring parts of Europe fall into the podzol and brown-earth groups, together with areas of skeletal soils in the uplands and of immature soils in the lower valleys and along the coasts.

There are three main divisions of British soils. In the first place there are those derived from the underlying solid rocks, either from the old resistant rocks of the north and west, or from the newer clays, marls and chalk of the south-eastern scarplands. In the second place there are soils derived from the extensive mantle of drift deposits which covers so much of Britain, including types of till, glacial loam, sands and gravels, brick-earths (somewhat akin to limon), Clay-with-flints in patches on the Downs, alluvium in the river flood-plains, and marine silts around the Wash. In the third place there are various soils derived from organic deposits, including mild humic soils, mild fen-peat and acid moorland peat.

The Soil Survey Research Board, under the aegis of the Agricultural Research Council, was set up in 1947 to coordinate soil surveys in Great Britain. The Soil Survey of England and Wales has its headquarters at Rothamsted, while the Soil Survey of Scotland is carried out by the Macaulay Institute near Aberdeen, founded in 1930. Soil surveys are being actively pursued in many

parts of the country; they are intended to obtain systematic information for purely scientific ends on the one hand, for practical purposes
on the other: for agriculture, forestry and the future planning of
land-use. Maps on various scales are being produced for various
parts of the country, together with accompanying *Memoirs*.

Within each of the major soil groups subdivisions are mapped as
soil series, each with a distinctive set of profile features developed on
a particular parent material and having a specific drainage condition. These series are generally named according to the locality
where they were first mapped; thus the *Winchester Series* occurs widely
on the Clay-with-flints on the chalklands of southern England.

LAND CLASSIFICATION

In the words of L. D. Stamp, 'The modern study of soils may be
described as the philosophical approach to the subject in that the
soil is studied in the first instance as an entity for its own sake and
not in relation to the use which may be made of it'. While the
geographer is not a pedologist, he should know something of the
processes whereby the soil-mantle has developed and is developing,
of the major types of soil and their potentialities for plant growth,
and of the detriment to the soil which may result from bad or
neglectful farming practices. He is interested in the soil both for its
own sake as a varied mantle which covers the solid rocks and contributes in no small way to the character of the landscape, and as the
medium in which plants grow, hence the scene of production,
directly or indirectly, of most of man's food and a large part of his
raw materials.

However, more important than soil classification to the geographer, and indeed to the planner, is *land classification*, which is of
course, closely associated with land-use.

L. D. Stamp proposed three major categories and ten types
of land classification, based on (i) the nature of the site (elevation,
slope and aspect), and (ii) the nature of the soil (its depth, water
condition and texture). The classification is listed in detail by
Stamp in *The Land of Britain : Its Use and Misuse* (1948). It is interesting to note that at that time 37·9 per cent of the total area of Britain
fell into the Major Category I, Good, and 35·2 per cent into the
Major Category III, Poor. For Scotland alone, as might be expected,
the respective figures were 17·5 and 70·8 respectively.

Other land classifications have been prepared and used in attempts
to judge the quality of land for possible agricultural use. Such
surveys are especially useful in the under-developed lands in connection with reclamation, irrigation and other developments.

SOIL EROSION

Many references have been made to the work of man in promoting the welfare of the soil—drainage, ploughing and ridging, manuring and composting, fertilizing, liming and marling. He carries out these various operations in order to maintain or even improve the quality of the soil on which his food supply largely depends.

On the other hand, it is possible for soil to deteriorate unless it is carefully cultivated. One mistake is to take too much of one substance out of a soil too quickly. But the most serious way in which man has adversely affected the soil-cover is by allowing the forces of nature to remove it more rapidly than the soil-forming processes can replace it; this is soil erosion, and its prevention is one of the major problems in the world.

Soil erosion may be divided into three main types: *wind erosion, sheet erosion* and *gully erosion*. While the chief causes of each type vary in detail, the fundamental reason for soil erosion is the removal of the natural vegetation which formerly bound the soil and held it in place. The ploughing of grasslands or the clearing of forest may start a train of events, the effects of which become progressively more serious (plates 102, 103).

Wind erosion The eroding and transporting power of the wind involves the actual removal of dry, unconsolidated material. Farmers at the beginning of this century ploughed up large areas of the grasslands of the Midwestern states of America, tempted by the accumulated fertility of these virgin soils and by a number of years which subsequently proved to be wetter than average. Later a series of drier years caused some of these marginal lands to be abandoned, but now devoid of their protective mat of grass vegetation. The wind was able to sweep the finer soil particles away, and the region became notorious as the 'Dust Bowl'. Many prosperous farms were ruined by the removal of their top-soil; others outside the Dust Bowl have been partially buried by material blown by the wind.

Only in recent years has the problem been tackled systematically. The main remedy has been to keep as much of the land as possible covered with vegetation ('cover-crops'), while only a limited strip is ploughed. Another method is the planting of long lines of trees as wind-breaks. The practice of dry-farming (the retention of a fine tilth over the surface to check evaporation) has perforce been abandoned in many areas.

Wind erosion sometimes operates in eastern England, less spectacularly than in the U.S.A., but its cumulative adverse effect is serious. The light sandy soils of western Cambridgeshire (developed on the Lower Greensand), parts of the fen-peat areas of northern Cambridgeshire and Lincolnshire, and the sandy soils of western Norfolk are all liable to 'blowing', particularly when there has been an unusually dry spring followed by strong winds in April and May. Especially liable are the soils broken down to provide a fine seed-bed for sugar-beet and turnips. In the heathland areas of Belgium and the Netherlands, clouds of fine sandy dust blow continuously from newly ploughed fields in spite of the planting of poplar wind-breaks to divide the land into squares, and surface movement is arrested only by deep ditches which reach the ground-water.

Sheet, rill and gully erosion These types of soil erosion are caused mainly by the action of water (pp. 102–5, 136). Sheet erosion is a general removal of the surface soil over a large area, as it slips gradually down the slopes. Rill and gully erosion are more localized, when a sudden rainstorm, producing a concentrated run-off, rips gashes into the land. Once again, the cause in each case is the re-moval of vegetation, particularly of trees with their widespread binding roots. These types of erosion are shown drastically in the Mississippi Basin. The Tennessee valley was inhabited mainly by poverty-stricken farmers using primitive methods, and as population increased they cleared the woods from the hill-sides, burning the lower growth, and then planted the bare slopes with maize. The soil soon lost its fertility and was then abandoned for a newly cleared area, so that the torrents tore away the unprotected soil and swept it into the Tennessee river, which flowed as a turgid brown stream into the Ohio, thence into the Mississippi. It has been calculated that on an average a metre-depth of soil has been removed from the Tennessee valley, leaving deep searing gullies or scoured rocky slopes. American statisticians state that 3000 million tons of top-soil are still blown or washed annually from their land.

These forms of soil erosion are widespread in the world—in the Mediterranean lands of Spain, Italy and Greece, in the monsoon lands of India and Ceylon, in Java and Sumatra, in the savanna lands of Africa. Shifting cultivation, 'row-cropping' instead of 'cover-cropping', over-grazing especially by goats, deforestation, burning, widespread clearing for over-ambitious schemes of planta-tion agriculture—all these enable rampant erosion to proceed. Even in Britain sheet erosion takes place where a field has been ploughed up and down the slope, and occasional downpours falling upon newly ploughed soil produce gullies.

As in the case of wind erosion, palliatives include the maintenance of an effective vegetation cover. The land reclamation schemes of peninsular Italy, Sicily and Cyprus involve extensive planting of trees and shrubs on the upper slopes; in Ceylon no jungle may be cleared above 1500 m (5000 ft). Contour ploughing, where arable land lies on a slope, is replacing up-and-down furrows. Terracing is used to intercept water as it flows down the gradient, either allowing it slowly to sink in or be removed by a drainage channel running across the slope. Gullies are dammed or filled with brush-wood to help hold the soil.

In the U.S.A. the Tennessee Valley Authority has tackled the whole problem in that region by an integrated programme, in-volving flood-control by means of a series of major and minor dams and storage lakes, the construction of power-stations using the water thus harnessed, and widespread social and material improve-ments. Soil-erosion measures, in addition to the all-important flood-control, included the reafforestation or regrassing of great areas (150 million trees were planted in the Valley). Improved agricultural practice involved the manufacture and distribution of fertilizers, the introduction of contour-ploughing, the sowing of 'cover-crop' plants, the diversification of agriculture instead of the former monoculture of maize or cotton, and a drive toward scientific agricultural education. The TVA acted as 'a multiple-purpose, long-term development agency', set up by the Federal Government, producing a whole coordinated plan to restore prosperity to an area of about 109,000 sq. km (42,000 sq. miles).

The Bureau of Plant Industry, Soils and Agricultural Engineer-ing (United States Department of Agriculture) is carrying out a 'national cooperative soil-survey'. Basic soil-maps, 'land-use capa-bility' maps, and 'erosion hazard' maps are produced on a county basis, together with interpretative data. This information is of immense practical value for farmers and agricultural agencies.

CHAPTER 20

Vegetation

THE living mantle of vegetation which covers most of the land surface of the world has been called the 'intermediate link' between physical geography on the one hand and economic and human geography on the other. The various plant associations owe their nature to the interaction of different climatic elements—heat, light, moisture and wind—with other factors, principally surface relief and the soil-cover. These conditions impart a special character and a distinctive appearance to a particular association of plants; it is possible to distinguish areas of the world which have broadly similar plant associations, called vegetation regions.

Man has altered so much of the 'natural vegetation', that is, the primeval plant-cover, that over most, if not all, of the earth's surface the term can hardly be used in its strictly correct sense. Probably the more remote parts of the rain-forest of the Amazon Basin and of the coniferous forest of northern Asia and America are primeval. Elsewhere the nature of the natural vegetation can only be deduced, often very speculatively. The original vegetation has been replaced in two ways.

Cultivated vegetation In the first place, it has been destroyed by man and subsequently replaced by crops and 'cultivated trees', as well as in relatively small areas by houses and factories, roads and airfields. The only relation which this deliberately planted vegetation has with the natural vegetation is it must be able to grow successfully under more or less the same physical conditions. Man can modify these conditions, as his technical ability increases, by such devices as irrigation or drainage, just as he is able to make the soil-cover more amenable to his requirements. Alternatively, he can modify the plants themselves, not only to increase the yield of a particular species, but to breed varieties which will withstand less favourable physical conditions, and so make it possible to extend their habitat.

The study of this cultivated vegetation forms an important part of economic geography. A 'predominant land-use map' distinguishes such categories as 'improved pasture', 'areas in which arable land occupies more than 90 per cent of the total', 'perennial tree-crops', and so on. This 'cultivated vegetation' covers much of Europe, the eastern and central parts of the U.S.A. and Canada, much of the Argentine and Uruguay in the Plate estuary area, a large part of India, Java and China, much of South Africa, and the former grass-lands of Australia and New Zealand.

Semi-natural vegetation Quite apart from cultivation, a considerable part of the land surface is covered with what can best be described as semi-natural vegetation. It consists of 'wild' plants in the sense that they have not been planted deliberately by man, but owe their existence to the greater or less, direct or indirect, influence of man and of animals. Some of the European heathlands, in which plants such as ling grow naturally, have developed from an original forest-cover. Centuries of shifting hoe-cultivation by the peoples of the savanna lands (p. 528) of Africa has modified the original vegetation so much that some authorities claim that savanna represents the replacement of a former extensive forest-cover. When the rain-forest areas of Indonesia, once cleared for cultivation, are allowed to revert, they are rapidly covered by a 'wild' growth of secondary forest with a dense undergrowth (p. 520). Not only man but animals—rabbits, red deer, goats—may considerably affect the plant-cover; if woodland is cleared, its natural regeneration may be impossible because of close grazing.

PLANT COMMUNITIES AND SUCCESSIONS

Plant communities A group of plants growing in some particular area, under certain physical conditions which satisfy them, is known as a 'plant community', which can be of different grades. The largest is sometimes called a *plant formation*, as for example, the Temperate Deciduous Forest, in which the distinctive plant-form is the deciduous tree. Within this major community is a grouping known as a *plant association*; an oak forest is a particular association within the formation of the Temperate Deciduous Forest. A *plant society* is a local community with special conditions and species. An ecologist distinguishes smaller, more detailed and more local communities in the course of his work.

Within any community is a variety of individual plants, but the strongest and most vigorous is known as the *dominant*. In an oak-wood the dominant is the oak tree, but other plants are found, such as hazel, wood anemones, primroses and mosses. In a heathland, ling (*Calluna vulgaris*) is dominant, but other plants include bilberry, dwarf gorse, juniper, heath grasses, mosses and lichens. In complex communities several strata occur, known as *layers*; there may be a *canopy-layer* made up of the crowns of the dominant trees, a *shrub-*

layer of bushes or weaker trees, a *field-* or *herb-layer* of small plants, and a *ground-layer* of mosses and lichens.

Plant communities, on whatever scale they are considered, are rarely defined, unless circumscribed by man's activities. The edge of a heath may be delimited by controlled burning, of a particular grassland community by a rabbit-proof fence. But most communities reveal gradual transitions as the physical conditions change.

Plant succession Much vegetation, particularly that which is semi-natural, reveals a gradual sequence of changes or phases over a period of time, even if the climate remains unaltered. An area of land, bare of vegetation after a volcanic eruption, a widespread fire, submergence by water or man's destructive activities, will experience a gradual succession of vegetation changes until the optimum for the conditions, known as the *climatic climax community*, establishes itself. When this is attained, there is no further alteration in the broad aspect of the vegetation, and a 'situation of equilibrium' is attained. This term climatic climax community is now preferred to natural vegetation.

One of the most striking examples of successional change is that of the small islands produced after the eruption of Krakatoa in 1883 (p. 75); the original vegetation was destroyed and the new islands consisted of sterile volcanic rocks. By 1886 eight species of plants had established themselves, and within fifty years the islands were covered with dense secondary forest, in which large trees are now slowly becoming dominant. The tendency seems to be the ultimate re-establishment of rain-forest, which will be the climatic climax community.

Recently, ecologists have sought to analyse the vegetation cover on a scientific basis in terms of developmental successions or sequences, known generally as *seres*. The whole series of communities which leads from a virgin land surface to a climatic climax community is termed a *prisere*. Each stage in this succession is a *seral community*, itself only temporary and affording conditions successively more favourable for colonization by more demanding communities: the gradual development of a soil cover, the increase in the humus content, the creation of shade beneath whose protection seedlings can develop, etc.

There are several types of prisere, depending on the physical character of the original conditions. Thus a *xerosere* originates on a dry surface, which may be either of rock (a *lithosere*) or of sand (a *psammosere*). Where the surface is of fresh water, such as a shallow lake, a *hydrosere* develops, or if of salt water (as on a tidal mud-flat) it is a *halosere*.

But not all priseres necessarily lead to a climax; indeed, in the world today very few seem to do so. A prisere may develop to a certain stage, then become arrested by some extraneous factor. If this is of a permanent character, the community is known as a *subclimax*. Thus the waterlogging of an undrained swamp, or the presence of poor sandy soils on a windswept plateau, will in each case prevent the sere from attaining a climax; the respective result will be a swamp and a heath, each an example of a subclimax. If, however, the factors which arrest or interrupt the succession are only temporary, a *subsere* will be initiated. Thus a forest fire or clear-felling of trees will initiate a subsere, and in due course the climax forest will ultimately re-establish itself.

One other type of community involves the interruption of climax vegetation by prolonged direct or indirect human interference. Many English uplands, formerly wooded, have been turned into grassland as the result of the initial clearing of the trees, followed by continuous grazing which has precluded the regeneration of the woodland. This is a *plagiosere*, leading to an ultimate community in equilibrium known as a *plagioclimax*. If, however, the interference ceases, a subsere will be established; as grazing on many hill-slopes has declined in recent years, the grassland has become colonized first by bracken, then by small shrubs, and then by birch and ash as subseral development proceeds.

This discussion of various types of plant succession emphasizes in more scientific terms what has been said above, that over much of the world man's influence has been so marked that true climatic climax (or 'natural') vegetation is indeed rare, and that much of the surface is occupied either with subclimax or plagioclimax vegetation, or with subseral development.

THE FACTORS CONTROLLING VEGETATION

The geographer is interested in the relationship between plants and the physical environments in which they grow. These environments vary, but the elements which help to distinguish them represent factors in the complex processes of plant life. These factors can be divided into three groups—*climatic* (the several climatic elements), *edaphic* (mainly soil conditions) and *physiographic* (relief, aspect and drainage). However, these various factors are closely interrelated and interdependent.

Two other groups are the *biotic factors*, the effect of organisms, and those due to *human activities*, the grazing of domestic animals, burning, the felling of forests and drainage schemes.

Climatic factors The two main elements on which plant

growth depends are rainfall and temperature, particularly their seasonal distribution. In other words, great significance results on the one hand from the length of the dry season, especially in low latitudes where temperatures are usually continuously favourable for plant growth, and on the other hand from the length and degree of severity of the cold season, especially in high latitudes where the winter temperatures are below the point at which plant growth can take place.

In a broad sense, rainfall determines the vegetation type; abundant rainfall tends to produce forest, light rainfall causes grassland, and meagre or deficient rainfall results in scrub and desert. Temperature, however, plays a large part in determining the actual constituents of the flora; there is the 'Boreal Forest' of lands near the Arctic Circle, the Cool Temperate Forest, the Warm Temperate Forest and the Tropical Rain-forest—all forests, but differing considerably in their nature and appearence largely because of the varied temperature conditions. This relationship between rainfall and temperature holds good for the most part, but there are exceptions. For example, the poleward limit of tree growth is set not by rainfall total but by temperature, since trees must have a growing season of at least three months with minimum temperatures of 6° C (43° F).

Temperature Other things being equal, low temperatures tend to result in slower plant growth and smaller size, while higher temperatures produce more luxuriant growth. With an increase in either latitude or altitude, assuming that the available moisture remains constant, there is a corresponding decrease in the size, luxuriance and rate of growth of the vegetation. Conversely, constant heat with ample humidity results in a profuse, rapid growth of tall vegetation; the writer has observed a bamboo stem grow 20 cm (8 in) in twenty-four hours and a banana leaf increase 5·7 cm (2·25 in) in length in the same period of time.

Of especial importance are 'critical' temperatures at particular times of the year and actual extremes of heat and cold. Freezing-point is one of the most crucial temperatures, since so many plants are vulnerable to frost. Another is the figure already mentioned, 6° C, since for most plants it seems that active growth cannot take place at temperatures much below this point. Not only actual extremes of temperature are important, but also 'accumulated temperatures' (p. 393), that is the duration of temperatures above a specific minimum figure, in other words, the length of the growing season. For most of the species of a temperate deciduous forest there must be at least six months with temperatures above the minimum for tree growth. A broad-leaved evergreen forest, such as is found in equatorial latitudes, experiences temperatures well above this minimum throughout the year, and so growth is continuous.

Various classifications of plants have been made on the basis of their temperature requirements; one such is as follows (fig. 209):

(i) *Megatherms* (e.g. palm) coldest month above 18° C

(ii) *Mesotherms* (e.g. olive) coldest month between 6° C and 18° C
warmest month over 22° C

(iii) *Microtherms* (e.g. oak) coldest month above 6° C
warmest month between 10° C and 22° C

(iv) *Hekistotherms*
(e.g. reindeer moss) warmest month less than 10° C

(These figures refer to monthly means).

Rainfall Moisture is vital to every plant, since it is by water absorbed through its roots that it receives in solution the foods it needs. Surplus water is transpired in the form of vapour through the minute pores (or *stomata*) which cover the leaves. Larger and more luxuriant types of vegetation, notably trees, need a constant supply of moisture during the growing season (p. 433).

Plants may be classified according to their water-economy. *Hydrophytes* are purely aquatic plants, both floating and submerged. *Hygrophytes* include those plants which live in an environment with a plentiful water-supply, such as many species of the Tropical Rainforest. *Mesophytes* have available a moderate water-supply, and most trees fall into this category. *Xerophytes* are adapted in various ways to dry conditions, seasonal or perennial; their modifications include long roots penetrating deeply in search of ground-water, small leaves protected against transpiration with thick skins or with the surfaces coated with wax or fine hairs, hard thorny buds and shoots, thick rough bark as in the case of the cork oak, tuberous roots, and even water-storage devices such as have the baobab tree and the multifarious cacti. The true xerophyte lives a slow, thrifty sort of life, with long dormant periods, and a short intense growing period in response to a sudden though brief supply of moisture. Many other plants possess various characteristics to enable them to survive a period of seasonal adversity; these are known as *tropophytes*. Thus leaf-shedding enables a plant to survive a period of seasonal adversity, due to either cold or drought. Such a plant behaves as a xerophyte at one season, as a hygrophyte at another.

Light Another climatic factor of importance to plant growth is the amount of available light, both its strength and its diurnal and seasonal duration. Chlorophyll, the green pigment in plants, absorbs energy from sunlight, and as a result carbohydrates are formed

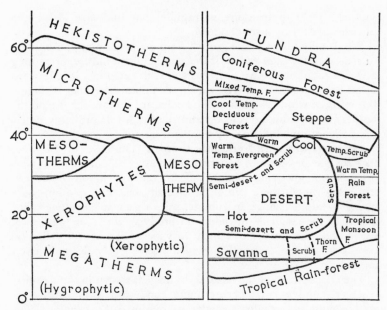

Fig. 209 *Diagrammatic representation of vegetation types in the northern hemisphere*

from carbon dioxide and water by a very complex and not fully understood process, and plant growth is sustained. Where light is weak, therefore, growth is lank and pale, as in the dim green light under the 'canopy' of a Tropical Rain-forest. On the other hand, excessive light, associated with high temperatures and transpiration, can destroy chlorophyll, and many species in tropical forests actually move their leaves during the course of the day, turning the edges toward the direct rays to minimize their effects (*heliotropism*).

An interesting result of environment is that within the polar circles continuous summer daylight induces a short-lived but extremely intensive plant growth, a characteristic of tundra vegetation (p. 532). The short days and weak light of winter in cool temperate latitudes is probably one factor which induces a deciduous habit in some trees, providing that the summer is sufficiently long to enable them to complete the cycle of leaf development, growth and fall in one season.

Wind The major influence of wind upon plants is that it removes the vapour transpired from the stomata; a strong wind means rapid transpiration. If a wind is warm and dry, this effect is very marked; the sirocco (p. 422) may cause trees and shrubs in southern Italy to transpire so rapidly that buds are killed and blackened, as if by a sudden frost.

Winds may exclude tree growth from exposed upland, mountain or coastal districts, or reduce it to a stunted, bent nature, presenting a wedge-shape to the wind, with its growth extending in a leeward direction. The scattered trees on the Pennines show very strikingly this effect of wind orientation. The poleward limit of shrubs, on the borders of the coniferous forest and the tundra, is probably determined more by wind than by temperature.

Physiographic and edaphic factors The relief of the land, its steepness or levelness, whether there are valleys and mountains or uniform low-lying plains, affects the vegetation-cover. On mountain ranges there is a great diversity in the vegetation-cover owing to rapid change in climate, soil, aspect and drainage, but generally the result of altitude is to produce 'zones' or 'girdles' of vegetation, the details of which vary according to the 'climatic base-level' of the mountains—which itself largely depends upon latitude (p. 484).

The effectiveness of the natural drainage is shown in the Tropics, where mangrove forest occurs under conditions of brackish tidal waterlogging, swamp-forest under conditions of fresh-water flooding, and true rain-forest where drainage is adequate. The high bogs on the moors of northern Germany are closely related botanically to the heaths which surround them, for both have the same poverty of mineral nutrients and the same high acidity; the bogs have a permanently damp surface while the heaths are dry. Should the bog dry out, naturally or artificially, heath will result.

Edaphic factors are those concerned with the characteristics of the soil-cover, since in this are rooted the majority of plants, and the plant-cover itself contributes to the processes which produce the mature soil (p. 491). There is no need to stress the influence of the soil-factor; compare the thin skeletal soils on steep mountains, the acid-peat soils of the moorlands, the sandy soils of the heaths and the coastal areas, the clay and alluvial soils of the lowlands. The ecologist is concerned with much more detailed edaphic differences; he studies an assemblage of species, and in an ecological survey of a grassland, for example, he finds considerable botanical differences due to slight variations in acidity and alkalinity, water content and mineral content.

WORLD VEGETATION TYPES

It is possible for the geographer, using the data collected by the botanist, to divide the surface of the earth into regions in which the 'wild' plant communities (that is, including both natural and semi-natural vegetation) are broadly identical (p. 510). The world's plant-covering does not consist 'of a mosaic of entirely dissimilar parts, but on the contrary ... it offers obvious similarities and relations' (Hardy). There have been many attempts to produce vegetation maps of the whole world, of individual continents (fig. 210), and even of smaller units; many difficulties are encountered.

In the first place, to outline arbitrarily these regions is in itself a difficult task. There are very few clean-cut lines in nature, so it is necessary to draw a boundary demarcating one vegetation region from the next through what is in effect a zone of transition. Such vegetation regions can only be broadly and generally defined. In the second place, there are many areas of the world whose vegetation has not yet been mapped and described, and certain assumptions must therefore be made. In the third place, where 'cultivated vegetation' is now widespread, the 'wild' vegetation has had to be deduced from surviving patches; in some areas these deductions are highly questionable. In the fourth place, the delineation of these regions has been made in terms of 'plant geography', or what has been called 'plant landscapes'; the geographer uses only the major plant communities as his regional basis. However, vegetation in some way summarizes the various physical features, and vegetation regions have been called 'climatico-botanical frames'. The broad relationship between climatic type, zonal soil group and vegetation region is shown in the following table:

Climate		Soil	Vegetation
Arctic		tundra	Tundra
Cool Temperate	cooler	podzols	Coniferous Forest
			Heathland
	warmer	brown-earths	Temperate Deciduous Forest
Cool Temperate (Continental)		chernozem prairie soils	Temperate Grassland
Cool Temperate	arid	chestnut-coloured soils	Arid Grassland
	very arid	saline soils	Salt desert plants
Mediterranean Warm Temperate		various (much terra rossa)	Warm Temperate Evergreen Woodlands
Hot Desert	semi-desert	red desert soils	Desert Scrub
	true desert	sandy, saline soils	Plantless Deserts
Tropical (summer rain)		red-earths lateritic soils	Savanna Grassland
Tropical (monsoon)		lateritic soils regur	Monsoon Forest

Climate	Soil	Vegetation
Equatorial	lateritic and associated forms	Tropical Rain-forest Swamp Forest Mangrove Swamp
Mountain (zoning according to altitude)	Skeletal	Zoning according to altitude, culminating in high moorland, mountain grassland, scrub, alpine flora, and plantless ice-cap and snowfield

FORESTS AND WOODLANDS

(i) **Tropical rain-forest** The tropical evergreen rain-forest (sometimes called *selva*) is confined to hot, humid areas in the vicinity of the Equator: the Amazon Basin, the coasts of central America, the Congo Basin, the West African coastlands, Malaysia, the Indonesian archipelago, and the lowland parts of southern Vietnam (plate 105).

The chief feature of the rain-forest is its profusion and variety; one estimate by an ecologist gave the plant population of forest in southern Malaya as about 64 million individuals per square mile (2·6 sq. km). Many thousands of different species are represented, and rarely is a 'stand' (a more or less continuous assemblage) of one tree found. A Dutch botanist has estimated that there are 30,000 different flowering species in Indonesia and 10,000 in Borneo (compared with 2000 species in the British Isles), while there are no less than 3000 separate species of trees. A large proportion of this flora is still unnamed and undescribed. Occasionally there may be small stands of such trees as ironwood and in Borneo of the camphor tree, but this is unusual.

The vegetation shows a distinct arrangement in layers. The large trees grow upward to the light for 45 m (150 ft) or more, with smooth branchless boles, supported by large 'plank-buttress' roots, and surmounted by a crown of broad evergreen leaves which forms a continuous 'canopy' or upper layer. The surface of this canopy is far from level, since the large rounded crowns of the biggest trees project. Among well-known trees are palms and the hard 'furniture woods', mahogany, rosewood, ebony, greenheart and ironwood, but many trees, both of hard- and softwood, are known only by their native or botanical names; some of these have made a post-war appearance in British timber-yards.

Below the canopy an 'intermediate layer' consists of smaller trees, tree-ferns, a maze of woody climbers known as *lianas* (often

104 Mangrove swamp
(Paul Popper)

105 Tropical Rain-forest in New Guinea
(Paul Popper)

106 Tropical swamp-forest in Guyana
This photograph was taken on the banks of the Potaro River, a tributary of the
Essequibo.

(Paul Popper)

107 Giant redwoods in Del Norte County, California
Paul Popper)

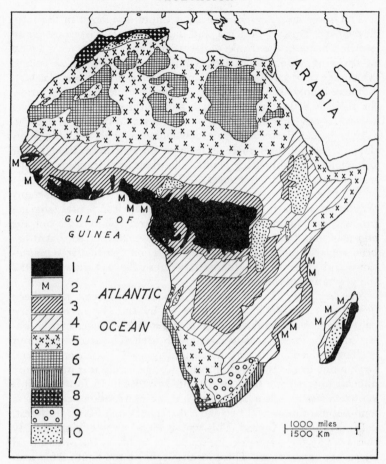

Fig. 210 *The vegetation regions of Africa*

1 Tropical evergreen rain-forest; **2** mangrove swamp; **3** tropical forest (with marked dry season); **4** savanna; **5** scrub and semi-desert vegetation; **6** hot desert —rock or sand, with scanty vegetation; **7** warm temperate rain-forest; **8** Mediterranean evergreen woodland; **9** temperate grassland; **10** mountain forest and grassland. Cultivated vegetation (e.g. the Nile delta and valley, and plantations) is not shown.

a hundred or more metres long and as thick as a man's arm), and innumerable *epiphytes* (plants growing on others, though not deriving their substance from them). One of the commonest lianas in Indonesia is the rattan, a species of climbing palm with thorny stems so long and strong that it is used by the native peoples in place of rope. Epiphytes include numerous varieties of orchids and ferns.

The lower layer, or *undergrowth*, is thinly developed in the rain-forest, and consists of ferns and large fleshy herbaceous plants growing between the trunks of the trees. The ground layer is a mass of decaying vegetable matter, with numerous *saprophytes* (plants which live on decaying matter) and scattered herbaceous plants.

There is no seasonal periodicity in the life-cycle of the vegetation, for there is little climatic change and so the general aspect never varies. At any time any plant, or even a single branch of a plant, may be in leaf, flower and fruit, new foliage may grow within a few days of leaf-fall, and there is little or no synchronization in the behaviour of neighbouring plants. The whole appearance is one of a monotonous uniform greenness.

The true rain-forest as just described is not as extensive as many vegetation maps indicate. Not only do environmental changes produce modifications such as swamp-forest (plate 105), but the efforts of man have destroyed the natural vegetation, so that it has been replaced both by cultivated vegetation (particularly plantations) and by semi-natural secondary forest. Fig. 211 shows in detail the vegetation of Sumatra.

Secondary forest Large areas of primary rain-forest have been destroyed either to obtain timber or by the system of shifting cultivation. A subseral sequence of vegetation changes can be observed as the forest seeks to re-establish itself. The secondary forest is characterized by its very dense undergrowth, rank herbage and large numbers of fast-growing softwoods; generally it is much more difficult to penetrate than virgin rain-forest itself. In Indonesia it is known as *bloekar*. Ultimately, if no further interruption occurs these semi-natural forests will become indistinguishable from rain-forest.

(ii) **Mangrove forest** This type is found on low-lying, muddy coasts where the water is calm inside the surf-belt, but where tidal effects are experienced. These forests occur on the South American coasts near the Amazon estuary, along the edge of the Niger delta, and along the east coasts of Sumatra and Borneo, extending inland in brackish swamps and lagoons. They grow where mud is being deposited, and so the belt gradually extends seaward as they help to further the accumulation of mud between their roots (plate 104).

The name 'mangrove' is given collectively to some twenty or thirty species of trees, not all botanically related, but with certain common physical characteristics, especially the ability to grow on tide-washed mud-flats. At high tide a mass of dull greenish-grey leathery foliage appears to float on the water, but at low tide one can see short stumpy trunks often supported by a maze of stilt-like aerial roots. Others send out horizontal roots under the mud, from which

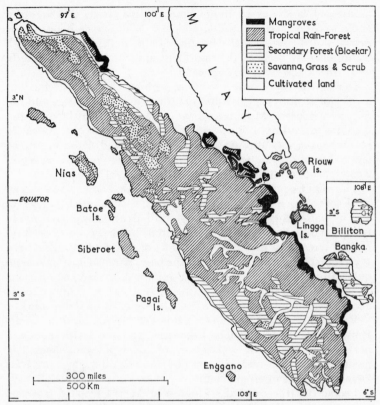

Fig. 211 *The vegetation regions of Sumatra*
After *Atlas van Tropisch Nederland.*
The system of shading does not apply to Malaya, which is left blank.

vertical roots grow above the surface, and others have horizontal roots which bend in loops. These complicated root-systems serve both to anchor the tree in the mud and, provided with breathing pores, as aerating organs.

The various species of mangrove are arranged in belts more or less parallel to the shore. On the seaward edge are the mangroves which can tolerate unstable mud and considerable tidal inundation; as the coast is built up by natural accretion, the 'pioneer' species advance seaward, producing a seral succession. On the landward side a mangrove swamp gradually changes into fresh-water swamp.

(iii) **Tropical monsoon forest** This is tropical forest which experiences a marked seasonal drought. It may occur both in true monsoonal lands, as in Indo-China, Burma, parts of Indonesia, India and northern Australia, and also in the lands lying on the margins of the equatorial climatic belt (p. 459), where it forms a transition zone between the rain-forest and the tropical grasslands, as in Africa and central America.

During the rainy season a monsoon forest looks somewhat like a rain-forest, although it is less luxuriant and more open, and has far fewer species. Most trees have a markedly deciduous habit, shedding their leaves during the dry season, when the associated herbaceous plants also die down. Epiphytes and lianas are much fewer than in the rain-forest. The ground layer consists of thickets of woody shrubs or, near the margins of the forest, of coarse grass.

The most frequently occurring trees are teak, bamboo and acacia. Teak forest has a single dominant species, the teak itself (known as *djati* in south-eastern Asia); the largest stands are now to be found in Burma and Siam. Its chief value lies in its hardness and durability; mature trees often provide 20 m (60 ft) of straight branchless bole, 1·5 m (5 ft) or more in diameter at the base. Bamboo is common and quick-growing and usually springs up when an area of monsoon forest, cleared for cultivation or for timber, has been subsequently abandoned. Bamboo thickets are well developed in Indo-China.

Varieties of acacia are widespread in the African forests, a species known as *pilang* is found in Indonesia, and the wattle grows in northern Australia. Several species of eucalypt (*gum*) flourish in northern Australia and in parts of Indonesia, though many trees have been deliberately planted.

Unlike the true tropical rain-forest areas, the tropical monsoon lands have long been densely populated. Many of the lowlands have been intensively cultivated for centuries and the natural vegetation over much of India, southern Burma, the deltas of Indo-China and the island of Java has long since vanished. In Africa, too, shifting cultivation and grazing have eaten into the forest margins.

(iv) **Tropical mountain forest** Many mountainous areas within tropical latitudes are forested, and a series of altitudinal zones can be distinguished as the vegetation changes with height. Some of these higher zones have coniferous and deciduous trees botanically similar to those in western Europe; oaks, chestnuts and pines are found in Sumatra and Borneo.

Some types of mountain forest are peculiar to low latitudes. *Moss-forest* occurs on the mountains of Ceylon, Malaya and Indonesia at heights of between 900 and 1500 m (3–5000 ft). The trees are covered with thick layers of moss and liverwort, dripping with moisture, and present an indescribably fantastic appearance. The writer has pushed his way to the forested summits of a number of Ceylonese peaks through this type of vegetation, perhaps the most unpleasant form of mountaineering in the world. Other types of mountain forest consist of brakes of tree-ferns. The mountain ranges of East Africa are swathed in 'a fantastic tangle of rotting vegetation —giant groundsel, lobelia and giant heath—all thickly covered in moss'; thus Eric Shipton described the slopes of Ruwenzori. The giant groundsel grows about 3·7m (12 ft) high on soft pulpy stems, with a sort of cabbage-leaf growth at the top, while the lobelia forms a tall cone of feathery leaves.

(v) **Thorn forest** In certain parts of the Tropics, notably in north-eastern Brazil (where it is known as the *Caatinga*), southern Mexico, Venezuela, the East African plateau, the southern Deccan and central Burma, a type of tropical woodland occurs which endures a long period of drought. It might almost be placed in the category of tropical scrub and semi-desert, but the tree growth is sufficiently continuous to merit the term 'tropical deciduous xerophytic woodland'. It indicates a transition stage between monsoon forest in which the dry season is marked but not prolonged, and scrub where the dry season is dominant. It forms a thick impenetrable mass of thorny jungle, from which rise occasional taller trees such as acacias and euphorbias. For seven months it presents a greyish, dead-looking, leafless tangle, intermingled with cacti, thorny lianas and a few hardy evergreen shrubs. All are protected in one way or another against drought conditions, either by storing water or by reducing transpiration to a minimum.

When the short concentrated rains occur, the vegetation bursts into intensive life. The dusty landscape becomes verdant, many trees develop brilliant blooms, and herbs and bulbous plants flower. When the period of rains, lasting some four months, has ended, the vegetation rapidly reverts to its dormant state.

(vi) **Warm temperate rain-forest** In some parts of the world occurs a type of forest favoured by a warm temperate climate, with a well distributed rainfall of 150–300 cm (60–120 in) and with temperatures ranging from 10° to 21° C. A dry season is either unknown or of brief duration. Such conditions are found in the eastern and south-eastern states of the U.S.A., especially along the Gulf Coast and in Florida, southern Brazil and the upper Paraná-Paraguay valleys, the Natal coastlands, north-eastern Burma, much of central and southern China, southern Japan, eastern Australia and the North Island of New Zealand. Similar forests grow on the slopes of mountain ranges in tropical latitudes.

In many ways these forests superficially resemble those of equatorial latitudes. Most of the trees are evergreen, but the increasing seasonal effect of winter induces an admixture of deciduous trees. Winter conditions are so mild, however, that there is no strong seasonal climatic rhythm and growth is fairly continuous. There is an extraordinary variety of trees, from the bamboo, palm, tulip-tree, camphor-tree, magnolia and camellia on the equatorward margins, to evergreen oaks, hemlock, locust, walnut, maple, dogwood, azalea and laurel on the poleward side. Conifers are widespread in some parts, mixed with deciduous trees, as in Georgia and Virginia where yellow and white pines grow, in Florida where cypress and evergreen oak occur, and in New Zealand where the kauri pine grows. Tree-ferns are common, forming dense 'fern-forest' in New Zealand. The undergrowth is usually thick, and the forests are often more difficult to penetrate than those of the Tropics. There is a great wealth of epiphytes, lianas, mosses and lichens festooning the trunks and branches.

Many of these original forests no longer exist, for the climatic conditions are very suitable for subtropical agriculture. In China, Japan and the 'Deep South' of the U.S.A., dense agricultural populations grow cotton, tobacco, soybean, mulberries and sugar-cane. Many other commercial products—kauri gum, camphor, yerba-maté, coca—are obtained.

'*Dry forests*' Some varieties of these warm temperate or subtropical forests are marked by a distinct dry season. In southern and eastern Australia are forests with a thin, scrubby undergrowth of dwarf acacia species from which rise giant evergreen eucalypts, also known as Australian gum-trees. These have tall, straight boles, with tattered barks and a flat crown of drooping greyish leaves, the edges of which are constantly turned to the sun. Some trees attain a height exceeding 90 m (300 ft), and the timber is of excellent quality. In southern Brazil forests of *araucaria pine* (popularly known as 'monkey puzzle') are separated by open spaces of coarse grass or scattered shrubs of yerba-maté. In the Gran Chaco of Paraguay–Argentina occur the *quebracho evergreen forests*, gnarled bushy trees with exceptionally hard wood growing from a thick undergrowth of evergreen shrubs. All these forests merge into grasslands, both tropical and temperate, and form part of the 'parkland' aspect of the grassland borders.

(vii) **Mediterranean evergreen woodland** The climatic type described (pp. 467–9) as 'Western Margin Warm Temperate' (commonly known as 'Mediterranean climate') was characteristically covered with open woodlands, most consisting of evergreen oak, cork oak and other oaks. There is now, however, very little real forest of evergreen oak, for much has been removed for domestic fuel, by the charcoal burners, and to make way for the cultivation of the characteristic Mediterranean products—both native fruits such as olive, fig and vine, and the imported citrus fruits. Widespread fires during the summer drought have also been a potent cause of deforestation.

The evergreen oak-forests have been replaced in part by evergreen aromatic shrubs, such as oleander, myrtle, heath, rosemary, lavender and a profusion of creepers, vines, herbaceous and bulbous plants, forming in places a dense thicket 2–3 m (8–10 ft) high. In the Mediterranean Basin this is given the name of the *maquis* (French) or *macchia* (Italian).

Elsewhere forests of conifers occur, such as pines, firs, cypresses and cedars. In Corsica the Corsican pine and the maritime pine flourish; the first of these, where there is some degree of forest management, grows into a fine tree which may reach a height of 45 m (150 ft) with a basal circumference of 6 m (20 ft), taking five or six centuries to attain its full development. The chief tree of the coastlands of the Mediterranean Basin is the Aleppo pine, found notably in Greece.

On the upland slopes, particularly from about 600–750 m (2000–2500 ft), are considerable woods of sweet chestnut associated with beech; the upland areas of eastern Corsica are known as the *Châtaigneraie* or *Castagniccia*. Another widespread tree, especially on stony slopes, is the wild olive.

Although the species are often different, the same general aspect and characteristics obtain in areas with similar climates in other parts of the world. The equivalent of the maquis in California is *chaparral*, and pines are represented by the Monterey cypress. There, too, grow the magnificent coast redwoods (*Sequoia sempervirens*) (plate 107), and the 'big-trees' (*S. gigantea*) of the Sierra Nevada, the latter now preserved in national parks; the famous 'General Sherman' tree, 83 m(272 ft) high, is 9·32 m (30 ft 7 in) in diameter at the base, and at 37 m (120 ft) high its diameter is still 5 m (17 ft). In western Australia two representatives of the eucalyptus family—jarrah and karri—form fine timber trees.

The basic characteristic of almost every plant, tree or shrub is that it is adapted in some way to the summer drought. It may have small, hard, evergreen leaves, compact woody stems, a thick rough bark, and long penetrating or fleshy bulbous roots. The dry evergreen form is suited to a slow thrifty growth throughout the year, with increased activity during the autumn rains; the herbaceous perennials and bulbous plants burst into a short-lived glory in early spring, to die down during the summer heat and drought.

(viii) **Cool temperate deciduous forest** The natural vegetation of much of central and western Europe, and of eastern North America south of the St Lawrence, was probably deciduous forest, but little if any of this primeval forest is left in these closely settled lands. In Europe for two thousand years the axe of the woodman has cleared the way for the plough, and the wood was needed for domestic building, for shipbuilding, and to burn for charcoal to meet the demands of the iron-smelters. Europe in general and England in particular became a land of scattered copse and maintained parks. In eastern North America three hundred years of vigorous clearance has produced a similar result.

A broadly similar forest, but much less affected by man, still occurs in the eastern Asiatic lands—in the Amur Basin, on the uplands of Manchuria, in Korea and northern Japan. There are also small stretches in southern Chile and Tierra del Fuego.

The response of the vegetation to a temperate moist climate, with its comparatively low winter temperatures, is a dormant season during which the trees are bare of leaves and a growing season during which the trees slowly build up wood. The variety of trees is not large, and stands of the same species are common. There are the beech woods of the English scarplands and central European uplands (also found more extensively in southern Chile), the now widely scattered oak woods, the mixed woods of poplar, elm, sycamore, chestnut, hornbeam and others, the 'damp woodland' of willows, poplar, alder and aspen, and the 'coppice woodland' of hazel, alder and hawthorn. The North American deciduous forest is much more varied and luxuriant, and includes, besides the trees mentioned, walnut, maple, hickory, magnolias, and conifers such as cedars and spruces, while an undergrowth of rhododendrons, azaleas, fuchsias and other flowering shrubs is well developed.

Probably the most magnificent area of these forests is in the Great Smoky Mountains National Park, linked by the Blue Ridge Parkway with Shenandoah National Park in eastern U.S.A. Here there are estimated to be 130 different native trees, with 1300 varieties, and about 100,000 hectares (250,000 acres) are virtually virgin forest. This type of forest merges southward into the Warm Temperate Forest of the southern states.

(ix) **Mixed deciduous-coniferous forest** The deciduous forest merges poleward (and up mountain slopes) into coniferous forest, forming in north-central Europe a transition zone known as 'mixed forest'. Even in the true coniferous forests the deciduous trees are represented by birch, alder and poplar; dwarf birch are found farther poleward than are the conifers. The common criterion taken to indicate the northern limit of mixed forests in Europe is the limit of the oak, a line running across central Sweden and more or less eastward from the Gulf of Finland.

(x) **Coniferous forest** The coniferous forest proper is found in high latitudes or on mountain-slopes, although coniferous species occur in many other vegetation regions. The Siberian name of *Taiga* is often applied. Their greatest development is in the northern hemisphere where the land-masses are at their broadest, and they extend right across the North American and Eurasiatic continents. In western Europe they extend southward to about 60° N, in eastern Asia to about 50° N, in eastern America as far south as 45° N.

The trees have to withstand hard climatic conditions—extremely cold winters, short cool summers, a light summer rainfall, and winter precipitation in the form of snow. They grow at a slow rate, they are frugal in their needs, their hard needle-shaped leaves reduce transpiration to a minimum, and the compact conical structure both helps their stability against the wind and prevents too heavy an accumulation of snow upon the branches. They have a big proportion of wood in comparison to leaves, and their timber is 'soft', while the evergreen foliage allows of the maximum utilization of sunlight throughout the short growing season.

There are not many species of tree, and they occur in great uniform stands. Pines predominate on dry, sandy soils, heavy spruce-forests on damper soils, larch (which is not an evergreen) on deep soils of fair quality, and various species of fir are widespread. The graceful silver birch helps to relieve the gloomy monotony of a dark spruce-forest. Much of the accessible woodland has been logged at least once—notably round the Baltic and along the St Lawrence valley, although a large part of the forests of Siberia and the Canadian Shield has been but little exploited. In countries such as Finland and Sweden, whose forests form a considerable part of their economy, cutting is carefully regulated, and as one area is cleared seedlings are planted. In the south of Finland, for example, a fir takes some fifty to sixty years to mature, in the centre about two centuries. Hence a long period of forest rotation has been evolved, and few of the commercial forests are ever 'clean-felled'—seedlings, saplings, young and mature trees grow side by side.

The undergrowth is thin and poor both in variety and quantity, partly because the surface layers of soil in which the smaller plants are rooted are subject to long hard freezing, and partly because of the thick layer of pine-needles inimical to most plant life. A few stunted shrubs, such as bilberry, cranberry, crowberry and dwarf birch, and numerous lichens and mosses, represent the ground vegetation.

The coniferous forest fades out northward into the Tundra, as trees become more stunted and wider spaced, with occasional clumps and individuals standing as gaunt outposts.

GRASSLAND

Grassland formations occur in areas which experience a season of prolonged drought, but which have a period of rainfall coinciding with the season of growth. The total rainfall is, however, generally inadequate for tree growth, except for scattered xerophytic types. The rainfall effectiveness (p. 432) is the main factor; in middle latitudes the rainfall total varies from about 30–60 cm (12–24 in), but in the Tropics, as a result of high evaporation, even 100 cm (40 in) is not enough for tree growth and grass or scrub is found.

The life-cycle of grass enables it to avoid the effects of drought by seeding and then dying down. The seeds lie in the mat of humus formed by the roots, and grow rapidly when stimulated by renewed rainfall.

Grassland landscapes are more varied than the name 'grass' might at first sight suggest. The botanist recognizes a wide range of grasses, known as *Gramineae*, distinguished by simple sheathing leaves, jointed tubular stems and flowering heads enclosed in scales; they include the familiar cultivated temperate turf and meadow grasses (pp. 536–7), desert and dune grasses, swamp grasses, alpine grasses, heath and moorland grasses, salt grasses and tall tropical grasses. As grasslands occur on the borders of both woodland and desert, the transition zones are scattered with trees and shrubs, usually possessing xerophytic characteristics.

It is doubtful whether any true natural grassland exists today. Much of it is either semi-natural, affected by centuries of grazing and burning or by liming, fertilizing and draining, or cultivated, where carefully selected grass seeds have been sown and the pastures laid down for some years. These areas are the scene of varied agricultural activity, where either cereals are grown or where animals are pastured.

(xi) **Tropical (savanna) grassland** It is difficult to draw a dividing line between tropical forest, tropical grassland, scrub and hot desert; the savanna partakes, to a varying degree, of the character of all four, with resulting wide differences in aspect. In some areas it forms a discontinuous cover of tufted grass, growing rapidly to a height of 2 m (6 ft) as a stiff yellow straw-grass crowned with silvery spikes. The tallest variety is elephant grass, often as much as

4·6 m (15 ft) high. Clumps of trees, such as palms, baobabs, acacias and ceiba, adapted to drought as in the Monsoon and Thorn Forests, grow in hollows where ground-water approaches the surface. The term 'park-savanna' is sometimes used to indicate this tree-dotted landscape (plate 108), and it is also called the 'high grass–low tree' type. Many trees are wedge- or umbrella-shaped, as a result of the strong winds that frequently blow over the open grasslands. Tubers, bulbs and hardy perennials burst into flower when the rains come.

As the desert border is approached, the grass becomes shorter and more tufted, with bare sand between and scattered thorny bushes; it is difficult to distinguish between 'desert grass savanna' and 'desert scrub'.

The savanna covers a great area in South America (on the Guiana plateau it is known as *llano*, in Brazil as *campo*); in Africa it is very widespread, both north and south of the rain-forest of the Congo Basin and West Africa and on the plateaus of East Africa; it occurs on the drier parts of the Deccan; and in Australia it extends north and east of the central desert.

These savanna lands are inhabited by numbers of animals, both herbivores and carnivores. Particularly in Africa it has been the scene of many native cultures based on hunting, animal rearing or shifting cultivation, and the direct or indirect effects of man must have reduced much former forest to savanna. Modern animal rearing, as in Brazil and Queensland, the extension of irrigation schemes, and the large-scale attempts to grow ground-nuts and coarse grains for feeding-stuffs have also greatly modified the vegetation cover.

In Java and Sumatra extents of coarse tussocky grass are known as *alang-alang*. While it has the appearance of a true savanna, it is the direct result of the repeated burning of forest. Once established it tends to maintain itself, since it forms a mattress of dead leaves which suppresses other plants.

(xii) **Temperate grassland** Extensive areas of grassland occur in the continental interiors of middle latitudes, where a rainfall total of *c.* 30–50 cm (12–20 in) is received mainly in early summer. These are known as *steppe* in Eurasia (extending from the *pusztas* of Hungary into the Ukraine and on eastward across Asia into Manchuria), as *prairie* in the central lowlands of North America, as *pampas* in the Argentine, as *veld* in Africa as far north as the Tropic of Capricorn because of the altitude of the plateau, and as *downs* in south-eastern Australia and eastern South Island, New Zealand.

The natural grassland forms a continuous cover of tufted grass, interspersed with bulbous and leguminous plants. The grass is dry and hardy, bluish-green in spring, yellow and straw-like in summer, and tipped with feathery spikes. In the drier parts of the Argentine pampas and of the downs of New Zealand, the grass grows in larger tufts with bare earth between. Trees are found only in hollows or along watercourses.

Little or no natural temperate grassland exists today. The better-watered parts have been ploughed up and form the world's wheat-lands, favoured by the gently rolling plains, the rich accumulated humus-content of the black-earth soils, and the climatic regime of early summer rain (growing season) and late summer drought and sunshine (ripening and harvesting season). Other parts have a pastoral economy, varying from the nomadic mode of life in the dry steppes of central Asia to the large-scale ranches of America and the sheep-runs of Australia and New Zealand. Rarely is the pasture unimproved; in Argentina the natural grass has been replaced by alfalfa, in New Zealand by British meadow grasses.

(xiii) **Mountain grassland** Irregular areas of grassland are found in nearly all mountainous regions, forming an altitudinal zone above the tree-line, the result of a short vegetative period. The position and nature of this 'grass girdle' varies with latitude, altitude and aspect, and a few examples must suffice.

In Ceylon coarse *patana* grassland covers much of the uplands above 1800 m (6000 ft), while in East Africa thick tufted grassland of a downland aspect extends upward from about 3000 m (10,000 ft), the limit of the bamboo forest. The intermont plateaus of South America at a height of 3700 m to 4900 m (12,000–16,000 ft) have a sparse cover of coarse 'cushions' of grass, separated by stony ground; these are the *puna*, the driest and poorest being the *puna brava*. The name of *pamir* is given to similar grasslands in the central Asiatic plateaus.

Grassland is not common in the Mediterranean lands because of the scorching summer drought, but in the higher parts of Corsica, Italy and Greece moisture from the melting snows starts its growth. On the alps of central Europe rich grasses grow rapidly, nurtured by the melting snow as the winter snow-line retreats up the slopes; to these upper pastures come the valley dwellers with their animals, a seasonal movement known as *transhumance*. The lower mountains of middle latitudes, such as those of Britain, have a varied grass-cover, notably the chalk downland grasses, the rough hill grazings and the grass moors (p. 538).

SCRUB AND DESERT

Many references have been made to a scrub vegetation, since it is a direct reflection of increasing aridity, both rainfall total and its seasonal distribution. In the Tropics the evergreen rain-forest degenerates through monsoon forest and thorn-forest into either savanna where grass predominates or a scrub of thorny shrubby plants. In the Mediterranean regions the maquis (p. 525) becomes increasingly desiccated, particularly on limestone soils; the maquis is a scrub, a subclimax form of evergreen xerophytic woodland, and broadly similar types are found in America and Australia. In the cold lands the forest cover thins out poleward and becomes more stunted, finally passing into tundra. All these transition zones are so gradual that the effects of man and animals—over-grazing, clearing, burning—can produce a progressive degeneration which has undoubtedly increased the area of scrubland at the expense of forest and grassland.

The poor grasslands and scrubs shade off into true desert, because of increasing drought, cold and also soil salinity. As de Martonne has said: 'The desert seems to be the goal towards which the gradual impoverishment of plant life is tending'.

(xiv) **Tropical scrub and semi-desert** The main plant in the tropical scrublands is the acacia, which is bare for most of the year. In some areas thorny succulents of the cactus family are common, while other plants include tubers, dry heath-like plants and desert grass. An occasional tropical downpour produces a short-lived burst of plant growth—shrubs and herbaceous plants blossom exotically for a brief season (as happened in Death Valley, California, in 1935, 1940 and 1947), and a carpet of grass springs up, soon to be scorched by the heat. Many more species can grow under these adverse conditions than is realized; a recent detailed survey of Death Valley discovered over 600 different species of plants.

Many varieties of this scrubland have broadly the same appearance. In New Mexico and Arizona are great areas of fantastic-looking cacti (as in the Organ Pipe Cactus National Monument); in Mexico are the yucca, cereus, agave and opuntia; in East Africa, particularly around the coast of the Eastern Horn, occur acacia, euphorbia, aloes and agave; varieties of acacia dominate in the Kalahari and Karroo country of south-western Africa; the *mulga* scrub on the borders of the Australian desert consists of dense thickets of acacia; the sharp *spinifex* (sometimes known as 'porcupine grass') is the main element in the scrubs to the north of the central Australian desert; and the *espinal* and *chañaral* occur in northern Argentina and central Chile. The chañaral is well-nigh impenetrable, for the chañar bush has large, vicious thorns. In the scrublands of Persia and Turkestan the common bush is the *saxaoul*, which resembles a pollarded willow without leaves.

(xv) **Tropical desert** The hot deserts are, except in the moving dune-belts, rarely completely without vegetation. Even the sand-desert bears occasional tamarisks and tufts of coarse, spiky grass, while the hamada and reg deserts (p. 277) have scattered clumps of leafless thorny perennials. Elsewhere dwarf salt-bushes, prickly prostrate plants, small brittle heath-like plants, such strange forms as the *welwitschia* of South-west Africa (which looks like a dead stump from which huge trailing leaves spread out, growing incredibly slowly from the base as the ends die), cacti, and in fact all the thorn bushes of the scrublands, are found, but much more stuntedly and sparsely. Most vegetation exists in a virtually dormant state, but after many years may burst into growth for a few days as the result of a rare downpour. Exceptional rains cause the more continuous vegetation cover of the margins to advance temporarily into the true desert, as happened in Australia in 1950–1, tempting the farmers to graze their flocks more widely.

(xvi) **Warm temperate scrub** Scrub is found in the drier parts of the warm temperate lands, where summer temperatures are quite high. In the Mediterranean lands the maquis and chaparral (p. 525) are kinds of evergreen scrub. On some arid calcareous soils an even poorer vegetation (*garigue*) may develop; this includes stunted evergreen oak, thorny aromatic shrubs, prostrate prickly plants and tuberous perennials, separated by bare rock.

Warm temperate scrub in Australia includes the *mallee* of dense thickets of dwarf eucalypts, 2 m or so in height, the *brigalow* of acacias, and the *mulga*, also of acacias but more open.

Another widespread scrub vegetation type is the *sage brush*, which consists of greyish heath-like shrubs up to 2 m in height, forming sometimes a dense continuous cover, sometimes scattered patches. Sage brush occurs widely in the Great Basin of Utah, the Colorado and Mexican plateaus, Patagonia, the Kalahari Desert, the Plateau of the Shotts in North Africa, the plateaus of central Asia, and the desert borderlands of Australia.

(xvii) **Cold deserts** Cold desert vegetation, or 'Barren Lands', occurs across the land-masses of the northern hemisphere more or less north of the Arctic Circle beyond the tree-line, where it is known as the *tundra*. It is also an altitudinal zone, lying between the upper limit of the alpine meadows and the permanent snow-line, and isolated 'alpine plants' grow among rock and scree. An example of this type is the *paramo* in the Andes.

The tundra is the result of excessively cold, strong winds, a permanently frozen subsoil (*permafrost*) (p. 254), and long alternate periods of darkness and light. The duration of plant growth is short, two months or less, but growth is then virtually continuous. The most characteristic aspect is a close growth of dwarf shallow-rooted shrubs such as crowberry, bilberry and bearberry rising from a carpet of mosses and lichens, with an occasional dwarf birch or willow above the general level. Patches of sedge and cottongrass grow in badly drained areas, giving rise to thick layers of acid peat. The south-facing slopes may exhibit in summer a short-lived splendour of flowering herbaceous plants.

CHAPTER 21

The Vegetation of the British Isles

AN analysis of a world-wide pattern of soil types involves extensive generalizations, the weakness of which was all too apparent when the soils of the British Isles were considered. The same holds true concerning vegetation; on a world or continental vegetation map these islands are covered by two or three types only, yet the plant-cover is extremely varied. This is the result partly of the wide range of relief, soil and climate, all remarkably diverse in so small a compass, and partly of the remarkable modifications produced by the long-continued activities of man and animals in a densely populated country. A most interesting line of research is the reconstruction of the plant-cover of the past, both in prehistoric times from an analysis of peat bogs and the plant remains which comprise them, and in historic times when the information from written records can be systematically pieced together. Leaving out of account the cultivated vegetation, most of the other vegetation of the British Isles is semi-natural. The original natural vegetation probably consisted of forest, fens and marshes over the lowlands, and moorlands on the uplands, although probably much of the present treeless hill-country, perhaps even the Pennines, was once wooded; roots and stumps are commonly found preserved in Pennine beat bogs.

The major plant communities comprise (i) woodland; (ii) scrubland; (iii) grassland; (iv) heathland; (v) moorland and bog; (vi) fen and marsh; (vii) maritime vegetation; and (viii) mountain vegetation.

(i) **Woodland** *Oak-forest* was probably the natural vegetation of the lowland areas, but by the eighteenth century this had largely been cleared (p. 526). Mature oak-forest is now rare, and most was deliberately planted in royal parks and on the estates of the great land-owners of the past. While the oak is dominant, the wych-elm, ash, maple and hornbeam are often present. A common association is oak and hazel, grown as 'coppice-with-standards', for long an important part of rural economy; the hazel trees are cut back hard to the base, and so produce a large number of stems, 6 m or more high, which provide poles for hurdles and fencing, while the oaks grow on as 'standards'. The details of the oak-woods—the shrub-,

108 Park savanna in Tanganyika
(Paul Popper)

109 Sand-dunes on the Lancashire coast between Southport and Ainsdale
(Paul Popper)

110, 111 The colonization of coastal mud-flats by vegetation
(Above) The development of mud-flats on which grows perennial rice-grass
(*Spartina townsendii*), along the Hampshire coast of the Solent.

(*J. A. Bailey*)

(Below) The mud-flats on the coast of the Solent at Lymington with the Isle of
Wight in the background.

(*Aerofilms Ltd*)

field- and ground-layers—vary with environment; oak-woods on sandy soils may have bramble and bracken, those on clays and loams the typical woodland flora of primrose, wood-sorrel and anemone, and those on constantly wet soils alders and water-loving plants.

Beech-woods are dominant on the chalk hills of south-eastern England, usually growing on loamy patches lying on the Chalk (such as the Clay-with-flints). Their chief need is good drainage, so that beeches rarely occur on the heavy clays where oaks are dominant. It seems certain that the beech, like the oak, is a true 'native' tree. Smaller, more scrubby ash-woods are found on calcareous soils, notably in the Pennine valleys, and alder-woods occur in damp localities, as on the margins of the Broads.

The characteristic Scots pine (*Pinus sylvestris*) is one of the dominant trees of the European coniferous forest, and probably once covered much of the Highlands of Scotland, both the floors of the glens and the lower hill-slopes up to the tree-line; in Inverness-shire the trees grow up to 600 m (2000 ft) above sea-level. But very few tracts of native pinewood survive. Birches (both the silver birch and the hairy birch) are British species, and often grow in association with pines. Continuous tracts of birch woodland are to be found in parts of Scotland.

The deliberate planting of quick-maturing conifers, both in areas not suitable for cultivation (such as heathland, drained moorland and hill-grazings) and on areas formerly covered by deciduous woodland, now perhaps derelict or scrub, has been the practice for a century or more, accelerated in recent years by the work of the Forestry Commission. The hill-slopes of the Highlands of Scotland, the English Lake District, the Pennines, North Wales and the North York Moors have been the scene of extensive planting. Conifers have also been planted on the heathlands of Dorset, in the New Forest, the central Weald, and the Breckland of East Anglia, and on coastal dunes, as at Ainsdale and Formby (plate 109) and on the shores of the Moray Firth. They can tolerate light sandy soils, indeed they help to 'fix' them, and they attain maturity reasonably quickly. While Scots pine is still used, exotic conifers have been extensively introduced, particularly Corsican pine, spruce (both Norway and Sitka varieties), larch (Japanese and European), and Douglas fir. Many small isolated groups of pine-trees are scattered about the countryside, particularly on heathlands, spread sub-spontaneously where seed has blown from neighbouring plantations.

(ii) **Scrubland** Scrub may grow in the form of 'undershrub' among woodland, producing a distinct layer, as hazel, blackthorn, sallow (commonly called 'palm'), briar and bramble grow among oaks, and as holly, yew and box occur among beeches. Where the trees have been felled or have died without replacement, or where arable and pasture lands have been abandoned, scrub bushes are dominant; this is *woodland scrub*.

Sometimes thickets of thorny scrub, including hawthorn, blackthorn, gorse and brambles, can form a more or less impenetrable growth, or *thicket scrub*. It occurs on limestone soils and on windswept hills, where conditions are unfavourable for tree growth.

Thus scrub may occur as a layer in a woodland, it may represent a stage in a succession which will lead to a forest climax, or it may represent a subclimax itself.

(iii) **Grassland** Grassland occupies a considerable part of the British Isles, replacing to a large extent what must have been deciduous forest. Part consists of permanent grassland, which was originally sown by man but has developed like a natural plant community, yet is subject to more or less continuous grazing. Part consists of upland grassland. Both of these categories have been improved in many areas, usually by liming, fertilizing, natural manuring and drainage. The cultivated grasses, sown mixtures fitting into the arable rotation and remaining down for a limited number of years ('short ley'), are outside this survey, although they are frequently allowed to revert to permanent pasture.

Botanically grasslands are complex. They include three main groups: *turf grasses* which when continuously cropped or mown make a close mat of fibrous roots with short, close blades; *meadow grasses* which send up taller shoots; and *tussock grasses* which form clumps and tussocks above the surface. There are many varieties of grass, depending mainly on the soil—whether they grow on acid, limy or neutral soils, on sandy or water-logged soils. In addition, most grassland contains a variety of leguminous plants such as clover, and miscellaneous flowering plants. The grasslands may be described under five heads.

(*a*) The *neutral grasslands* form most of the permanent grassland of the lowlands, are generally enclosed in fields, and are constantly 'improved' in quality by good farming practice. They are dominated by perennial rye-grass (*Lolium perenne*), which with white clover (*Trifolium repens*) makes the best pasture. Other grasses include common meadow-grass (*Poa pratensis*) and timothy (*Phleum pratense*).

(*b*) The *basic grasslands* occur on chalk and limestone terrain. The dominant grasses are sheep's fescue (*Festuca ovina*) and red fescue (*Festuca rubra*); well-grazed downland pasture, now much reduced in extent, consists of a delightful springy turf of these grasses.

(c) The *acid grasslands*, growing on the shallow base-deficient soils which have developed on the old siliceous rocks of north-western Britain, form the better 'hill-grazings' of the English Lake District, North Wales, the Pennines and the Scottish Highlands, wherever they are well-grazed and well-drained. The dominant grasses are the common bent (*Agrostis tenuis*) and the sheep's fescue. Unfortunately another common plant of this bent–fescue grassland is the bracken fern, which is a menace to many hill-farmers, as it spreads rapidly and so destroys the pasture.

(d) *Moor grasses* develop on the peaty soils, poorly drained and acid, dominated by the mat-grass or 'white bent' (*Nardus stricta*), which forms a tough unpalatable grass of little grazing value. In autumn and winter it is bleached almost white. Wavy hair-grass (*Deschampsia flexuosa*) is also common. On damper peat soils the grassland is dominated by purple moor grass (*Molinia caerulea*), especially on the gritstone moorlands of the southern Pennines.

(e) Some of the characteristic heathlands are covered in parts with grass, so that they may be termed *grass-heaths*. Heavily grazed or burnt heath may, in fact, develop into grassland, while conversely if grazing ceases the characteristic heathland plants will re-establish themselves. Wavy hair-grass and fescue are the chief grasses.

(iv) **Heathland** Heathland occurs on coarse sandy or gravelly soils, as in the centre of the Weald, in the Hampshire Basin, in Dorset and in the Breckland. The soils are poor since they are heavily leached, and they are usually underlain by a hardpan (p. 493). A thin, rather dry layer of acid peat, formed from lichens and mosses, covers the surface. The dominant plant is the ling (*Calluna vulgaris*), and associated forms include bilberry, dwarf gorse and, on south-facing sunny slopes, the purple bell-heather (*Erica cinerea*). In damper areas the cross-leaved heath (*Erica tetralix*) is common, while in districts where burning or grazing has kept the ling from full development a more open grass-heath is found.

Scattered trees are not uncommon; silver birch grows sporadically, and in places where it occurs in conjunction with dwarf oak, the trees are so numerous as to justify the name 'oak-birch heath', as in parts of the Weald and the Hampshire Basin. On the heathlands are to be found all stages between oak-birch heath and true *Calluna* heath; undisturbed heath may be colonized by an oak-birch association, and over-grazed or over-cut oak-birch heath will revert to true heath. The factor most inimical to tree growth is the hardpan which prevents the penetration of tree roots; if this is well developed the heath may well form a stable association.

In some parts of the British hills between about 300 m and 900 m

(1000–3000 ft) occur 'upland heaths', sometimes incorrectly called 'moors'. Some of the 'moors' in the northern Pennines, much of the North York Moors, the Irish hills, and the 'grouse moors' of the Scottish Highlands come into this category. Many of these are systematically burnt, particularly where preserved for grouse, and in the Scottish Highlands they are grazed by the red deer. This upland heath has extended at the expense both of former woodlands (regeneration of which is prevented in the Highlands by the red deer) and also, with the decline of sheep-farming, of upland pasture.

(v) **Moorland and bog** Used in its strict sense, moorland refers to an upland where acid peat has accumulated to an appreciable thickness, usually under damp conditions (p. 501). The characteristic vegetation includes bog-moss (*Sphagnum*), cotton-grass and reeds; purple moor-grass is common in parts and ling occurs on the margins. The wetter parts form peat bogs.

The moors and bogs are found for the most part in the damper western areas of the British Isles, although much has now been drained. They include the cotton-grass and bilberry moors of the gritstone Pennines, the higher parts of Dartmoor, the reed-moors of the North-west Highlands of Scotland and the Irish Wicklow Hills, the grass-moors of the Southern Uplands of Scotland, the 'raised-bogs' of the central plain of Ireland, and the swathing 'blanket-bogs'. The last cover vast areas of Connemara and Mayo in Ireland, and of western Scotland, notably the Moor of Rannoch, crossed by the railway line between Crianlarich and Fort William.

(vi) **Fen and marsh** The term *fen* is restricted to a water-logged area in which peat is accumulating, but in which the ground-water is markedly alkaline so that the peat is not acid, as it is in the bogs (p. 501). The largest area in Britain was the Fenland of northern Cambridgeshire and southern Lincolnshire, although only a few 'preserved' fragments (such as Wicken Fen) now remain after centuries of drainage and reclamation. The low-lying lands of the Vale of Pickering were similarly covered with fen, and small areas occur in the English Lake District and round the shores of Lough Neagh in Northern Ireland. Pond weeds, reeds, rushes, various grasses, and such shrubs as sallow, osier, willow and alder are the typical plants, together with a large number of water-loving plants. The term *carr* is applied to such a fen of scrubby bushes.

A *marsh* is an area where there may be some variation in the height of the water-table, where rivers overflow and where much silt (p. 490) is deposited; the soil is therefore largely inorganic. Such are the silt-lands of the northern Fen District. A *swamp* is a very generally used term, but should apply to land that is permanently inundated.

(vii) **Maritime vegetation** The main vegetation groups associated with the coasts are the salt-marsh communities (plates 110 and 111), and those of the foreshore, sand-dunes and shingle beaches. These plants are of interest not only to the botanist but also to the student of land-forms, for the development of vegetation is a major factor in the reclamation of lowland coasts, both naturally and artificially (p. 303). Some plants help to trap particles of mud and then bind the mud-banks together, others 'fix' the loose sands.

Salt-marsh communities, occupying marginal parts of tidal estuaries and bays, occur in zones from the lowest tide-mark to the dry land, and as the marsh is reclaimed the zones advance seaward. Sea-grass (*Zostera*), marsh-samphire (*Salicornia*) and then sea manna-grass (*Puccinellia*) which forms salt pastures, succeed each other landward. Then come familiar flowering plants—sea-lavender, thrift and many more.

An interesting plant is the perennial rice grass (*Spartina townsendii*), first reported in this country in Southampton Water in 1870. It is a strong-growing perennial grass, probably a hybrid by origin, which thrives on deep loose mud upon which *Salicornia* could not take hold. No other species can compete with it, and none can compare in helping the natural reclamation of land from the sea. It has spread rapidly in Southampton Water, in the other creeks of the Solent, and in Poole Harbour. Within the last thirty years, however, the *Spartina* swards have shown progressive signs of 'die-back', for reasons not wholly understood.

The sand-dunes are dominated by marram-grass (*Ammophila arenaria*), often deliberately planted (p. 301), while other sand grasses include the sea couch-grass (*Agropyron junceum*) and a dune variety of red fescue (*Festuca rubra*). The fixed dunes in due course are colonized by an increasing variety of plants and ultimately by heathland species (plate 109).

(viii) **Mountain vegetation** While uplands cover a large part of western and northern Britain, the greater part of their vegetation has already been mentioned—the deliberately planted conifers on the lower slopes, the acid grasslands of the hill-grazings, the upland heaths, moors and bogs. Covering the highest areas is a group of plants known as 'arctic-alpine', of great interest to the botanist; these include mosses, lichens and small prostrate rosette-like alpine plants, some of considerable rarity and limited occurrence.

CHAPTER 22

Conclusion

THE geographer seeks to describe the diverse features of the earth's surface, to explain if possible how these features have come to be what they are, and to discuss how they influence the distribution of man with his multifarious activities. This book has been concerned more particularly with the various aspects of man's physical surroundings. For convenience these aspects have been dealt with one by one. In the first place we described the materials which compose the crust of the earth, the internal forces which have affected this crust, and the external forces of earth-sculpture which have produced and are producing a series of gradual changes.

In turn were discussed the configuration of the seas and oceans, the relief of the sea-floor, and the character and movements of the waters of the oceans. Then we examined the physical processes within the surrounding atmosphere, in terms both of the weather, which is such a familiar and vital element in our daily lives, and of climate, summarizing this in terms of climatic types. Next we passed on to the soil-cover, and finally described the mantle of vegetation which clothes much of the land-surface. Throughout the distinctive facts and factors were surveyed in a series of organized classifications, but their mutual inter-relationships were stressed.

The nature and detail involved in any geographical description and analysis depends upon the extent under consideration—whether, for example, one is dealing with major units such as the Mediterranean Basin or the plateau of Tibet, or with minor units such as Wirral or the Vale of Pewsey. To these is commonly applied the term *region*. The basis or criterion of definition of any region may be primarily that of structure and relief, for example, the Scandinavian Plateau or the Central Massif of France. It may be that of climate; compare the stamp put upon the Mediterranean lands with that upon the Monsoon lands by reason of their widely differing climatic regimes. Or it may be that of vegetation, as in the case of the coniferous forests of North America or the savanna lands of Africa. But perhaps the most convincing regions, certainly in the case of the smaller ones, are those whose identity is recognized, consciously or unconsciously, by their own inhabitants; notable examples of such regions are the *pays* of France. The chief problem is how to define them. Their actual boundaries, of whatever order of magnitude or whatever type, are seldom clear-cut; they are more often zones of transition than sharply defined lines.

The geographer, therefore, faced with the task of presenting in an ordered account the mass of material which he may have accumulated, tries to demarcate his units, one by one, on some convenient scale, and then goes on to explain the characteristics of the landscape of each—both the physical landscape and the 'humanized' or 'cultural' landscape. The features of the latter result from the direct or indirect, constructive or destructive, impact or impress of man upon his natural surroundings. It is important to realize that there is a constant interplay between the various forces and factors of man and his environment, giving different results; it is the study of this interplay that is the very essence of modern geography. In the end, there is a 'sequence of geographies', interwoven with and based on the complex physical environment, and shot through with the threads of man's activities. As H. C. Darby has said, 'The landscape as we see it today represents a collection of legacies from the past'. This book has attempted to analyse the various legacies bequeathed by physical geography. A systematic inter-related analysis of the physical features of any region is an essential prelude to any survey of its human and economic geography.

One example must suffice to illustrate this principle. The writer once sought to portray systematically the essential geographical character of a small area in the north-east of Belgium, known to the people of that country as the *Kempenland*, or in French *la Campine*. It was necessary to examine first the geological features—the great thickness of Lower Pliocene sands and gravels on the surface, over-lying the Cretaceous rocks, and below those again, at considerable depth, the Upper Carboniferous rocks containing workable Coal Measures. Physically the region consists of a low plateau, bordered on the east by the sharply-defined edge of the Meuse valley and on the south by the Demer valley, but dipping gently away westward and northward to the Scheldt estuary and the Dutch polder-lands. The region experiences a mean annual rainfall of from 60–75 cm, mean temperatures of from 3° C (January) to 18° C (July), and frequent strong winds.

The soils developed on the sands are of a dry podzol-type, acid in character and deficient in nutrients, and only a small amount of humus forms, derived from the matted fibrous remains of mosses, lichens and woody heath plants. Commonly there is an impermeable pan, sometimes consisting of humus compounds, sometimes of sand-grains or gravel cemented by the ferric salts deposited from percolating solutions.

With this physical basis in mind, it was possible to describe the main geographical features of the heathland which covers most of the Kempen plateau. While in past centuries much of the region must have been under woodland, the clearance schemes of religious houses, the slow but gradual widening of the perimeter of arable land and of pasture around the villages, and the wholesale cutting for fuel by the armies which repeatedly fought over the Low Countries, took their toll. Once the woodland had been cleared, natural regeneration became increasingly difficult, for the clearing allowed the rapid destruction of the mild humus which forms under deciduous woodland, and its replacement by a thin acid layer in which heathland associations flourish. Moreover, regeneration was often prevented by the close grazing of sheep, goats and rodents which destroyed the young seedlings. The net result of the interaction of the direct or indirect activities of man with this physical basis was that by the end of the eighteenth century the Kempenland was covered with almost continuous heathland. The most characteristic feature is the extensive area of ling (*Calluna vulgaris*), commonly the dominant species, although elsewhere mixed heath, including broom, whin, juniper and bell-heather, is to be found. Where burning or grazing has kept the ling from full development there is a more open grass-heath, while scattered silver birch and dwarf-oak may form scrubby thickets. In the neighbourhood of the many small lakes and meres, aquatic and heath communities come into close association; there are all transitions between bog, where permanently wet soil conditions occur, the intermediate 'wet-heath' and the true *Calluna* heath.

After discussing these features, it was possible to examine the various methods by which the heathland has been improved or modified—the planting of conifers, the laying down of pasture, the use of irrigation, the exploitation of the deep coalfield, the building of factories and the housing-estates where live their workers, and the resulting enormous increase of population, proportionally greater than in any other part of Belgium during this century. All these topics are inter-dependent, and form part of a single theme, that is, the increasingly profitable use by Belgium of an area described in the early nineteenth century as consisting of 'sterile wastes even as the sands of the sea-shore'. And just as it is impossible to survey adequately the characteristics of any region without considering how these have come to be what they are, it proved to be as important to examine how the Kempenlanders utilized their environment in the past and the changes they produced in so doing, as it was to describe the facts of the present-day geography.

Much of the satisfaction derived from making this regional study of the Kempenland was due to the necessity of visiting the area and studying systematically its features *in the field*. Ideally a geographer should visit and see for himself hot deserts, ice-caps and glaciers, rift-valleys and deltas, volcanoes and coral reefs. Not very long ago if a man wished to be regarded as a geographer it was necessary for him to travel widely, to be an explorer in the usually accepted sense of the word, and to help fill in some 'blank on the map'. Today few of us are fortunate enough to travel as widely as we would like. Nevertheless, we can still be explorers, if only we try to look with understanding at our own surroundings, limited though they may seem to be, and to take every opportunity to widen them. Many a young British geographer owes much to the Youth Hostels Association and to the Field Studies Council. It is indeed difficult to overemphasize the contribution which a student can make to his study from his own experience. Fieldwork and personal observation, the re-exploration of a perhaps familiar area with a more-seeing eye, is a most vitalizing stimulus, the most useful contribution to reality in geographical study. Wherever a student lives, he can look at and try to explain the phenomena of his environment. He can seek to acquire an understanding eye for country, and so enrich his descriptive and explanatory powers. York Powell wrote in his preface to Grant Allen's *County and Town in England*: 'It was a pleasant thing to go a walk with him. The country was to him a living being, developing under his eyes, and the history of its past was to be discovered from the conditions of its present.' If fieldwork is necessary for the historical geographer, it is indispensable for the physical geographer, who needs as his first and basic item of equipment a pair of stout boots.

But no one can go everywhere, even if these stout boots were indeed 'seven league' ones. His other main item of equipment is the map, which is a mine of information, in that it is a precise and accurate depiction of spatial distributions and relationships. The maps a physical geographer is required to handle may be divided broadly into two groups.

In the first place there is the small-scale atlas map. Many of the chapters in this book have necessarily been concerned with distributions on a world or continental scale: the very patterns of land and ocean, shields and fold-ranges, continental shelves and ocean deeps, ocean currents and major wind-systems, tropical forests and temperate grasslands. It is necessary for the student to have a good atlas constantly at his side for immediate and constant reference, from which he can give precise location to the feature with which he is concerned, and from which he can draw his sketch-maps.

In the second place, there is the large-scale topographical map. Such a map can be used, of course, for the essentially practical object of finding one's way, whether on foot, on a cycle or by car. The ability to use map and compass is a vital part of the technique of the walker among the British hills, especially in bad weather when low-lying cloud may descend almost to the valley floor. The geographer can use the topographical map to supplement and corroborate observation in the field; indeed, fieldwork should be planned carefully in advance with a map to avoid desultory wanderings, and the map will be a guide to methodical study as one walks over the ground. Moreover, it provides much precise and exact information which can be used as a basis, as a starting point, for further interpolation. The student can plot and add to it the features he has actually observed. If he intends to visit the Craven district of the West Riding of Yorkshire, he will be well repaid if he studies beforehand the relevant sheet of the current Ordnance Survey One-inch Series, or the corresponding sheets of the 1:25,000 Series. To these maps, or to tracings from them, he may then add, when he is actually in the field, the scars, dry-valleys, sink-holes, clint pavements and peat bogs, and other interesting detail.

But in addition the topographical map is a clear, accurate source of special information. The geographer can make selective tracings of significant contours and of drainage patterns. In recent years increasing attention has been paid to morphometric techniques, that is, the utilization in various cartographical and diagrammatic ways of statistical information about the earth's surface, either provided by published topographical maps, or actually measured in the field. This is one aspect of the advance of *quantification* generally in geographical method. Most of these morphometric techniques are concerned with slope and altitude; to the patterns of distribution provided by the dimensions of length and breadth is added the dimension of height, from the map in the form of contour-lines and spot-heights, from fieldwork by levelling or the use of various kinds of altimeter. This information may be classified, interpreted and presented by various devices: profiles, slope analysis, area-height diagrams, hypsometric curves, clinographic curves, height-range diagrams, and many more. In the words of R. J. Chorley, 'Quantitative techniques, supported by statistical analysis, provide a standardized, rigorous, conservative and object-ive framework for the investigation of many of the problems of earth science. . . .' It is important to realize, however, that morpho-metric methods alone may be inadequate; quite apart from the danger of applying 'high-power methods to low-power data', it may sometimes appear that the objectivity claimed for quantifi-cation is illusory, that elements of subjectivity may enter into the work, and that there is no substitute for lengthy and painstaking fieldwork illuminated by flashes of qualitative inspiration and intuition. As Chorley goes on to say, '. . . these techniques and analyses are only an adjunct to, and not a substitute for, the initial qualitative stage of any investigation'.

With training the geographer can also deduce and interpret infor-mation from a map, as he looks behind the conventional symbols. C. C. Carter wrote in the preface to his delightful book, *Land-forms and Life*, 'Even if it [the map] cannot present a complete image of an actual scene or a full record of human activities, yet some of its subject-matter emerges conspicuous and certain, while for the rest, like an artist's picture, it presents a background that is often brimful of suggestion'. If the geographer is studying chalklands, he will do well to pore over the One-inch maps of the Chilterns or the South Downs; if he is concerned with volcanoes he may be able to refer to the 1:100,000 sheet on which Vesuvius appears, published by the Italian *Istituto Geografico Militare*. As C. E. Montague wrote in one of his charming essays, 'Thus does the large-scale map woo the sus-ceptible mind'. The interpretation of a topographical map is not easy, and it needs much practice to enable the landscape to be visualized in all its aspects. This practice is an integral and most worth-while part of the training of a geographer, and the careful study of the map is secondary only to the study of the ground itself.

Index

The following abbreviations are used: I. Island or Isle; L. Lake; Mts Mountains; R. River. Plates are indicated by **bold** numerals. A reference to a figure is indicated by the page number in *italics*.